1613609234

AF616385

WITHDRAWN FROM
STOCK

Mechanics of Quasi-Brittle Materials and Structures

HERMES Science Publications
8, quai du Marché-Neuf
75004 Paris

ISBN 2-86601-729-3

Cataloging in Publication Data: Electre-Bibliographie

Mechanics of Quasi-Brittle Materials and Structures
Pijaudier-Cabot, Gilles*Bittnar, Zdenek*Gérard, Bruno, Ed. –
Paris: Hermès Science Publications, 1999
ISBN 2-86601-729-3
RAMEAU: matériaux: propriétés mécaniques
construction: stabilité
durée de vie (ingénierie)
DEWEY: 620.2: Mécanique de l'ingénieur. Matériaux.
Matériaux

Mechanics of Quasi-Brittle Materials and Structures

A Volume in Honour of
Professor Zdenek P. Bazant 60th Birthday

Editors

Gilles Pijaudier-Cabot
Ecole Normale Supérieure de Cachan
& Institut Universitaire de France

Zdenek Bittnar
Czech Technical University

Bruno Gérard
Direction des Etudes et Recherches
Electricité de France

HERMES

Serveur web : http://www.editions-hermes.fr
http://www.hermes-science.com

The cover picture is adapted from the computed distribution of damage in the double edge notch specimen tested by Nooru Mohamed at Delft University of Technology (1992). The computation has been performed at the Laboratoire de Mécanique et Technologie (ENS de Cachan-CNRS-Université Pierre et Marie Curie) by Stéphanie Fichant.

Contents

Foreword

This volume, honouring Zdenek P. Bazant on his 60th birthday, features most of the papers presented at the Workshop on Mechanics of Quasi-Brittle Materials and Structures. This Workshop, in honour of Zdenek, was held during March 27-28, 1998, at his alma mater, the Czech Technical University in Prague (CVUT). It was organised in collaboration with the Laboratoire de Mécanique et Technologie at Ecole Normale Supérieure de Cachan, France, and sponsored by Electricité de France, Stavby silnic a zcleznic, Stavby mostri Prague, and Vodni stavby Bohemia.

Born in Prague on December 10, 1937, Zdenek studied engineering at Czech Technical University in Prague (CVUT), receiving the degree of Civil Engineer (Ing.) in 1960 (with a straight A record, first in class). In 1963, he obtained a Ph.D. in mechanics from the Czechoslovak Academy of Sciences, and in 1966 a postgraduate diploma in physics from Charles University, both in Prague. In 1967, he attained habilitation at CVUT as « Docent » in concrete structures. Through all his graduate studies, he was employed full-time as a design engineer and construction supervisor; he designed six bridges, one of them a prestressed box girder over Jizera near Korenov, noteworthy for its highly curved spans (each with a 30° central angle). During 1964-1967, he conducted research in polymer-fibber composites in the Klokner Institute of CVUT, served as adjunct assistant professor, and continued consulting in structural engineering.

In the fall of 1966, the French government awarded Zdenek an ASTEF fellowship for a six-month visit of CEBTP, Paris. Zdenek feels lucky that his advisor and mentor was the famous Robert L'Hermite. Since that time, Zdenek has always maintained close contacts with French researchers and has had a number of French assistants and collaborators (G. Pijaudier-Cabot, J. Mazars, C. Huet, L. Granger, E. Becq-Giraudon, Y. Berthaud, F.-J. Ulm and others). As a result of his stay in France, Zdenek started his activity in RILEM.

The year 1967 was critical for Zdenek. He left his native land and moved to America. After spending two years on visiting appointments at the University of Toronto and the University of California, Berkeley, he joined in September 1969 the faculty of Northwestern University as Associate Professor of Civil Engineering. He became full Professor in 1973 and was named to the W.P. Murphy distinguished professorial chair in 1990. During 1981-88, he was the Director of the Center for Geomaterials, and during 1974-78 and 1992-96, he was the Structural Engineering Coordinator.

Zdenek has made lasting contributions to mechanics of solids and concrete engineering which received wide attention (as documented by his extraordinarily high citation index, now running about 550 annually). Since 1958, he published over 380 research papers in refereed journals. In 1991, he published (with L. Cedolin) an important book on Stability of Structures, which is the first to cover systematically stability problems of fracture, damage and inelastic behaviour, and has been acclaimed in reviews by leading mechanics experts. Bazant's latest book (with J. Planas, 1998) on Fracture and Size Effect is the first to present a systematic theory of size effects in quasibrittle failure, and his book (with M. Kaplan, 1996) on Concrete at High Temperatures is the first to systematically treat mathematical modelling in this field.

Bazant is by now well known for his size effect law and the nonlocal concept for strain-softening materials. Until 1984, the observed size effects on structural strength were explained by Weibull's statistical theory, but this changed after Bazant showed theoretically, and verified experimentally, that for quasibrittle failures preceded by large stable crack growth (as observed in concrete, rock masses, tough composites, sea ice and other quasibrittle materials), the size effect is caused mainly by the release of energy stored in the structure. He introduced the size-effect method to identify non-linear fracture characteristics (adopted as RILEM Recommendation). He was the first to demonstrate, beginning in 1976, that finite element codes that model distributed cracking by means of strain-softening stress-strain relations are plagued by spurious mesh sensitivity, ill-posedness and localisation, and lack size effect. His simple remedy, the energy based crack band model, found wide use in industry and is being introduced in commercial codes (e.g., DIANA, SBETA). As a more general remedy, he pioneered, beginning in 1983, the nonlocal continuum models, as well as gradient models for damage localisation, and later justified them physically by microcrack interactions.

In his lab, Bazant generated an extensive experimental basis for the quasibrittle size effect. He extended the size effect law to rate dependence (discovering the reversal of softening to hardening after a sudden increase of loading rate), to compression failures (columns, borehole breakout, fiber laminates) and to bending fractures of sea ice plates. He showed that, for quasibrittle materials, Paris' law for fatigue crack growth requires a size effect correction. He extended the plastic « strut-and-tie » model for failures of reinforced concrete structures (such as diagonal shear) to size effect by incorporating quasibrittle fracture mechanics. He also elucidated the size effect and fracture mechanics aspects of quasibrittle compression failures, particularly reinforced concrete columns and microbuckling kink bands in unidirectional fiber composites. He demonstrated further how the previously accepted Weibull-type statistical strength theory of size effect can be extended to nonlocality. Bazant also produced a series of progressively more powerful non-linear triaxial constitutive models for concrete and soils. Extending G.I. Taylor's idea from plasticity to damage, he developed the microplane constitutive model for concrete and soils, which is used in some large codes (EPIC) and is proving more realistic than the classical plasticity-type models. In this context, he found a new and more efficient (21-point) Gaussian integration formula for a spherical surface (published in a mathematics journal, it has also been used in computational chemistry and radiation problems).

Furthermore, Bazant solved the three-dimensional elastic stress singularity and edge angle for crack-surface intersections, and the singularity at the tip of a conical notch or inclusion. He derived conditions of localisation into ellipsoidal domains and layers; clarified the thermodynamic basis of the criterion of stable post-bifurcation path; demonstrated bifurcation and crack arrest occurring in systems of parallel cooling or shrinkage cracks; derived consistent micropolar continuum approximation for buckling of regular lattices; demonstrated and quantified spurious wave reflection and diffraction due to a changing finite element size (which found implications in atomic lattice studies and in geophysics, and was recently republished in a special volume of most important papers by the American Society for Exploration Geophysics). In 1971, Bazant clarified the correlation among three-dimensional continuum stability theories associated with different finite strain measures, such as Green's, Biot's and Hencky's, which had hitherto been thought to be in conflict. This, for example, showed the Engesser's and Haringx's formulas for shear buckling to be equivalent. He formulated a new finite strain tensor with compression-tension symmetry giving a close approximation to Hencky's but easier to compute.

Extending the work of Trost, Bazant, in 1972, formulated and rigorously proved the age-adjusted effective modulus method, which allowed approximately solving the system of integral equations for ageing creep effects in concrete structures by a single quasi-elastic analysis. This method became standard, embodied in American (ACI) and European (CEB-FIP) recommendations and featured in many books. As consultant to the Nuclear Reactor Safety Division of Argonne National Laboratory, he developed thermodynamically based models for creep, hygrothermal effects, coupled heat and mass transport and pore pressure in concrete, widely used to analyse nuclear accident scenarios. He formulated the solidification theory for creep of concrete which treats short-term ageing as a volume growth of a non-ageing constituent (cement gel) in the pores of cement stone, and the microprestress theory which describes long-term ageing and cross-couplings with diffusion processes (drying, heating) by relaxation of self-equilibrated prestress in the microsctructure generated chemically and by water adsorption. He explained various phenomena in creep of concrete by surface thermodynamics of water adsorption in gel pores. He elucidated various stochastic aspects of concrete creep, developed a Latin hypercube sampling approach to assess the effect of uncertainty of creep parameters on structures, often used in design of sensitive structures, and conceived a Bayesian model for updating these predictions based on short-time measurements. Adapting the concept of ergodicity, he formulated a spectral method for determining the effects of random environmental humidity and temperature on an ageing structure. He clarified creep and shrinkage effects on nuclear reactor containments. Bazant's efficient exponential step-by-step algorithm for concrete creep (1971), based on converting an integral-type to rate-type creep law, has found use in various finite element codes. Bazant's contributions to creep, humidity effects and their statistical analysis are important for improving durability of infrastructures as well as for designing more daring structures with high-performance concretes.

Bazant is a member of the National Academy of Engineering (elected in 1996, he was cited for *contributions to solid mechanics, particularly structural stability and size effects in fracture*). He received honorary doctorates (Dr.h.c.) from Czech Technical University, Prague (1991) and Universität Karlsruhe, Germany (1998). In 1996, the Society of Engineering Science awarded him the Prager Medal, given for *outstanding contributions to solid mechanics*. In 1997, ASME awarded him the W.R. Warner Medal, which honors *outstanding contributions to the permanent literature of engineering; cited for important contributions to solid mechanics, focusing on the size-effect law for failure of brittle structures, modeling of material damage from softening, local and nonlocal concepts, stability and propagation of fracture and damage in material and thermodynamic concepts associated with stability of non-elastic structures*. In 1996, ASCE awarded him the Newmark Medal (which honors *a member who, through contributions to structural mechanics, has helped substantially to strengthen the scientific base of structural engineering; cited for fundamental contributions to the understanding of constitutive behaviour of structural materials, non-linear fracture mechanics and stability of structures*). Other honors include: 1975 L'Hermite Medal from RILEM (cited for *brilliant developments in mechanics of materials, thermodynamics of creep and stability theory, bridging experimental and theoretical research*); Huber Research Prize (1976), T.Y. Lin Award (1977) and Croes Medal (1997) from ASCE; Guggenheim (1978), Ford Foundation (1967), JSPS (Japan 1995), Kajima Foundation (Tokyo 1987), NATO Senior Scientist (France 1988) Fellowships; A. von Humboldt Award (Germany 1989); 1991 National Science Council of China (Taiwan) Lectureship Award, 1992 Best Engineering Book of the Year Award (Association of American Publishers), Meritorious Publication Award (1992) from Structures Engineers Association, Medal of Merit (1992) (for advances in mechanics) from Czech Society for Mechanics; Outstanding New Citizen from Metropolitan Chicago Citizenship Council (1976); and 1990 Gold Medal from Building Research Institute of Spain (cited for outstanding achievements in the fields of structural engineering and mechanics of concrete). He was elected an Honorary Member of that Institute (1991), of Czech Society of Civil Engineers (Prague 1991) and of Czech Society for Mechanics (1992), and a Fellow of American Academy of Mechanics, American Society of Mechanical Engineers (ASME), American Society of Civil Engineers (ASCE), American Concrete Institute (ACI) and RILEM (International Union of Research in Materials & Structures, Paris).

Zdenek has been very active in engineering societies. He was, (1991-93), the first president of the International Association. for Fracture Mechanics of Concrete Structures (IA-FraMCoS), incorporated in Illinois. In 1993, he was president of the Society of Engineering. Science. During 1983-94, he was division coordinator in International Association for Structures Mechanics in Reactor Technology (IA-SMiRT). He has been an inspiring leader and determined organizer, forming new committees in several societies and producing (with several committees he chaired) influential state-of-art reports. He served, (1988-94), as Editor-in-chief of ASCE Journal of Engineering Mechanics. He is a Regional Editor of the International Journal of Fracture, and a member of editorial boards of 14 other journals. He chaired the ACI Committee on Fracture Mechanics, Concrete Structures Division of the International Association for Structural Mechanics in Reactor Technology, ASCE-EMD Programs Committee and ASCE-EMD Committee on Properties of Materials. In RILEM, he currently chairs a committee on creep and a committee on scaling of failure. He organised and chaired IUTAM Prager Symposium (Evanston 1983), 4th RILEM International Symposium on Concrete Creep (Evanston 1986); FraMCoS1 (Breckenridge 1992); and co-organized and co-chaired NSF Workshop on High Strength Concrete (Chicago 1979), NSF Symposium on Concrete Creep (Lausanne 1980), AFOSR Workshop on Localisation (Minneapolis 1987), France-U.S. Workshop on Strain Localisation and Damage (Paris-Cachan 1988), RILEM 5th International Symposium on Concrete Creep (Barcelona, 1993), NSF-Eur. Union Workshop on Quasi-Brittle Materials in Prague (1994), etc. An Illinois Registered Structural Engineer (S.E.), he has been consultant for many firms and, during 1974-1996, has served as staff consultant on nuclear reactor structures to Argonne National Laboratory.

Zdenek P. Bazant comes from an old family of engineers and intellectuals. Zdenek is the fifth generation civil engineer in the line of Bazant's. His grandfather Zdenek Bazant was professor of structural mechanics at the Czech Technical University in Prague (CVUT), where he served as the dean and rector, and was member of the Czechoslovak Academy of Sciences. His father Zdenek J. Bazant was the chief engineer of Lanna, the largest construction firm in pre-war Czechoslovakia, and then for thirty years professor of foundation engineering at CVUT and a widely sought consultant. Zdenek's wife Iva, whom he married in Prague in 1967 (just two days before leaving for America), works as a physician in a State of Illinois hospital. Their son Martin, with a doctorate in physics from Harvard University, just started teaching at M.I.T., and their daughter Eva pursues graduate studies in public health at Columbia University.

Most of all, Zdenek always emphasises the great help in research he received from his outstanding doctoral students (40 completed Ph.D.'s so far). He is proud that 18 of them became professors (in the USA, France, Spain, Turkey, Japan, Korea, Taiwan, etc.). Five became deans, five directors of research institutes. Others distinguished themselves in industry.

Zdenek has many human qualities which are appreciated by all the people he has been working with. He has always cared for his students and co-workers. He believes that being a professor does not only mean being successful in research and teaching. It also means helping co-workers at developing their own original way of thinking and assisting them in finding, depending on their interests, the best place for the future. In other words, Zdenek knows that advising does not stop at the end of a Ph.D. defense or a stay at Northwestern. It is almost a life-time effort and Zdenek has always been up to it, collaborating with many former students on new research problems for years after they left Northwestern.

For his co-workers, he is not only an outstanding scientist and an experienced advisor, but also a very much appreciated friend. We are all looking forward to celebrate many more Zdenek's birthdays in the future!

Gilles PIJAUDIER-CABOT
ENS de Cachan
& Institut Universitaire de France

Zdenek BITTNAR
Czech Technical University

Bruno GÉRARD
Direction des Etudes et Recherches
Electricité de France

Introduction

How can R&D help to manage with "aging assessment" of concrete structures for electric power generation?

Within the framework of the large french national equipment program for power generation, Electricité de France has developped during the last forty years advanced capabilities in civil engineering to design hydraulic and nuclear power plants. Today, up to 200 dams and 58 nuclear units are operated in France. A number of major concrete dams are older than 30 years ; the average expected residual life time of nuclear plants is close to 20 years, refering to the design life time of 40 years. A key question for EDF is now:

How long can we continue to operate existing power units with the same high level safety requirements?

The answer obviously depends on the maintenance costs: maintenance has to be adapted to aging consequences, such as loss of structural integrity or loss of containement capabilities. This is why a better understanding of aging phenomena appears as a main target for EDF: What are the key physical phenomena ? How do they combine and do their kinetics change in case of coupling?

The main time-dependent aging mechanisms affecting large concrete structures have already been pointed out: creep and shrinkage of reactor containment, reinforment corrosion for cooling towers and water supplier equipments, swelling due to alkali-aggregate reaction (AAR) fo certain dams and chemical attack by leaching for long term waste disposal containers.

In most cases, safety assessment requires numerical simulation to predict long-term behavior of structures under complex mechanical loadings combined to environmental aging. The reliability of the prediction strongly depends on the performance of the models, i.e., (i) capability to account for the key physical mechanisms, (ii) validation in the required range of operating conditions and industrial materials, and (iii) numerical robustness when integrated in large scale computer codes.

To ensure models reliability, EDF recently decided to support a large scientific program conducted by the R&D Division. The general purpose of the project is to provide expertise and technical support in the process of decision making: When and how to repair? How to optimise materials and design rules for future plants?

The main targets of this project are to understand and explain degradations, to validate existing models and propose new ones if needed, to derive optimised repairing criteria for existing plants and design rules including aging effects for plants yet to be designed.

In this project, a number of key topics referring to *mechanics of quasi-brittle materials and structures* have been identified. A large range of competences is required to progress in this field, and a number of them are already developped by research teams all over the world. That is why EDF strongly support network organisations for joint focused R&D programs, and is proud to sponsor this symposium in honor of Professor Z.P. Bazant, regarding his great scientific contribution to industrial problems in structural mechanics and durability.

Marc LASNE
EDF-R&D Division

Chapter 1

Mechanics of Material Failure

Towards an Universal Theory for Fracture of Concrete

Jan G.M. van Mier

Delft University of Technology
Faculty of Civil Engineering and Geo-Sciences
P0 Box 5048, 2600 GA Delft, The Netherlands

J.Vanmier@ct.tudelft.nl

ABSTRACT. *Determination of fracture parameters of concrete and other brittle disordered materials like rocks and non-transformable ceramics, is influenced by severe boundary condition and size effects. From the observation that boundary conditions affect the post-peak behaviour of concrete it can be concluded that softening is not a material property (at the macro-level). Moreover, size effects not only confirm this observation, but reinforce the idea that the fracturing of each structure must be regarded as a unique process from microfracturing at the scale of the individual aggregates in the material to large scale crack propagation. Although similarities may exist between structures loaded under specific boundary conditions and of varying size, no unique (macroscopic) law seems capable of describing all the observed differences. Instead, a meso-level approach where effects caused by the heterogeneity of the material are directly incorporated, and where the entire fracture process is mimicked in detail seems to be capable of effectively simulating the fracture behaviour. The disadvantage is the time consuming procedure, which makes direct practical applications rather unwieldy. For future developments in the field, however, the more detailed meso-approach may be helpful when new concretes emerge.*

KEY WORDS: *Boundary Conditions, Concrete, Fracture Process, Size Effects, Uniaxial Tension, Uniaxial Compression, Modelling.*

Introduction

Fracture mechanics research of concrete has rapidly expanded over the past few decades, in particular after the pioneering work of, among others, Kaplan [KAP61] and Hillerborg and co-workers [HIL 76]. The idea is simple: develop a tool to predict the fracturing of large scale concrete structures, which can be used by practical engineers to solve design problems. The intellectual effort is demanding, but non-the-less very rewarding. The demand of an effective theory with predictive qualities has a large influence on how such a theory (or model, or computer simulation technique) should be developed. The parameters that are used in the theory should be uniquely defined, and one should be capable of measuring them in a simple and straightforward manner. Inverse and indirect techniques should be avoided as much as possible because in general several 'correct' solutions can be found. Thus, bias should be avoided under all circumstances. In principle one should descend all the way to the level of fundamental particles making up atoms and molecules. Geometrical considerations are the most important aspect in such approaches, and of course the laws describing the interactions between the elementary particles. For engineering, obviously such an approach is too far fetched. Such an approach, which indeed is universal, could be labelled as a cosmological theory. The interesting question to be posed is what size scales should be included to develop a reliable theory which operates at the macro-level, and which is based on knowledge of microstructural processes taking place in the material at a more fundamental level. Or stated differently, is it sufficient to resort to meso/macro level approaches, or should we span the complete range micro/ meso/macro in order to come to universal theories with some predictive power.

Basic to such an approach is understanding all elements of fracture processes in materials and structures. Important for progress in the field seems to accept that a macroscopic test to measure macroscopic fracture parameters is **not** a test that yields direct information on the properties of the material under consideration, but rather an experiment on a small scale structure with its own specific size and boundary conditions. Accepting the analogue made in Figure 1 between materials testing at the macroscopic level and the structural level is to my opinion essential to progress in the field. The size variation goes even farther, i.e. from the extreme small (sub-atomic level) to the size of the real structure built in practice as argued before.

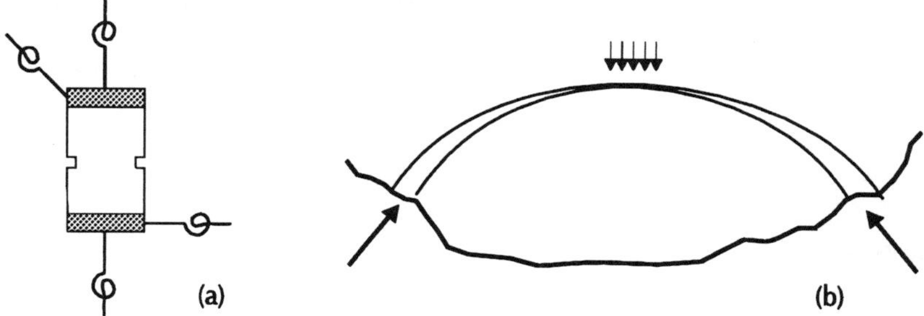

Figure 1. *Testing at the macroscopic material level (a) and structural level (b). In both cases boundary and size effects affect the measured response significantly*

For practical engineering, however, the length scales to be considered are limited to three distinct levels, namely the micro-, meso- and macro-level. For concrete this means that we have to consider the fracture response from the level of the cement structure up to the level where, through homogenisation, the material can be considered as a continuum. At the largest level contradictions will however always arise. The so-called 'fixed' material structure, defined by the size and geometry of the constituting materials (aggregate and cement, and in some cases fibres) will have limited effect on the measured response if the specimen is larger than the so-called representative volume element (RVE), which is normally considered to be at least five times the size of the largest aggregate particle in normal weight gravel concrete. There seems to be no problem, until we recognise that the fracture process in such particle composites brings in a new length scale, namely the size of the critical crack. At the end of the fracture process, the length of such cracks will be of the order of the size of the specimen (or structure) under consideration. Because the process is a continuous process with steadily increasing crack length, there is no way of circumventing the problem by increasing the size of the specimen. Note that the above does not necessarily imply that crack growth is stable. Unstable jumps may occur during the growth process. The important point made here is that there is a 'fixed' material structure defined by the aggregate and cement structure, and that there is a continuously changing crack structure. The cracks can however only increase in size. The dilemma we are facing must be clear by now. Continuum assumptions can be made with respect to the 'fixed' material structure, but size and boundary condition effects become important as the crack size increases relative to the specimen or structure size.

Now let us first consider two specific loading cases, namely uniaxial tension and uniaxial compression. In uniaxial tension, which will be treated in section 1, the tensile strength and fracture energy depend on the size and boundary rotations allowed in the test. In uniaxial compression (cf. section 2), friction between loading platens and specimen size have a significant effect on the measured compressive strength and fracture energy as well. In both cases no unique number or shape of post-peak softening diagram can be given. In the sections following hereafter, it is argued that a meso-level model may help to understand the boundary and size effects observed at the macro-level, but is at the same time hampered by the same experimental difficulties as are experienced at the macro-level. As a matter of fact, the model parameters and fracture properties seem even harder to determine. Including the micro-level, and thus spanning three scale-levels, might be a promising approach. The way out of the maze is, however, not straightforward, but seems essential for future engineering applications of fracture mechanics.

1. Size and boundary effects in uniaxial tension

In a uniaxial tension test on concrete, a small specimen (of size larger than the RVE defined by the largest aggregate particle) is glued between two loading platens and pulled in displacement control until no load can be transferred anymore. For practical reasons it is important to know where the crack will develop. This makes the definition of a control loop for the servo-mechanism needed in the displacement controlled experiment more easy. Displacement controlled testing is needed for a full record of the softening branch, and thus for measuring the complete fracture energy. The fracture energy is defined as the area under the total stress-elongation diagram. Often corrections are made for the pre-peak energy, but this is not considered correct. Experience teaches us that specimens are already fractured before the maximum load is reached, and consequently, the energy consumed in these pre-peak crack processes should be included in the total fracture energy.

When the specimen size, the control loop, the measuring length, the electronics and the hydraulics are all set-up, the test can be conducted. The complete procedures are explained in detail in [MIE 97a]. The definition of the specimen size and shape, as well as the translational and rotational freedom in the loading platens are now important. In Figure 2, two examples of stress-displacement diagrams are shown, viz. for uniaxial tensile tests between fixed (Figure 2a) and freely rotating loading platens (Figure 2b) respectively. The specimens are 100 mm long cylinders of 100 mm diameter. At half length a 5 mm deep circumferential notch was sawed in the specimens. Clearly, the average stress-deformation diagrams are quite different, and seem affected by the boundary conditions in each experiment. When the loading platens are fixed, i.e., when they are forced to translate parallel to one another during the entire experiment, a distinct plateau is observed in the softening curve (see Figure 2a). In contrast, a very smooth curve is measured when freely rotating loading platens are used (Figure 2b). The difference is caused by allowing certain stress-redistributions in the test with fixed loading platens, which cannot occur when rotations are allowed.

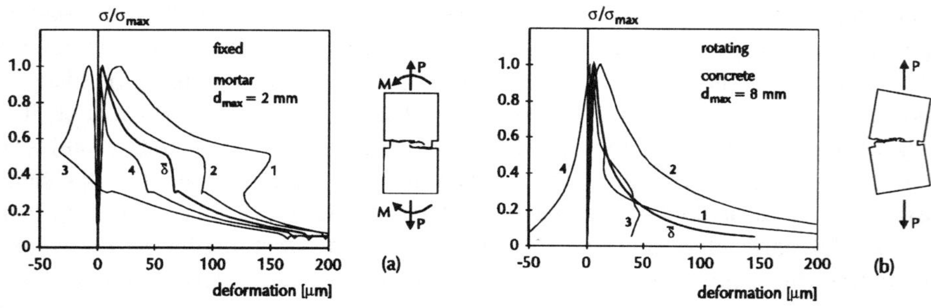

Figure 2. *Average and local stress-deformation diagrams under uniaxial tension: (a) fixed loading platens, and (b) freely rotating loading platens, after [MIE 97a]*

In the case of fixed platens, the crack will start to grow at one side of the specimen, dictated by the local variations in material properties caused by the heterogeneous material structure. Because the loading platens are kept parallel to each other, a bending moment will develop, which arrests the crack to further propagation. This causes the bump in the average stress-deformation diagram. Note that in Figure 2a, crack initiation is near LVDT no. 1, which registers the local deformation at one side of the cylinder. The opposite LVDT no. 3 registers unloading during opening at no. 1. The two other LVDTs are somewhere in between of these two extremes. At the end of the plateau in the softening diagram, suddenly large deformations are measured at LVDT no. 3. This means that a second crack develops from the other side of the specimen. Thus, a system of two interacting cracks develops, which causes the fracture energy to increase in comparison to the case with freely rotating loading platens. In the latter case, the local deformation measurements point towards crack initiation and continuous growth from a single point along the circumference. No stress-redistributions can now occur. Detailed measurements on two different concretes and two different sandstones have shown that the fracture energy can be about 40 % larger when fixed loading platens are used. At the same time it must be recognised that more crack surface develops in the fixed test as a consequence of the stress-redistributions during crack growth. The result from the test between the freely rotating loading platens is related to the growth of a single crack. Therefore the fracture energy in this test seems to be a lower bound. A complication that arises here is, however, the compressive zone which develops in the specimen opposite to the crack initiation point. The implications of this are not clear at all. In conclusion, it must be stated that a uniaxial tension test is not a pure test as usually assumed. Therefore the term 'direct tension' test should be abandoned altogether.

Related to the above boundary effect is the size effect. When specimens of different sizes are tested in displacement control, differences in fracture energy are measured. This was, for example, observed by Ferro [FER 95] and more recently by Van Vliet [VLI 98]. Figure 3 is taken from [VLI 98], and demonstrates clearly the effect of size on fracture energy, which seems to be in agreement with findings by

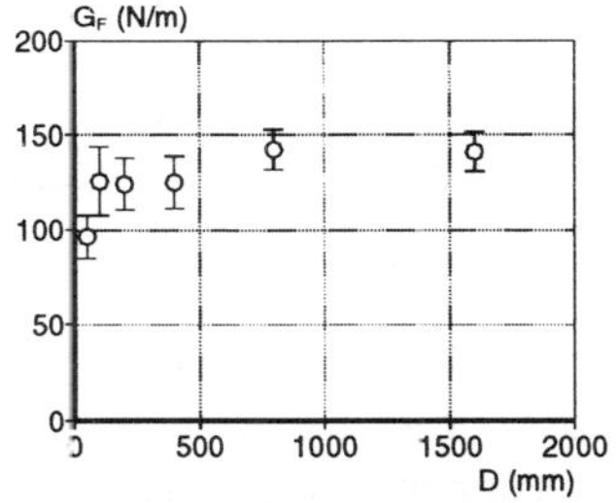

Figure 3. *Effect of specimen size on tensile fracture energy, after [VLI 98]*

Ferro. Note, however, that the value of the fracture energy must be determined from a test which has been pulled till the maximum crack opening. If this has not been done, certain assumptions must be made to extrapolate the softening curve to the maximum opening. The cause for the size effect is not clear, but in analogy to the boundary effect discussed above, the reason will probably be found in a higher crack density and the ability to redistribute stresses differently when the specimen size increases.

The results of Figures 2 and 3 illustrate that a direct measurement of the tensile fracture energy is not possible. The results from the tensile tests must be translated to an effective fracture energy, and even more important to an effective shape of the softening curve for structural analysis, which means that the total area of crack surface at any stage of the fracture process must be determined, as well as any stress redistributions that occur. In view of the very heterogeneous and complicated fracture process in uniaxial tension, i.e., a process from distributed microcracking to localised macro cracking [MIE 97a], this is not a simple and straightforward task, not in the last place because direct detection of internal cracks is virtually impossible. Development of acoustic techniques seems essential for progress.

Next to the size and boundary effect on fracture energy, it should be mentioned that an effect on strength exists as well. Rotating loading platens lead to lower tensile strength and an increased scatter in comparison to fixed loading platens, see [MIE 94]. Also, specimens of smaller sizes yield higher tensile strength, with an increased variability when the size comes close to the RVE, see [VLI 98].

2. Size effect and boundary restraint in uniaxial compression

The line of reasoning for uniaxial compression is almost identical to that of section 1. Again, size and boundary effects have a significant influence on strength and compressive fracture energy. The interpretation of the results is, however, much more complicated. In the first place, localisation of deformations is very clear in uniaxial tension. In uniaxial compression it was demonstrated in 1984 for the first time that localisation of deformations occurs in uniaxial compression tests between brushes as well, [MIE 84]. Later this was challenged by some authors, e.g. [VON 92], but the results from an extensive round robin test on the nature of the compressive strain softening diagram, organised by the RILEM committee 148 SSC Strain Softening of Concrete, clearly confirmed the findings of 1984, see [MIE 97] and [VLI 96]. Secondly, frictional effects seem to play a more prominent role in compressive fracture. Meso-mechanical analysis of compressive failure indicates that tensile microfracturing precedes the development of shear transfer in cracks and the growth of shear bands. From experiments it is known that the restraint between loading platen and specimen has a significant effect on the peak strength and the shape of the softening diagram. When high friction steel platens are used in a test, the stress-strain diagrams tend to be more ductile as shown in Figure 4. In that figure, stress-strain diagrams for normal (4a,b) and high strength concrete (4c,d) are shown, both for tests between high friction steel platens (4a,c) and low friction teflon platens (4b,d). In the same figures, the effect of specimen slenderness can be seen. The specimens were prisms with a constant cross-section of 100 x 100 mm^2, but with a varying slenderness between 0.25 and 2.0, (i.e., variation of specimen height between 25 and 200 mm). Clearly, the slenderness has an effect on the post-peak brittleness, which becomes even more prominent when high strength concrete is tested [MAR 95] or when the slenderness is further increased [JAN 97]. In all, these results confirm the findings of Van Mier [MIE 84] and Kotsovos [KOT 83] on the effect of slenderness and boundary restraint respectively. Note that it was shown by Bazant [BAZ 87] that a simple series model suffices to describe the post-peak localization which comes from the tests where slenderness is varied.

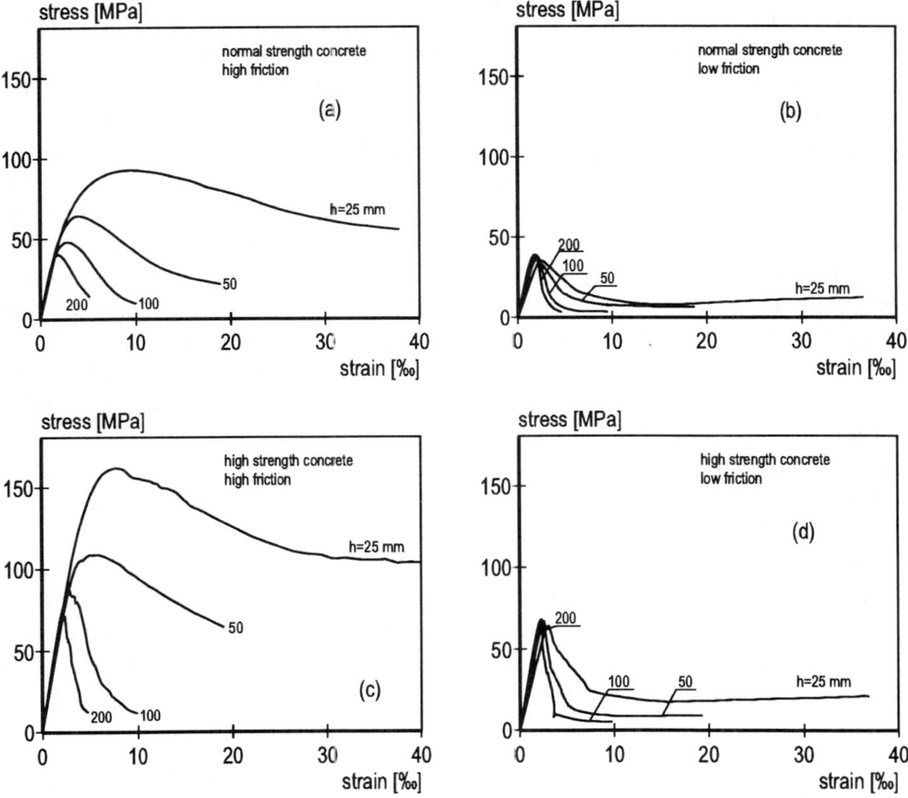

Figure 4. *Stress-strain curves under uniaxial compression showing the effect of boundary restraint and specimen slenderness, after [MIE97b]*

Another conclusion from Figure 4 is that the uniaxial compressive strength is dependent on slenderness when high friction platens are used, whereas slenderness independent results are found when teflon (low-friction) platens are used. The effect is caused by the frictional restraint at the top and bottom surface of the specimens. The restraint produces triaxially confined volumes in the specimens, which may be larger or smaller relative to the total specimen volume. It is well known that the strength of concrete under triaxial compressive stress is substantially higher than under uniaxial compressive stress. In all, it is not known which stress-strain diagram should be used for the analysis of reinforced concrete structures. Virtually any size and shape of the softening diagram can be obtained. Interaction with structural engineers seems essential to come to a diagram and a measurement method that will yield the most optimal result.

3. Meso-level modelling

The various size/scale effects and boundary influences can be modelled with simple meso-level models. The assumption made in such approaches is that much of the behaviour of concrete depends on the relatively large heterogeneity of the material in relation to the dimensions of the specimen. The heterogeneity is directly incorporated into the finite element model and the fracture laws to be incorporated are assumed to be more simple than those used at the macro-level (i.e., after homogenization). Different types of models can be used for meso-level modelling of concrete fracture. One of the first was based on the finite element method, [ROE 85]. Three phases were distinguished, namely aggregate, bond and matrix material. The aggregate and matrix phases were modelled by using either plane stress or brick elements (depending whether the model is 2D or 3D), whereas the interface behaviour was modelled by means of interface springs. A similar approach was adopted by Vonk [VON 92], but his model was developed on the basis of UDEC, i.e., the distinct element method which was originally made for modelling the behaviour of fractured rock [CUN 71]. Vonk also used interface elements to model the interactions between aggregates and matrix. In both these models, softening was incorporated as a fracture law at the meso-level. The reason is that the aggregate and matrix material and the bond zones are heterogeneous materials in their own right, but now at a smaller scale. Thus, as argued also in the introduction, concrete must be regarded as a material spanning several size scales. Because of the inherent heterogeneity of the matrix and bond zones, softening must be included in the meso-level models. In general, the matrix can be regarded as a 2 mm mortar because only the largest aggregates are included in the discretization. A 2 mm mortar clearly exhibits softening behaviour, see for example in [PET 81].

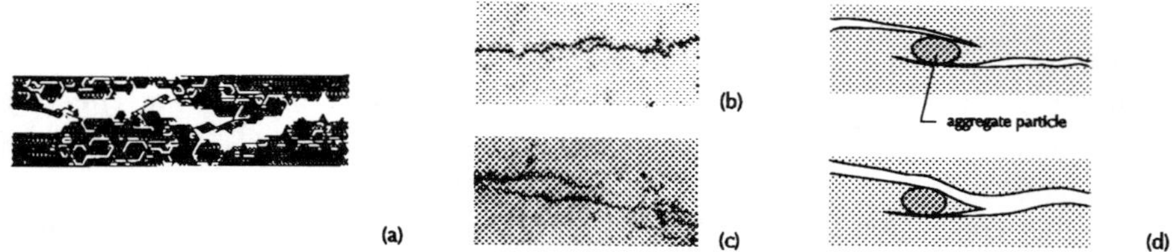

Figure 5. *Crack face bridging through crack overlaps from lattice analysis after [SCH 92] (a), and from experiments [MIE 97a] (b,c) and schematic representation of the overlap mechanism [MIE 91] (d)*

Another approach comes from statistical physics [HER 89], and was recently applied to concrete [SCH 92]. In this approach the material is discretized as a network of brittle breaking linear elements. The elements can be bars, springs or beams in all kinds of geometrical configurations. The lattice can be two-dimensional or three-dimensional. Heterogeneity is included by projecting the material structure over the lattice, for example the particle structure of concrete can be superimposed on a regular or random triangular lattice, [SCH 93, VER 97]. One example showing bridging in tensile fracture from a lattice analysis is shown in Figure 5a. Experimental proof of crack overlaps as an important bridging mechanism, which explains the softening phenomenon in concrete, is shown in Figures 5b and c for 2 mm mortar and 12 mm lytag concrete respectively. The average crack opening in the analysis and the experiments is the same, namely 100 μm.

The lattice model can be used to explain boundary condition effects in tension and compression, under global shear loading, and many mechanisms at the meso-level are an automatic outcome of the simulations. In fact, based on assumptions of a meso-level fracture law, the complete stress-strain curve (or stress-crack opening curve in the case of uniaxial tension) is computed. The surprising feature of all this is that a simple tensile fracture criterion at the meso-level seems to suffice. Even global shear failure seems a natural outcome from this, and compressive and shear failure seems to be a tensile phenomenon at a smaller size scale.

By means of the meso-level models, size effect on strength and fracture energy can be simulated as well. Again it is important to keep the scale of the material the same in all specimens. Thus the fineness of the aggregate structure, and the finite element mesh projected on that material structure are constant. The outer dimensions of the specimens, however, vary. Using such a simple approach, size effects can be computed, see for example [VLI 98], [RIE 91]. The advantage of a simple meso-level analysis is that the fracture mechanisms and stress-redistributions can be visualised. In addition, non-homogeneously distributed material properties from casting procedures, which are normally assumed identical for specimens of different sizes, and from drying, can be included in the analysis, and the true reason for the variation of fracture strength and fracture energy with structure size can be elucidated.

4. Macroscopic size-effect laws

Quite opposite to the physics-based approach of the previous section is a mechanics model where results from macroscopic fracture tests on specimens of varying size are used directly to derive size effects laws. Examples of such efforts are [BAZ 97], [CAR 94], [ARS 95]. Obviously, refinements such as non-homogeneities caused by casting procedures, eigen-stresses from hydration and non-uniform drying in specimens of different sizes, and local stress-redistributions during the crack growth process can not be treated in a continuum approach. This means that the material as well as the crack growth process are homogenised. They are regarded as global processes that can be smeared out over a volume or an area. In this respect it should be mentioned that softening according to the original ideas of Hillerborg in the fictitious crack model still contains some remnants of continuum ideas. The localisation occurs in longitudinal direction, i.e., in the direction of the applied load (see Figure 6a), but a close observation of tensile tests shows that during the steep part of the softening curve, the fracture zone traverses the specimens cross section (see Figure 6b). In the fictitious crack model it is assumed that the localization zone develops uniformly over the specimens cross-section. Physically this is not correct, as the tensile diagrams of Figure 2 clearly show that **localized crack growth starts from a single point along the specimens circumference.**

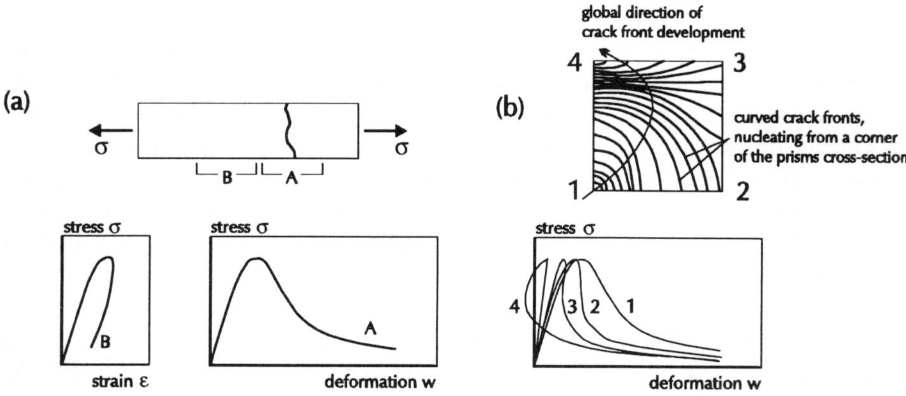

Figure 6. *Localized fracture zone in a tensile bar and separation of response in a pre-peak stress-strain curve and post-peak stress-crack opening curve (a) and propagation of the fracture zone through the specimens cross-section during the steep part of the softening curve, after [MIE97a]*

Essentially this means that the weakest link in combination with the largest stress concentration determines where the crack localization **point** will be located. This corresponds to the ideas presented in the previous section, but is difficult to include this in models that are based on homogenisation ideas.

5. The role of standard testing

It will be clear that the above sketched approach to the size effect will take considerable research effort. It is considered essential, however, because many different concretes are introduced in practice nowadays, and testing for size effects on all these different new mixtures seems a hughe task as well. With a meso-level model, the material aspects can be dealt with in a straightforward manner. A problem that hampers progress in this field, but this is true for both the macroscopic and mesoscopic approaches, is that many available and new experimental results are hard to compare. In that respect there is a need for a well defined standard test, both for uniaxial tensile and uniaxial compressive fracture. The idea is that through such a standard test, which must indeed be simple and easy to interpret, future experimental data will be more simple to compare. Such standard tests will have to be carried out parallel to the real fracture study that is done by any researcher. In addition, when the test is simple, and relates to current standard test practice, it will be more easily accepted and will pave the way for practical applications of fracture mechanics theories. This means that a test has to be selected which is preferentially not related to any theory existing in the field of fracture mechanics of concrete to date, but at the same time should be a test which gives results that can be translated to accommodate all existing theories. In this respect, it should be mentioned that true universal theories exist in particle and atomic physics only, i.e., theories where the behaviour is computed from fundamental properties of the smallest components. Such properties should indeed be true physical constants. We all know that this cannot be achieved at the macro-level where too many factors affect the observed response of structures as shown above. The question remains whether descending to the meso-level is sufficient, or whether crossing two size scales is needed, i.e., from macro to meso and subsequently from meso to micro, in order to obtain a model with some predictive qualities. Universality (which means according to the Concise Oxford dictionary "applicable to all cases") can perhaps then be obtained. For cases where no frictional restraint in cracks occurs, some progress has been made in the past years, but much work is still needed for the frictional cracks.

6. Conclusion and future outlook

In this paper a strong plea is made for the further development of meso-level models for fracture of concrete. Introduction of new binders and aggregates for application in concrete, makes that the mechanical behaviour of the materials can vary widely. Moreover, the meso-level models are capable of describing size and boundary effects. In tensile fracture, a simple meso-level strength based criterion seems to suffice, and basic mechanisms like bridging by crack overlaps and crack branching can be simulated. In compressive fracture frictional restraint in cracks is involved. The cracks seem to have developed primarily under local tensile stress, and the frictional phenomenon adds complications to such an approach. In spite of the fact that many problems still have to be solved, the meso-level models are generally capable of simulating correctly the effect of boundary restraint on the global failure mode in a uniaxial compression test. The meso-level approach requires powerful computers, which seems the limiting factor at present. This means that research should not only focus on meso-level materials science models for concrete, but in view of short term practical applications should include macroscopic approaches as well.

7. References

[ARS 95] Arslan, A. and Ince, R., The neural network-based analysis of size effect in concrete fracture, in *Fracture Mechanics of Concrete Structures, Proc. FraMCoS-2*, ed. F.H. Wittmann, AEDIFICATIO Publ., Freiburg, 693-706, 1995.

[BAZ 89] Bazant, Z.P., Identification of strain-softening constitutive relation from uniaxial tests by series coupling model for localization, *Cem. Conc. Res.*, 19, 973-979, 1989.

[BAZ 97] Bazant, Z.P., Scaling of quasi-brittle fracture: Asymptotic analysis, *Int. J. Fracture*, 83, 19-40, 1997.

[CAR 94] Carpinteri, A., Fractal nature of material microstructure and size effects on apparent mechanical properties, *Mech. Mater.*, 89-101, 1994.

[CUN 71] Cundall, P.A., A computer model for simulating progressive large scale movements in blocky rock systems, in *Proc. ISRM Symposium*, Nancy, France, 1 (II-8), 1971.

[FER 95] Ferro, G., Effetti di scala sulla resistenza a trazione dei materiali, *PhD thesis*, Politecinco di Torino, 1994.

[KAP 61] Kaplan, M.F., Crack propagation and the fracture of concrete, *J. Am. Conc. Inst.*, 58(11), 1961.

[JAN 97] Jansen, D. and Shah, S.P., Effect of length on compressive strain softening of concrete, *J. Engng. Mech. (ASCE)*, 123, 25-35, 1997.

[HER 89] Herrmann, H.J., Hansen, H. and Roux, S., Fracture of disordered. elastic lattices in two dimensions, *Phys. Rev. B*, 39, 637-648, 1989.

[HIL 76] Hillerborg, A., Modeer, M. and Petersson, P.-E., Analysis of Crack Formation and Crack Growth in Concrete by means of Fracture Mechanics and Finite Elements, *Cem. Conc. Res.*, 6. 773-782, 1976.

[KOT 83] Kotsovos, M.D., Effect of testing techniques on the post-ultimate behaviour of concrete in compression, *Mater. Struct. (RILEM)*, 16, 3-12, 1983.

[MAR 97] Markeset, G., High strength concrete phase 3E - SP4 - Comments on size dependency and brittleness of HSC, *SINTEF Structures and Concrete*, Trondheim, Norway, February 1995, 23 p.

[MIE 84] Van Mier, J.G.M., Strain Softening of concrete under multiaxial loading conditions, *PhD thesis*, Eindhoven University of Technology, 1984.

[MIE 91] Van Mier, J.G.M., Mode I fracture of concrete: Discontinuous crack growth and crack interface grain bridging, *Cem. Conc. Res.*, 21, 1-16, 1991.

[MIE 94] Van Mier, J.G.M., Vervuurt, A. and Schlangen, E., Boundary and size effects in uniaxial tensile tests: A numerical and experimental study, in *Fracture and Damage in Quasi-Brittle Structures*, ed. Z.P. Bazant, Z. Bittnarr, M. Jirásek and J. Mazars, E&FN Spon, 289-302, 1994.

[MIE 97a] Van Mier, J.G.M. *Fracture Processes of Concrete - Assessment of Material Parameters for Fracture Models*, CRC Press Inc., Boca Raton (FL), USA, 1997.

[MIE 97b] Van Mier, J.G.M., Shah, S.P., Arnaud, M., Balayssac, J.P., Bascoul, A., Choi, S., Dasenbrock, D., Ferrara, G., French, C., Gobbi, M.E., Karihaloo, B.L., König, G., Kotsovos, M.D., Labuz, J., Lange-Kornbak, D., Markeset, G., Pavlovic, M.N., Simsch, G., Thienel, K-C., Turatsinze, A., Ulmer, M., Van Geel, H.J.G.M., Van Vliet, M.R.A., Zissopoulos, D., Strain-Softening of Concrete in Uniaxial Compression - Report of the Round-Robin Test carried out by RILEM TC 148SSC, *Mater. Struct. (RILEM)*, 30(198), 195-209, 1997.

[PET 81] Petersson, P.-E., Crack growth and development of fracture zones in plain concrete and similar materials, *Report TVBM-1006*, Division of Building Materials, Lund University, Sweden, 1981.

[RIE 91] Riera, J.D. and Rocha, M.M., On size effects and rupture of non-homogeneous materials, in *Fracture processes in Concrete, Rock and Ceramics*, ed. J.G.M. van Mier, J.G. Rots and A. Bakker, E&FN Spon, 451-460, 1991.

[ROE 85] Roelfstra, P.E., Sadouki, H. and Wittmann, F.H., Le béton numérique (Numerical Concrete), *Mater. Struct. (RILEM)*, 18, 327, (1985).

[SCH 92] Schlangen, E. and Van Mier, J.G.M., Experimental and numerical analysis of micro-mechanisms of fracture of cement-based composites, *Cem. Conc. Composites*, 14, 105-118, (1992).

[SCH 93] Schlangen, E., Experimental and numerical analysis of fracture processes in concrete, *PhD thesis*, Delft University of Technology, The Netherlands, (1993).

[VER 97] Vervuurt, A., Interface Fracture in Concrete, *PhD thesis*, Delft University of Technology, (1997).

[VLI 96] Van Vliet, M.R.A. and Van Mier, J.G.M., Experimental investigation of concrete fracture under uniaxial compression, *Mech. Coh. Frict. Mater.*, 1, 115-127, 1996.

[VLI 98] Van Vliet, M.R.A. and Van Mier, J.G.M.,Experimental investigation of size effect in concrete under uniaxial tension, in *Proc. FraMCoS-3*, Gifu, Japan, AEDIFICATIO Publishers, 1998 (in press).

[VON 92] Vonk, R.A., Softening of concrete loaded in compression, *PhD thesis*, Eindhoven University of Technology, 1992.

Strength Scaling Law for Elastic Materials with Interacting Defects

Christian Huet

Swiss Federal Institute of Technology Lausanne
Department of Materials Science, Laboratory of Construction Materials
MX G Ecublens, CH-1015 Lausanne, Switzerland

christian.huet@epfl.ch

ABSTRACT. *It is shown that scaling laws for the nominal strength of elastic bodies, possibly heterogeneous, may be explained by the influence of preexisting microcracks in the framework of Linear Elastic Fracture Thermodynamics. Non-interacting microcracks provide two asymptotes of the Bazant scaling law, including the one generally attributed to plasticity effects. Interacting microcracks provide the full scaling law of Bazant including the transitory part of the curve.*

KEY WORDS: *Strength, Scaling-laws, Materials, Defects, Bazant.*

Introduction

In the recent past, scaling law has become the subject of increasing interest to the community of materials and structural scientists and engineers, which has also given rise to some controversies. We are pleased, in this Anniversary volume, to acknowledge the prominent contributions of Zdenek Bazant in this area as in many other fields, demonstrated by his numerous publications on the subject, listed in the recent book by [BAZ 98]. We show here that the main features of the now celebrated Bazant scaling law can be derived in the framework of Linear Elastic Fracture Thermodynamics, which stands as a particular case of the general Fracture Thermodynamics formalism that we used for dissipative heterogeneous bodies with cracks and microcracks in [HUE 94, 95, 96, 97].

1. Thermodynamic rate equations for crack growth

We consider a body – possibly heterogeneous and anisotropic – submitted to quasistatic boundary conditions with negligible volume forces and we make use of the following notation:

t : time,
x : coordinate vector of a material point,
ξ: displacement vector,
$\dot{\psi} = \frac{\partial \psi}{\partial t}$: time derivative of any quantity ψ,
U: total internal energy of the body,
P: density of the external surface tractions.

For the monothermal case, defined by a uniform temperature T_0 on ∂D_0 we denote:

Φ^{ε}: the global free energy,

Φ^{σ}: minus the free enthalpy (quasi-free enthalpy) defined by:

$$\Phi^{\varepsilon} = U - T_0 S$$
$$\Phi^{\sigma} = \int_D \sigma : \varepsilon \, dV - \Phi^{\varepsilon} \qquad [1]$$

For a dissipative material of any kind, combination of the universal balance equations for energy and entropy in primitive global form make it possible to define the overall potential energy Ψ^{ε} and the overall complementary energy Ψ^{σ} by:

$$\Psi^{\varepsilon} = \Phi^{\varepsilon} - \int_{D_0} F^d \cdot \xi \, dV - \int_{\partial D_{0\sigma}} P^d \cdot \xi \, d\Sigma \qquad [2]$$

$$\Psi^{\sigma} = \Phi^{\sigma} - \int_{\partial D_{0\xi}} P \cdot \xi^d \, d\Sigma = -\Psi^{\varepsilon} \qquad [3]$$

similar to the elastic case. Using a virtual dissipative identity in the global forms provides the governing equation for crack growth as:

$$R_F^k(\ldots, \dot{a}^k, \ldots) + \frac{\partial \underset{\approx}{\Psi^{\varepsilon}}}{\partial a^k} = 0 \qquad [4]$$

or, equivalently:

$$R_F^k(\ldots, \dot{a}^k, \ldots) - \frac{\partial \underset{\approx}{\Psi^{\sigma}}}{\partial a^k} = 0 \qquad [5]$$

where the $R_F^k(\ldots, \dot{a}^k, \ldots)$ are the crack resistance forces associated, in the expression of the dissipation for the specific material, to each crack geometry parameter a^k, [HUE 94, 95, 96, 97]. Owing to the second principle, they must be rate dependent.

Introducing the tensor energy release rate G^k defined for the general dissipative case by:

$$G^k = -\frac{\partial \Psi^\varepsilon}{\partial \underset{\approx}{a}^k} = \frac{\partial \Psi^\sigma}{\partial \underset{\approx}{a}^k} \qquad [6]$$

Equations [4] and [5] generalize the classical Griffith criterion which is restituted in the elastic case for a single crack with constant crack resistance force, thus defining the so-called Linear Elastic Fracture Mechanics (LEFM).

3. Crack Compliance Tensor and Crack Energy

For an elastic body made of a homogeneous material with any compliance tensor S and submited to static uniform boundary conditions (σ_0-SUBC), the strain is homogeneous and given by:

$$\varepsilon^h = S : \sigma_0 \qquad [7]$$

For the same body with defects in the homogeneous matrix, the apparent strain can then be written in the form:

$$\varepsilon^{app} = \varepsilon^h + \Delta\varepsilon \; = \; \varepsilon^h = S : \sigma_0 + \Delta\varepsilon = S_\sigma^{app\,c} : \sigma_0 \qquad [8]$$

where $S_\sigma^{app\,c}$ is the overall compliance tensor of the cracked body in σ_0-SUBC. This yields:

$$\Delta\varepsilon \; = \; (S_\sigma^{app\,c} - S) : \sigma_0 \; = \; H : \sigma_0 \qquad [9]$$

with H a crack compliance functions tensor given by:

$$H = S_\sigma^{app\,c} - S \qquad [10]$$

This gives Ψ^σ in the form:

$$\Psi^\sigma \; = \; \Phi^\sigma \; = \frac{V}{2} \; (S + H) : \sigma_0 : \sigma_0 \; = \; \Psi_0^\sigma + \Delta\Psi^\sigma \qquad [11]$$

where Ψ_0^σ is the complementary energy of the homogeneous body without defects in the form of cracks or voids, [KAC 94]. The second Equation [11] shows that seeking explicit expressions for H, for $S_\sigma^{app\,c}$ or for $\Delta\Psi^\sigma$ for a microcracked body with homogeneous matrix are equivalent problems. This makes it possible to apply the formalism developed in [HUE 97] for the overall properties of multicracked bodies.

4. Crack density tensors and Strength Scaling Law for Non-Interacting cracks

When seeking explicit expressions of the generalized energy release rates for specific elastic materials, formulae of the kind derived by [MAU 92] and [KAC 94] for $\Delta\Psi^{\sigma}$ may be used. These formulae are based on tensorial variables of the form, for the kth defect in an isotropic matrix:

$$\lambda^k = \lambda^k_{\alpha\beta}\, n^{k\alpha} \otimes n^{k\beta} \qquad [12]$$

where $n^{k\alpha}$ are orientation vectors, while $\lambda^k_{\alpha\beta}$ are functions of the geometric parameters defining the dimensions and shape of the defects. Other terms with tensor products of higher orders in the $n^{k\alpha}$ may also be involved, together with cross-products in case of interacting defects. For an anisotropic matrix, products of the form $n^{k\alpha} \otimes n^{k\beta}$ are replaced by expressions of the form $n^{k\alpha} \otimes B \otimes n^{k\beta}$ where B is the crack compliance tensor (symmetric and of the second rank) relating the average $< b >^{k\alpha}$ of the defect surface displacement to a uniform traction vector applied on the defect surface $\Gamma^{k\alpha} = \partial D^{k\alpha}$ in the absence of external loading:

$$< [\xi]^+_- >^{k\alpha} \; = < b >^{k\alpha} \; = B.(\sigma_0 . n^{k\alpha}) \qquad [13]$$

As shown by Kachanov, this yields the anisotropy induced by a set of non-interacting cracks in an isotropic matrix to be an orthotropic one, with principal axes in the proper directions of the crack density tensor defined by:

$$\alpha = \frac{1}{A} \sum_k \; (l^2 n \otimes n)^k \qquad [14]$$

In an anisotropic matrix, another crack density tensor is needed. For non-interacting cracks, it is defined by:

$$\beta_\alpha = \frac{1}{A} \sum_k \; (ln \otimes B \otimes n)^{k\alpha} \qquad [15]$$

The form of the dependence of the scalar invariant function $\Delta\Psi^{\sigma}$ upon these parameters will thus involve a finite set of the proper basic invariants of these last and of joint invariants obtained from their tensor products with the stress tensor σ_0. From this, the size effects, boundary conditions effects and scaling laws may easily be derived by taking the partial derivatives of $\Delta\Psi^{\sigma}$ in terms of those of the defect parameters that are changing with the growth of the defect.

For non-interacting defects, the resulting $\Delta\Psi^{\sigma}$ will be obained by addition of the $\Delta\Psi^{\sigma k}$. For non-interacting straight microcracks in 2D, this gives, [KAC 94]:

$$\Delta\Psi^{\sigma} = \sum_{k=1}^{N} \Delta\Psi^{\sigma k} = \frac{\pi}{E_0}(\sigma_0 . \sigma_0):\alpha \qquad [16]$$

We consider an unnotched body with fixed size. For the growth of the kth crack in its initial direction, this gives the crack growth criterion in the form:

$$G_{\sigma}^{k}(t) = \frac{1}{2}\frac{\partial \Delta\Psi^{\sigma}}{\partial l^{k}} = R_F^{l^k} \qquad [17]$$

with:

$$R_F^{l^k} = G_c \quad for\ \dot{l}^{\,k} > 0 \quad ; R_F^{l^k} = 0 \quad for\ \dot{l}^{\,k} < 0 \qquad [18]$$

which characterizes LEFM for G_c being a constant. For uniaxial loading Equation [16] gives:

$$\sigma_N = \frac{C_0}{\sqrt{l^k}} \qquad [19]$$

which is a minimum for the largest crack of a given orientation, C_0 being a constant depending on the orientation of the crack. In fact, the strength of an unnotched body of a given size with non-interacting distributed microcracks is governed by the largest crack projection perpendicular to the load direction, [KAC 94] thus explaining the Rankine criterion found in numerical simulations, [WAN 94, 96].

We consider notched bodies with varying sizes having a set $\left\{l^i, i \neq k\right\}$ of initial cracks with fixed lengths while l^k with k fixed denotes a single notch with size proportional to the size of the body. We can, up to another constant written $1/d_0$, replace l^k by the size d of the body, which is varied in the scaling law. This gives:

$$\sigma_N = \frac{C_0}{\sqrt{\dfrac{d}{d_0}}} \quad ; \quad d > d_0 \qquad [20]$$

$$\sigma_N = C_0 \quad ; \quad d \leq d_0 \qquad [21]$$

which are the two asymptotes of the well known [BAZ 93] scaling law, observed in experimental data in many cases, [BAZ 98].

5. Crack density tensors and Strength Scaling law for Interacting cracks

For denser distributions, the cracks are interacting and the coupling between cracks is involved. For interacting cracks in an isotropic matrix, an expression of the apparent compliance was given by [MAU 92] in the form:

$$S_{\sigma}^{app\,c} = S_{\sigma} + \frac{1}{2A} \sum_{i,k} l^i l^k \Xi^{ik} \qquad [22]$$

where Ξ^{ik} is an appropriate tensor of the fourth rank expressing the coupling between the ith and kth cracks and of the form:

$$\Xi_{pqrs}^{ik} = \Omega_{qr}^{ik} n_p^i n_s^k + \Omega_{ps}^{ik} n_q^i n_r^k \qquad [23]$$

with Ω^{ik} an appropriate tensor of the second rank expressing the transmission of average stresses from one crack to another. For a growth of the kth crack in its initial direction, this gives the energy release rate in the form:

$$G_{\sigma}^{l^k}(t) = \frac{1}{2} \frac{\partial \underset{\sim}{S}_{\sigma\alpha}^{app\,c}}{\partial l_{\alpha}^k} : \sigma_0 : \sigma_0 = \frac{\sigma_0 \otimes \sigma_0}{2A} \sum_i l^k \Xi_{pqrs}^{ik} + \frac{1}{4A} \sum_i l^i \Xi_{pqrs}^{ii} \qquad [24]$$

For uniaxial loading, this gives in a critical state:

$$G_{\sigma}^{k}(t) = \frac{(\sigma_0)^2}{2A} [\ (\sum_i l^i \Xi_{1111}^{ik} l^i) + \frac{1}{4A} \Xi_{1111}^{kk} l^k\ ] = G_c \qquad [25]$$

Thus, the nominal strength σ_N has the form:

$$\sigma_N = \frac{C_0}{\sqrt{1 + C_1 l^k}} \qquad [26]$$

where C_0 and C_1 are functions of the l^i's for $i \neq k$.

Considering again that the set $\{l^i, i \neq k\}$ is a set of initial cracks with fixed lengths in the material, while l^k is a notch with size proportional to the size of the body which, up to another constant $1/d_0$, may be replaced by the characteristic size d of the body varied in the scaling law, we finally obtain:

$$\sigma_N = \frac{C_0}{\sqrt{1 + \frac{d}{d_0}}} \qquad [27]$$

in which the celebrated Bazant scaling law can be recognized. Since, in most real materials, interacting cracks do exist, even at a very small scale, this may explain the almost universal character of the Bazant law.

Let us emphasize the fact that here, the Bazant scaling law is derived using LEFM concepts only. Thus it is not necessary to call for Non-Linear Fracture Mechanics (NLFM) concepts (which correspond to non-constant G_c) to give a theoretical justification of this scaling law.

7. Conclusion

We have shown that the scaling laws observed experimentally in many situations may find a theoretical explanation even in the framework of linear elastic Fracture Thermodynamics for Materials with distributed microcracks, as always found at some scale in real materials and in particular in concrete, [HUE 93, 97].

Non-interacting cracks give the two asymptotes of the Bazant scaling law while interacting cracks with interaction terms limited to the second order give this law in its full expression.

In fact, there exist many other kinds of coupling which may give similar results. For instance, in the interaction of cracks with inclusions of voids, it may be expected that cross terms of the form $r^i l^k$ will be involved, where r^i is the radius of the ith grain, giving similar scaling law for the nominal strength of a notched body in terms of the size of the body at constant granulometry of the material for a notch proportional to the body size. This may explain why Bazant's scaling law applies very well to granular composites like concrete since microscopic observation shows that most of the cracks in concrete are very near grain-matrix interface, and that distributed voids of various sizes do exist, [HUE 93.]

The results obtained here are for static boundary conditions, for which the traction vector is prescribed on the whole boundary. Attention should be paid to the fact that, as shown in [HUE 90, 93, 97], the magnitude and even the trend of size effects are highly coupled with the boundary conditions. This has to be taken into account when trying to generalize the present results.

Some extensions of these results to the viscoelastic case, with and without aging, may be found in [HUE 98], with numerical simulation examples in [GUI 98].

Acknowledgements

Partial support for this work from the Swiss National Fonds for Scientific Research is gratefully acknowleged. This paper was completed while the author was a visiting guest of Prof. Ostoja-Starzewski at the Institute of Paper Science and Technology and the Georgia Institute of Technology in Atlanta, whose kind hospitality is also acknowleged.

References

[BAZ 93] BAZANT, Z.P., Scaling laws in Mechanics of Failure, *J. Eng. Mech.-ASCE*, 119(9), pp. 1828-1844, 1993

[BAZ 98] BAZANT, Z.P., PLANAS, J., *Fracture and Size Effect in Fracture and other Quasi-Brittle Materials*, CRC Press, London, 1998.

[GUI 98] GUIDOUM, A., HUET, C., CECOT, C., BENGOUGAM, A., 2D and 3D Finite-Element micromechanical computations of the long term development of internal stresses and deterioration in random viscoelastic granular composites with chemical solidification, climate sensitivity and microcracks. In E. Oñate and S.R. Idelsohn (Eds.), *Computational Mechanics,* New Trends and Applications, *Proceedings of IVth WCCM Congress,* Buenos-Aires, June 1998 ©CIMNE, Barcelona, Spain, Compact disk, 1998.

[HUE 90] HUET, C., Application of variational concepts to size effects in elastic heterogeneous bodies, *J. of the Mechanics and Physics of Solids*, vol. 38, pp. 813-841, 1990.

[HUE 93] HUET, C., *Micromechanics of Concrete and Cementitious Materials*, Presses Polytechniques et Universitaires Romandes, Lausanne, 1993.

[HUE 94] HUET, C., Some aspects of the application of fracture mechanics to concrete dams, In E. Bourdarot, J. Mazars and V. Saouma (Eds), *Dam Fracture and Damage.* Balkema, Rotterdam, 79-89, 1994.

[HUE 95] HUET, C., A continuum thermodynamics approach for studying microstructural effects on the non-linear fracture behaviour of concrete seen as a multicracked granular composite material, In F. H. Wittmann (Ed.), *Fracture Mechanics of Concrete and Concrete Structures,* Aedificatio, Freiburg, 1995, pp. 1089-1108, 1995.

[HUE 96] HUET, C., A continuum thermodynamics approach for size effects in the failure of concrete type materials and structures, In A. Carpinteri (Ed.), *Size-Scale Effects in the Failure Mechanisms of Materials and Structures*, Spon, London, pp. 259-276, 1996.

[HUE 97] HUET, C., An integrated micromechanics and statistical continuum thermodynamics approach for studying the fracture behaviour of microcracked heterogeneous materials with delayed response, *Engineering Fracture Mechanics*, Special Issue, 58, 5-6, pp. 459-556, 1997.

[HUE 96] HUET, C., Dissipative continuum and fracture thermodynamics, energy release and hierarchical bounds for overall properties of damaging viscoelastic composites with dissolution, solidification and microcracks.In E. Oñate and S.R. Idelsohn (Eds), *Computational Mechanics,* New Trends and Applications, *Proceedings of IVth WCCM Congress,* Buenos-Aires, June 1998, ©CIMNE, Barcelona, Spain, Compact disk, 1998.

[KAC 94] KACHANOV, M., Elastic Solids with many cracks and related problems, In J. W. Hutchinson and T. Y. Hu (Eds) *Advances in Applied Mechanics,* Vol. 3, pp. 259-445, 1994.

[MAU 92] MAUGE, C., KACHANOV, M., Interacting arbitrarily oriented cracks in anisotropic matrix. Stress intensity factor and effective moduli, *Int. J. of Fracture,* 58, R69-R74, 1992.

[WAN 94] WANG, J., NAVI, P., HUET, C., Numerical study of granule influence on the crack propagation in concrete, In Z.P. Bazant (Ed.), *Fracture Mechanics of Concrete Structures*, Elsevier, Amsterdam, pp. 373-378, 1994.

[WAN 96] WANG, J., NAVI, P., HUET, C., Numerical analysis of crack propagation in tension specimens of concrete considered as a 2D multicracked granular composite, *Materials and Structures,* 30, pp. 11-21, 1996.

Isotropic and Anisotropic Damage Models for Concrete Fracture

René de Borst

Koiter Institute Delft
Faculty of Civil Engineering and Geosciences
Delft University of Technology
NL-2600 GA Delft, The Netherlands

R.deBorst@ct.TUDelft.nl

ABSTRACT. *Isotropic and anisotropic damage formulations for concrete fracture are reviewed, including the classical fixed and rotating smeared crack models and more refined approaches based on the microplane concept. Higher-order gradients are introduced to avoid the boundary value problem becoming ill-posed at the onset of softening. For an infinite onedimensional bar dispersion analyses are carried out to examine the effect of using gradients of different strain measures or internal variables.*

KEY WORDS: *Smeared Cracking, Rotating Crack Model, Fixed Crack Model, Microplane Model, Damage, Failure, Fracture, Localisation, Regularisation, Dispersion Analysis.*

1. Introduction

Computational modelling of concrete structures started in the late 1960s with the landmark papers of Ngo and Scordelis [NGO 67] and Rashid [RAS 68], in which the discrete and smeared crack approaches were introduced. The following three decades have seen many refinements, both for the discrete crack models and for the smeared crack models. The two major deficiencies of the discrete crack approach, namely that crack propagation was restricted to interelement boundaries and that it caused a continuous change in the topology of the mesh, were remedied by the introduction of automatic remeshing at the crack tip [ING 95], and by a priori inserting interface elements in the finite element discretisation [ROT 91]. However, neither of these solutions removes both disadvantages. Smeared crack models have proven to be more flexible in the sense that, in principle, arbitrary crack propagation, including curved cracks, can be simulated since no topological constraints exists. Nevertheless, experience has shown that this advantage is less rigorous than it would seem, since it appeared rather difficult to simulate curved crack paths by smeared representations [BOR 86, ROT 91, FEE 95]. Indeed, discrete approaches have proven to be more successful, either when lattice type approaches were used [SCH 93], or when finite element methods with remeshing were employed (e.g. [ING 95]) based on pure linear elastic fracture mechanics, or van Vroonhoven and de Borst [VRO 97], who employed a hybrid approach, incorporating concepts of fracture mechanics as well as damage mechanics.

In the early 1990s it became apparent that the failure of the smeared approach to properly predict curved crack paths is rooted in the fact that a smeared concept inevitably introduces *strain* softening into the constitutive model, which at a certain level of loading causes a loss of well-posedness of the incremental boundary value problem. This ill-posedness creates an infinite number of solutions [BEN 88, BOR 93], from which a numerical method selects the solution with the smallest energy dissipation that is available in the finite dimensional solution space. In the limit of an infinitely dense mesh, solutions are computed which predict failure without energy dissipation, thus rendering the solution physically meaningless. Early solutions as fracture-energy models in their various forms [PIE 81, BAZ 83b, WIL 94], provide a partial solution for the mesh densification problem, but fail to repair the mesh bias issue, i.e. they still predict crack propagation along the direction of the grid lines.

A rigorous solution is the introduction of higher-order continuum models. The first models that were applied to fracture in concrete were nonlocal damage models [PIJ 87, BAZ 88a] and gradient plasticity models [BOR 92, BOR 96b]. Fully nonlocal approaches, in which spatially averaged quantities are employed in the constitutive models, are computationally unwieldy and are not believed to have potential for large-scale computations of concrete structures. The gradient approaches are more promising, but the gradient plasticity model suffers from the drawback that there is an internal boundary between the elastic and plastic domain, which necessitates a smooth interpolation of the plastic strain field. The required C^1-continuity is believed to reduce the ability of the gradient plasticity model to simulate curved crack propagation accurately, although globally proper directions of crack propagation were computed, independent of the discretisation. Gradient damage approaches, first introduced in a computationally feasible format in [PEE 96a], do not necessarily require a higher-order continuity of the interpolants of the strain field or the damage field and, as was recently shown in [PEE 98], are capable of simulating curved cracks.

This contribution will review recent developments of gradient damage approaches. We shall start by local, isotropic damage formulations, and then extend the formulation to anisotropic models, including various forms of smeared crack concepts, such as the fixed crack model, the rotating crack model [COP 80] and the microplane models [BAZ 84]. Then, starting from fully nonlocal damage models, we will develop isotropic and anisotropic gradient damage models. Specifically, we will derive a gradient smeared crack model, and we will indicate how gradient microplane models can be developed [KUH 98]. Some theoretical considerations using dispersion analyses for one-dimensional infinite media conclude the presentation to bring out some of the salient differences between the various types of gradient enhancements.

2. Standard Damage Models

2.1. *Isotropic Damage Models*

The basic structure of constitutive models that are set up in the spirit of damage mechanics is simple. We have a total stress-strain relation, which for the case of isotropic damage evolution reads

$$\sigma_{ij} = [(1-\omega_1)G(\delta_{ik}\delta_{jl} + \delta_{il}\delta_{jk} - 2/3\delta_{ij}\delta_{kl}) + (1-\omega_2)K\delta_{ij}\delta_{kl}]\varepsilon_{kl} \quad [1]$$

with G the virgin shear modulus and K the virgin bulk modulus, which are degraded by the scalar damage variables ω_1 and ω_2, respectively. A further simplification can be achieved if it is assumed that the secant shear stiffness and bulk moduli, $(1-\omega_1)G$ and $(1-\omega_2)K$, degrade in the same manner during damage growth. Essentially, this means that Poisson's ratio ν remains constant throughout the damage process and we have

$$\sigma_{ij} = (1-\omega)D^{e}_{ijkl}\varepsilon_{kl} \quad [2]$$

with ω the damage variable which grows from zero to one (at complete loss of integrity) and D^{e}_{ijkl} the fourth-order elastic stiffness tensor:

$$D^{e}_{ijkl} = \frac{E}{2(1+\nu)}(\delta_{ik}\delta_{jl} + \delta_{il}\delta_{jk}) + \frac{E\nu}{(1+\nu)(1-2\nu)}\delta_{ij}\delta_{kl} \quad [3]$$

with E the virgin Young's modulus. The total stress-strain relation [2] is complemented by a damage loading function f, which reads:

$$f = f(\tilde{\varepsilon}, \tilde{\sigma}, \kappa) \quad [4]$$

with $\tilde{\varepsilon}$ and $\tilde{\sigma}$ scalar-valued functions of the strain and the stress tensors, respectively, and κ the only remaining scalar history variable. The damage loading function f and the rate of the history variable, $\dot{\kappa}$, have to satisfy the discrete Kuhn-Tucker loading-unloading conditions:

$$f \leq 0 \quad , \quad \dot{\kappa} \geq 0 \quad , \quad f\dot{\kappa} = 0 \quad [5]$$

We first consider the case that the damage loading function does not depend on $\tilde{\sigma}$. Then,

$$f(\tilde{\varepsilon}, \kappa) = \tilde{\varepsilon} - \kappa \quad [6]$$

The parameter κ starts at a damage threshold level κ_{i} and is updated by the requirement that during damage growth $f = 0$. Damage growth occurs according to an evolution law such that

$$\omega = \omega(\kappa) \quad [7]$$

which can be determined from a uniaxial test. For instance, for a uniaxial (tensile) stress-strain diagram, where the tensile strength $f_{\mathrm{ct}} = E\kappa_{\mathrm{i}}$ is followed by a linear descending branch up to a strain κ_{u} at which the load-carrying capacity is exhausted (and thus $\omega = 1$), the evolution law reads:

$$\omega(\kappa) = \frac{\kappa_{\mathrm{u}}(\kappa - \kappa_{\mathrm{i}})}{\kappa(\kappa_{\mathrm{u}} - \kappa_{\mathrm{i}})} \quad [8]$$

Stress-based isotropic damage formulations can be elaborated by omitting $\tilde{\varepsilon}$ from the damage loading function f. Then, the following damage relation ensues

$$f(\tilde{\sigma}, \kappa) = \tilde{\sigma} - \kappa \quad [9]$$

still subjected to the Kuhn-Tucker loading-unloading conditions and equipped with an evolution relation for the damage variable $\omega = \omega(\kappa)$. Stress-based damage formulations bear some resemblance to plasticity approaches, with the notable difference that in plasticity the elastic stiffness moduli remain unchanged, while in an isotropic damage formalism they degrade in an isotropic fashion. However, as we will see in Section 4, for monotonic loading conditions and uniaxial stressing both models can be made identical. In a first step a damage strain $\varepsilon_{ij}^{\mathrm{d}}$ can be defined:

$$\varepsilon_{ij}^{\mathrm{d}} = \omega \varepsilon_{ij} \quad [10]$$

so that Eq. [2] changes into

$$\sigma_{ij} = D_{ijkl}^{\mathrm{e}}(\varepsilon_{kl} - \varepsilon_{kl}^{\mathrm{d}}) \quad [11]$$

In a fashion similar to the total equivalent strain $\tilde{\varepsilon}$, scalar measures for the equivalent stress, $\tilde{\sigma}$, and for the damage strain, $\tilde{\varepsilon}^{\mathrm{d}}$ can be defined. In fact, if the same definition is applied for $\tilde{\sigma}$, $\tilde{\varepsilon}$ and $\tilde{\varepsilon}^{\mathrm{d}}$, these quantities can be uniquely related for any given hardening/softening characteristics. Accordingly, the loading function [9] can also be cast in the following format:

$$f(\tilde{\varepsilon}^{\mathrm{d}}, \kappa) = \tilde{\varepsilon}^{\mathrm{d}} - \kappa \quad [12]$$

from which it becomes apparent that now $\kappa = \kappa(\tilde{\varepsilon}^{\mathrm{d}})$, quite similar to plasticity approaches. In the part where we compare the properties of the different damage approaches, this identity will be taken as point of departure for the evolution of the internal variable κ, or more precisely, we shall use

$$f(\tilde{\sigma}, \kappa) = \tilde{\sigma} - \kappa(\tilde{\varepsilon}^{\mathrm{d}}) \quad [13]$$

2.2. *Anisotropic Damage Models*

While isotropic damage models have been used successfully for describing progressive crack propagation, their disadvantage is that possible compressive strut action is eliminated. This is a drawback especially for the analysis of reinforced concrete members. Directional dependence of damage evolution can be incorporated by degrading the Young's modulus E in a preferential direction. When, for plane-stress conditions, distinction is made between the global x, y-coordinate system and a local n, s-coordinate system, a simple loading function in the local coordinate system would be

$$f(\varepsilon_{nn}, \kappa) = \varepsilon_{nn} - \kappa \quad [14]$$

with ε_{nn} the normal strain in the local n, s-coordinate system, subject to the standard Kuhn-Tucker loading-unloading conditions. The secant stiffness relation now reads

$$\boldsymbol{\sigma}_{\mathrm{ns}} = \mathbf{D}^{\mathrm{s}}_{\mathrm{ns}} \boldsymbol{\varepsilon}_{\mathrm{ns}}, \quad [15]$$

with $\boldsymbol{\sigma}_{\mathrm{ns}} = [\sigma_{nn}, \sigma_{ss}, \sigma_{ns}]^{\mathrm{T}}$, $\boldsymbol{\varepsilon}_{\mathrm{ns}} = [\varepsilon_{nn}, \varepsilon_{ss}, \gamma_{ns}]^{\mathrm{T}}$ and $\mathbf{D}^{\mathrm{s}}_{\mathrm{ns}}$ defined as

$$\mathbf{D}^{\mathrm{s}}_{\mathrm{ns}} = \begin{bmatrix} \dfrac{(1-\omega_1)E}{1-(1-\omega_1)\nu^2} & \dfrac{(1-\omega_1)\nu E}{1-(1-\omega_1)\nu^2} & 0 \\ \dfrac{(1-\omega_1)\nu E}{1-(1-\omega_1)\nu^2} & \dfrac{E}{1-(1-\omega_1)\nu^2} & 0 \\ 0 & 0 & (1-\omega_2)G \end{bmatrix} \quad [16]$$

with $\omega_1 = \omega_1(\kappa)$ and $\omega_2 = \omega_2(\kappa)$. The factor $1 - \omega_2$ represents the degradation of the shear stiffness and can be identified with the traditional shear retention factor β [SUI 73]. It is emphasised that because of the choice of a preferential direction in which damage takes place, the damage variables ω_1 and ω_2 have an entirely different meaning than those that were introduced in the basic isotropic formulation of Eq. [1].

If we introduce ϕ as the angle from the x-axis to the n-axis, we can relate the components of $\boldsymbol{\varepsilon}_{\mathrm{ns}}$ and $\boldsymbol{\sigma}_{\mathrm{ns}}$ to those in the global x, y-coordinate system via the standard transformation matrices $\mathbf{T}_\varepsilon$ and $\mathbf{T}_\sigma$:

$$\boldsymbol{\varepsilon}_{\mathrm{ns}} = \mathbf{T}_\varepsilon(\phi)\boldsymbol{\varepsilon}_{\mathrm{xy}} \quad [17]$$

and

$$\boldsymbol{\sigma}_{\mathrm{ns}} = \mathbf{T}_\sigma(\phi)\boldsymbol{\sigma}_{\mathrm{xy}} \quad [18]$$

Using Eqs [17] and [18], the damage loading function [14] can be written in terms of the strain components ε_{xx}, ε_{yy} and γ_{xy} of the global x, y-coordinate system:

$$f = \varepsilon_{xx} \cos^2 \phi + \varepsilon_{yy} \sin^2 \phi + \gamma_{xy} \sin \phi \cos \phi - \kappa \tag{19}$$

Similarly, we obtain the following secant stress-strain relation instead of Eq. (15):

$$\boldsymbol{\sigma}_{\mathrm{xy}} = \mathbf{T}_{\sigma}^{-1}(\phi)\, \mathbf{D}_{\mathrm{ns}}^{\mathrm{s}} \mathbf{T}_{\varepsilon}(\phi) \boldsymbol{\varepsilon}_{\mathrm{xy}} \tag{20}$$

Eqs [19] and [20] incorporate the traditional fixed crack model and the rotating crack model. The only difference is that in the fixed crack model the inclination angle ϕ is fixed when the major principal stress first reaches the tensile strength ($\phi = \phi_0$), while in the rotating crack concept ϕ changes such that the n-axis continues to coincide with the major principal stress direction. This difference has profound consequences when deriving the tangential stiffness, especially with regard to the shear term.

The above framework also allows for incorporation of constitutive models that are based on the microplane concept. As an example, we shall consider a microplane model based on the so-called kinematic constraint, which implies that the normal and tangential strains on a microplane that is labelled α, can be derived by a simple projection of the global strains $\boldsymbol{\varepsilon}_{\mathrm{xy}}$ similar to Eq. [17]:

$$\boldsymbol{\varepsilon}_{\mathrm{ns}}^{\alpha} = \mathbf{T}_{\varepsilon}(\phi^{\alpha}) \boldsymbol{\varepsilon}_{\mathrm{xy}} \tag{21}$$

The stresses on this microplane can be derived in a fashion similar to Eq. [15]:

$$\boldsymbol{\sigma}_{\mathrm{ns}}^{\alpha} = \mathbf{D}_{\mathrm{ns}}^{\alpha} \boldsymbol{\varepsilon}_{\mathrm{ns}}^{\alpha} \tag{22}$$

with $\mathbf{D}_{\mathrm{ns}}^{\alpha}$ given by:

$$\mathbf{D}_{\mathrm{ns}}^{\alpha} = \begin{bmatrix} (1-\omega_{\mathrm{N}}^{\alpha})E_{\mathrm{N}} & 0 & 0 \\ 0 & 0 & 0 \\ 0 & 0 & (1-\omega_{\mathrm{T}}^{\alpha})E_{\mathrm{T}} \end{bmatrix} \tag{23}$$

where the initial stiffness moduli E_{N} and E_{T} are functions of Young's modulus, Poisson's ratio and a weight parameter (e.g. [BAZ 88b]). The damage parameters $\omega_{\mathrm{N}}^{\alpha}$ and $\omega_{\mathrm{T}}^{\alpha}$ for the normal stiffness and the shear stiffness are functions of the history parameters $\kappa_{\mathrm{N}}^{\alpha}$ and $\kappa_{\mathrm{T}}^{\alpha}$ in standard fashion: $\omega_{\mathrm{N}}^{\alpha} = \omega_{\mathrm{N}}^{\alpha}(\kappa_{\mathrm{N}}^{\alpha})$ and $\omega_{\mathrm{T}}^{\alpha} = \omega_{\mathrm{T}}^{\alpha}(\kappa_{\mathrm{T}}^{\alpha})$. The main departure from the fixed crack model as outlined above is the fact that we need two damage loading functions on each microplane α:

$$f_{\mathrm{N}}^{\alpha} = \varepsilon_{nn}^{\alpha} - \kappa_{\mathrm{N}}^{\alpha} \tag{24}$$

and

$$f_{\mathrm{T}}^{\alpha} = \gamma_{ns}^{\alpha} - \kappa_{\mathrm{T}}^{\alpha} \tag{25}$$

each subject to the standard Kuhn-Tucker loading-unloading conditions. Finally, the stresses in the global x, y coordinate system are recovered by summing over all the microplanes and by transforming them in a standard fashion according to Eq. [18]. With Eqs [21] and [22], we finally arrive at:

$$\boldsymbol{\sigma}_{\mathrm{xy}} = \sum_{\alpha=1}^{n} w^{\alpha} \mathbf{T}_{\sigma}^{-1}(\phi^{\alpha}) \mathbf{D}_{\mathrm{ns}}^{\alpha} \mathbf{T}_{\varepsilon}(\phi^{\alpha}) \boldsymbol{\varepsilon}_{\mathrm{xy}} \qquad [26]$$

with n the chosen number of microplanes and w^{α} the weighting factors.

Attention is drawn to the fact that the second row of $\mathbf{D}_{\mathrm{ns}}^{\alpha}$ consists of zeros. This is because in the microplane concept only the normal stress and the shear stress are resolved on each microplane. The normal stress parallel to this plane therefore becomes irrelevant. Furthermore, attention is drawn to the fact that here we use a relatively simple version of the microplane model, namely one in which no splitting in volumetric and deviatoric components is considered [BAZ 84]. Nevertheless, more sophisticated microplane models, e.g. [BAZ 88b], which incorporate such a split, can be captured by the same formalism [KUH 98]. It is finally noted that the microplane is very similar to the multiple fixed-crack model [BOR 85, BOR 87], except for the fact that the multiple fixed-crack model has been formulated in terms of a strain decomposition in the sense of the stress-based damage model of Eq. [13].

For the fixed crack model differentiation of Eq. [20] yields the tangential stress-strain relation needed in an incremental-iterative procedure which utilises the Newton-Raphson method:

$$\dot{\boldsymbol{\sigma}}_{\mathrm{xy}} = \mathbf{T}_{\sigma}^{-1}(\phi_0)(\mathbf{D}_{\mathrm{ns}}^{\mathrm{s}} - \Delta\mathbf{D}_{\mathrm{ns}}) \mathbf{T}_{\varepsilon}(\phi_0) \dot{\boldsymbol{\varepsilon}}_{\mathrm{xy}} \qquad [27]$$

with $\mathbf{D}_{\mathrm{ns}}^{\mathrm{s}}$ given by Eq. [16] and

$$\Delta\mathbf{D}_{\mathrm{ns}} = \begin{bmatrix} d_{11} & 0 & 0 \\ \nu d_{11} & 0 & 0 \\ d_{31} & 0 & 0 \end{bmatrix} \qquad [28]$$

with

$$d_{11} = \frac{\partial\omega_1}{\partial\kappa} \frac{\partial\kappa}{\partial\varepsilon_{nn}} \frac{E(\varepsilon_{nn} + \nu\varepsilon_{ss})}{(1-(1-\omega_1)\nu^2)^2} \quad \text{and} \quad d_{31} = \frac{\partial\omega_2}{\partial\kappa} \frac{\partial\kappa}{\partial\varepsilon_{nn}} G\gamma_{ns}$$

$\partial\kappa/\partial\varepsilon_{nn} = 1$ upon loading and zero otherwise. We observe that the local material tangential stiffness matrix $\mathbf{D}_{\mathrm{ns}}^{\mathrm{s}} - \Delta\mathbf{D}_{\mathrm{ns}}$ generally becomes non-symmetric.

The fact that in the rotating smeared crack model the local coordinate system of the crack and the principal axes of stress and strain coincide throughout the entire deformation process, implies that the secant stiffness matrix $\mathbf{D}_{\mathrm{ns}}^{\mathrm{s}}$ relates principal stresses to principal strains, and that a secant shear stiffness becomes superfluous. Consequently, there is only one remaining damage parameter, $\omega_1 = \omega$, and we have

$$\mathbf{D}_{\mathrm{ns}}^{\mathrm{s}} = \begin{bmatrix} \dfrac{(1-\omega)E}{1-(1-\omega)\nu^2} & \dfrac{(1-\omega)\nu E}{1-(1-\omega)\nu^2} & 0 \\ \dfrac{(1-\omega)\nu E}{1-(1-\omega)\nu^2} & \dfrac{E}{1-(1-\omega)\nu^2} & 0 \\ 0 & 0 & 0 \end{bmatrix} \qquad [29]$$

instead of expression [16]. Using Eq. [29], differentiation of the secant stiffness relation in global coordinates yields [BAZ 83a, WIL 86]:

$$\dot{\boldsymbol{\sigma}}_{\mathrm{xy}} = \mathbf{T}_{\sigma}^{-1}(\phi)(\mathbf{D}_{\mathrm{ns}}^{\mathrm{s}} - \Delta\mathbf{D}_{\mathrm{ns}})\mathbf{T}_{\varepsilon}(\phi)\dot{\boldsymbol{\varepsilon}}_{\mathrm{xy}} \qquad [30]$$

with $\Delta\mathbf{D}_{\mathrm{ns}}$ now given by:

$$\Delta\mathbf{D}_{\mathrm{ns}} = \begin{bmatrix} d_{11} & 0 & 0 \\ \nu d_{11} & 0 & 0 \\ 0 & 0 & d_{33} \end{bmatrix} \qquad [31]$$

with

$$d_{11} = \frac{\partial\omega}{\partial\kappa}\frac{\partial\kappa}{\partial\varepsilon_{nn}}\frac{E(\varepsilon_{nn}+\nu\varepsilon_{ss})}{(1-(1-\omega)\nu^2)^2} \quad \text{and} \quad d_{33} = -\frac{\sigma_{nn}-\sigma_{ss}}{2(\varepsilon_{nn}-\varepsilon_{ss})}$$

The tangential stress-strain relation of the microplane model can be cast in the same formalism as that of the rotating and the fixed crack models. Indeed, from linearisation of Eq. [26] we obtain

$$\dot{\boldsymbol{\sigma}}_{\mathrm{xy}} = \sum_{\alpha=1}^{n} w^{\alpha}\mathbf{T}_{\sigma}^{-1}(\phi^{\alpha})(\mathbf{D}_{\mathrm{ns}}^{\alpha} - \Delta\mathbf{D}_{\mathrm{ns}}^{\alpha})\mathbf{T}_{\varepsilon}(\phi^{\alpha})\dot{\boldsymbol{\varepsilon}}_{\mathrm{xy}} \qquad [32]$$

with $\mathbf{D}_{\mathrm{ns}}^{\alpha}$ given by Eq. [23] and $\Delta\mathbf{D}_{\mathrm{ns}}^{\alpha}$ defined as

$$\Delta\mathbf{D}_{\mathrm{ns}}^{\alpha} = \begin{bmatrix} d_{11}^{\alpha} & 0 & 0 \\ 0 & 0 & 0 \\ 0 & 0 & d_{33}^{\alpha} \end{bmatrix} \qquad [33]$$

with

$$d_{11}^{\alpha} = \frac{\partial\omega_{\mathrm{N}}^{\alpha}}{\partial\kappa_{\mathrm{N}}^{\alpha}}\frac{\partial\kappa_{\mathrm{N}}^{\alpha}}{\partial\varepsilon_{nn}^{\alpha}} E_{\mathrm{N}}\varepsilon_{nn}^{\alpha}$$

where $\partial\kappa_{\mathrm{N}}^{\alpha}/\partial\varepsilon_{nn}^{\alpha} = 1$ if $f_{\mathrm{N}}^{\alpha} = 0$ and zero otherwise, and

$$d_{33}^{\alpha} = \frac{\partial\omega_{\mathrm{T}}^{\alpha}}{\partial\kappa_{\mathrm{T}}^{\alpha}}\frac{\partial\kappa_{\mathrm{T}}^{\alpha}}{\partial\gamma_{ns}^{\alpha}} E_{\mathrm{T}}\gamma_{ns}^{\alpha}$$

while $\partial\kappa_{\mathrm{T}}^{\alpha}/\partial\gamma_{ns}^{\alpha} = 1$ if $f_{\mathrm{T}}^{\alpha} = 0$ and zero otherwise.

3. Enhanced Damage Models

In a nonlocal generalisation, the equivalent strain $\tilde{\varepsilon}$ is normally replaced by a spatially averaged quantity in the damage loading function [PIJ 87, BAZ 88a]:

$$f(\bar{\varepsilon}, \kappa) = \bar{\varepsilon} - \kappa \qquad [34]$$

where the nonlocal strain $\bar{\varepsilon}$ is computed from:

$$\bar{\varepsilon}(\mathbf{x}) = \frac{1}{V_{\mathrm{r}}(\mathbf{x})} \int_V \psi(\mathbf{s}) \tilde{\varepsilon}(\mathbf{x}+\mathbf{s})\, \mathrm{d}V \quad , \quad V_{\mathrm{r}}(\mathbf{x}) = \int_V \psi(\mathbf{s})\, \mathrm{d}V \qquad [35]$$

with $\psi(\mathbf{s})$ a weight function, e.g., the error function, and $\mathbf{s}$ a relative position vector pointing to the infinitesimal volume $\mathrm{d}V$. In this formulation all the other relations remain local: the Kuhn-Tucker loading-unloading conditions [5], the local stress-strain relation [2] and the dependence of the damage variable ω on the history variable κ, Eq. [7].

Alternatively, the locally defined history parameter κ may be replaced in the damage loading function f by a spatially averaged quantity:

$$f(\tilde{\varepsilon}, \bar{\kappa}) = \tilde{\varepsilon} - \bar{\kappa} \qquad [36]$$

where the nonlocal history parameter $\bar{\kappa}$ follows from:

$$\bar{\kappa}(\mathbf{x}) = \frac{1}{V_{\mathrm{r}}(\mathbf{x})} \int_V \psi(\mathbf{s}) \kappa(\mathbf{x}+\mathbf{s})\, \mathrm{d}V \qquad [37]$$

where ψ typically has a different form than the weight function in [35]. This alternative formulation requires some slight modifications with respect to some of the other governing equations. The local stress-strain relation [2] and the damage evolution law [7] remain unaltered, but the Kuhn-Tucker loading-unloading conditions must be rephrased as:

$$f \leq 0 \quad , \quad \dot{\bar{\kappa}} \geq 0 \quad , \quad f \dot{\bar{\kappa}} = 0 \qquad [38]$$

3.1. *Isotropic Gradient Damage Models*

Nonlocal constitutive relations can be considered as a point of departure for constructing gradient models, although we wish to emphasise that the latter class of models can also be defined directly by supplying higher-order gradients in the damage loading function. Yet, we will follow the first-mentioned route to underline the connection between integral and differential type nonlocal models. Again, this can either be done by expanding the kernel $\tilde{\varepsilon}$ of the integral in [35] in a Taylor series, or by expanding the history parameter κ in [37] in a Taylor series. We will first consider the expansion of $\tilde{\varepsilon}$ and then we will do the same for κ. If we truncate after the second-order terms and carry out the integration implied in [35] under the assumption of isotropy, the following relation ensues:

$$\bar{\varepsilon} = \tilde{\varepsilon} + g\nabla^2\tilde{\varepsilon} \qquad [39]$$

where g is a material parameter of the dimension length squared. It can be related to the averaging volume and then becomes dependent on the precise form of the weight function ψ. For instance, for a one-dimensional continuum and taking

$$\psi(s) = \frac{1}{\sqrt{2\pi}l}\, \mathrm{e}^{-s^2/2l^2} \qquad [40]$$

we obtain $g = {}^1\!/_2 l^2$. Here, we adopt the phenomenological view that $\sqrt{g}$ reflects the length scale of the failure process which we wish to describe macroscopically.

Formulation [39] has a severe disadvantage when applied in a finite element context, namely that it requires computation of second-order gradients of the local equivalent strain $\tilde{\varepsilon}$. Since this quantity is a function of the strain tensor, and since the strain tensor involves first-order derivatives of the displacements, third-order derivatives of the displacements have to be computed, which would necessitate C^1-continuity of the shape functions. To obviate this problem, Eq. [39] is differentiated twice and the result is substituted again into Eq. [39]. Again, neglecting fourth-order terms then leads to

$$\bar{\varepsilon} - g\nabla^2\bar{\varepsilon} = \tilde{\varepsilon} \qquad [41]$$

When $\bar{\varepsilon}$ is discretised independently and use is made of the divergence theorem, a C^0-interpolation for $\bar{\varepsilon}$ suffices [PEE 96a].

In a fashion similar to the derivation of the gradient damage models based on the averaging of the equivalent strain $\tilde{\varepsilon}$, we can elaborate a gradient approximation of [37], i.e., by developing κ into a Taylor series. For an isotropic, infinite medium and truncating after the second term, we then have [BOR 96a]:

$$\bar{\kappa} = \kappa + g\nabla^2\kappa \qquad [42]$$

Since the weight functions for the different gradient formulations may be quite different, the gradient parameter g may also be very different for the various formulations. For instance, the gradient parameter g of [42] may differ considerably from that in [39] or [41]. According to Eq. [7], the damage variable ω is a function of the internal variable κ, and therefore, the differential equation [42] can be replaced by [BOR 96a, COM 98]:

$$\bar{\omega} = \omega + g\nabla^2\omega \qquad [43]$$

where $\bar{\omega}$ is a spatially averaged damage field, similar to $\bar{\varepsilon}$ or $\bar{\kappa}$.

For stress-based isotropic damage models, we replace the loading function [13] by

$$f(\tilde{\sigma}, \kappa) = \tilde{\sigma} - \kappa(\bar{\varepsilon}^{\,d}) \quad [44]$$

with

$$\bar{\varepsilon}^{\,d} = \tilde{\varepsilon}^{\,d} + g\nabla^2\tilde{\varepsilon}^{\,d} \quad [45]$$

It is noted that the set of Eqs [44]-[45] formally becomes identical to the gradient-enhanced plasticity model [BOR 92]. Of course, the internal variable $\tilde{\varepsilon}^{\,d}$ has a completely different meaning than the equivalent plastic strain $\tilde{\varepsilon}^{\,p}$ in plasticity approaches, in the sense that $\tilde{\varepsilon}^{\,d}$ does not correspond to a permanent strain and that, accordingly, all strain is recoverable.

3.2. *Anisotropic Gradient Damage Models*

Anisotropic gradient damage models can be developed in a manner similar to isotropic gradient damage models. We take as point of departure the damage loading function [14] in the local n, s-coordinate system, where we replace the normal strain ε_{nn} by its nonlocal equivalent:

$$f(\bar{\varepsilon}_{nn}, \kappa) = \bar{\varepsilon}_{nn} - \kappa \quad [46]$$

or, identically

$$f = \bar{\varepsilon}_{xx}\cos^2\phi + \bar{\varepsilon}_{yy}\sin^2\phi + \bar{\gamma}_{xy}\sin\phi\cos\phi - \kappa \quad [47]$$

instead of Eq. [19]. We can now apply the averaging process to each of the (nonlocal) strain components $\bar{\varepsilon}_{xx}$, $\bar{\varepsilon}_{yy}$ and $\bar{\gamma}_{xy}$. In the spirit of Eq. [41] they have to satisfy

$$\begin{aligned} \bar{\varepsilon}_{xx} - g\nabla^2\bar{\varepsilon}_{xx} &= \varepsilon_{xx} \\ \bar{\varepsilon}_{yy} - g\nabla^2\bar{\varepsilon}_{yy} &= \varepsilon_{yy} \\ \bar{\gamma}_{xy} - g\nabla^2\bar{\gamma}_{xy} &= \gamma_{xy} \end{aligned} \quad [48]$$

or written in a vector format

$$\bar{\boldsymbol{\varepsilon}}_{\mathrm{xy}} - g\nabla^2\bar{\boldsymbol{\varepsilon}}_{\mathrm{xy}} = \boldsymbol{\varepsilon}_{\mathrm{xy}} \quad [49]$$

After solving the set of Helmholtz equations [49], $\bar{\boldsymbol{\varepsilon}}_{\mathrm{xy}}$ is substituted into the damage loading function [47] to determine whether loading occurs, and then into the secant stress-strain relation to solve for the stresses $\boldsymbol{\sigma}_{\mathrm{xy}}$.

The tangential stress-strain relation attains a slightly different format due to the fact that the nonlocal strain components must be discretised independently. For a gradient-enhanced version of the fixed crack model, we obtain upon linearisation of Eq. [20]

$$\dot{\boldsymbol{\sigma}}_{\mathrm{xy}} = \mathbf{T}_\sigma^{-1}(\phi_0)\mathbf{D}^{\mathrm{s}}_{\mathrm{ns}}\mathbf{T}_\varepsilon(\phi_0)\dot{\boldsymbol{\varepsilon}}_{\mathrm{xy}} - \mathbf{T}_\sigma^{-1}(\phi_0)\Delta\mathbf{D}_{\mathrm{ns}}\mathbf{T}_\varepsilon(\phi_0)\dot{\bar{\boldsymbol{\varepsilon}}}_{\mathrm{xy}} \quad [50]$$

with

$$\Delta \mathbf{D}_{\rm ns} = \begin{bmatrix} d_{11} & 0 & 0 \\ 0 & 0 & 0 \\ d_{31} & 0 & 0 \end{bmatrix} \qquad [51]$$

where

$$d_{11} = \frac{\partial \omega_1}{\partial \kappa} \frac{\partial \kappa}{\partial \bar{\varepsilon}_{nn}} E \varepsilon_{nn} \quad \text{and} \quad d_{31} = \frac{\partial \omega_2}{\partial \kappa} \frac{\partial \kappa}{\partial \bar{\varepsilon}_{nn}} G \gamma_{ns}$$

$\partial \kappa / \partial \bar{\varepsilon}_{nn} = 1$ upon loading and zero otherwise.

For the gradient-enhanced rotating crack model, we obtain upon linearisation of Eq. [20]

$$\dot{\boldsymbol{\sigma}}_{\rm xy} = \mathbf{T}_\sigma^{-1}(\phi) \mathbf{D}_{\rm ns}^{\rm s} \mathbf{T}_\varepsilon(\phi) \dot{\boldsymbol{\varepsilon}}_{\rm xy} - \mathbf{T}_\sigma^{-1}(\phi) \Delta \mathbf{D}_{\rm ns} \mathbf{T}_\varepsilon(\phi) \dot{\bar{\boldsymbol{\varepsilon}}}_{\rm xy} \qquad [52]$$

but now $\Delta \mathbf{D}_{\rm ns}$ is given by

$$\Delta \mathbf{D}_{\rm ns} = \begin{bmatrix} d_{11} & 0 & 0 \\ 0 & 0 & 0 \\ 0 & 0 & d_{33} \end{bmatrix} \qquad [53]$$

with

$$d_{11} = \frac{\partial \omega}{\partial \kappa} \frac{\partial \kappa}{\partial \bar{\varepsilon}_{nn}} E \varepsilon_{nn} \quad \text{and} \quad d_{33} = -\frac{\sigma_{nn} - \sigma_{ss}}{2(\varepsilon_{nn} - \varepsilon_{ss})}$$

In a gradient-enhanced microplane model [KUH 98], the single damage loading condition [46], or equivalently Eq. [47], must be replaced by two loading functions for each microplane:

$$f_{\rm N} = \bar{\varepsilon}_{nn}^\alpha - \kappa_{\rm N}^\alpha \quad \text{and} \quad f_{\rm T} = \bar{\gamma}_{ns}^\alpha - \kappa_{\rm T}^\alpha \qquad [54]$$

where $\bar{\varepsilon}_{nn}^\alpha$ is obtained by solving Eq. [49] for $\bar{\varepsilon}_{xx}$, $\bar{\varepsilon}_{yy}$ and $\bar{\varepsilon}_{xy}$ followed by a transformation to the coordinate system of microplane α via the matrix $\mathbf{T}_\varepsilon(\phi^\alpha)$. The tangential stiffness relation can then be derived to be of a form similar to the fixed and rotating crack models. Indeed, upon linearisation of Eq. [26] we obtain

$$\dot{\boldsymbol{\sigma}}_{\rm xy} = \sum_{\alpha=1}^{n} w^\alpha \mathbf{T}_\sigma^{-1}(\phi^\alpha) \left[\mathbf{D}_{\rm ns}^\alpha \mathbf{T}_\varepsilon(\phi^\alpha) \dot{\boldsymbol{\varepsilon}}_{\rm xy} - \Delta \mathbf{D}_{\rm ns}^\alpha \mathbf{T}_\varepsilon(\phi^\alpha) \dot{\bar{\boldsymbol{\varepsilon}}}_{\rm xy} \right] \qquad [55]$$

with $\mathbf{D}_{\rm ns}^\alpha$ given by Eq. [23] and $\Delta \mathbf{D}_{\rm ns}^\alpha$ defined as

$$\Delta \mathbf{D}_{\rm ns}^\alpha = \begin{bmatrix} d_{11}^\alpha & 0 & 0 \\ 0 & 0 & 0 \\ 0 & 0 & d_{33}^\alpha \end{bmatrix} \qquad [56]$$

with

$$d_{11}^{\alpha} = \frac{\partial \omega_{\mathrm{N}}^{\alpha}}{\partial \kappa_{\mathrm{N}}^{\alpha}} \frac{\partial \kappa_{\mathrm{N}}^{\alpha}}{\partial \bar{\varepsilon}_{nn}^{\alpha}} E_{\mathrm{N}} \varepsilon_{nn}^{\alpha}$$

with $\partial \kappa_{\mathrm{N}}^{\alpha} / \partial \bar{\varepsilon}_{nn}^{\alpha} = 1$ if $f_{\mathrm{N}}^{\alpha} = 0$ and zero otherwise, and

$$d_{33}^{\alpha} = \frac{\partial \omega_{\mathrm{T}}^{\alpha}}{\partial \kappa_{\mathrm{T}}^{\alpha}} \frac{\partial \kappa_{\mathrm{T}}^{\alpha}}{\partial \bar{\gamma}_{ns}^{\alpha}} E_{\mathrm{T}} \gamma_{ns}^{\alpha}$$

$\partial \kappa_{\mathrm{T}}^{\alpha} / \partial \bar{\gamma}_{ns}^{\alpha} = 1$ if $f_{\mathrm{T}}^{\alpha} = 0$ and zero otherwise.

We now introduce the interpolation matrices $\mathbf{N}_{\mathrm{a}}$ and $\mathbf{N}_{\bar{\varepsilon}}$ for the displacements $\mathbf{u}$ and the nonlocal strains $\bar{\boldsymbol{\varepsilon}}_{\mathrm{xy}}$, respectively:

$$\mathbf{u} = \mathbf{N}_{\mathrm{a}} \mathbf{a} \quad , \quad \bar{\boldsymbol{\varepsilon}}_{\mathrm{xy}} = \mathbf{N}_{\bar{\varepsilon}} \bar{\boldsymbol{\varepsilon}} \qquad [57]$$

The strains and the gradients of the nonlocal strains are derived in a standard manner as

$$\boldsymbol{\varepsilon} = \mathbf{B}_{\mathrm{a}} \mathbf{a} \quad , \quad \nabla \bar{\boldsymbol{\varepsilon}}_{\mathrm{xy}} = \mathbf{B}_{\bar{\varepsilon}} \bar{\boldsymbol{\varepsilon}} \qquad [58]$$

so that a consistent incremental-iterative scheme is written as

$$\begin{bmatrix} \mathbf{K}_{\mathrm{aa}} & \mathbf{K}_{\mathrm{a}\bar{\varepsilon}} \\ \mathbf{K}_{\bar{\varepsilon}\mathrm{a}} & \mathbf{K}_{\bar{\varepsilon}\bar{\varepsilon}} \end{bmatrix} \begin{bmatrix} \mathrm{d}\mathbf{a} \\ \mathrm{d}\bar{\boldsymbol{\varepsilon}} \end{bmatrix} = \begin{bmatrix} \mathbf{f}_{\mathrm{ext}} - \mathbf{f}_{\mathrm{int}} \\ \mathbf{f}_{\bar{\varepsilon}} - \mathbf{K}_{\bar{\varepsilon}\bar{\varepsilon}} \bar{\boldsymbol{\varepsilon}} \end{bmatrix} \qquad [59]$$

with $\mathrm{d}\mathbf{a}$ and $\mathrm{d}\bar{\boldsymbol{\varepsilon}}$ the iterative corrections to the nodal values of the displacements and the nonlocal strains, respectively. In Eq. [59] the following definitions for the sub-matrices have been employed

$$\mathbf{K}_{\mathrm{aa}} = \int_V \sum_{\alpha=1}^{n} w^{\alpha} \mathbf{B}_{\mathrm{a}}^{\mathrm{T}} \mathbf{T}_{\sigma}^{-1} \mathbf{D}_{\mathrm{ns}}^{\alpha} \mathbf{T}_{\varepsilon} \mathbf{B}_{\mathrm{a}} \, \mathrm{d}V$$

$$\mathbf{K}_{\bar{\varepsilon}\bar{\varepsilon}} = \int_V (\mathbf{N}_{\bar{\varepsilon}}^{\mathrm{T}} \mathbf{N}_{\bar{\varepsilon}} + g \mathbf{B}_{\bar{\varepsilon}}^{\mathrm{T}} \mathbf{B}_{\bar{\varepsilon}}) \, \mathrm{d}V$$

$$\mathbf{K}_{\mathrm{a}\bar{\varepsilon}} = -\int_V \sum_{\alpha=1}^{n} w^{\alpha} \mathbf{B}_{\mathrm{a}}^{\mathrm{T}} \mathbf{T}_{\sigma}^{-1} \Delta \mathbf{D}_{\mathrm{ns}} \mathbf{T}_{\varepsilon} \mathbf{N}_{\bar{\varepsilon}} \, \mathrm{d}V$$

$$\mathbf{K}_{\bar{\varepsilon}\mathrm{a}} = -\int_V \mathbf{N}_{\bar{\varepsilon}}^{\mathrm{T}} \mathbf{B}_{\mathrm{a}} \, \mathrm{d}V$$

where $n = 1$ and $w^{\alpha} = 1$ for the fixed and rotating crack models. For the right-hand side vectors we have

$$\mathbf{f}_{\mathrm{int}} = \int_V \mathbf{B}_{\mathrm{a}}^{\mathrm{T}} \boldsymbol{\sigma} \, \mathrm{d}V$$

$$\mathbf{f}_{\bar{\varepsilon}} = \int_V \mathbf{N}_{\bar{\varepsilon}}^{\mathrm{T}} \boldsymbol{\varepsilon} \, \mathrm{d}V$$

and $\mathbf{f}_{\text{ext}}$ the standard external load vector that arises from body forces and boundary tractions.

4. Dispersion Analyses

In the isotropic gradient-enhanced damage formulations of Section 3.1, different variables have been considered for regularising the governing set of field equations. In Section 3.2 on anisotropic gradient-enhanced damage models, the regularisation by means of the nonlocal strain $\bar{\varepsilon}$ [PEE 96a] has been used exclusively. Indeed, this regularisation has proven to be theoretically sound [PEE 96b], computationally robust and versatile in the sense that curved crack propagation can be simulated [PEE 98]. Below we shall carry out dispersion analyses for the various gradient models and use these results to bring out their typical properties, weaknesses and strengths.

We consider an infinite one-dimensional medium, so that the equations of motion, the kinematic equations and the stress-strain relation reduce to

$$\frac{\partial \sigma}{\partial x} = \rho \frac{\partial^2 u}{\partial t^2} \quad [60]$$

$$\varepsilon = \frac{\partial u}{\partial x} \quad [61]$$

and

$$\sigma = (1-\omega) E \varepsilon \quad [62]$$

complemented by a model-dependent definition of the nonlocal quantity. For the loading function, we assume that momentarily $f=0$ throughout the bar, so that we have a linear comparison solid in the sense of Hill [HIL 58]. Combination of Eqs [60]-[62] leads to

$$c_{\text{e}}^{-2} \frac{\partial^2 u}{\partial t^2} = (1-\omega) \frac{\partial^2 u}{\partial x^2} - \varepsilon \frac{\partial \omega}{\partial \kappa} \frac{\partial \bar{\varepsilon}}{\partial x} \quad [63]$$

with $c_{\text{e}} = \sqrt{E/\rho}$ the one-dimensional elastic bar velocity.

We first consider the 'explicit' gradient damage model of Eq. [39], which for the one-dimensional case reduces to

$$\bar{\varepsilon} = \varepsilon + g \frac{\partial^2 \varepsilon}{\partial x^2} \quad [64]$$

Substitution of Eq. [64] into Eq. [63] yields

$$c_{\text{e}}^{-2} \frac{\partial^2 u}{\partial t^2} = \left(1 - \omega - \varepsilon \frac{\partial \omega}{\partial \kappa}\right) \frac{\partial^2 u}{\partial x^2} - g \varepsilon \frac{\partial \omega}{\partial \kappa} \frac{\partial^4 u}{\partial x^4} \quad [65]$$

We now substitute a small harmonic disturbance

$$\delta u = \hat{u}\, \mathrm{e}^{\mathrm{i}k(x-ct)} \qquad [66]$$

with $\hat{u}$ the amplitude, k the wave number and c the propagation speed of the disturbance, into Eq. [65] with $\varepsilon = \varepsilon_0$ and $\omega = \omega_0$, which characterises a homogeneous state. The result is given by [PEE 96b]:

$$c = c_{\mathrm{e}} \sqrt{1 - \omega_0 - \varepsilon_0 \left(\frac{\partial \omega}{\partial \kappa} \right)_0 (1 - gk^2)} \qquad [67]$$

For the *linear* softening relation of Eq. [8] the above expression reduces to

$$c = c_{\mathrm{e}} \sqrt{\frac{h}{h+E} \left(1 - \frac{\kappa_{\mathrm{u}}}{\varepsilon_0} gk^2 \right)} \qquad [68]$$

with h the softening modulus: $h = -f_{\mathrm{ct}}/\kappa_{\mathrm{u}}$. Real wave speeds are obtained if the term under the square root is nonnegative, which implies that there is a critical wave number k_{crit} where we have a stationary wave. The corresponding critical wave length is given by

$$\lambda_{\mathrm{crit}} = 2\pi \sqrt{g\, \kappa_{\mathrm{u}} / \varepsilon_0} \qquad [69]$$

which corresponds to the width of the localisation zone. We observe that for progressive damage, λ_{crit} gradually decreases to a minimum value of $\lambda_{\mathrm{crit}} = 2\pi\sqrt{g}$ for $\varepsilon_0 = \kappa_{\mathrm{u}}$.

The latter observation is in contrast with the 'implicit' gradient model of Eq. [41]. For the one-dimensional case we now have

$$\bar{\varepsilon} = \varepsilon + g \frac{\partial^2 \bar{\varepsilon}}{\partial x^2} \qquad [70]$$

instead of Eq. [64]. We cannot directly substitute Eq. [70] into Eq. [63], because $\bar{\varepsilon}$ is defined in an implicit manner. For this reason we consider Eqs [63] and [70] simultaneously, and substitute the perturbations [66] and

$$\delta \bar{\varepsilon} = \hat{\varepsilon}\, \mathrm{e}^{\mathrm{i}k(x-ct)} \qquad [71]$$

We obtain for the propagation speed of the perturbation

$$c = c_{\mathrm{e}} \sqrt{1 - \omega_0 - \varepsilon_0 \left(\frac{\partial \omega}{\partial \kappa} \right)_0 (1 + gk^2)^{-1}} \qquad [72]$$

For linear softening Eq. [72] reduces to

$$c = c_{\mathrm{e}} \sqrt{\frac{h}{h+E} \left(1 - \frac{\kappa_{\mathrm{u}}}{\varepsilon_0} \frac{gk^2}{1 + gk^2} \right)} \qquad [73]$$

and the critical wave length now reads

$$\lambda_{\text{crit}} = 2\pi\sqrt{g\,(\kappa_{\text{u}}\,/\,\varepsilon_0 - 1)} \tag{74}$$

Similar to the 'explicit' gradient model, we observe that for progressive damage, λ_{crit} decreases. A striking difference is that for $\varepsilon_0 = \kappa_{\text{u}}$, λ_{crit} reduces to zero, which implies that in the limiting case of complete stiffness degradation, the width of the localisation zone reduces to zero, i.e. we recover a line crack as one would expect physically.

Next, we consider the gradient damage model that is obtained by replacing the history parameter κ by its nonlocal equivalent $\bar{\kappa}$, cf. Eq. [42], which for the one-dimensional case reads:

$$\bar{\kappa} = \kappa + g\,\frac{\partial^2 \kappa}{\partial x^2} \tag{75}$$

Applying the perturbations [66] and

$$\delta\kappa = \hat{\kappa}\,\mathrm{e}^{ik(x-ct)} \tag{76}$$

to Eqs [63] and [75], we obtain for the velocity

$$c = c_{\text{e}}\sqrt{1 - \omega_0 - \varepsilon_0\left(\frac{\partial\omega}{\partial\kappa}\right)_0 (1 - gk^2)^{-1}} \tag{77}$$

which implies that the gradient constant g should now have a negative sign for expression (77) to be meaningful. Then, the result is fully identical to that of the 'implicit' strain-based gradient model (70), including the expression for the critical wave length.

Finally, we examine the stress-based gradient damage model of Eq. [44]. In a one-dimensional context Eq. [45] is replaced by

$$\bar{\varepsilon}^{\text{d}} = \varepsilon^{\text{d}} + g\,\frac{\partial^2 \varepsilon^{\text{d}}}{\partial x^2} \tag{78}$$

while, considering Eq. [11], the one-dimensional stress-strain relation reads

$$\sigma = E(\varepsilon - \varepsilon^{\text{d}}) \tag{79}$$

Combining the equation of motion [60], the kinematic relation [61], the stress-strain relation [79] and definition [78] for the nonlocal damage strain $\bar{\varepsilon}^{\text{d}}$, we obtain after some algebraic manipulations

$$c_{\text{e}}^{-2}\,\frac{\partial^2 u}{\partial t^2} = \frac{h}{h+E}\,\frac{\partial^2 u}{\partial x^2} + \frac{gh}{h+E}\left(\frac{\partial^4 u}{\partial x^4} - c_{\text{e}}^{-2}\,\frac{\partial^4 u}{\partial x^2 \partial t^2}\right) \tag{80}$$

where $h = \partial\kappa/\partial\bar{\varepsilon}^{\text{d}}$, which becomes a (negative) constant for linear softening. Not surprisingly, Eq. [80] is formally identical to gradient plasticity [SLU 93] and so are the expressions for the propagation speed

$$c = c_e \sqrt{\frac{h(1-gk^2)}{h+E-hgk^2}} \qquad [81]$$

and the critical wave length for stationary waves

$$\lambda_{crit} = 2\pi\sqrt{g} \qquad [82]$$

Indeed, since in this one-dimensional dispersion analysis we consider a linear comparison solid, the coincidence is complete. Note, that in contrast to the other three gradient damage approaches, the critical wave length attains a finite value irrespective of the strain level. Apparently, the choice of the regularising field variable can have a major influence in addition to the choice of the formalism, i.e. plasticity or damage [HUE 94], or the choice of a differential *vs* an integral type of nonlocal model [PEE 96b].

5. References

[BAZ 83a] BAZANT, Z.P., "Comment on orthotropic models for concrete and geomaterials", *ASCE J. Eng. Mech.,* vol. 109, p. 849-865, 1983.

[BAZ 83b] BAZANT, Z.P., OH, B., "Crack band theory for fracture of concrete", *RILEM Mat. Struct.,* vol. 16, p. 155-177, 1983.

[BAZ 84] BAZANT, Z.P., GAMBAROVA, P., "Crack shear in concrete: Crack band microplane model", *ASCE J. Struct. Eng.,* vol. 110, p. 2015-2036, 1984.

[BAZ 88a] BAZANT, Z.P., PIJAUDIER-CABOT, G., "Nonlocal continuum damage, localization instability and convergence", *ASME J. Appl. Mech.,* vol. 55, p. 287-293, 1988.

[BAZ 88b] BAZANT, Z.P., PRAT, P., "Microplane model for brittle plastic material. I. Theory & II. Verification", *ASCE J. Eng. Mech.,* vol. 114, p. 1672-1702, 1988.

[BEN 88] BENALLAL, A., BILLARDON, R., GEYMONAT, G., "Some mathematical aspects of the damage softening problem", *Cracking and Damage,* Elsevier, Amsterdam and London, p. 247-258, 1988.

[BOR 85] de BORST, R., Nauta, P., "Non-orthogonal cracks in a smeared finite element model", *Eng. Comput.,* vol. 2, p. 35-46, 1985.

[BOR 86] de BORST, R., Non-linear analysis of frictional materials, dissertation, Delft University of Technology, 1986.

[BOR 87] de BORST, R., Smeared cracking, plasticity, creep and thermal loading - a unified approach", *Comp. Meth. Appl. Mech. Eng.,* vol. 62, p. 89-110, 1987.

[BOR 92] de BORST, R., MUHLHAUS, H.-B., Gradient-dependent plasticity: Formulation and algorithmic aspects. *Int. J. Num. Meth. Eng.,* vol. 35, p. 521-539, 1992.

[BOR 93] de BORST, R., SLUYS, L.J. MUHLHAUS, H.-B., PAMIN, J., "Fundamental issues in finite element analysis of localisation of deformation, *Eng. Comput.,* vol. 10, p. 99-122, 1993.

[BOR 96a] de BORST, R., BENALLAL, A., HEERES, O.M., "A gradient-enhanced damage approach to fracture", *J. de Physique IV,* vol. C6, p. 491-502, 1996.

[BOR 96b] de BORST, R., PAMIN, J., "Gradient plasticity in numerical simulation of concrete cracking", *Eur. J. Mechanics: A/Solids,* vol. 15, p. 295-320, 1996.

[COM 98] COMI, C., "Computational modelling of gradient-enhanced damage in quasi-brittle materials", *Mech. Coh.-frict. Mat.,* in press.

[COP 80] COPE, R.J., RAO, P.V., ClARK, L.A., NORRIS, P., "Modelling of reinforced concrete behaviour for finite element analysis of bridge slabs", *Numerical Methods for Non-Linear Problems,* Pineridge Press, Swansea, vol. 1, p. 457-470, 1980.

[FEE 95] FEENSTRA, P.H., de BORST, R., "A plasticity model for mode-I cracking in concrete", *Int. J. Num. Meth. Eng.,* vol. 38, p. 2509-2529, 1995.

[HIL 58] HILL, R., "A general theory of uniqueness and stability in elastic-plastic solids", *J. Mech. Phys. Solids,* vol. 7, p. 236-249, 1958.

[HUE 94] HUERTA, A., PIJAUDIER-CABOT, G., "Discretization influence on the regularization by two localization limiters", *ASCE J. Eng. Mech.,* vol. 120, p. 1198-1218, 1994.

[ING 95] INGRAFFEA, A., "Topology-controlled modeling of linear and nonlinear 3D crack propagation in geomaterials", *Fracture of Brittle Disordered Materials: Concrete, Rock and Ceramics,* E. & F.N. Spon, London, p. 301-318. 1995.

[KUH 98a] KUHL, E., RAMM, E., On the linearization of the microplane model. *Mech. Coh.-frict. Mat.,* in press.

[KUH 98b] KUHL, E., RAMM, E., de BORST, R., "Anisotropic gradient damage with the microplane model", *Computational Modelling of Concrete Structures,* Balkema, Rotterdam-Boston, p. 103-112, 1998.

[NGO 67] NGO, D., SCORDELIS, A.C., "Finite element analysis of reinforced concrete beams", *J. Amer. Concr. Inst.,* vol. 64, p. 152-163, 1967.

[PEE 96a] PEERLINGS, R.H.J., de BORST, R., BREKELMANS, W.A.M., de VREE, J.H.P., "Gradient-enhanced damage for quasi-brittle materials", *Int. J. Num. Meth. Eng.,* vol. 39, p. 3391-3403, 1996.

[PEE 96b] PEERLINGS, R.H.J., de BORST, R., BREKELMANS, W.A.M., de VREE, J.H.P., SPEE, I., "Some observations on localisation in non-local and gradient damage models", *Eur. J. Mech./A: Solids,* vol. 15, p. 937-953, 1996.

[PEE 98] PEERLINGS, R.H.J., de BORST, R., BREKELMANS, W.A.M., GEERS, M.G.D., "Gradient-enhanced modelling of concrete fracture", *Mech. Coh.-frict. Mat.,* in press.

[PIE 81] PIETRUSZCZAK, S., MROZ, Z., "Finite element analysis of deformation of strain softening materials", *Int. J. Num. Meth. Eng.,* vol. 17, p. 327-334, 1981.

[PIE 87] PIJAUDIER-CABOT, G., BAZANT, Z.P., "Nonlocal damage theory", *ASCE J. Eng. Mech.,* vol. 113, p. 1512-1533, 1987.

[RAS 68] RASHID, Y.R., "Analysis of prestressed concrete pressure vessels", *Nucl. Eng. Des.,* vol. 7, p. 334-344, 1968.

[ROT 91] ROTS, J.G., "Smeared and discrete representations of localized fracture", *Int. J. Fracture,* vol. 51, p. 45-59, 1991.

[SCH 91] SCHLANGEN, E., Experimental and numerical analysis of fracture processes in concrete, dissertation, Delft University of Technology, 1993.

[SLU 93] SLUYS, L.J., de BORST, R., MUHLHAUS, H.-B., "Wave propagation, localization and dispersion in a gradient-dependent medium", *Int. J. Solids Structures,* vol. 30, p. 1153-1171, 1993.

[SUI 73] SUIDAN, M., SCHNOBRICH, W.C., "Finite element analysis of reinforced concrete", *ASCE J. Struct. Div.,* vol. 99, p. 2109-2122, 1973.

[VRO 97] van VROONHOVEN, J.C.W., de BORST, R., "Uncoupled numerical method for fracture analysis", *Int. J. Fracture,* vol. 84, p. 175-190, 1997.

[WIL 84] WILLAM, K.J., "Experimental and computational aspects of concrete fracture", *Proc. Int. Conf. Computer Aided Analysis and Design of Concrete Structures,* Pineridge Press, Swansea, vol. 1, p. 33-70, 1984.

[WIL 86] WILLAM, K.J., PRAMONO, E., STURE, S., "Fundamental issues of smeared crack models", *Proc. SEM/RILEM Int. Conf. Fracture of Concrete and Rock,* Springer-Verlag, Heidelberg-Berlin-New York, p. 142-157, 1986.

Comments on Microplane Theory

Milan Jirásek

Laboratory of Structural and Continuum Mechanics
Swiss Federal Institute of Technology
LSC-DCC, EPFL, CH-1015 Lausanne, Switzerland

ABSTRACT. *This paper discusses several aspects of constitutive modeling based on the microplane concept. The common idea is a systematic and consistent application of the principle of virtual work. The first part focuses on microplane damage models derived from the principle of energy equivalence. The second part addresses theoretical problems related to the extension of microplane models into the large-strain range. Finally, certain considerations regarding the symmetry of Cauchy stress tensor lead to the idea of a micropolar (Cosserat-type) microplane model.*

KEY WORDS: *Microplane Model, Anisotropic Damage, Large Strain, Micropolar Continuum.*

1. Introduction

Most of traditional constitutive models have a tensorial character in the sense that they establish a direct relationship between the strain tensor and the stress tensor; the relationship must satisfy certain requirements of frame indifference (independence of the choice of a particular coordinate system). An appealing alternative approach can be based on the microplane concept, motivated by the slip theory for metals [Tay38, BB49]. Microplane models work with stress and strain *vectors* on a set of planes of various orientations (so-called microplanes). The basic constitutive laws are defined on the level of the microplane and must be transformed to the level of the material point using certain relations between the tensorial and vectorial components. The most natural choice would be to construct the stress and strain vectors on each microplane by projecting the corresponding tensors, i.e., by contracting the tensors with the vector normal to the plane. However, it is impossible to use this procedure for both stress and strain and still satisfy a general law relating the vectorial components on every microplane. The original slip theory for metals worked with stress vectors as projections of the stress tensor; this is now called the *static constraint*. Most versions of the microplane model for concrete and soils have been based on the *kinematic constraint*, which defines the strain vector on an arbitrary microplane by projecting the strain tensor.

The development of microplane models for concrete started in the eighties when Bažant and coworkers proposed the first version for tensile fracturing [BG84, BO85] and then generalized it to arbitrary stress states including triaxial compression [BP88]. The model was implemented into a nonlocal finite element code [BO90] and applied to analyses of compression failure [BO92a], behavior under cyclic loading [BO92b], and other problems. Theoretical aspects also received considerable attention: Carol et al. separated the constitutive equations into two independent parts representing damage and rheology [CBP91], and they revised the model such that a computationally efficient, fully explicit implementation became possible [CPB92]. The detection of a pathological behavior under large tensile strains [Jir93] motivated the development of a modified version based on the so-called stress-strain boundaries [BXP96]. Currently, a large-strain generalization of the microplane model is being investigated [BAC$^+$98].

The aim of the present paper is to stimulate further development of the microplane family of models by commenting on certain theoretical aspects related to the application of the principle of virtual work (PVW). Section 2 focuses on the microplane characterization of damage and Section 3 presents a particular version of the microplane damage model. A consistent derivation of the stress-evaluation formula valid for large strains is given in Section 4. Finally, Section 5 outlines an extension of the microplane theory to the area of micropolar continua.

2. Anisotropic Damage

2.1. *Principle of Energy Equivalence*

Advanced damage models are frequently based on the principle of energy equivalence [CS79]. This principle works with the effective stress, $\tilde{\sigma}$, and effective strain, $\tilde{\varepsilon}$, which form a pair of work-conjugate tensors. The effective quantities reflect the conditions in the undamaged material between cracks, voids, etc., and in the simplest case of an elastic damage model, they are linked by the generalized Hooke's law

$$\tilde{\boldsymbol{\sigma}} = \boldsymbol{D}_e : \tilde{\boldsymbol{\varepsilon}} \qquad [1]$$

where $\boldsymbol{D}_e$ is the elastic stiffness tensor. Progressive loss of material integrity due to the initiation and propagation of microdefects is taken into account by the relations between the effective quantities and the macroscopically observed ones. The effective stress is expressed as $\tilde{\boldsymbol{\sigma}} = \boldsymbol{M} : \boldsymbol{\sigma}$, where $\boldsymbol{M}$ is a fourth-order damage effect tensor and $\boldsymbol{\sigma}$ is the apparent (macroscopically observed) stress tensor. To be consistent with [CB97], we shall work with the inverse of the damage effect tensor, $\boldsymbol{\beta} = \boldsymbol{M}^{-1}$, and we shall write the relationship between the apparent stress and the effective stress as

$$\boldsymbol{\sigma} = \boldsymbol{\beta} : \tilde{\boldsymbol{\sigma}} \qquad [2]$$

The principle of energy equivalence states that the complementary energy of the damaged material under the apparent stress is equal to the complementary energy of the virgin (undamaged) material under the effective stress, i.e.,

$$\frac{1}{2}\boldsymbol{\sigma} : \boldsymbol{C} : \boldsymbol{\sigma} = \frac{1}{2}\tilde{\boldsymbol{\sigma}} : \boldsymbol{C}_e : \tilde{\boldsymbol{\sigma}} \qquad [3]$$

where $\boldsymbol{C}$ is the secant (unloading) compliance tensor of the damaged material, and $\boldsymbol{C}_e = \boldsymbol{D}_e^{-1}$ is the elastic compliance tensor. Alternatively, one could use the principle of complementary virtual work, stating that the work of virtual stress on the actual strain must be equal to the work of virtual effective stress on the effective strain, for any virtual stress state satisfying [2]. Substituting $\delta\boldsymbol{\sigma} = \boldsymbol{\beta} : \delta\tilde{\boldsymbol{\sigma}}$ into

$$\boldsymbol{\varepsilon} : \delta\boldsymbol{\sigma} = \tilde{\boldsymbol{\varepsilon}} : \delta\tilde{\boldsymbol{\sigma}} \qquad [4]$$

and taking into account the independence of variations, we obtain an expression for the effective strain,

$$\tilde{\boldsymbol{\varepsilon}} = \boldsymbol{\varepsilon} : \boldsymbol{\beta} = \boldsymbol{\beta}^T : \boldsymbol{\varepsilon} \qquad [5]$$

which is dual to [2]. It is tacitly assumed that $\boldsymbol{\beta}$ exhibits minor symmetry, i.e., $\beta_{ijkl} = \beta_{jikl} = \beta_{ijlk}$. Major symmetry is in general not required.

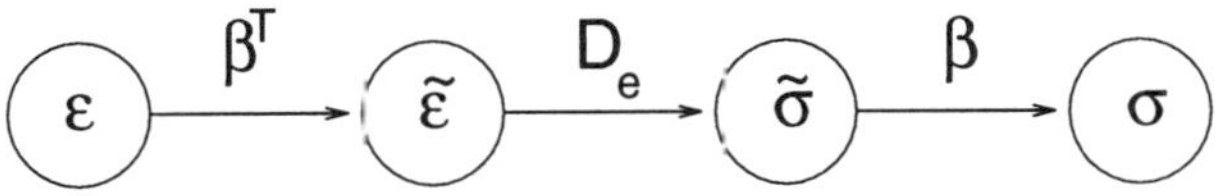

Figure 1. *Damage model based on energy equivalence—tensorial approach*

The structure of the model is schematically depicted in Fig. 1. Combining equations [1], [2], and [5] we can construct the total stress-strain relations

$$\boldsymbol{\sigma} = \boldsymbol{\beta} : \tilde{\boldsymbol{\sigma}} = \boldsymbol{\beta} : \boldsymbol{D}_e : \tilde{\boldsymbol{\varepsilon}} = \boldsymbol{\beta} : \boldsymbol{D}_e : \boldsymbol{\beta}^T : \boldsymbol{\varepsilon} = \boldsymbol{D} : \boldsymbol{\varepsilon} \qquad [6]$$

where

$$D = \beta : D_e : \beta^T \qquad [7]$$

is the damaged material stiffness tensor. Inverting [7] we obtain an expression for the damaged compliance tensor

$$C = D^{-1} = \beta^{-T} : D_e^{-1} : \beta^{-1} = M^T : C_e : M \qquad [8]$$

2.2. *Microplane Description of Damage*

The key ingredient of any damage model is an evolution law for the tensor characterizing the current state of damage. In the case of a simple isotropic damage model, the damage effect tensor is a scalar multiple of the unit fourth-order tensor, and the dependence of a scalar damage parameter on the applied strain can be identified from experiments relatively easily. However, for general anisotropic damage models the evolution laws are difficult to develop. In standard tensorial theories, the damage effect tensor is usually expressed in terms of another tensor, $\boldsymbol{\omega}$, which characterizes damage more directly and can have a lower order than $\boldsymbol{M}$. However, the choice of the (nonlinear) function $\boldsymbol{M}(\boldsymbol{\omega})$ is to a large extent arbitrary, and a tensorial evolution law for $\boldsymbol{\omega}$ is still needed. Instead of constructing complex tensorial laws, it is possible to exploit the microplane concept and characterize the evolution of damage on each microplane separately [CBP91, Fic96, CB97]. Carol and Bažant [CB97] applied this idea to the model based on the principle of energy equivalence. The general structure of the theory is illustrated by the schematic diagram in Fig. 2. Each of the strain and stress tensors is associated with a family of microplane vectors, denoted by the corresponding Latin letter.

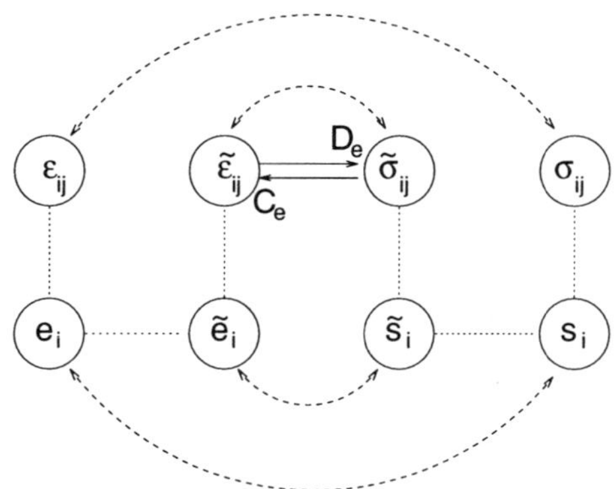

Figure 2. *Damage model based on energy equivalence—microplane approach*

Quantities located symmetrically with respect to the vertical axis in Fig. 2 are work-conjugate; this is emphasized in the diagram by double arrows. Each of the four work-conjugate pairs renders an expression for virtual work. As all the expressions must be identical, the PVW allows one to construct three independent relations. The six links represented by the dotted lines in Fig. 2 form three pairs of work-conjugate relations. In each pair, one relation has to be postulated, and the other can be derived from the PVW or from the principle of complementary virtual work.

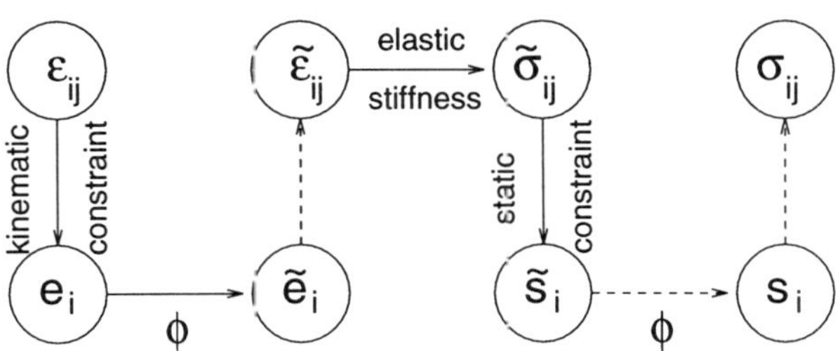

Figure 3. *Structure of microplane damage model—stiffness version*

Now, our goal is to construct a continuous path linking the strain tensor with the stress tensor. One possible scheme is indicated in Fig. 3. It exploits the kinematic constraint for the apparent strain,

$$\boldsymbol{e} = \boldsymbol{\varepsilon} \cdot \boldsymbol{n} \qquad [9]$$

and the static constraint for the effective stress,

$$\tilde{\boldsymbol{s}} = \tilde{\boldsymbol{\sigma}} \cdot \boldsymbol{n} \qquad [10]$$

in which $\boldsymbol{n}$ is the unit normal to the microplane.

The virtual work equality for the apparent quantities,

$$\boldsymbol{\sigma} : \delta\boldsymbol{\varepsilon} = \frac{3}{2\pi}\int_\Omega \boldsymbol{s} \cdot \delta\boldsymbol{e} \, \mathrm{d}\Omega \qquad [11]$$

sets the work of stress tensor on virtual strain equal to the work of microplane stresses on virtual microplane strains (both expressions taken per unit volume of the material). The work of the microcomponents is integrated over all possible microplane orientations, represented by the unit hemisphere, Ω, and it is divided by the volume of the unit hemisphere, $2\pi/3$. Substituting relation $\delta\boldsymbol{e} = \delta\boldsymbol{\varepsilon} \cdot \boldsymbol{n}$ that follows from the kinematic constraint and taking into account that the virtual strain tensor $\delta\boldsymbol{\varepsilon}$ is symmetric but otherwise arbitrary, we obtain an integral formula for the apparent stress tensor,

$$\boldsymbol{\sigma} = \frac{3}{2\pi}\int_\Omega (\boldsymbol{s} \otimes \boldsymbol{n})_{sym} \, \mathrm{d}\Omega \qquad [12]$$

where $\otimes$ is the symbol for the direct product of tensors, and $(.)_{sym}$ denotes the symmetric part of a tensor. Similarly, the complementary virtual work equality for the effective quantities,

$\boldsymbol{e} = \boldsymbol{\varepsilon} \cdot \boldsymbol{n}$	$\tilde{\boldsymbol{e}} = \phi \boldsymbol{e}$	$\tilde{\boldsymbol{s}} = \tilde{\boldsymbol{\sigma}} \cdot \boldsymbol{n}$
$\boldsymbol{\sigma} : \delta\boldsymbol{\varepsilon} = \frac{3}{2\pi} \int_\Omega \boldsymbol{s} \cdot \delta\boldsymbol{e} \, \mathrm{d}\Omega$	$\boldsymbol{s} \cdot \delta\boldsymbol{e} = \tilde{\boldsymbol{s}} \cdot \delta\tilde{\boldsymbol{e}}$	$\delta\tilde{\boldsymbol{\sigma}} : \tilde{\boldsymbol{\varepsilon}} = \frac{3}{2\pi} \int_\Omega \delta\tilde{\boldsymbol{s}} \cdot \tilde{\boldsymbol{e}} \, \mathrm{d}\Omega$
$\boldsymbol{\sigma} = \frac{3}{2\pi} \int_\Omega (\boldsymbol{s} \otimes \boldsymbol{n})_{sym} \, \mathrm{d}\Omega$	$\boldsymbol{s} = \phi \tilde{\boldsymbol{s}}$	$\tilde{\boldsymbol{\varepsilon}} = \frac{3}{2\pi} \int_\Omega (\tilde{\boldsymbol{e}} \otimes \boldsymbol{n})_{sym} \, \mathrm{d}\Omega$

Figure 4. *Basic equations—stiffness version*

$$\delta\tilde{\boldsymbol{\sigma}} : \tilde{\boldsymbol{\varepsilon}} = \frac{3}{2\pi} \int_\Omega \delta\tilde{\boldsymbol{s}} \cdot \tilde{\boldsymbol{e}} \, \mathrm{d}\Omega \qquad [13]$$

combined with the static constraint $\delta\tilde{\boldsymbol{s}} = \delta\tilde{\boldsymbol{\sigma}} \cdot \boldsymbol{n}$ leads to an integral formula for the effective strain tensor,

$$\tilde{\boldsymbol{\varepsilon}} = \frac{3}{2\pi} \int_\Omega (\tilde{\boldsymbol{e}} \otimes \boldsymbol{n})_{sym} \, \mathrm{d}\Omega \qquad [14]$$

The simplest form of the microplane damage law is

$$\tilde{\boldsymbol{e}} = \phi \boldsymbol{e} \qquad [15]$$

where ϕ is a parameter characterizing damage on the microplane level. This parameter is a scalar, but its value on each microplane is in general different, and so the overall damage effect is anisotropic. The PVW written in terms of the microplane components,

$$\frac{3}{2\pi} \int_\Omega \boldsymbol{s} \cdot \delta\boldsymbol{e} \, \mathrm{d}\Omega = \frac{3}{2\pi} \int_\Omega \tilde{\boldsymbol{s}} \cdot \delta\tilde{\boldsymbol{e}} \, \mathrm{d}\Omega \qquad [16]$$

leads to the law linking the effective and apparent microplane stresses,

$$\boldsymbol{s} = \phi \tilde{\boldsymbol{s}} \qquad [17]$$

since the virtual microplane strains have to satisfy the condition $\delta\tilde{\boldsymbol{e}} = \phi \delta\boldsymbol{e}$, but otherwise they can be considered as independent on each microplane (i.e., in the present context they are not subjected to any kinematic constraint).

The version of the microplane model summarized in Fig. 4 (and called here the *stiffness version*) exactly corresponds to the original proposal of Carol and Bažant [CB97]. It is also possible to develop a dual model, schematically depicted by the diagram in Fig. 5 and described by the equations in Fig. 6. This so-called *compliance version* is based on the static constraint for apparent stresses, $\boldsymbol{s} = \boldsymbol{\sigma} \cdot \boldsymbol{n}$, kinematic constraint for effective strains, $\tilde{\boldsymbol{e}} = \tilde{\boldsymbol{\varepsilon}} \cdot \boldsymbol{n}$, and a microplane damage law in the compliance form $\tilde{\boldsymbol{s}} = \psi \boldsymbol{s}$. If the new parameter ψ is set equal to $1/\phi$, this microplane law is fully equivalent to [17].

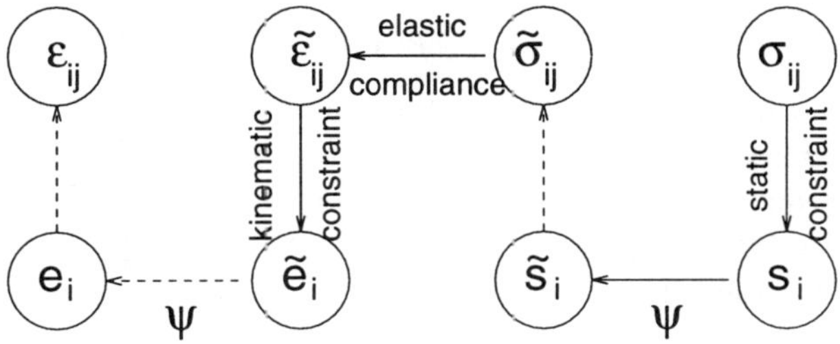

Figure 5. *Structure of microplane damage model—compliance version*

2.3. *Damage Tensor*

Combining the basic equations from Figs. 4 or 6, we can construct direct links between the apparent and the effective tensorial components. For the stiffness version (Figs. 3 and 4) we obtain

$$\begin{aligned}\tilde{\varepsilon} &= \frac{3}{2\pi}\int_\Omega (\tilde{e}\otimes \boldsymbol{n})_{sym}\,\mathrm{d}\Omega = \frac{3}{2\pi}\int_\Omega (\phi \boldsymbol{e}\otimes \boldsymbol{n})_{sym}\,\mathrm{d}\Omega = \\ &= \frac{3}{2\pi}\int_\Omega (\phi\varepsilon\cdot \boldsymbol{n}\otimes \boldsymbol{n})_{sym}\,\mathrm{d}\Omega = \frac{3}{2\pi}\left[\varepsilon\cdot\int_\Omega (\phi\boldsymbol{n}\otimes\boldsymbol{n})\,\mathrm{d}\Omega\right]_{sym} = (\boldsymbol{\phi}\cdot\varepsilon)_{sym}\end{aligned} \quad [18]$$

where

$$\boldsymbol{\phi} = \frac{3}{2\pi}\int_\Omega \phi\boldsymbol{n}\otimes\boldsymbol{n}\,\mathrm{d}\Omega \quad [19]$$

is a symmetric second-order tensor characterizing damage from the macroscopic point of view. An analogous equation,

$$\boldsymbol{\sigma} = (\boldsymbol{\phi}\cdot\tilde{\boldsymbol{\sigma}})_{sym} \quad [20]$$

can be derived for the tensors of apparent and effective stress. This means that the general relations [2] and [5] assume a special form, with components of the fourth-order tensor $\boldsymbol{\beta}$ given by

$\tilde{\boldsymbol{e}} = \tilde{\varepsilon}\cdot\boldsymbol{n}$	$\tilde{\boldsymbol{s}} = \psi\boldsymbol{s}$	$\boldsymbol{s} = \boldsymbol{\sigma}\cdot\boldsymbol{n}$
$\tilde{\boldsymbol{\sigma}} : \delta\tilde{\varepsilon} = \frac{3}{2\pi}\int_\Omega \tilde{\boldsymbol{s}}\cdot\delta\tilde{\boldsymbol{e}}\,\mathrm{d}\Omega$	$\delta\boldsymbol{s}\cdot\boldsymbol{e} = d\tilde{\boldsymbol{s}}\cdot\tilde{\boldsymbol{e}}$	$\delta\boldsymbol{\sigma} : \varepsilon = \frac{3}{2\pi}\int_\Omega \delta\boldsymbol{s}\cdot\boldsymbol{e}\,\mathrm{d}\Omega$
$\tilde{\boldsymbol{\sigma}} = \frac{3}{2\pi}\int_\Omega (\tilde{\boldsymbol{s}}\otimes\boldsymbol{n})_{sym}\,\mathrm{d}\Omega$	$\boldsymbol{e} = \psi\tilde{\boldsymbol{e}}$	$\varepsilon = \frac{3}{2\pi}\int_\Omega (\boldsymbol{e}\otimes\boldsymbol{n})_{sym}\,\mathrm{d}\Omega$

Figure 6. *Basic equations—compliance version*

$$\beta_{ijkl} = \frac{1}{4}\left(\phi_{ik}\delta_{jl} + \phi_{il}\delta_{jk} + \phi_{jk}\delta_{il} + \phi_{jl}\delta_{ik}\right) \qquad [21]$$

where ϕ_{ij} are components of the second-order tensor $\boldsymbol{\phi}$, and δ_{ij} is Kronecker delta. This particular form of $\boldsymbol{\beta}$ exhibits not only minor but also major symmetries. Consequently, $\boldsymbol{\beta} = \boldsymbol{\beta}^T$ and [7] can be rewritten as $\boldsymbol{D} = \boldsymbol{\beta} : \boldsymbol{D}_e : \boldsymbol{\beta}$.

Substituting [21] into [7], it is possible to derive an expression for the components of the damaged stiffness tensor,

$$\begin{aligned} D_{ijkl} &= D^e_{pqrs}\beta_{ijpq}\beta_{klrs} = \qquad [22] \\ &= \frac{1}{4}\left(D^e_{mjnl}\phi_{im}\phi_{kn} + D^e_{imnl}\phi_{jm}\phi_{kn} + D^e_{mjkn}\phi_{im}\phi_{ln} + D^e_{imkn}\phi_{jm}\phi_{ln}\right) \end{aligned}$$

For illustration, consider a two-dimensional model with the effective stress-strain law corresponding to isotropic plane stress elasticity. In principal coordinates of the damage tensor $\boldsymbol{\phi}$, equation [18] in engineering notation reads

$$\begin{Bmatrix} \tilde{\varepsilon}_{11} \\ \tilde{\varepsilon}_{22} \\ \tilde{\gamma}_{12} \end{Bmatrix} = \begin{bmatrix} \phi_1 & 0 & 0 \\ 0 & \phi_2 & 0 \\ 0 & 0 & (\phi_1+\phi_2)/2 \end{bmatrix} \begin{Bmatrix} \varepsilon_{11} \\ \varepsilon_{22} \\ \gamma_{12} \end{Bmatrix} \qquad [23]$$

where ϕ_1 and ϕ_2 are the principal values of $\boldsymbol{\phi}$. The damaged stiffness can be described by the matrix

$$\boldsymbol{D} = \frac{E}{1-\nu^2}\begin{bmatrix} \phi_1^2 & \nu\phi_1\phi_2 & 0 \\ \nu\phi_1\phi_2 & \phi_2^2 & 0 \\ 0 & 0 & (1-\nu)(\phi_1+\phi_2)^2/8 \end{bmatrix} \qquad [24]$$

Note that the present formulation differs from the one advocated in [CB97]. Carol and Bažant suggested to write the components of the fourth-order tensor $\boldsymbol{\beta}$ in the form

$$\beta_{ijkl} = w_{ik}w_{jl} \qquad [25]$$

where $\boldsymbol{w}$ is the tensorial square root of $\boldsymbol{\phi}$, i.e., a symmetric second-order tensor for which $\boldsymbol{w}\cdot\boldsymbol{w} = \boldsymbol{\phi}$. They linked the apparent and the effective tensors by the relationships

$$\tilde{\boldsymbol{\varepsilon}} = \boldsymbol{w}\cdot\boldsymbol{\varepsilon}\cdot\boldsymbol{w}, \qquad \boldsymbol{\sigma} = \boldsymbol{w}\cdot\tilde{\boldsymbol{\sigma}}\cdot\boldsymbol{w} \qquad [26]$$

which means that [23] would be replaced by

$$\begin{Bmatrix} \tilde{\varepsilon}_{11} \\ \tilde{\varepsilon}_{22} \\ \tilde{\gamma}_{12} \end{Bmatrix} = \begin{bmatrix} \phi_1 & 0 & 0 \\ 0 & \phi_2 & 0 \\ 0 & 0 & \sqrt{\phi_1\phi_2} \end{bmatrix} \begin{Bmatrix} \varepsilon_{11} \\ \varepsilon_{22} \\ \gamma_{12} \end{Bmatrix} \qquad [27]$$

However, [25] can be considered only as an approximation. The exact relation between $\boldsymbol{\beta}$ and $\boldsymbol{\phi}$ is given by formula [21], which agrees with formula [86] from [CB97]. In the principal coordinates of $\boldsymbol{\phi}$ we have $\beta_{1212} = \beta_{1221} = (\phi_1+\phi_2)/4$, while equation [25] would give $\beta_{1212} = w_{11}w_{22} = \sqrt{\phi_1\phi_2}$ and $\beta_{1221} = w_{12}w_{21} = 0$.

The microplane theory consistently derived from the kinematic constraint for the apparent quantities and the static constraint for the effective quantities is correctly described by [18] and [20] rather than by [26], and it naturally leads to the so-called *sum-type symmetrization*. The coefficient multiplying the shear strain in the principal damage plane, $(\phi_1 + \phi_2)/2$, is given by the *arithmetic average* of principal values of $\boldsymbol{\phi}$. On the other hand, if we do not strictly adhere to the variational derivation, we can use the *product-type symmetrization* implied by [26], and then the coefficient multiplying the shear strain, $\sqrt{\phi_1 \phi_2}$, corresponds to the *geometric average* of principal values of $\boldsymbol{\phi}$. As shown in [CB97], such an approach leads to elegant formulas for the damaged stiffness tensor.

Let us turn attention to the compliance version (Figs. 5 and 6). In terms of the symmetric tensor

$$\boldsymbol{\psi} = \frac{3}{2\pi} \int_{\Omega} \psi \boldsymbol{n} \otimes \boldsymbol{n} \, \mathrm{d}\Omega = \frac{3}{2\pi} \int_{\Omega} \phi^{-1} \boldsymbol{n} \otimes \boldsymbol{n} \, \mathrm{d}\Omega \qquad [28]$$

the relationship between the apparent and effective strains and stresses can be described by equations

$$\boldsymbol{\varepsilon} = (\boldsymbol{\psi} \cdot \tilde{\boldsymbol{\varepsilon}})_{sym}, \qquad \tilde{\boldsymbol{\sigma}} = (\boldsymbol{\psi} \cdot \boldsymbol{\sigma})_{sym} \qquad [29]$$

and the damaged compliance matrix can be expressed according to [8], in which $\boldsymbol{M} = \boldsymbol{\beta}^{-1}$ is the damage effect tensor with components

$$M_{ijkl} = \frac{1}{4} (\psi_{ik}\delta_{jl} + \psi_{il}\delta_{jk} + \psi_{jk}\delta_{il} + \psi_{jl}\delta_{ik}) \qquad [30]$$

Note that even if we set $\psi = \phi^{-1}$, the tensor $\boldsymbol{\psi}$ defined in [28] is *not* the inverse of $\boldsymbol{\phi}$ from [19].

3. Microplane Damage Model

Based on the general theoretical framework, outlined in the previous section, a specific version of a microplane damage model has been developed and implemented into a finite element code. The law governing the evolution of damage on the microplane level has been postulated in the form

$$\phi = f(e_{max}) \qquad [31]$$

where e_{max} is the largest value of the equivalent microplane strain ever reached in the previous history of the material. The equivalent microplane strain, e, is a scalar measure of the microplane strain vector, in general defined as

$$e = \sqrt{\langle e_N \rangle^2 + c\, \boldsymbol{e}_T \cdot \boldsymbol{e}_T} \qquad [32]$$

where

$$e_N = \boldsymbol{n} \cdot \boldsymbol{e} = \boldsymbol{n} \cdot \boldsymbol{\varepsilon} \cdot \boldsymbol{n} \qquad [33]$$

is the normal microplane strain,

$$e_T = e - e_N n \tag{34}$$

is the shear microplane strain, and McAuley brackets $\langle . \rangle$ denote the "positive part of" operator. In [32], c is a nonnegative parameter that controls the influence of the shear microplane strain on the damage process. The specific form of the function f from [31] controls the shape of the resulting stress-strain curve. Acceptable results have been obtained with a function inspired by the exponential softening law,

$$f(e_{max}) = \begin{cases} 1 & \text{if} \quad e_{max} \leq e_p \\ \sqrt{\dfrac{e_p}{e_{max}} \exp\left(-\dfrac{e_{max} - e_p}{e_f - e_p}\right)} & \text{if} \quad e_{max} \geq e_p \end{cases} \tag{35}$$

where e_p is a parameter controlling the elastic limit, and $e_f > e_p$ is another parameter controlling ductility.

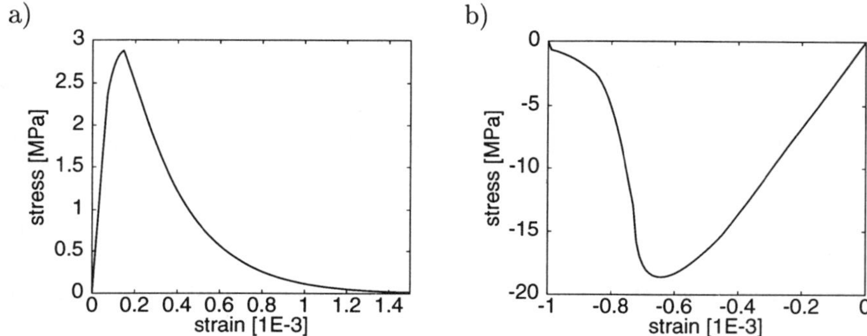

Figure 7. *Uniaxial stress-strain curve: a) tension, b) compression*

The best numerical results have been obtained with the compliance version (equations [28]–[30]) and using $c = 0$, i.e., with the equivalent microplane strain defined simply as the positive part of the normal microplane strain, $e = \langle e_N \rangle$. Fig. 7a shows the macroscopic stress-strain diagram for uniaxial tension, constructed with parameters $E = 34$ GPa, $\nu = 0.2$, $c = 0$, $e_p = 59 \times 10^{-6}$, and $e_f = 250 \times 10^{-6}$. The resulting macroscopic tensile strength is approximately 2.9 MPa. Note that this value is higher than the "microplane strength", $Ee_p = 2.0$ MPa, which corresponds to the elasticity limit. Even after the degradation process has started on the microplane normal to the direction of loading, the overall response exhibits hardening because most of the microplanes are still in the elastic range. The stress-strain diagram for uniaxial compression in Fig. 7b reveals that the compressive strength of the material is 18.6 MPa, which is much too low compared to the tensile strength. The compressive failure is too brittle, with a rapid stress drop in the post-peak range. This phenomenon is due to the fast deterioration of microplanes parallel (or almost parallel) to the direction of loading. A similar problem with overestimated damage due to lateral positive strains under uniaxial compression is typical of damage models driven by the equivalent positive strain. A possible remedy is to reduce the equivalent strain if the loading is compressive. In the present study, the equivalent strain has been redefined as

$$e = \frac{\langle e_N \rangle}{1 - \dfrac{m}{Ee_p}\sigma_{kk}} \qquad [36]$$

where m is a nonnegative parameter that controls the sensitivity to the volumetric pressure, σ_{kk} is the trace of the stress tensor, and the scaling by Ee_p is introduced in order to render the parameter m nondimensional. Under compressive stress states (characterized by $\sigma_{kk} < 0$), the denominator in [36] is larger than 1, and the equivalent strain is reduced, which also leads to a reduction of damage. The stress evaluation algorithm is no more explicit because the equivalent strain now depends on the volumetric pressure, which is not known in advance. However, as this dependence is rather weak, a good estimate is usually obtained with σ_{kk} taken from the last converged state, and after the evaluation of the new stress tensor the initial estimate can be corrected.

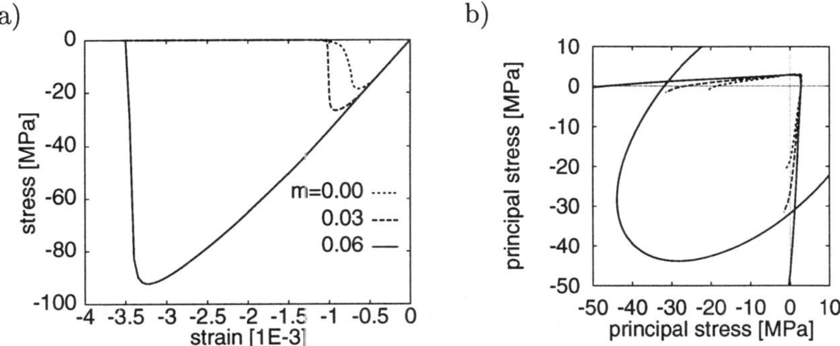

Figure 8. *Effect of parameter m on a) uniaxial stress-strain curve, b) biaxial failure envelope*

The effect of parameter m on the compressive stress-strain diagram is shown in Fig. 8a. By adjusting this parameter, it is possible to increase the compressive strength of the model and prevent sudden failure under relatively low compressive stress. Nevertheless, the model is primarily designed for the simulation of tensile failure. In its present simple form, it does not provide a sufficiently realistic description of compressive failure. If a more general model is required, it is possible to combine the present microplane damage model with a plasticity model describing the behavior under compression. This is schematically shown in Fig. 8b, which depicts the biaxial failure envelope obtained with the microplane damage model and a plastic yield surface of the Drucker-Prager type.

4. Microplane Theory For Large Strains

In recent years, Bažant and coworkers started investigating the potential of microplane models in the domain of large strain [BXP96, BAC$^+$98]. The most difficult part of the problem is of course the formulation of realistic microplane constitutive laws. They have to be carefully designed, based partially on intuition, and verified by comparison to experiments. A thermodynamically consistent framework for the development of microplane constitutive laws shall be described in a future publication. The present paper focuses on the other ingredients of microplane models—the kinematic constraint and the formula for macroscopic stress. It advocates a specific choice of the microplane stress and strain components and proposes a consistent derivation of the relations that link them to the macroscopic tensors. The aim is to preserve objectivity in a simple and elegant manner.

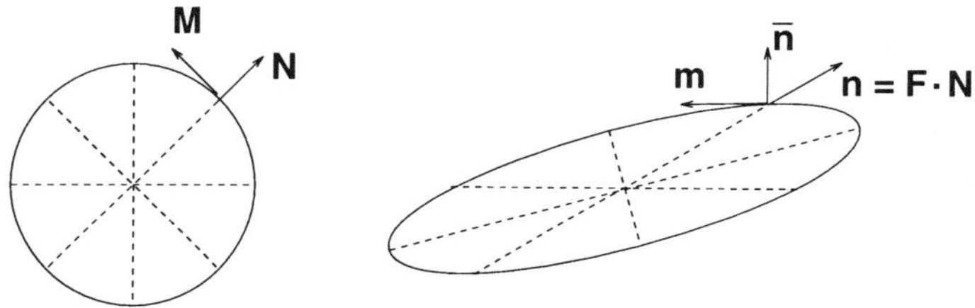

Figure 9. *Initial and deformed microplane base vectors*

4.1. *Kinematic Constraint and Microplane Strains*

On the macroscopic level, the deformation is fully described by the deformation gradient, $\boldsymbol{F}$. When characterizing the deformation on the microplane level, it is natural to observe the change of the initial orthonormal basis, consisting of the unit microplane normal $\boldsymbol{N}$ and unit vectors $\boldsymbol{M}$ and $\boldsymbol{L}$ tangential to the microplane, which are transformed by the deformation process into $\boldsymbol{n} = \boldsymbol{F}\cdot\boldsymbol{N}$, $\boldsymbol{m} = \boldsymbol{F}\cdot\boldsymbol{M}$, and $\boldsymbol{l} = \boldsymbol{F}\cdot\boldsymbol{L}$; see Fig. 9. Vectors $\boldsymbol{m}$ and $\boldsymbol{l}$ remain in the microplane but vector $\boldsymbol{n}$ is no longer normal to the microplane. Defining the unit normal vector to the deformed microplane,

$$\bar{\boldsymbol{n}} = \frac{\boldsymbol{m}\times\boldsymbol{l}}{|\boldsymbol{m}\times\boldsymbol{l}|} = \frac{\boldsymbol{F}^{-T}\cdot\boldsymbol{N}}{|\boldsymbol{F}^{-T}\cdot\boldsymbol{N}|} \qquad [37]$$

we can characterize the normal component of the deformation on the microplane level by the projection of $\boldsymbol{n}$ onto the normal direction,

$$\bar{\lambda}_N = \boldsymbol{n}\cdot\bar{\boldsymbol{n}} = \frac{\boldsymbol{n}\cdot(\boldsymbol{m}\times\boldsymbol{l})}{|\boldsymbol{m}\times\boldsymbol{l}|} = \frac{J\,\mathrm{d}A}{\mathrm{d}a} \qquad [38]$$

where $J = \det \boldsymbol{F}$ is the Jacobian of the deformation gradient, $\mathrm{d}A$ is the infinitesimal microplane area in the initial configuration, $\boldsymbol{m} \times \boldsymbol{l}$ is the vector product of $\boldsymbol{m}$ and $\boldsymbol{l}$, and $\mathrm{d}a = |\boldsymbol{m} \times \boldsymbol{l}|\,\mathrm{d}A$ is the infinitesimal microplane area in the deformed configuration. Quantity $\bar{\lambda}_N$ characterizes the relative thickness of a layer of material parallel to the given microplane. An equivalent expression is

$$\bar{\lambda}_N = \boldsymbol{n} \cdot \bar{\boldsymbol{n}} = \frac{\boldsymbol{n} \cdot \boldsymbol{F}^{-T} \cdot \boldsymbol{N}}{|\boldsymbol{F}^{-T} \cdot \boldsymbol{N}|} = \frac{\boldsymbol{N} \cdot \boldsymbol{F}^{T} \cdot \boldsymbol{F}^{-T} \cdot \boldsymbol{N}}{|\boldsymbol{F}^{-T} \cdot \boldsymbol{N}|} = \frac{1}{|\boldsymbol{F}^{-T} \cdot \boldsymbol{N}|} \qquad [39]$$

Another possible measure of deformation on the microplane level is the stretch of a fiber initially normal to the microplane,

$$\begin{aligned}\lambda_N &= |\boldsymbol{n}| = \sqrt{\boldsymbol{n} \cdot \boldsymbol{n}} = \sqrt{(\boldsymbol{F} \cdot \boldsymbol{N}) \cdot (\boldsymbol{F} \cdot \boldsymbol{N})} = \sqrt{\boldsymbol{N} \cdot \boldsymbol{F}^{T} \cdot \boldsymbol{F} \cdot \boldsymbol{N}} = \\ &= \sqrt{\boldsymbol{N} \cdot (\boldsymbol{I} + 2\boldsymbol{E}) \cdot \boldsymbol{N}} = \sqrt{1 + 2E_{NN}} \qquad [40]\end{aligned}$$

where $\boldsymbol{E}$ is the Green's Lagrangian strain tensor, and $E_{NN} = \boldsymbol{N} \cdot \boldsymbol{E} \cdot \boldsymbol{N}$ is its normal component with respect to the local coordinate system. The above defined measures $\bar{\lambda}_N$ and λ_N have the meaning of stretches (Fig. 10), i.e., they equal 1 in the undeformed configuration. Either of them can be transformed into strain-type measures that equal 0 in the undeformed configuration, e.g., $\bar{\varepsilon}_N^{(1)} = \bar{\lambda}_N - 1$ (Biot strain), $\bar{\varepsilon}_N^{(2)} = (\bar{\lambda}_N^2 - 1)/2$ (Green's Lagrangian strain), or $\bar{\varepsilon}_N^{(0)} = \ln \bar{\lambda}_N$ (Hencky logarithmic strain), and $\varepsilon_N^{(1)} = \lambda_N - 1$, $\varepsilon_N^{(2)} = (\lambda_N^2 - 1)/2$, or $\varepsilon_N^{(0)} = \ln \lambda_N$. Note that $\varepsilon_N^{(2)} = (\lambda_N^2 - 1)/2 = E_{NN} = \boldsymbol{N} \cdot \boldsymbol{E} \cdot \boldsymbol{N}$ is the projected Green's Lagrangian strain but the other strain measures are in general different from the projections of the corresponding strain tensors.

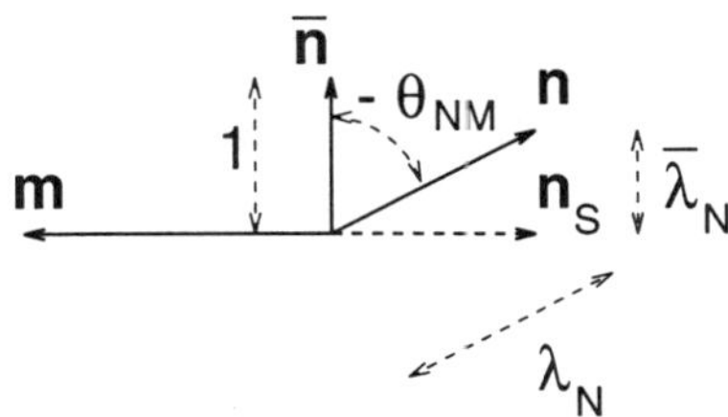

Figure 10. *Microplane deformation measures*

The shear deformation on the microplane level can be represented by the projection of the deformed normal $\boldsymbol{n}$ onto the deformed microplane, i.e., by the vector $\boldsymbol{n}_T = \boldsymbol{n} - \bar{\lambda}_N \bar{\boldsymbol{n}}$. An alternative choice are the changes of the initially right angles between $\boldsymbol{n}$ and $\boldsymbol{m}$ and between $\boldsymbol{n}$ and $\boldsymbol{l}$,

$$\theta_{NM} = \arcsin \frac{\boldsymbol{n} \cdot \boldsymbol{m}}{|\boldsymbol{n}|\,|\boldsymbol{m}|} = \arcsin \frac{2E_{NM}}{\sqrt{(1 + 2E_{NN})(1 + 2E_{MM})}} \qquad [41]$$

$$\theta_{NL} = \arcsin \frac{\boldsymbol{n} \cdot \boldsymbol{l}}{|\boldsymbol{n}|\,|\boldsymbol{l}|} = \arcsin \frac{2E_{NL}}{\sqrt{(1+2E_{NN})(1+2E_{LL})}} \quad [42]$$

where $E_{NM} = \boldsymbol{N}\cdot\boldsymbol{E}\cdot\boldsymbol{M}$, etc., are again components of the Green's Lagrangian strain tensor in the local coordinate system aligned with the microplane. Note that if the material is subjected to uniform volumetric expansion or contraction described by the deformation gradient $\boldsymbol{F} = \lambda \boldsymbol{I}$ where $\boldsymbol{I}$ is the unit second-order tensor, we have $\lambda_N = \bar{\lambda}_N = \lambda$ and $\boldsymbol{n}_T = \boldsymbol{0}$, $\theta_{NM} = \theta_{NL} = 0$.

If the volumetric-deviatoric split is required for proper formulation of the microplane constitutive laws, the deformation gradient can be decomposed as

$$\boldsymbol{F} = \left(J^{1/3}\boldsymbol{I}\right)\cdot\left(J^{-1/3}\boldsymbol{F}\right) \quad [43]$$

where the first term on the right-hand side represents purely volumetric expansion or contraction and the second term corresponds to isochoric deformation. The volumetric part can be characterized by the volumetric stretch $\lambda_V = J^{1/3}$ or by strain-like quantities $\varepsilon_V^{(1)} = \lambda_V - 1$, $\varepsilon_V^{(2)} = (\lambda_V^2 - 1)/2$, or $\varepsilon_V^{(0)} = \ln \lambda_V$. For J close to 1 we have $J^{1/3} = 1 + (J-1)/3 + O\left((J-1)^2\right)$, and so $\varepsilon_V \approx (J-1)/3$ for any of the above definitions.

In the standard small-strain microplane model, it is usually assumed that the normal microplane traction depends only on the normal and volumetric strain and that the shear traction in a certain direction ($\boldsymbol{m}$ or $\boldsymbol{l}$) depends in the elastic regime only on the corresponding shear strain and in the inelastic regime also on the normal or volumetric strain that determines the frictional resistance of the microplane. For large strains, the situation is much more complex. First of all, the difference between the initial orthogonal basis and the deformed basis is not negligible, and so it is necessary to specify, whether we work with covariant or contravariant components of the shear strains and stresses. If a hyperelastic microplane law is used in the elastic regime, covariant components of strains are work-conjugate with contravariant components of stresses, but it does not seem to be reasonable to assume that the directions $\boldsymbol{m}$ and $\boldsymbol{l}$ are decoupled, because then, shear strain in one direction would produce shear traction having a different direction. Thus, the shear components should not be treated separately anymore.

Furthermore, it is questionable whether the normal traction can be taken as independent of the shear strains. For example, the stretch measure λ_N characterizes the extension of a *fiber* initially perpendicular to the microplane. If λ_N grows, this fiber is extended and, before it breaks, it should produce a tensile traction in the direction of the deformed normal, $\boldsymbol{n}$. But we could alternatively think of *layers* of material that are parallel to the microplane. The layers can experience a very large relative slip that corresponds to the plastic shear strain. The situation in the direction normal to the microplane is then better characterized by $\bar{\lambda}_N$. It might easily happen that the layer is compressed in the normal direction, so that $\bar{\lambda}_N$ is smaller than 1 but λ_N is larger than 1 due to the effect of shear; see Fig. 10b. For this "layer analogy" it seems more reasonable to assume that the traction in the direction of the normal to the microplane, $\bar{\boldsymbol{n}}$, depends on a strain measure derived from $\bar{\lambda}_N$. As shear on the microplane has such an important effect, it is not clear whether it is possible to design a reasonable law for the normal components that does not depend at all on the shear strains.

4.2. *Stress Evaluation Formula and Objectivity*

The formula for evaluation of the macroscopic stress tensor can be rigorously derived from the PVW if the microplane stresses have a clearly defined physical meaning. We suggest to use Cauchy tractions $\boldsymbol{t}$, which represent the actual forces acting on the microplane per unit deformed area. Once these tractions are determined from the microplane constitutive laws, they can be integrated over all microplane orientations to obtain the equivalent macroscopic stress tensor. The integration formula can be constructed by setting the virtual work done by the macroscopic stress equal to the virtual work done by the microplane tractions for any virtual change of the state of deformation.

An arbitrary virtual change of deformation is described by a virtual increment $\delta\boldsymbol{F}$ of the deformation gradient. Consider an elementary volume of material that occupies a ball of radius $\mathrm{d}R$ in the initial configuration. The virtual work done by the macroscopic stress on the elementary ball is

$$\delta W_{mac} = \frac{4\pi(\mathrm{d}R)^3}{3}\boldsymbol{P} : \delta\boldsymbol{F} \qquad [44]$$

where $\boldsymbol{P}$ is the first Piola-Kirchhoff stress tensor.

Now let us express the virtual work done by the microplane tractions. The elementary ball is after deformation transformed into an ellipsoid; see Fig. 9. Without affecting the virtual work expression, we can fix the center of both the initial ball and the deformed ellipsoid to be at the origin. For any unit vector $\boldsymbol{N}$, the end point of the position vector $\boldsymbol{N}\mathrm{d}R$ is located at the surface of the elementary ball and can be associated with a facet $\mathrm{d}A$ representing a microplane with initial normal $\boldsymbol{N}$. After deformation, this point moves to $\boldsymbol{n}\mathrm{d}R = \boldsymbol{F}\cdot\boldsymbol{N}\mathrm{d}R$ and the area of the elementary facet changes to $\mathrm{d}a = |\boldsymbol{m}\times\boldsymbol{l}|\,\mathrm{d}A = J\bar{\lambda}_N^{-1}\,\mathrm{d}A$. From the specification of the physical meaning of microplane tractions, $\boldsymbol{t}$, it follows that the force acting on the elementary facet is $\boldsymbol{t}\,\mathrm{d}a$. After a virtual change of the deformation state, the point of application of this elementary force moves by $\delta\boldsymbol{n}\mathrm{d}R = \delta\boldsymbol{F}\cdot\boldsymbol{N}\mathrm{d}R$ and the work done by the microplane traction is $\boldsymbol{t}\cdot\delta\boldsymbol{F}\cdot\boldsymbol{N}\mathrm{d}R\,\mathrm{d}a$. By integrating over all elementary facets, we obtain the "microscopic" virtual work

$$\delta W_{mic} = \int_a \boldsymbol{t}\cdot\delta\boldsymbol{F}\cdot\boldsymbol{N}\,\mathrm{d}a\,\mathrm{d}R = \int_a \boldsymbol{t}\otimes\boldsymbol{N}\,\mathrm{d}a : \delta\boldsymbol{F}\mathrm{d}R \qquad [45]$$

It should be emphasized that the domain of integration a is the surface of the *deformed* elementary ellipsoid. To simplify the numerical integration, we transform the integration domain to the initial elementary sphere A and then, making use of symmetry, to the unit hemisphere Ω:

$$\begin{aligned} \delta W_{mic} &= \int_a \boldsymbol{t} \otimes \boldsymbol{N} \, \mathrm{d}a : \delta \boldsymbol{F} \mathrm{d}R = \int_A \boldsymbol{t} \otimes \boldsymbol{N} J \bar{\lambda}_N^{-1} \, \mathrm{d}A : \delta \boldsymbol{F} \mathrm{d}R = \\ &= 2J \int_\Omega \boldsymbol{t} \otimes \boldsymbol{N} \bar{\lambda}_N^{-1} \, \mathrm{d}\Omega : \delta \boldsymbol{F} (\mathrm{d}R)^3 \end{aligned} \quad [46]$$

Setting $\delta W_{mac} = \delta W_{mic}$ for any $\delta \boldsymbol{F}$ and taking into account the independence of the components of $\delta \boldsymbol{F}$, we obtain an expression for the first Piola-Kirchhoff stress tensor,

$$\boldsymbol{P} = \frac{3J}{2\pi} \int_\Omega \boldsymbol{t} \otimes \boldsymbol{N} \bar{\lambda}_N^{-1} \, \mathrm{d}\Omega \quad [47]$$

which can be transformed into a formula for the Cauchy stress tensor,

$$\boldsymbol{\sigma} = \frac{1}{J} \boldsymbol{P} \cdot \boldsymbol{F}^T = \frac{3}{2\pi} \int_\Omega \boldsymbol{t} \otimes \boldsymbol{N} \cdot \boldsymbol{F}^T \bar{\lambda}_N^{-1} \, \mathrm{d}\Omega = \frac{3}{2\pi} \int_\Omega \boldsymbol{t} \otimes \boldsymbol{n} \bar{\lambda}_N^{-1} \, \mathrm{d}\Omega \quad [48]$$

A very appealing feature of the present approach is that it is intrinsically objective. No polar decomposition of the deformation gradient is required and no objective stress rate has to be defined. Objectivity is achieved by relating all quantities on the microplane level to the local coordinate system that travels with the microplane. If the material undergoes rigid rotation, the microplane strains remain constant, same as the traction components with respect to the (rotating) microplane coordinate system. Denoting the spin tensor as $\boldsymbol{W}$, we have $\dot{\boldsymbol{t}} = \boldsymbol{W} \cdot \boldsymbol{t}$, $\dot{\boldsymbol{n}} = \boldsymbol{W} \cdot \boldsymbol{n}$, $\dot{J} = 0$, and $\mathrm{d}|\boldsymbol{m} \times \boldsymbol{l}|/\mathrm{d}t = 0$. Differentiation of [48] gives

$$\dot{\boldsymbol{\sigma}} = \frac{3}{2\pi} \int_\Omega (\dot{\boldsymbol{t}} \otimes \boldsymbol{n} + \boldsymbol{t} \otimes \dot{\boldsymbol{n}}) \bar{\lambda}_N^{-1} \, \mathrm{d}\Omega \quad [49]$$

and taking into account that

$$\dot{\boldsymbol{t}} \otimes \boldsymbol{n} + \boldsymbol{t} \otimes \dot{\boldsymbol{n}} = \boldsymbol{W} \cdot \boldsymbol{t} \otimes \boldsymbol{n} + \boldsymbol{t} \otimes \boldsymbol{W} \cdot \boldsymbol{n} = \boldsymbol{W} \cdot \boldsymbol{t} \otimes \boldsymbol{n} + \boldsymbol{t} \otimes \boldsymbol{n} \cdot \boldsymbol{W}^T \quad [50]$$

we obtain

$$\dot{\boldsymbol{\sigma}} = \boldsymbol{W} \cdot \boldsymbol{\sigma} + \boldsymbol{\sigma} \cdot \boldsymbol{W}^T \quad [51]$$

which is the correct formula for the rate of Cauchy stress due to rigid rotation.

Formula [51] has been derived only for illustration. The suggested microplane approach works directly with the total values, not with rates, and it provides the exact update of Cauchy stress for arbitrarily large rotations.

4.3. *Symmetry of Cauchy Stress Tensor*

A puzzling point is that the Cauchy stress tensor evaluated from [48] is not necessarily symmetric. The reason is quite simple. We have tested the virtual work equality $\delta W_{mac} = \delta W_{mic}$ for an arbitrary virtual change of the deformation gradient, which has 9 independent components. This means that the virtual change includes 3 virtual rigid body rotations, which yield 3 moment equilibrium equations for the elementary material volume. In terms of macroscopic components, these equations are equivalent to the symmetry conditions for the shear components of Cauchy stress. However, the moment equilibrium conditions are not necessarily satisfied by the microplane tractions, and this results into loss of symmetry of the equivalent macroscopic stress tensor. If we ignore the violation of moment equilibrium, we can replace the stress tensor by its symmetric part. This procedure is implicitly contained in the standard microplane theory.

Let us denote material coordinates (identical with spatial coordinates in the initial configuration) by $\boldsymbol{X}$, and let $\boldsymbol{x}(\boldsymbol{X}) = \boldsymbol{X} + \boldsymbol{u}(\boldsymbol{X})$ be the coordinates of material points in the deformed configuration. Presenting the variation of the displacement field in the form

$$\delta \boldsymbol{u} = \delta \boldsymbol{\varepsilon} \cdot \boldsymbol{x} + \delta \boldsymbol{\omega} \cdot \boldsymbol{x} \tag{52}$$

where $\delta \boldsymbol{\varepsilon}$ is a symmetric tensor and $\delta \boldsymbol{\omega}$ is a skew-symmetric tensor, we can express the variation of the deformation gradient as

$$\delta \boldsymbol{F} = \frac{\partial \delta \boldsymbol{x}}{\partial \boldsymbol{X}} = \frac{\partial \delta \boldsymbol{u}}{\partial \boldsymbol{X}} = \delta \boldsymbol{\varepsilon} \cdot \frac{\partial \boldsymbol{x}}{\partial \boldsymbol{X}} + \delta \boldsymbol{\omega} \cdot \frac{\partial \boldsymbol{x}}{\partial \boldsymbol{X}} = \delta \boldsymbol{\varepsilon} \cdot \boldsymbol{F} + \delta \boldsymbol{\omega} \cdot \boldsymbol{F} \tag{53}$$

Virtual work of the macroscopic stresses is (alternatively to [44]) given by

$$\delta W_{mac} = J \frac{4\pi (\mathrm{d}R)^3}{3} \boldsymbol{\sigma} : \delta \boldsymbol{\varepsilon} \tag{54}$$

Substituting [53] into the expression for virtual work of microplane tractions [46] yields

$$\delta W_{mic} = 2J \int_{\Omega} \boldsymbol{t} \otimes \boldsymbol{n} \bar{\lambda}_N^{-1} \, \mathrm{d}\Omega : \delta \boldsymbol{\varepsilon} (\mathrm{d}R)^3 + 2J \int_{\Omega} \boldsymbol{t} \otimes \boldsymbol{n} \bar{\lambda}_N^{-1} \, \mathrm{d}\Omega : \delta \boldsymbol{\omega} (\mathrm{d}R)^3 \tag{55}$$

Now we compare [54] to [55] for an arbitrary virtual change. If we restrict our attention to changes for which $\delta \boldsymbol{\omega} = \boldsymbol{0}$ and take into account the symmetry of $\delta \boldsymbol{\varepsilon}$, we conclude that $\boldsymbol{\sigma}$ must be equal to the symmetric part of the right-hand side of [48], i.e.,

$$\boldsymbol{\sigma} = \frac{3}{4\pi} \int_{\Omega} (\boldsymbol{t} \otimes \boldsymbol{n} + \boldsymbol{n} \otimes \boldsymbol{t}) \bar{\lambda}_N^{-1} \, \mathrm{d}\Omega \tag{56}$$

This is the procedure exploited in the standard microplane theory. However, for virtual changes with $\delta \boldsymbol{\varepsilon} = \boldsymbol{0}$ but $\delta \boldsymbol{\omega} \neq \boldsymbol{0}$ we obtain (due to the skew-symmetry of $\delta \boldsymbol{\omega}$) the condition that $\int_{\Omega} \boldsymbol{t} \otimes \boldsymbol{n} \bar{\lambda}_N^{-1} \, \mathrm{d}\Omega$ must be symmetric. For small strains small strains and small displacements we can neglect the difference between the initial and deformed basis and set $\boldsymbol{t} = \boldsymbol{s} = \sigma_N \boldsymbol{n} + \sigma_{NM} \boldsymbol{m} + \sigma_{NL} \boldsymbol{l}$. The symmetry conditions then read

$$\int_\Omega [\sigma_{NM}(n_i m_j - m_i n_j) + \sigma_{NL}(n_i l_j - l_i n_j)] \,\mathrm{d}\Omega = 0; \quad \text{for} \quad i \neq j \qquad [57]$$

These three conditions have been ignored by the standard microplane theory.

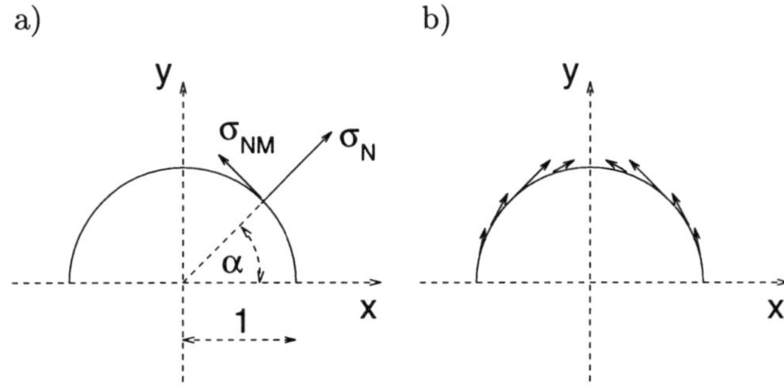

Figure 11. *Microplane stresses for small strain theory in two dimensions*

Some insight into the problem can be gained if we reduce the model to two dimensions. The unit hemisphere is replaced by the unit hemicircle, and the microplane base vectors are $\boldsymbol{n} = (\cos\alpha, \sin\alpha)^T$ and $\boldsymbol{m} = (-\sin\alpha, \cos\alpha)^T$ where $\alpha \in \langle 0, \pi \rangle$ is the angle between the microplane normal and the positive x-axis; see Fig. 11a. As $n_1 m_2 - m_1 n_2 = \cos^2\alpha + \sin^2\alpha = 1$, condition [57] written for $i = 1$ and $j = 2$ reads

$$\int_0^\pi \sigma_{NM} \mathrm{d}\alpha = 0 \qquad [58]$$

The integral on the left-hand side represents one half of the moment produced by the shear microplane tractions σ_{NM} with respect to the center of the unit circle. This moment vanishes as long as the material remains linear elastic. In that case, the microplane tractions are symmetric with respect to the principal strain axes. If we use the principal coordinate system (Fig. 11b), we have $\sigma_{NM}(\alpha) = -\sigma_{NM}(\pi - \alpha)$ and the integral in [58] is indeed zero. The microplane tractions remain symmetric in the inelastic regime if the principal axes do not rotate. Similar symmetry arguments can be used for the full three-dimensional model. We conclude that, for elastic processes and for inelastic processes with fixed principal axes, conditions [57] are always satisfied. However, for loading paths with rotating principal directions, these conditions are in general violated.

5. Micropolar Microplane Model

A fully consistent theory can be developed by extending the standard microplane model to a Cosserat-type formulation [CC09]. For the sake of simplicity, we restrict ourselves to small strains but a generalization covering large strains would be straightforward.

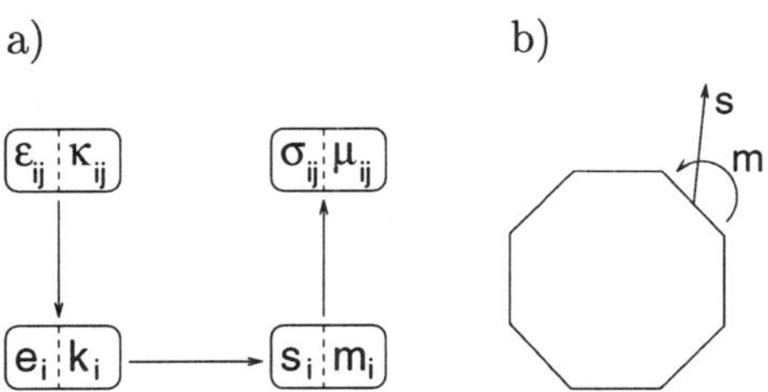

Figure 12. *a) Structure of micropolar microplane model, b) microplane tractions and moments*

Recall that micropolar theories enrich the standard continuum description by the fields of local rotations, $\boldsymbol{\omega}(\boldsymbol{x})$, microcurvatures, $\boldsymbol{\kappa}(\boldsymbol{x})$, and couple stresses, $\boldsymbol{\mu}(\boldsymbol{x})$. In the unconstrained Cosserat theory [Gün58], the tensors of stress and strain are in general not symmetric. The constitutive equations specify the relationship between the strains and microcurvatures on one side and the stresses and couple stresses on the other side. Applying the microplane concept, we introduce the microplane counterparts of the kinematic and static quantities according to the schematic diagram in Fig. 12a. Note that, in this section, $\boldsymbol{m}$ denotes the microplane couple stresses and not a unit vector tangential to the microplane. In the kinematically constrained version of the model, the microplane strain and curvature vectors are obtained by projecting the corresponding tensors:

$$\boldsymbol{e} = \boldsymbol{\varepsilon} \cdot \boldsymbol{n}, \qquad \boldsymbol{k} = \boldsymbol{\kappa} \cdot \boldsymbol{n} \tag{59}$$

The microplane stress and couple-stress vectors, $\boldsymbol{s}$ and $\boldsymbol{m}$, are then computed from the microplane constitutive laws. They have the meaning of tractions and moments acting on elementary facets; see Fig. 12b. Finally, the extended PVW

$$\boldsymbol{\sigma} : \delta\boldsymbol{\varepsilon} + \boldsymbol{\mu} : \delta\boldsymbol{\kappa} = \frac{3}{2\pi} \int_\Omega (\boldsymbol{s} \cdot \delta\boldsymbol{e} + \boldsymbol{m} \cdot \delta\boldsymbol{k}) \, \mathrm{d}\Omega \tag{60}$$

yields the evaluation formulas for the stress tensor,

$$\boldsymbol{\sigma} = \frac{3}{2\pi} \int_\Omega \boldsymbol{s} \otimes \boldsymbol{n} \, \mathrm{d}\Omega \tag{61}$$

and for the couple-stress tensor,

$$\boldsymbol{\mu} = \frac{3}{2\pi} \int_\Omega \boldsymbol{m} \otimes \boldsymbol{n} \, \mathrm{d}\Omega \tag{62}$$

6. Conclusions

The first part of the paper has been concerned with microplane damage models based on the principle of energy equivalence. The theoretical basis originally proposed in [CB97] has been reformulated and generalized. It has been demonstrated that a consistent variational approach leads to the sum-type symmetrization of the damage effect tensor. The abstract general framework has been specialized into a particular formulation intended for anisotropic damage modeling of concrete fracture. The compliance version seems to give better numerical results than the stiffness version.

In the second part, several new ideas related to the large-strain extension of microplane models have been advanced. Attention has been focused on the derivation of the stress evaluation formula from the principle of virtual work. As an alternative to the procedure proposed in [BAC+98], an approach leading directly to the Cauchy stress tensor has been advocated. Its salient feature is that it is intrinsically objective. No polar decomposition of the deformation gradient is required, and no objective stress rate has to be defined. Objectivity (frame invariance) is achieved by relating all quantities on the microplane level to a local coordinate system that travels with the microplane. A natural generalization of this approach leads to a micropolar (Cosserat-type) version of the microplane model.

Acknowledgment

Financial support of the Swiss Committee for Technology and Innovation under project CTI.3201.1 is gratefully acknowledged.

References

[BAC+98] Z. P. Bažant, M. D. Adley, I. Carol, M. Jirásek, B. Rohani, S. Akers, and F. Caner. Large-strain generalization of microplane model. *In preparation*, 1998.

[BB49] S. B. Batdorf and B. Budianski. A mathematical theory of plasticity based on the concept of slip. Technical Note 1871, National Advisory Committee for Aeronautics, Washington, D.C., 1949.

[BG84] Z. P. Bažant and P. Gambarova. Crack shear in concrete: Crack band microplane model. *Journal of Structural Engineering, ASCE*, 110:2015–2035, 1984.

[BO85] Z. P. Bažant and B.-H. Oh. Microplane model for progressive fracture of concrete and rock. *Journal of Engineering Mechanics, ASCE*, 111:559–582, 1985.

[BO90] Z. P. Bažant and J. Ožbolt. Nonlocal microplane model for fracture, damage, and size effect in structures. *Journal of Engineering Mechanics, ASCE*, 116:2485–2505, 1990.

[BO92a] Z. P. Bažant and J. Ožbolt. Compression failure of quasibrittle material: Nonlocal microplane model. *Journal of Engineering Mechanics, ASCE*, 118:540–556, 1992.

[BO92b] Z. P. Bažant and J. Ožbolt. Microplane model for cyclic triaxial behavior of concrete. *Journal of Engineering Mechanics, ASCE*, 118:1365–1386, 1992.

[BP88] Z. P. Bažant and P. Prat. Microplane model for brittle plastic materials. I: Theory, II: Verification. *Journal of Engineering Mechanics, ASCE*, 114:1672–1702, 1988.

[BXP96] Z. P. Bažant, Y. Xiang, and P. C. Prat. Microplane model for concrete. I: Stress-strain boundaries and finite strain. *Journal of Engineering Mechanics, ASCE*, 122(3):245–254, 1996.

[CB97] I. Carol and Z. P. Bažant. Damage and plasticity in microplane theory. *International Journal of Solids and Structures*, 34:3807–3835, 1997.

[CBP91] I. Carol, Z. P. Bažant, and P. Prat. Geometric damage tensor based on microplane model. *Journal of Engineering Mechanics, ASCE*, 117:2429–2448, 1991.

[CC09] E. Cosserat and F. Cosserat. *Théorie des corps déformables.* A. Herrman et Fils, Paris, 1909.

[CPB92] I. Carol, P. Prat, and Z. P. Bažant. New explicit microplane model for concrete: Theoretical aspects and numerical implementation. *International Journal of Solids and Structures*, 29:1173–1191, 1992.

[CS79] J. P. Cordebois and F. Sidoroff. Anisotropie élastique induite par endommagement. In *Comportement mécanique des solides anisotropes*, number 295 in Colloques internationaux du CNRS, pages 761–774, Grenoble, 1979. Editions du CNRS.

[Fic96] S. Fichant. *Endommagement et anisotropie induite du béton de structures. Modélisations approchées.* PhD thesis, E.N.S. de Cachan, Université Paris 6, Cachan, France, 1996.

[Gün58] W. Günther. Zur Statik und Kinematik des Cosseratschen Kontinuum. *Abh. Braunschweig Wiss. Ges.*, 10:195–213, 1958.

[Jir93] M. Jirásek. *Modeling of fracture and damage in quasibrittle materials.* PhD thesis, Northwestern University, 1993.

[Tay38] G. I. Taylor. Plastic strain in metals. *J. Inst. Metals*, 62:307–324, 1938.

A Visco-Damage Model for the Tensile Behavior of Concrete at Moderately High Strain-Rates

Luigi Cedolin — Pietro Bianchi — Athena Ratti

Department of Structural Engineering
Politecnico di Milano,
P. Leonardo da Vinci 32
20133 Milano, Italia

cedolin@stru.polimi.it

ABSTRACT. *A simple rate-dependent fracture model for concrete under moderately high strain rates is presented. The model consists of a fracturing element in parallel with a Maxwell element with rate-history dependent viscosity. The viscosity is assumed to decrease with the damage accumulated by the material, estimated through the ratio between the absorbed and the recoverable energy densities. The model is validated through the simulation of recent experimental data obtained at Ispra Joint Research Center with the use of a split Hopkinson bar. The specimens are concrete cubes of 20 cm side and 25 mm maximum aggregate size, and the strain rates are 1* s^{-1} *and 10* s^{-1}*. The resulting stress-strain diagrams show, for the higher strain rate, a four-fold increase of the tensile strength of the material.*

KEY WORDS: *Concrete, Damage, Fracture, Tension, Softening, Strain-rate, Viscosity.*

Introduction

Strain-rate effects in concrete are a direct consequence of material viscosity. They are particularly relevant in the case of the tensile strength, which governs cracking and shear behavior and is instrumental for the transfer of bond stresses to reinforcing steel. They influence also the stress-strain relation of the material, including the post-peak softening regime.

Early experimental studies [KOM 70, HEI 77, BIR 71] have shown that both tensile strength and initial elastic modulus increase with strain rate. Less clear were the conclusions of these studies about the strain at the peak stress, because they involve a strain measurement. More recent experiments [KOR 80, REI 86] with a split Hopkinson bar also indicated that the strain at peak stress was increasing with the strain rate. In these experiments the stress was measured with strain gages mounted on the elastic bars, while the strain in concrete was measured as average strain between the opposite faces of the specimen.

The stress-strain relation for concrete at high strain rates cannot be determined, however, with direct measurements, because the specimen length (at least four times the maximum aggregate size) is such that the strain (and the stress) is not uniform along the specimen itself. The problem is complicated by the fact that, due to strain-softening, the deformation tends to localize into a narrow portion of the specimen.

It appears then, that the only procedure which may yield a stress-strain relation for concrete is an inverse modeling technique applied to the strain measurements on the elastic bars of a split Hopkinson bar apparatus. The objective of this paper is to propose a one-dimensional semi-physical model which fits the results of a recent experimental investigation conducted at the Joint Research Center of Ispra [CAD 97], with the use of an apparatus especially constructed for handling large specimens.

1. Reference experimental results

A comprehensive experimental study [CAD 97] of concrete behavior under moderately high strain rates has currently been completed at the Joint Research Center of Ispra, Italy. Use has been made of a split Hopkinson bar [HOP 13, KOL 53] whoch isschematically represented in Figure 1a.

It consists of two coaxial elastic bars between which the specimen is inserted. A tensile pulse applied to the left end of the incident bar (Figure 1a) propagates in the right direction, passing through the specimen and the transmitting bar. In order to minimize partial reflections of the pulse at the interfaces, the material used for the bars is an aluminum alloy, which has an impedance which is close to that of concrete. The pulse in the elastic bars is measured by strain gages 1 and 2 (Figure 1a), mounted respectively on the incident and on the transmitting bar. The bars must be long enough that any reflection from their extremities will not reach the specimen before its failure.

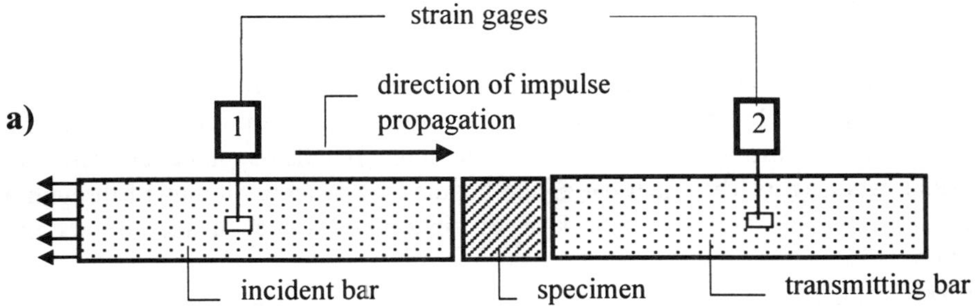

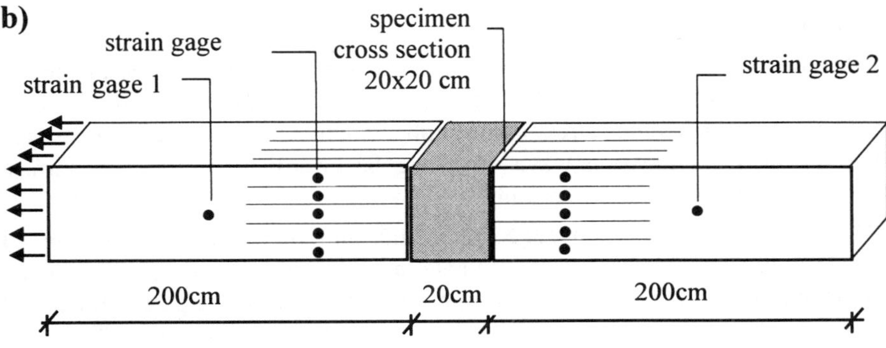

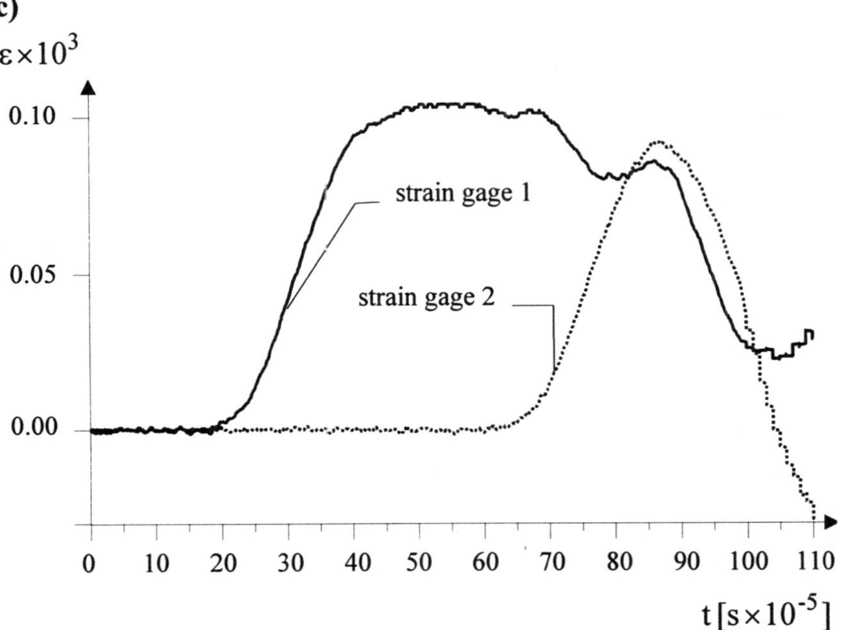

Figure 1. *(a) Split Hopkinson bar scheme; (b) J.R.C. Ispra equipment; (c) incident and transmitted pulses*

The experimental set up at Ispra is particularly powerful and refined, and it allows one to test at moderately high strain rates specimens with a large cross section (square of 20 cm side). The transmission of lateral restraining forces at the interfaces is minimized by cutting the solid bar into a bundle of smaller and laterally flexible bars (Figure 1b). Each bar of the bundle is separately monitored through a separate strain gage in order to check the distribution of normal stresses in the cross section.

The mechanical impulse at the left end of the incident bar is generated by the sudden release of the energy stored in high strength (100 m long) steel cables. Depending on the amount of the prestressing, the apparatus can produce loading rates up to 10 s^{-1}.

The type of signal recorded by strain gage 1 on the incident bar is represented in Figure 1c. It has a rising part with an almost linear loading rate, a plateau, which corresponds to the maximum stress of the applied impulse, and a decreasing part due to the reflected compression wave travelling backwards from the free surface generated by the fracture of the specimen.

The signal recorded by strain gage 2 on the transmitting bar is also represented in Figure 1c. It has a rising part which terminates at the amplitude corresponding to the dynamic tensile strength of the concrete specimen. The decreasing part of the signal is interrupted before the reflected wave coming from the free end reaches the strain gage.

It was verified experimentally [CAD 97] that the distribution of strains in the bundle of small bars is not too disuniform, so that the strain gages 1 and 2 can be assumed to give an average behavior of the concrete cross section.

In this paper we will analyze data corresponding to uniaxial behavior at strain rates of 1 s^{-1} and 10 s^{-1}, which are of the order of magnitude of those influencing the overall structural behavior. As already said, for these strain rates direct measurements of the average strain directly on the specimen are meaningless, because of strain localization. So, the only signals used to calibrate the constitutive relation are those recorded by strain gages 1 and 2.

2. Rate-dependent fracture models

Tensile behavior of concrete is characterized, after the peak stress, by strain softening induced by internal damage, and dependent also on the geometry and boundary conditions of the specimen. It can be macroscopically described by a smeared crack model, which treats the cracked zone as a continuum which obeys a softening stress-strain relation. Due to the uniaxial nature of the loading and to the one-dimensional measuring apparatus, we will restrict our attention to one-dimensional models.

Under transient loading conditions the stress-strain relation becomes strain-rate dependent. The inclusion of this dependence into the constitutive relation has the advantage of regularizing the solution after the onset of strain softening [NEE 88] and of introducing implicitly a length scale [SLU 92] which insures mesh independence to finite element simulations of the strain profile.

The rate dependence can be easily included in these models through a damper element in parallel with an elastic-fracturing element. The problem is, however, the identification of the viscosity of the damper from available test data.

A first choice was to express the viscosity of the damper element as a function of strain and strain rate. We will see in the next sections that this choice did not allow us to properly model the available experimental results, and that a rate-history dependence appears to be needed.

2.1. *Rate-dependent viscosity*

The one-dimensional representation of the model is shown in Figure 2a. The values of the parameters E_1, E_2, η_2 of the standard solid model (Kelvin type) are chosen in such a way that the stiffening effect due to the viscoelastic behavior at the early stages of loading is reproduced. When the stress reaches the static strength f_t', a fracturing element comes into play, in parallel with a damper of variable viscosity m.

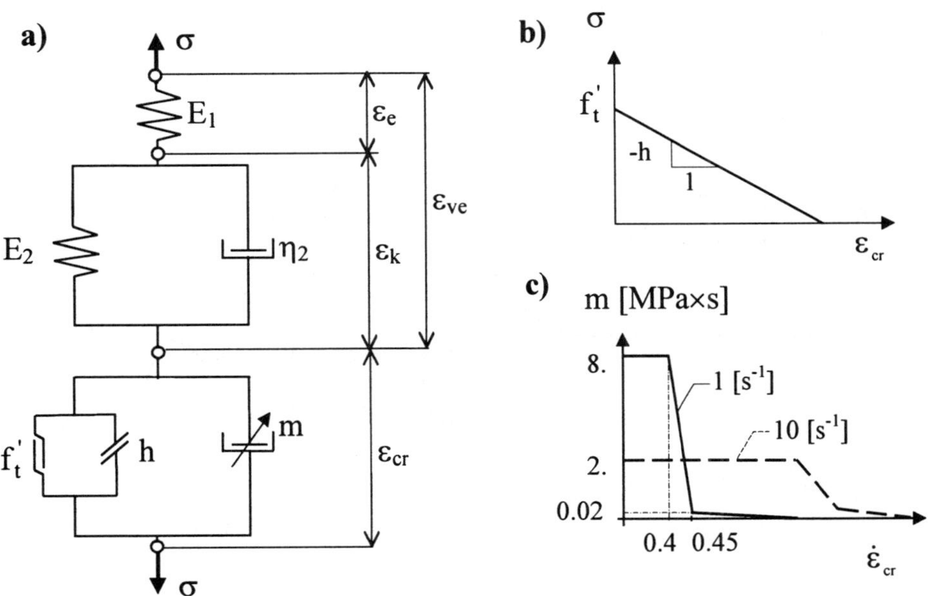

Figure 2. *(a) One-dimensional model with rate-dependent viscosity; (b) assumed fracturing law; (c) assumed laws of variation of m*

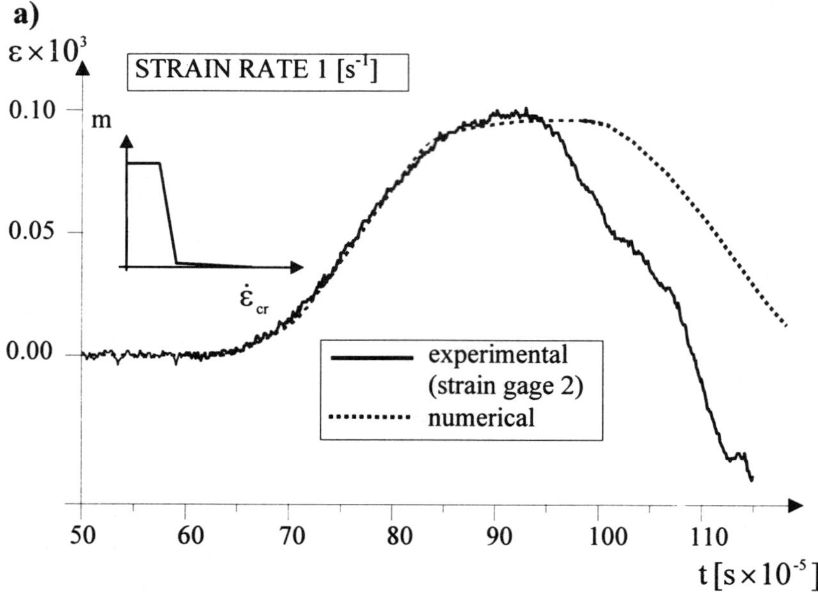

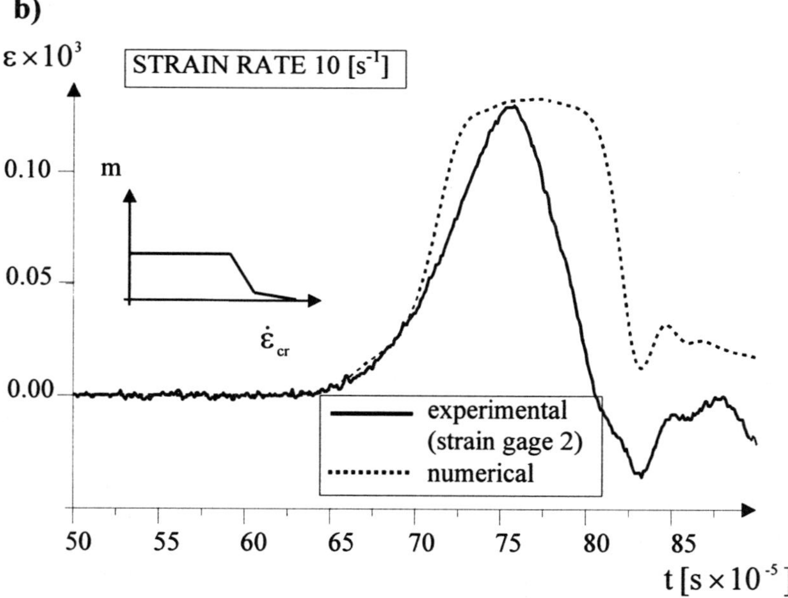

Figure 3. *Numerical simulation of the strain measured on the transmitting bar for different strain rates: (a) 1 s^{-1}; (b) 10 s^{-1}*

The fracturing element is characterized by the diagram of stress versus fracturing strain ε_{cr}, which is rate independent and, as a first approximation, has been assumed linear with slope –h (Figure 2b). The viscosity of the damper has been assumed variable either with ε_{cr} or $\dot{\varepsilon}_{cr}$. The results of the simulations with various values of h and with different laws of variation for m (Figure ac) indicated however that with the same set of parameters, the model could not fit the recorded signals for different strain rates. Figures 3a and 3b show the best approximations we could get of the strains measured on the transmitting bar for strain rates of the order of, 1 s^{-1} and 10 s^{-1} respectively, with, however, totally different laws of variation of m in the two cases.

More elaborate laws for h and m stemming from the activation energy theory for the rate of bond rupture [BAZ 93] were not successful either [BIA 97].

The conclusion of all these simulations was that a proper modeling of the experimental results required much less viscosity for higher strain rates, and this could not be achieved with rate-history independent models.

2.2. *Rate-history dependent viscosity*

A rate-history dependent model can have a much simpler representation (Figure 4a). In this model, the element with variable tangent modulus E_1 obeys the quasi-static stress-strain relation of concrete (Figure 4b). Connected in parallel is a Maxwell element with E_2 = constant and variable viscosity $\eta_2 = \tau_2 E_2$.

2.2.1. *Assumed quasi-static stress-strain relation*

The analytical representation adopted for the quasi-static stress-strain relation (Figure 4b) is:

– for $\varepsilon < \varepsilon_p$

$$\sigma_1 = \frac{s\xi - \xi^2}{1 + (s+2)\xi} \qquad [1]$$

in where $\xi = \dfrac{\varepsilon}{\varepsilon_p}$, $s = \dfrac{E_0}{f_t}\varepsilon_p$, $\varepsilon_p = 0.1 \times 10^{-3}$

– for $\varepsilon_p < \varepsilon < \varepsilon_u$

$$\sigma_1 = a\varepsilon^2 + b\varepsilon + c \qquad [2]$$

in where the coefficients a, b, c are determined by the conditions that for $\varepsilon = \varepsilon_p$, σ_1 = f_t and $d\sigma_1/d\varepsilon = 0$, and that for $\varepsilon = \varepsilon_u = 12\varepsilon_p$, $\sigma_1 = f_u = f_t/10$ (Figure 4b).

– for $\varepsilon > \varepsilon_u$ an exponential decay of the stress has been assumed.

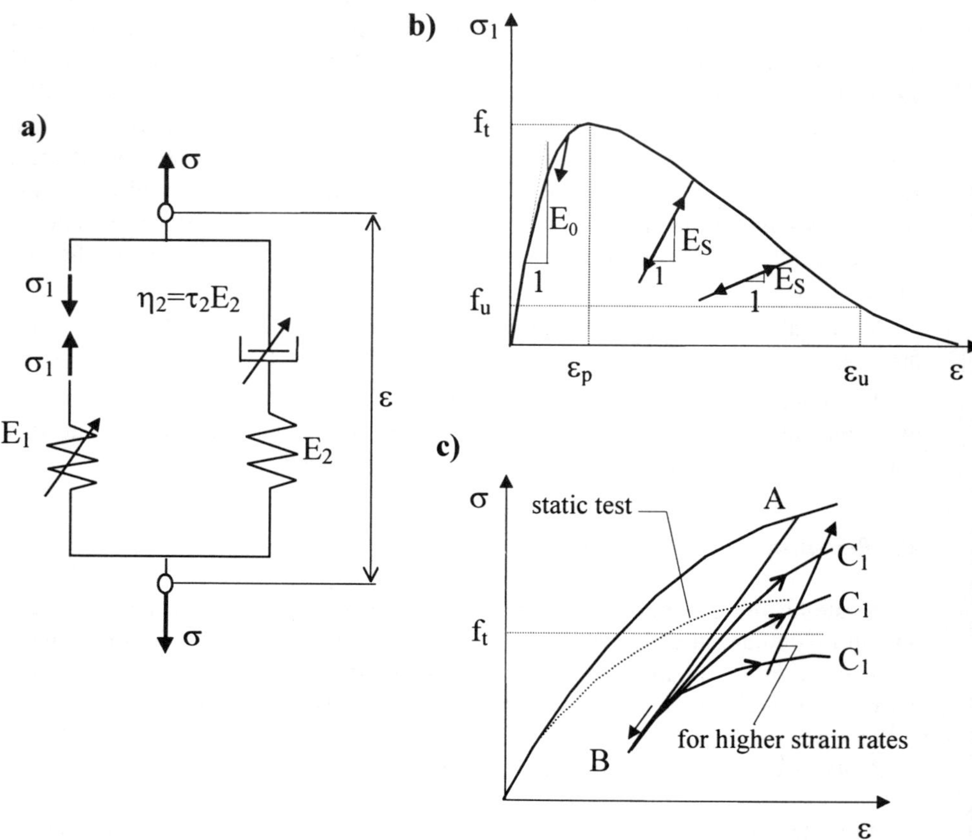

Figure 4. *(a) Rate-history dependent model; (b) quasi-static stress-strain relation and unloading modulus; (c) unloading-reloading path for different strain rates*

2.2.2. *Damage related viscosity*

The law of variation of the viscosity of the damper is assumed now to be a function of the rate history, in the sense that the higher the overstress due to strain rate effects, the higher the damage of the material, and, consequently, the smaller the value of the viscosity. An expression which satisfies this requirement is given by:

$$\eta_2 = \eta_0 \left(\frac{W_1}{W_2} \right)^{\frac{3}{2}} \qquad [3]$$

in which η_0 is the viscosity of the undamaged material, W_1 is the recoverable energy density upon unloading along the quasi-static stress-strain relation and W_2 is the specific work absorbed by the system at the current stress state (Figure 5). Figure 5a refers to pre-peak stress states, and Figure 5b refers to post-peak stress states.

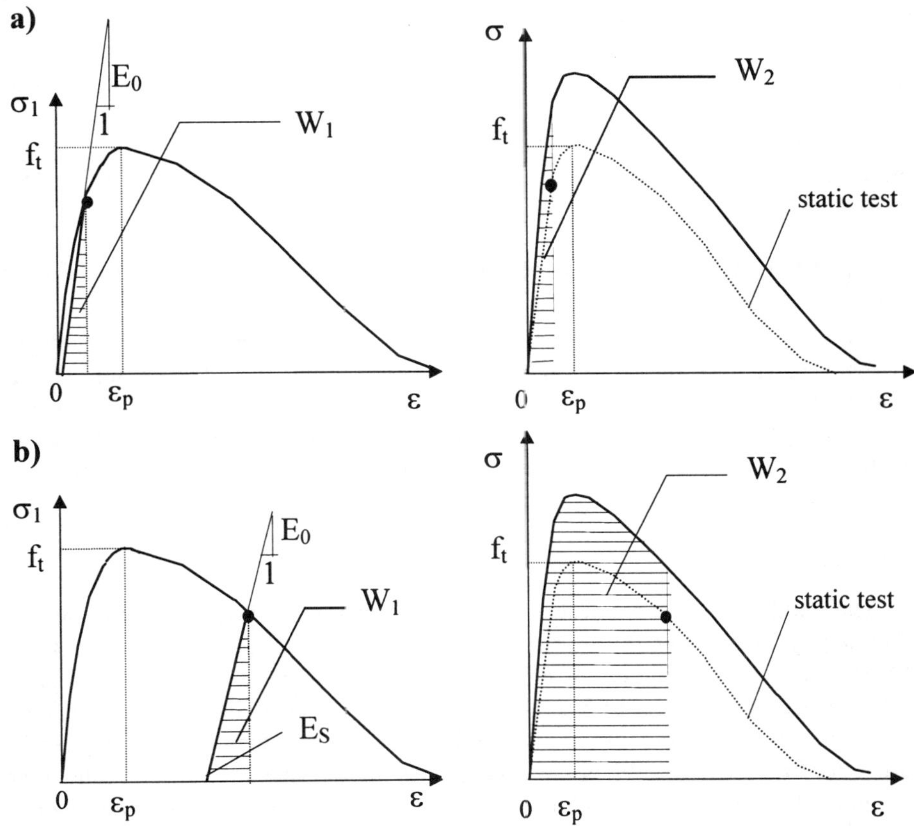

Figure 5. *Representation of recoverable (W_1) and absorbed (W_2) energies for (a) pre-peak states and (b) post-peak stress states*

The ratio W_1/W_2 keeps diminishing during the loading process, to a small extent in the pre-peak region in which W_2 is slightly higher than W_1, more rapidly in the post-peak range, in which the absorbed energy W_2 becomes much higher than the recoverable one W_1 and may then be considered indicative of the accumulation of damage.

2.2.3. *Unloading-reloading criteria*

For the element representing the quasi-static behavior, unloading in the pre-peak region is assumed to occur according to the initial elastic modulus E_0 (elasto-plastic behavior, Figure 4b). In the post-peak region, the unloading modulus E_s must reflect the development of internal damage which gradually reduces the slope (Figure 4b). The adopted assumptions are:

$$E_s = E_0 \qquad \text{for } 0 < \varepsilon < \varepsilon_p \qquad [4]$$

$$E_s = E_0\sqrt{\frac{\sigma_1}{f_t}} \qquad \text{for } \varepsilon > \varepsilon_p \qquad [5]$$

The reloading has been assumed to occur according to the same modulus of unloading.

For the damper, unloading and reloading occur with the same value of the viscosity reached at the instant of first unloading (point A in Figure 4c). This is equivalent to admitting that no damage due to strain rate is exercised on the material in the unloading-reloading process. The reloading, would then bring back the stress at the levels reached before unloading only with a comparable strain rate (Figure 4c).

2.2.4. *Mesh sensitivity*

The model has been checked for mesh sensitivity. Four different meshes obtained by subdividing the specimen into n columns of elements with n = 13, 20, 40, 80 have been considered. At the left end of the incident bar (Figure 6a) a pulse with a constant strain rate of 5 s^{-1} is applied for a time interval of 1.28×10^{-4} s, after which the stress is held constant at a level which is higher than the dynamic tensile strength of the material. When the pulse passes through the specimen, it causes the columns of elements closer to the left boundary to reach first the softening regime. Figure 6b shows that the strain localization in a zone of finite width is fairly independent from the mesh adopted.

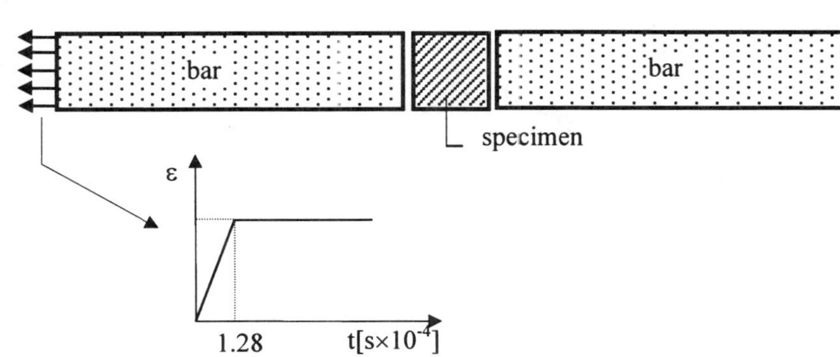

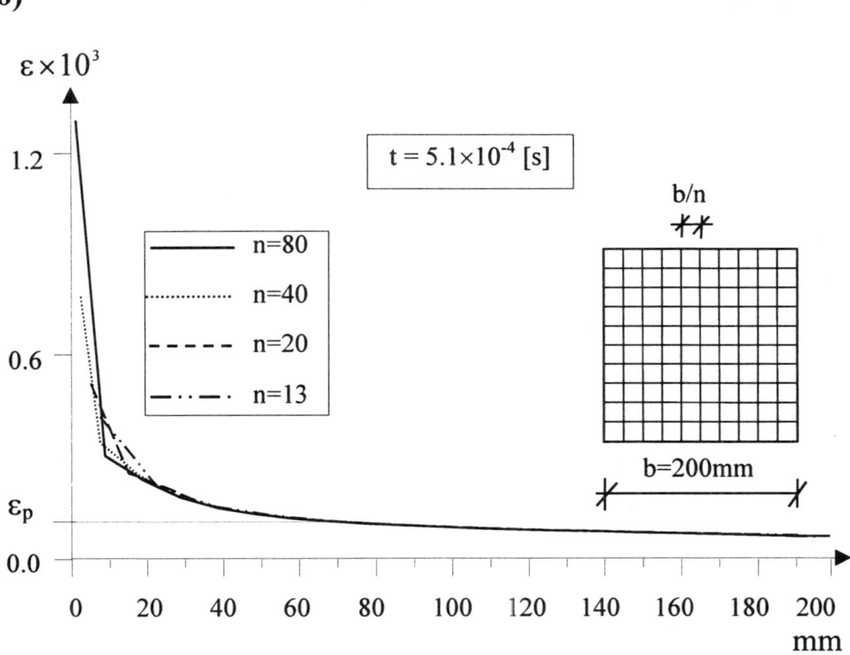

Figure 6. *(a) Loading pulse; (b) strain localization along the specimen at the instant* $t = 5.1 \times 10^{-4}$ *s*

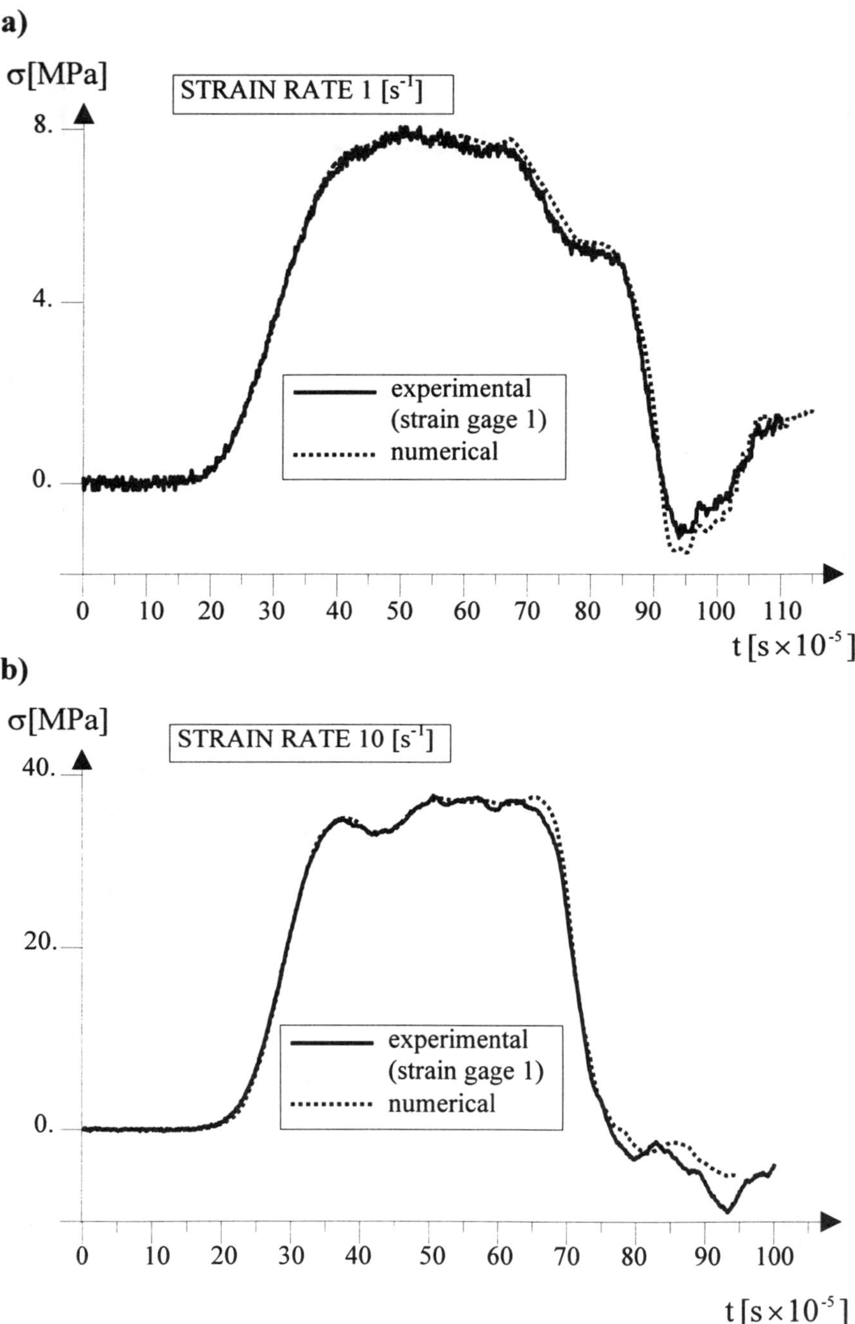

Figure 7. *Simulations of the pulse passing through the incident bar strain station for strain rates (a) 1 s^{-1} and (b) 10 s^{-1}*

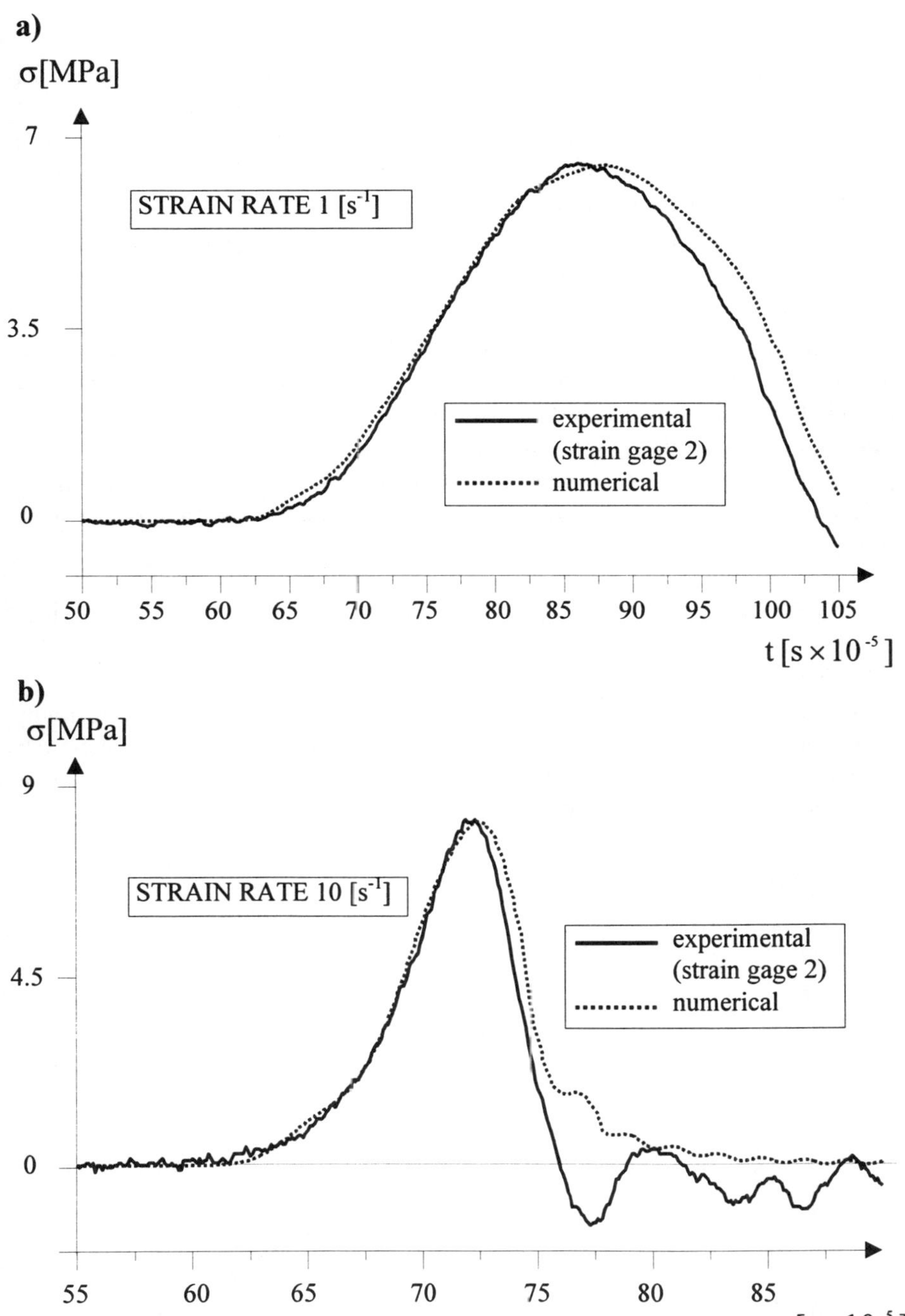

Figure 8. *Simulations of the pulse passing through the transmitting bar strain station for strain rates (a) 1 s^{-1} and (b) 10 s^{-1}*

2.2.5. *Numerical simulation*

The model has been applied to concrete cubes of side 20 cm which had been cast using Portland cement type 425. The concrete mix had a cement-sand-gravel ratio of 1:2.5:2.3 (all by weight) and a water cement ratio of 0.5. The maximum gravel size was 25 mm. The specimenns were cured at 95% relative humidity for 90 days, then kept sealed until a few days before the test.

The mechanical properties after 300 days were: compressive strength $f_c' = 58.86 \text{MPa}$; tensile strength $f_t' = 3.21 \text{MPa}$; elastic modulus $E_0 = 48.6 \times 10^3$ MPa.

The applied impulse at the left extremity of the incident bar has been chosen in such a way [BIA 97] that the resulting impulse on the incident bar (strain station 1) is accurately modeled for the two strain rates investigated (Figures 7a, b).

The simulations of the impulse recorded on the transmitting bar (strain station 2) is represented in Figures 8a and 8b for, respectively, strain rates 1 s^{-1} and 10 s^{-1}. Differently from Figure 3, the numerical values adopted in the two simulations have been identical, and given by:

$$E_2 = 138 \times 10^3 \text{ MPa}; \quad \eta_0 = 48.3 \text{ MPa} \times \text{s} \quad (\tau_0 = 3.5 \times 10^{-4} \text{ s}) \qquad [6]$$

One can also see that the transmitted impulse is well approximated.

The effect of moderately high strain rates on the resulting stress-strain relations is represented in Figure 9. There is a two-fold increase of strength passing from strain rates 0 s^{-1} to 1 s^{-1}, and a four-fold increase from 0 s^{-1} to 10 s^{-1}. The strain at peak stress has, instead, a more limited growth.

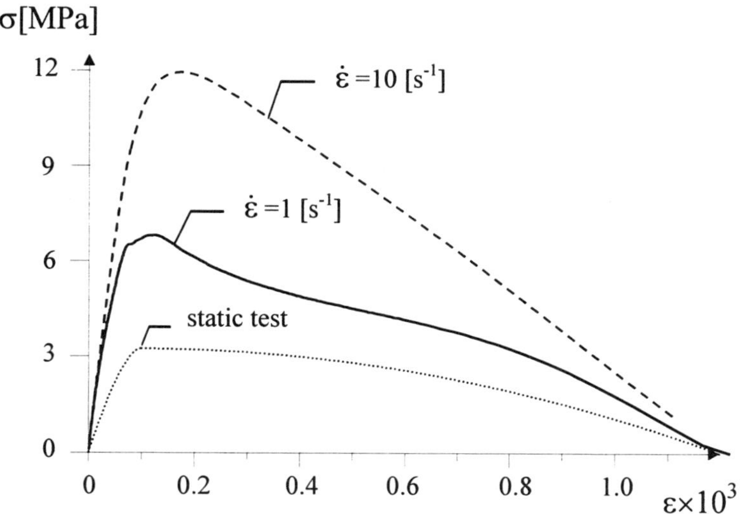

Figure 9. *Resulting stress-strain relations for different loading rates*

3. Conclusions

The smeared crack model for concrete under uniaxial tension can be generalized for transient loading conditions through the inclusion, in parallel with an elastic-plastic-fracturing element, of a Maxwell element with variable viscosity. The viscosity of the damper cannot be assumed as a function of strain rates, but, on the contrary, depends on the accumulated damage. A dependence on the relative value of absorbed energy with respect to the recoverable energy has proved to yield a very accurate fit of experimental results obtained through a split Hopkinson bar. The strain rates studied ($1\ s^{-1}$ and $10\ s^{-1}$) show a remarkable influence of strain rate on the dynamic tensile strength.

The results are relative to a normal concrete ($f_c' = 58.86\text{MPa}$) with 25 mm aggregate size and 95% relative humidity. The modeling of the effect of humidity content on the dynamic behavior is presented in a separate paper [CED 98].

Acknowledgments

The authors are indebted to Drs. Ezio Cadoni and Kamel Labibes for their patience and help in the interpretation of the experimental data. They also want to thank professors Zdenek P. Bazant and Umberto Perego for their valuable comments on the proposed model.

4. References

[BAZ 93] BAZANT, Z.P., "Current status and advances in the theory of creep and interaction with fracture", in Creep and Shrinkage of Concrete, *Proceedings of the 5th International RILEM Symposium*, E&FN SPON, 1993.

[BIA 97] BIANCHI, P., RATTI, A., Una legge costitutiva viscoplastica per il calcestruzzo sotto carichi impulsivi, dissertation, Politecnico di Milano, 1997.

[BIR 71] BIRKIMER, D.L., LINDEMANN, R., "Dynamic tensile strength of concrete materials", *A.C.I. Journal*, vol. 68, n° 1, 1971.

[CAD 97] CADONI, E., LABIBES, K., ALBERTINI, C., SOLOMOS, G., BEVILACQUA, G., BROGNERI, E., DELZANO, G., MURAROTTI, M., SCHABEL, W., Mechanical Response in Tension of Plain Concrete in a Large Range of Strain Rates, Technical Note I.97.194, J.R.C. Ispra, 1997.

[CED 98] CEDOLIN, L., ALBERTINI, C., BERRA, M., "The influence of humidity content on dynamic tensile strength of concrete", submitted for publication.

[HEI 77] HEILMANN, H.G., HILSDORF, H., FINSTERWALDER, K., *Festigkeit und Verformung von Beton unter Zugspannungen*, Deutscher Ausschuss für Stahlbeton, Heft 203, 1977.

[HOP 14] HOPKINSON, B., "A method of measuring the pressure produced in the detonation of high explosives or by the impact of bullets", *Phil. Trans. Roy. Soc.*, London, Series A, vol. 213, n° 10, p. 437-456, 1914.

[KOL 53] KOLSKY, H., *Stress waves in solid*, Clarendon Press, Oxford, 1953.

[KOM 70] KOMLOS, K., "Investigation of rheological properties of concrete in uniaxial tension", *Material Prüfung*, vol. 12, n° 9, p. 300-304, 1970.

[KOR 80] KORMELING, H.A., ZIELINSKI, A.J., REINHARDT, H.W., Experiments on concrete under single and repeated uniaxial tensile loading, Report n° 5-80-3, Stevin Laboratory, Delft University of Technology, 1980.

[NEE 88] NEEDLEMAN, A., "Material rate dependence and mesh sensitivity on localization problems", *Comp. Meth. Appl. Mech. Eng.*, vol. 67, p. 69-86, 1988.

[REI 86] REINHARDT, H.W., KORMELING, H.A., ZIELINSKI, A.J., "The split Hopkinson bar, a versatile tool for the impact testing of concrete", *Materiaux et constructions*, vol. 19, n° 109, p. 55-63, 1986.

[SLU 92] SLUYS, L.J., Wave propagation, localization and dispersion in softening solids, doctoral dissertation, Delft University, 1992.

Size Effect in Design of Fastenings

Rolf Eligehausen — Joško Ozbolt

Institut für Werkstoffe im Bauwessen
University of Stuttgart
70550 Stuttgart, Germany

ozbolt@iwb-uni.stuttgart.de

ABSTRACT. *In the present paper the failure mechanism and size effect on the concrete cone pull-out resistance is reviewed and studied. The influence of material and geometrical parameters on the failure mode and size effect is investigated in detail. In the numerical studies the smeared crack finite element analysis, based on the microplane material model for concrete, was used. Both, experimental and numerical results show that there is a strong size effect on the nominal pull-out strength. It is demonstrated that besides the embedment depth the scaling of the head of the stud as well as the scaling of the concrete member influence the nominal strength and the size effect.*

KEY WORDS: *Concrete, Pull-out, Fastening Technique, Microplane Model, Finite Elements, Size Effect.*

1. Introduction

In engineering practice headed anchors are often used to transfer loads into reinforced concrete members. Experience, a large number of experiments, as well as numerical studies for anchors of different sizes, confirm that fastenings are capable to transfer a tension force into a concrete member without using reinforcement.

The simplest fastening case, that will be considered in the present paper in more detail, is a single headed stud anchor which transfers a tensile force into a large unreinforced concrete block. Provided the steel strength of the stud is high enough, a headed stud subjected to a tensile load normally fails by pulling out a concrete cone. A typical concrete cone observed in experiments is shown in Figure 1. The failure is due to the failure of concrete in tension by forming a circumferential crack [ELI 97]. In the test shown in Figure 1, the concrete block has been reinforced with a surface reinforcement, however, this reinforcement did not influence the formation of the cone.

Figure 1. *Typical pull-out concrete cone obtained in the tests [ELI 97]*

To better understand the crack growth and to predict the pull-out failure load of headed studs for different embedment depths, a number of experimental and theoretical studies have been carried out [OTT 81], [BOD 85], [BAL 86], [ELI 85, 88, 89, 90, 92] [OZB 95]. From these activities, it can be said that the experimental results for headed anchors show a significant size effect on the pull-out failure strength. Furthermore, it has been shown that numerical finite element studies based on macroscopical constitutive models, according to the conventional plasticity or elasticity (strength) theory, are not capable to predict the behaviour of anchors as observed in the experiments [ELI 89, 90], [OZB 95]. Therefore, more sophisticated numerical fracture analyses need to be carried out in which the employed computational model should account for the concrete strength as well as for the equilibrium between the structural energy release rate and concrete energy consumption capacity. The pull-out from a concrete block does not represent a simple mode-I failure case. Consequently, to model the entire load history correctly, the concrete behaviour needs to be represented realistically, not only for dominant tensile loads, but also for general three-dimensional combinations of stresses and strains.

In numerical studies, the smeared fracture finite element analysis is often used. An important aspect in this approach is that it should not exhibit spurious mesh sensitivity [BAZ 86], [BOR 89]. This kind of the mesh dependency is a consequence of the localisation of damage into a volume whose size is mesh dependent. To assure the objectivity of the analysis with respect to the size of the finite elements a so called localisation limiter need to be used. Principally, there are two approaches available. The simplest one is the crack band approach [BAZ 83]. In this approach, the material constitutive law is adapted to the element size such that the energy consumption capacity of the material for different element sizes remains approximately constant and equal to the concrete fracture energy G_F. More general, but from the computational point of view more time consuming methods, are so called higher order methods (Cosserat continuum, nonlocal continuum of integral or gradient type [PIJ 87], [BOR 92], [OZB 96]).

In the present study, the failure mechanism and related size effect on the concrete cone pull-out capacity are reviewed and additional studies are carried out. The test data are compared with numerical results which are obtained using a finite element code that is based on a smeared type of the fracture analysis and mixed constraint microplane model for concrete (computer code ***MASA*** [OZB 97]). The influence of the material and geometrical parameters on the size effect is investigated. Besides the pull-out geometries which are typically used in the engineering practice, where the head size of the stud is not scaled proportional to the embedment depth, studs of different sizes with proportional scaling of the head are investigated as well. Finally, the size and scaling of the concrete block is studied.

2. Experimental evidence on the size effect

The resistance of the headed stud relies only on the concrete cone tensile resistance (no reinforcement). Therefore, to design safe and economical structures, it is important to fully understand the failure mechanism and to know how the variation of the material and geometrical properties influence the pull-out failure capacity. The first experiments in which the size effect on the concrete cone pull-out strength has systematically been investigated were performed by Bode [BOD 85]. The embedment depth was varied between 40 and 150 mm. Subsequently, a number of experiments were carried out by [ELI 97]. The embedment depth was varied between h_{ef} = 130 and 520 mm. However, the size of the concrete specimen were varied only approximately. Therefore, further experiments were performed by [ELI 92], with the variation of the embedment depth between h_{ef} = 50 mm and 450 mm (see Figure 1). In these experiments, the concrete properties have been kept constant, while the main diameters of the concrete specimen were scaled proportionally to the embedment depth. The distance between the support reaction and the outer edges of the member was kept constant and equal to 100 mm (see Figure 1). Furthermore, in the tests, the size of the heads was done such that the pressure under the head was approximately constant for different embedment depths. In one test series, the diameter of the head was scaled proportionally to the embedment depth, which leads to a smaller concrete pressure under the head at failure with increasing embedment depth.

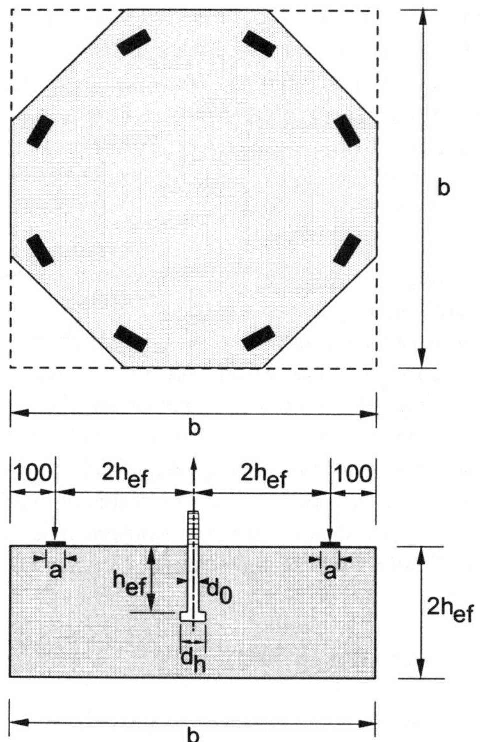

Figure 2. *Typical pull-out test geometry for the headed stud anchors [ELI 92]*

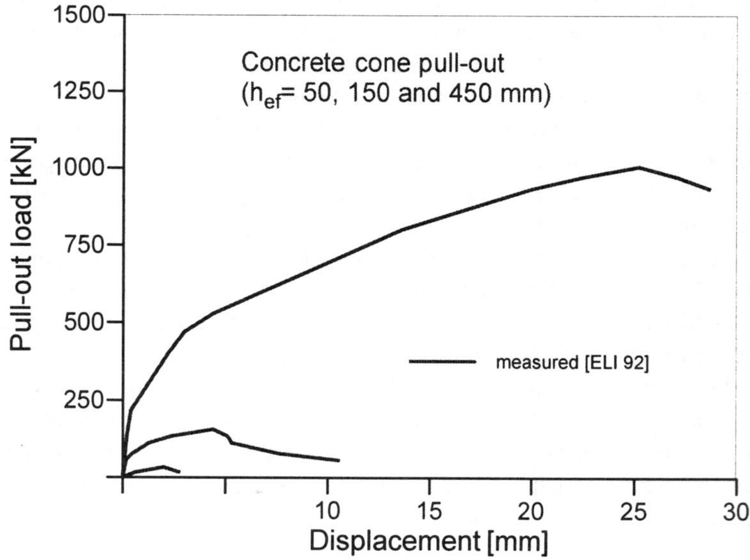

Figure 3. *Typical pull-out load-displacement curved measured in the experiments [ELI 92]*

The headed anchor is pulled out from a concrete block by controlling the displacement at the loaded anchor end. Figure 3 shows typical pull-out load-displacement (L-D) curves obtained from the experiments for three different embedment depths [ELI 92]. As can be seen, although in the test members no reinforcement was present, the curves exhibit a relatively ductile response. It has been observed that the major part of the displacement is due to the extremely large compressive deformations under the head of the stud.

The nominal strength for a number of experimental results are summarised in Figure 4. The measured nominal pull-out strengths normalised to the concrete cube compressive strength $f_{cc} = 33$ MPa (normalising factor $= (33/f_{cc})^{1/2}$), for the size range $h_{ef} = 40$ to 520 mm, are plotted versus the embedment depth in a log-log scale. The strength is calculated as the ultimate load divided by the area of a circle with a radius equal to the embedment depth i.e. $\sigma_N = P_U/(h_{ef}^2\pi)$. Comparing the test data with the ACI no size effect design formula [ACI 80], it is obvious that the experimental results for larger embedment depths are lower than the predicted values. In the same figure, the design formula, which is based on linear elastic fracture mechanics (LEFM), is also plotted [CEB 90]. This curve exhibits a good agreement with the experimental results for the whole size range. The fact that the experimental results agree with the LEFM based design formula means that the size effect on the nominal pull-out strength is strong since this formula predicts the maximal possible size effect.

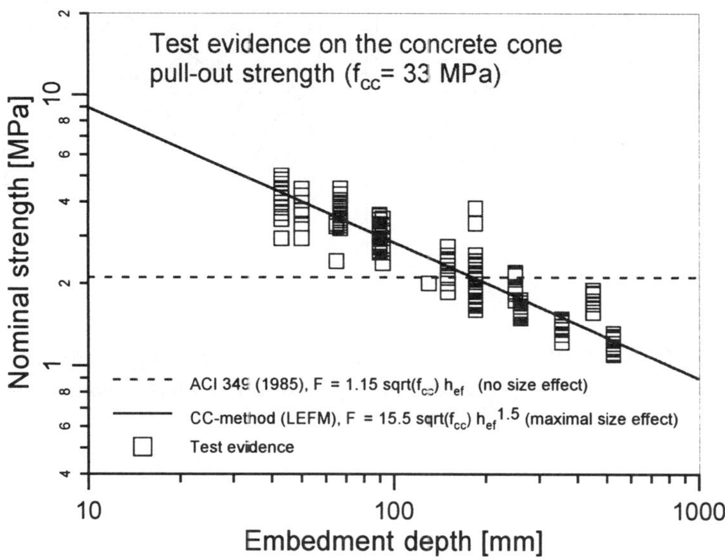

Figure 4. *Pull-out of the headed studs - summary of experimental results for the nominal pull-out strength and comparison with design formulas*

To find out the reason for the size effect, the crack development as well as the distribution of the stresses along the crack surface were measured in test with h_{ef} = 130 mm, 350 mm and 520 mm [ELI 89]. Figure 5 shows the position of the strain gages for the anchor with an embedment depth of h_{ef} = 520 mm. They were placed along the assumed concrete cone crack surface which was verified in a previous test. In Figure 6a the strains normal to the crack surface at 30% and 90% of the ultimate load are plotted. Cracking started at about 25% of the peak load at the head. At 90% of the ultimate load the crack length reaches approximately 35% of the total crack length at failure. The corresponding stresses perpendicular to the crack surface are plotted in Figure 6b. The test data clearly show a stable crack growth, *i.e.*, with increase of the crack length the load also increases and reaches the maximum value at a critical crack length of approximately $l_{cr} = 0.4l_{tot}$. As well known from LEFM, if the structure exhibits a stable crack growth, the size effect on the ultimate resistance is strong and the nominal strength is more sensitive on the variation of the fracture energy than on the variation of the concrete tensile strength [OZB 95].

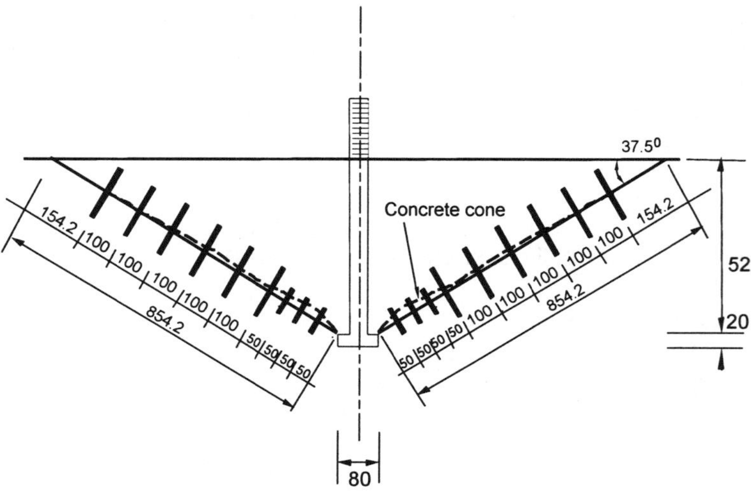

Figure 5. *Distribution of strain gages along the concrete cone surface in order to measure the strain and stress distribution for different load stages, h_{ef} = 520 mm [ELI 89]*

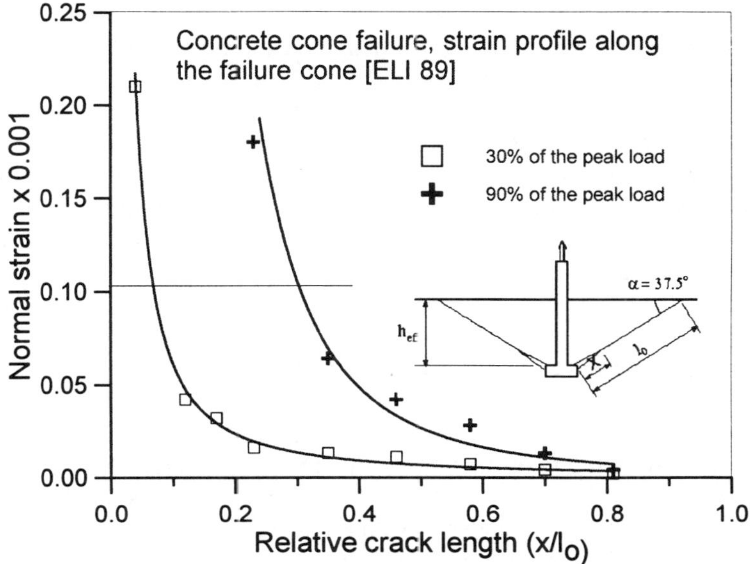

Figure 6a. *Distribution of strains along the concrete cone surface at 30% and 90% of the ultimate load, h_{ef} = 520 mm [ELI 89]*

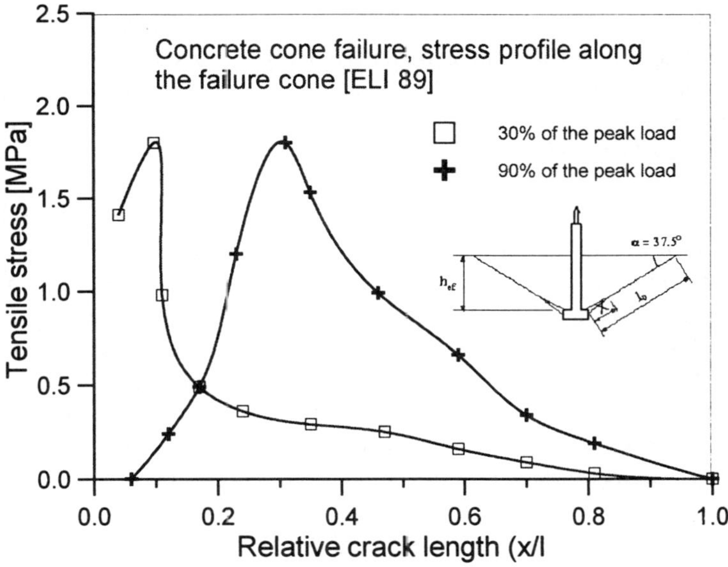

Figure 6b. *Distribution of stresses along the concrete cone surface at 30% and 90% of the ultimate load, h_{ef} = 520 mm [ELI 89]*

The embedment depths investigated in the tests cover most of the practical applications. As shown in Figure 4, for this range, the size effect on the nominal pull-out strength is strong and close to the LEFM prediction formula. The question is, whether the same trend may be expected for larger embedment depths. Furthermore, in the experiments, the influence of the concrete fracture properties (tensile strength and concrete fracture energy) and geometry on the pull-out capacity was not systematically investigated. Therefore, to study the influence of the above parameters and to better understand the experimental results, further numerical studies are carried out.

3. Numerical studies - failure mechanism and size effect

In the numerical analysis three different series of geometries are investigated. Two of them were tested by Eligehausen [ELI 89] [ELI 92] (see Figure 2). The third geometry is taken from the RILEM TC 90 Round robin proposal [RIL 90]. The numerical results are first compared with the test data. Subsequently, the influence of the geometry, *i.e.,* the scaling of the test specimen as well as the influence of material properties on the pull-out resistance is investigated.

The spatial discretization is performed by axisymmetrical finite elements. Four node quadrilateral elements with four integration points (linear strain field) are used [OZB 97]. The cracking and damage phenomena are modelled by employing the smeared crack approach. As a constitutive law for concrete, the mixed formulation of the microplane model was adopted [OZB 97]. To assure the objectivity of the analysis and to prevent so-called spurious mesh sensitivity, two approaches were alternatively used: (1) crack band approach [BAZ 83] and (2) nonlocal integral approach based on the interaction of microcracks [OZB 96].

The typical specimen geometry was the same as in the experiments (see Figure 2). In all calculations pulling of the anchor was performed by prescribing displacements at the bottom of the stud. Contact between steel stud and concrete existed only under the head of the stud. To account for the restraining effect of the embedded anchor, the displacements of the concrete surface along the steel stud in the vicinity of the head were fixed in direction perpendicular to the load direction. The supports were fixed in both vertical (loading) and horizontal direction. The distance between the support and the anchor was assumed to be $2h_{ef}$(see Figure 2) so that an unrestricted formation of the failure cone was possible.

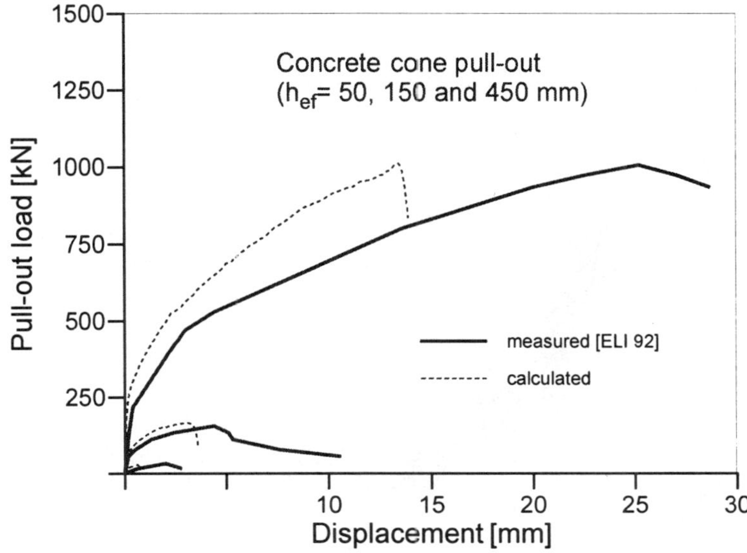

Figure 7. *Measured and calculated load-displacement curves for three embedment depths: h_{ef} = 50, 150 and 450 mm*

3.1. *Verification of the numerical results*

To confirm that the numerical analysis is able to predict the failure load and failure mode correctly, the numerical results for the geometry tested in [ELI 92] (see Figure 2) are compared with the test data. The analysis was carried out for three different embedment depths: h_{ef} = 50, 150 and 450 mm. The size of the head was relatively small ($d_h = 0.25h_{ef}$ for the smallest embedment depth h_{ef} = 50 mm) and it was scaled such that the compressive stresses under the head of the stud at peak load was approximately the same for all three embedment depths. Concrete properties were adopted as the average values measured in the experiment (Young's modulus $E = 29000$ MPa, Poisson's ratio $\nu = 0.20$, uniaxial compressive strength $f_c = 26$ MPa, uniaxial tensile strength $f_t = 2.9$ MPa and concrete fracture energy $G_F = 0.15$ MPa).

The calculated and measured pull-out load-displacement (L-D) curves for h_{ef} = 50, 150 and 450 mm are plotted in Figure 7. As can be seen, the calculated peak loads agree well with the measured failure loads. Furthermore, the corresponding peak displacements and shape of the calculated curves fit the test data reasonably well. Similar as observed in the experiments, the numerical results show a relatively ductile response. Due to the confinement of the concrete under the head of the stud, the compressive stresses and strains in all three directions are large. Depending on the head size, the compressive stresses may be up to 12 times larger then the uniaxial compressive strength of concrete. Consequently, the strains as well as the displacement in the loading direction are also relatively large.

a)

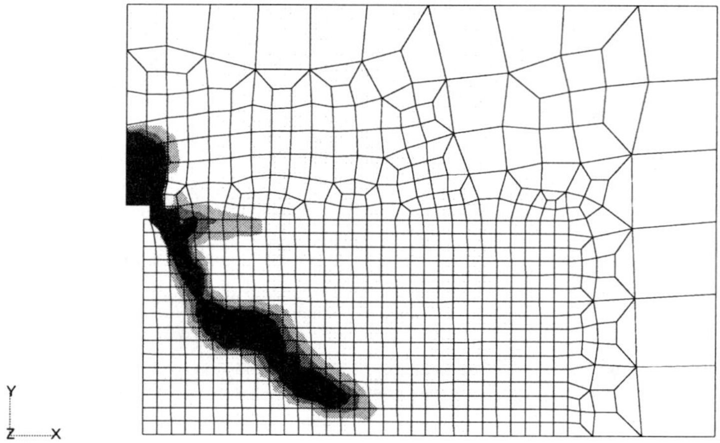

b)

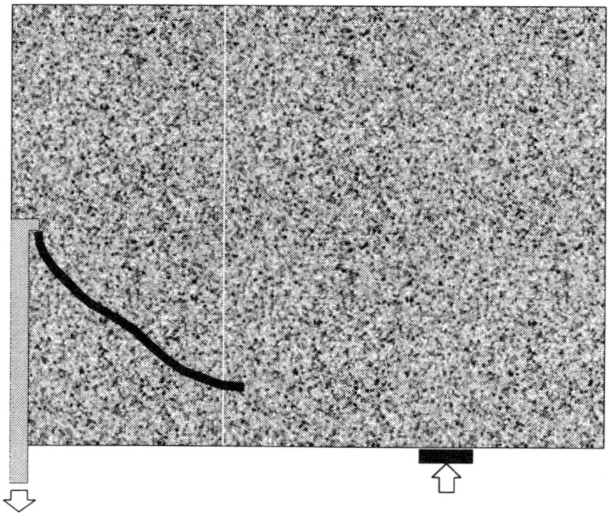

Figure 8. *Crack pattern in the post peak regime mm (approximately 70% of the ultimate load) for h_{ef} = 150: a) calculated, b) measured [ELI 89]*

The crack pattern for h_{ef} = 150 mm short after reaching peak load is shown in Figure 8a. The damage (crack) is plotted in terms of strains in the loading direction (ε_{yy}). For comparison, the corresponding crack pattern observed in the experiment is shown as well (see Figure 8b). As can be seen, under the stud head the crack propagates rather steep which is due to the shearing of the concrete. By subsequent loading the crack starts to deviate from the vertical direction and at failure it propagates under an average angle of approximately $\alpha \approx 35^0$.

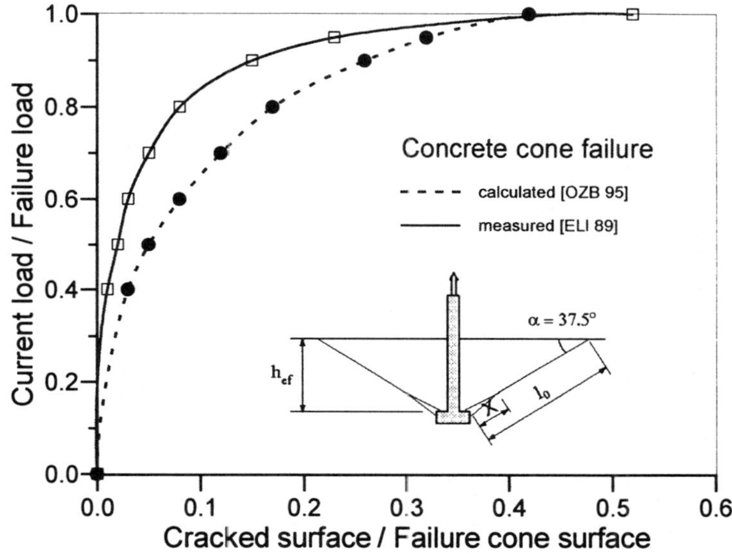

Figure 9. *Relative crack length as a function of the embedment depth [OZB 95]*

Due to the relatively small head, high compressive stresses are generated under the head leading to large displacements of the head. The concrete is sheared off. After the shearing zone propagates deep enough into the concrete, the pull-out crack starts to propagate under a much smaller angle α in failure mode that is approximately of mode-I type. Due to the large displacement, the actual embedment depth and thus the concrete cone failure load is reduced [FUR 94]. The slope of the failure cone and the failure load is influenced by the head size, which will be shown later.

In Figure 9, the relative crack length for h_{ef} = 150 mm is plotted as a function of the ratio between actual load to failure load [OZB 95]. For comparison, the test data are plotted as well [ELI 89]. As can be seen, as observed in the experiments, the numerical results show that before the ultimate load is reached a stable crack propagation takes place.

In Figure 10 the nominal strength, calculated from the numerically and experimentally obtained failure loads is plotted as a function of the embedment depth. The power fit of the calculated data results in an exponent $\beta = 1.62$ ($\sigma_N = A\, h_{ef}^{\beta}$, were A is a geometry and material dependent constant). The agreement between calculated and test data is good since the fit of the test data gives almost the same power function. Obviously, for the investigated size range the numerical and test results indicate a significant size effect on the nominal pull-out capacity. The nominal strength increases as a function of the embedment depth with an exponent which is much smaller than 2 ($\beta = 2$ means no size dependency).

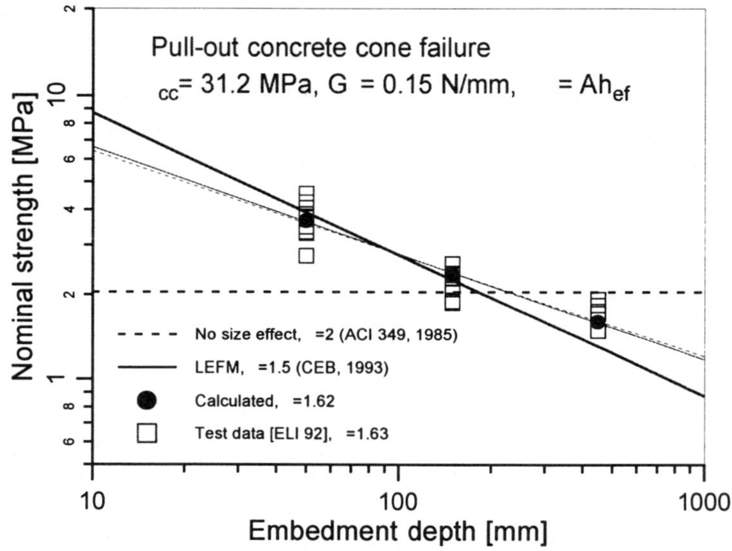

Figure 10. *Nominal strength as a function of the embedment depth - test data, calculated data and different prediction formulas*

The above comparison between numerical and experimental data shows that the axisymmetrical finite element analysis is able to realistically simulate the pull-out failure mechanism and size effect.

3.2. *Influence of the concrete fracture properties on failure load*

The pull-out concrete cone resistance relies on the concrete fracture properties. Therefore, it is important to know how macroscopical fracture properties of concrete influence the pull-out resistance. A numerical analysis was carried out for the geometry proposed by RILEM TC 90 [RIL 90]. A parameter study for headed stud specimens with h_{ef} = 450 mm was performed as follows: (1) for constant G_F = 0.08 N/mm, the tensile strength was varied from 2.4 to 3.6 MPa and (2) for constant f_t = 2.8 MPa, the concrete fracture energy was changed from 0.08 to 0.14 N/mm. The calculated nominal pull-out strengths are plotted in Figure 11 as a function of the tensile strength and fracture energy, respectively. As can be seen, the nominal strength is practically independent of the tensile strength (Figure 11a). However, Figure 11b shows approximately a square root dependency between the nominal pull-out strength and the concrete fracture energy. The same result has been found by Eligehausen [ELI 89], by analytical studies based on the LEFM and by tests on headed studs pulled out from a glass specimen [SAW 94].

The above results clearly show that the pull-out of headed stud concrete cracking is an important aspect of the resistance mechanism. Namely, in contrast to a number of structures which rely only on the material strength this type of structures relies also on the energy consumption capacity of concrete.

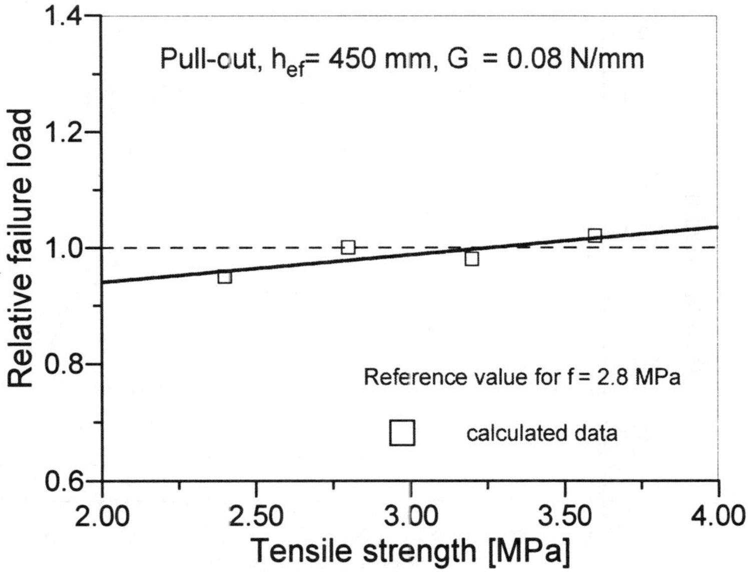

Figure 11a. *Nominal pull-out strength as a function of concrete tensile strength [OZB 95]*

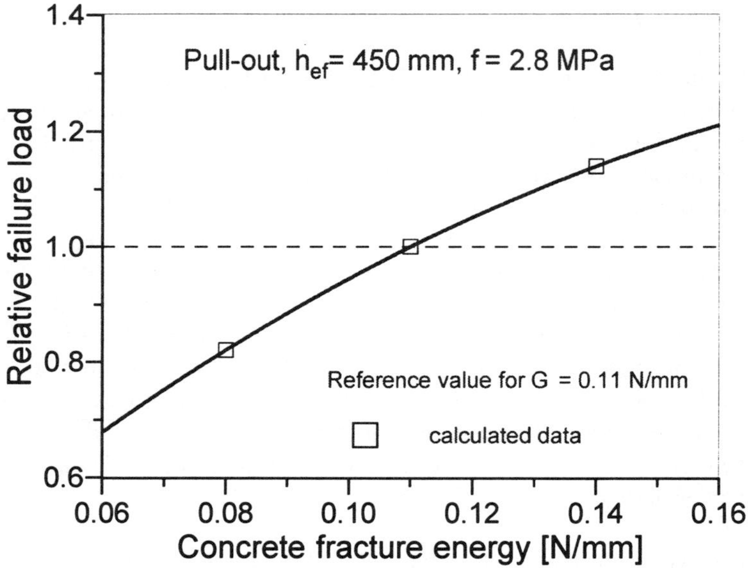

Figure 11b. *Nominal pull-out strength as a function of concrete fracture energy [OZB 95]*

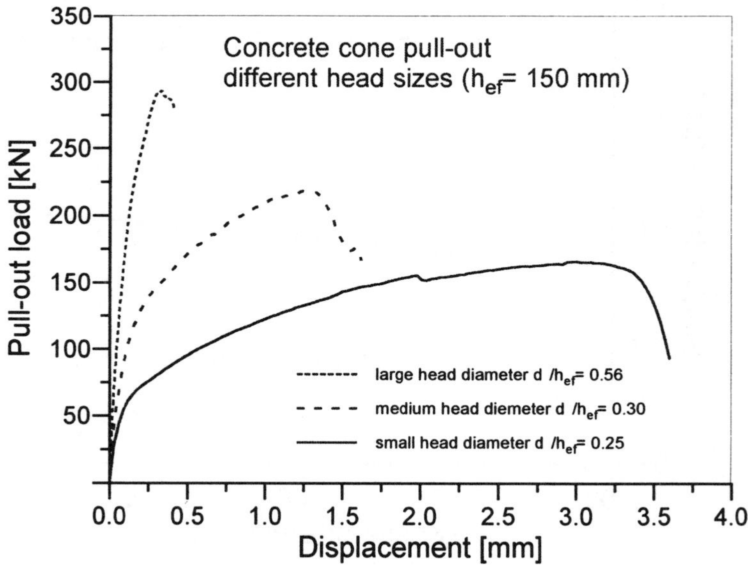

Figure 12. *Calculated L-D curves for three different head sizes: small, medium and large*

3.3. *Influence of the head geometry on failure mode and failure load*

The influence of the head size on the pull-out response and failure mode was studied for the geometry shown in Figure 2 [ELI 92]. The material properties and geometry were the same as in the experiment. For a constant embedment depth (h_{ef} = 150 mm) the size of the head was varied. Three different head sizes were investigated: small ($d_h = 0.22h_{ef}$), medium ($d_h = 0.30h_{ef}$) and large ($d_h = 0.56h_{ef}$).

The calculated L-D curves are plotted in Figure 12. A large head exhibit a stiffer response and a higher pull-out capacity than a small head. There are two reasons for this. The first is due to the fact that a larger head generates a larger concrete cone surface for the same embedment depth and therefore a higher pull-out capacity and stiffer response. The second reason, which is important for relatively small heads, is due to the shearing effect which influences the failure mode and reduces the embedment depth. To demonstrate this, in Figure 13 the crack patterns for three different head sizes are shown. The figure shows that in the case of a small head, there is a strong shearing effect. The crack starts to propagate almost vertically and before the mode-I failure is activated the embedment depth is reduced. By increasing the head the shearing is less pronounced and for a relatively large head there is almost no shearing with a crack growth which is for the whole load history approximately of mode-I type. A similar influence of the head size on the load-displacement behaviour and the failure load was found in tests by Furhe [FUR 94].

3.4. *Size effect*

3.4.1. *General*

As shown above, the series of experimental data as well as the numerical results for headed anchors indicate that there is a significant size effect on the pull-out capacity. To investigate the phenomena in more detail in the past, a number of theoretical studies have been performed [ELI 97]. Besides already available results, additional studies are carried out here.

In Figure 14, the nominal pull-out strength is plotted as a function of the embedment depth. The failure loads are normalised to a cube compressive concrete strength f_{cc} = 33 MPa. For most of the plotted test data, the size of the head was relatively small ($d_h/h_{ef} = 0.20 - 0.25$) and it was scaled such that the pressure under the head of the stud at peak load was approximately constant for all embedment depths. As can be seen, the test data agree well with the design formula based on LEFM, which means that for the test size range there is a strong size effect on the pull-out capacity. To check whether the same tendency is to be expected for larger embedment depths, a numerical analysis [OZB 95] was carried out for the size range h_{ef} = 50 to 2700 mm. In this analysis, the geometry proposed by RILEM TC 90 was used [RIL 90].

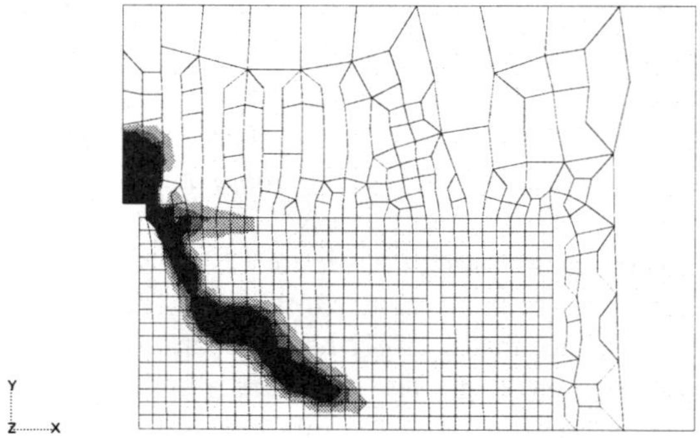

b)

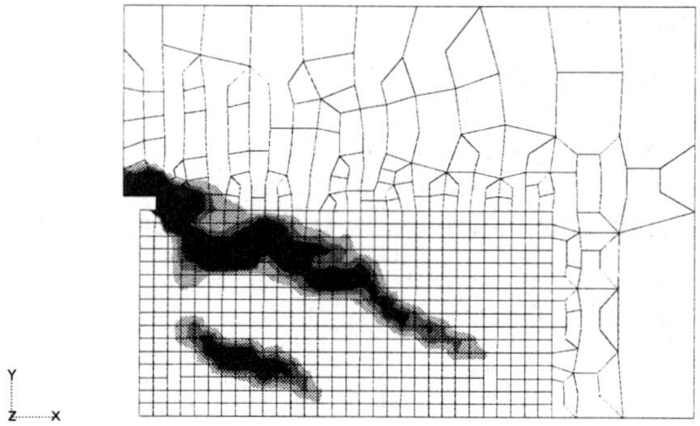

c)

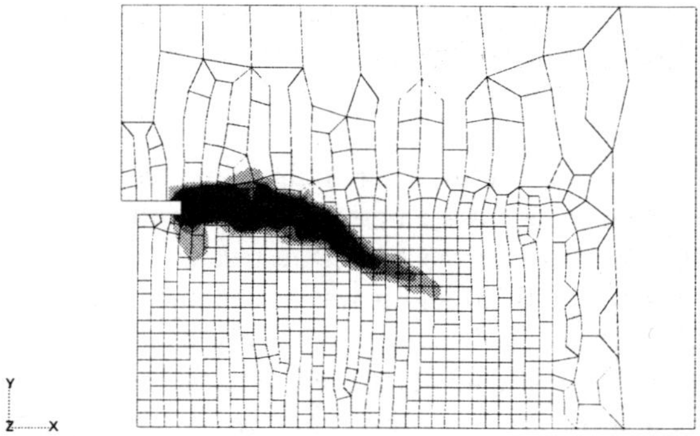

Figure 13. *Calculated crack patterns for three different head sizes: a) small, b) medium and c) large*

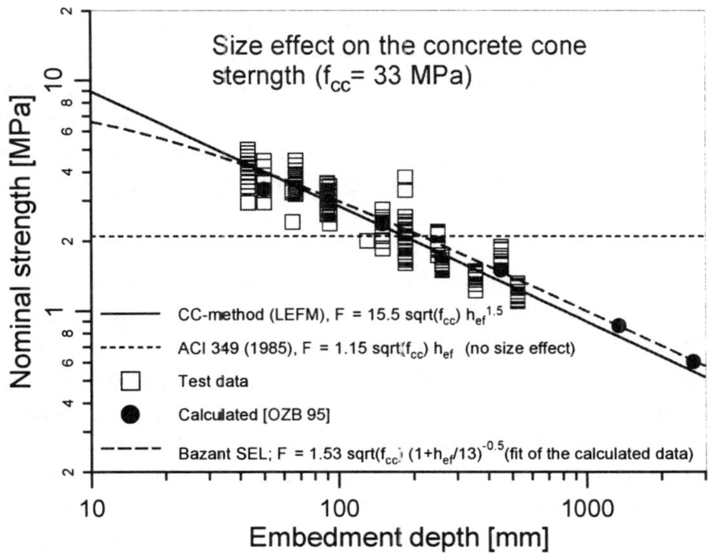

Figure 14. *Nominal pull-out strength as a function of the embedment depth for a broad size range (10-3000 mm) - test data, calculated data and size and no size effect prediction formulas*

The ratio head diameter to embedment depth was $d_h/h_{ef} = 0.3$ for all h_{ef}. The numerical results shown in Figure 14 are fitted with Bazant's size effect formula [BAZ 84]. As can be seen, the value of the so-called characteristic embedment depth $h_{ef,0}$ is relatively small ($h_{ef,0} \approx 13$ mm) which is a clear indicator for a relatively strong size effect. Namely, $h_{ef,0} \approx 13$ mm means that already for an embedment depth larger then 13 mm, the size effect should be closer to the LEFM prediction than to the prediction according to the theory of plasticity. It can be seen that the CEB design formula based on LEFM agrees well with Bazant's size effect formula in the whole size range. Obviously, according to the numerical results for embedment depths larger than available from the current test data a strong size effect may be expected as well.

The numerical results show that for $h_{ef} \to \infty$ the nominal strength coincides with the LEFM prediction and Bazant's size effect formula (see Figure 14, [OZB 95]). This means that there is a strong localisation of damage and stable crack growth. To confirm this in Figure 15, the calculated relative critical crack length at ultimate load is plotted against the embedment depth [OZB 95]. The relative critical crack length is measured as the ratio between the crack length at peak load and the maximum crack length of the concrete cone, under the assumption that the final crack is inclined with an average angle of $\alpha = 37.5^0$. As can be seen in Figure 15, for smaller embedment depths (h_{ef} = 50 to 150 mm), the relative critical length is approximately equal to 0.4. However, for $h_{ef} \to \infty$ it approaches a constant value of approximately 0.25. This confirms that even for a very large structure, the crack length at peak load is approximately proportional to the embedment depth, which implies stable cracking and consequently strong size effect [OZB 95]. The experimental data for the critical crack length [ELI 89] confirm the numerical results, although as aforementioned, these measurements exist only for embedment depths up to 520 mm.

The calculated crack patterns for specimen with relatively small proportionally scaled heads at peak load for small (h_{ef} = 50 mm) and larger (h_{ef} = 150 mm) embedment depth are shown in Figure 16. The dark zones indicate the localisation of damage (cracks). Comparing the crack length at peak load (critical crack length) for smaller and larger embedment depth, it can be seen that the relative crack length is larger for smaller embedment depth. Furthermore, the average angle α of the concrete cone is slightly larger for smaller embedment depths, *i.e.*, the crack tends to propagate under a larger slope. However, if the specimen head is not scaled proportionally, the opposite tendency is observed (compare Figure 13a with Figure 16a), which was also observed in experiments [ELI 92].

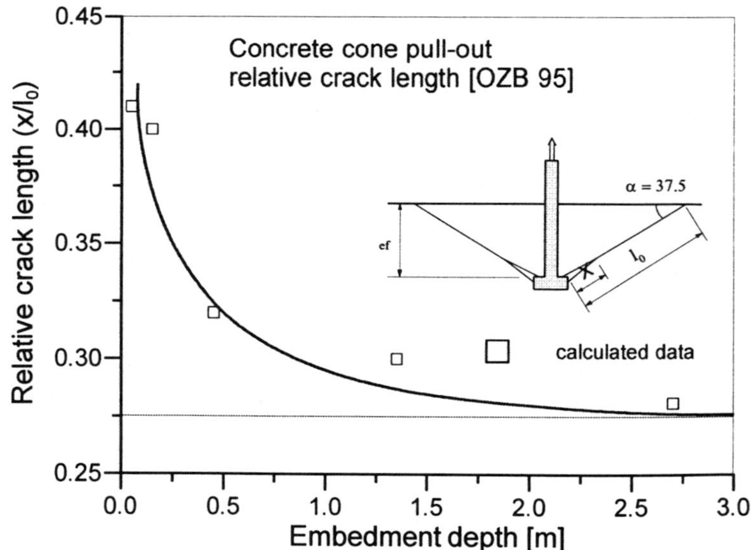

Figure 15. *Calculated relative crack length at peak load [OZB 95]*

a)

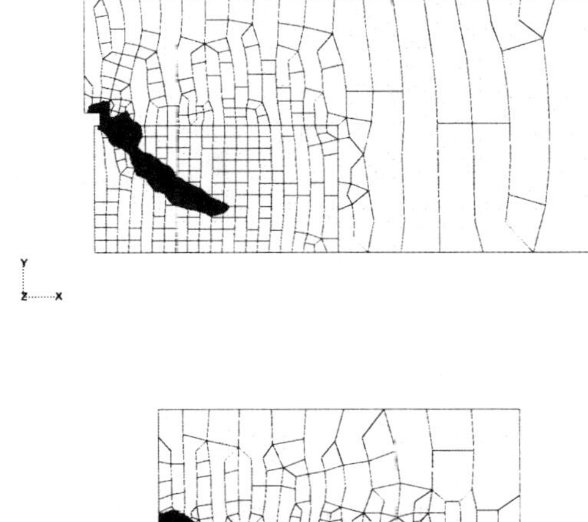

b)

Figure 16. *Calculated crack patterns short after the peak load (approximately $0.95P_U$) for: a) h_{ef} = 50 mm and b) h_{ef} = 150 mm*

3.4.2. *Influence of the head size*

To investigate the influence of the size and scaling of the head of the stud on the size effect, three different series of calculations were performed. In all series the embedment depth was scaled proportionally and the specimen size (geometry) was scaled the same as in the experiment, *i.e.,* not proportional [ELI 92]. The size (diameter) of the head of the stud was first taken relatively small (for the smallest size, h_{ef} = 50 mm, $d_h = 0.25h_{ef}$), and it was scaled such that the ultimate pressure under the head was approximately constant for all three embedment depths [ELI 92]. The results of this study are already compared with the experimental data (see Figure 10). Furthermore, the size of the head was taken as $d_h = 0.30h_{ef}$ and it was scaled proportional to the embedment depth. Finally, in the third series of calculations the head was relatively large ($d_h = 0.56h_{ef}$) and proportional to the embedment depth.

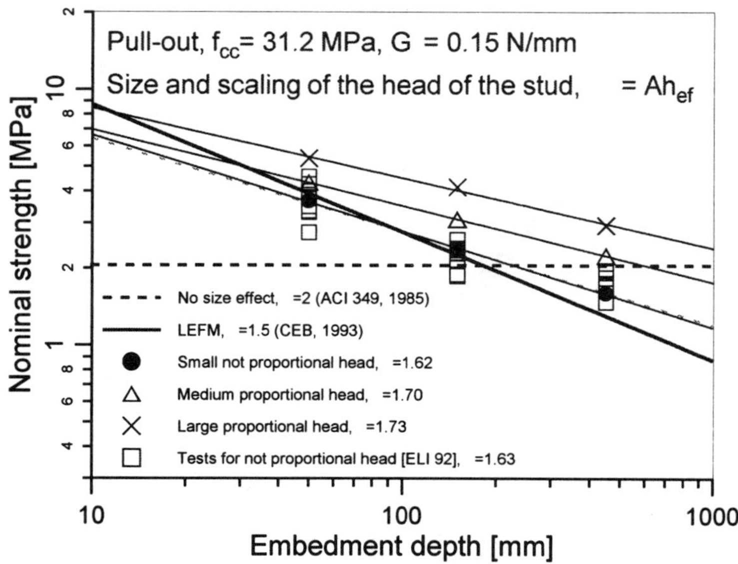

Figure 17. *Size effect on the nominal pull-out strength as a function of the head size diameter*

The results of the study are shown in Figure 17. For comparison, the test data for the first series of head geometry as well as the prediction according to the design formula based on LEFM are also plotted. As can be seen, the numerical results as well as the test data slightly deviate from the LEFM prediction which predicts the maximal size effect ($\beta = 1.5$). The power fit of the calculated and test data for small heads yields to an exponent of approximately β= 1.6. Furthermore, it is obvious that with the increase of the head diameter the nominal strength increases for all embedment depths. However, it is clear that for larger heads the size effect tends to be less pronounced. Namely, for smaller heads β is approximately equal to 1.6 and for relatively large heads equal to 1.7.

The explanation for the fact that with increase of the head size the size effect tends to be lower can be found in a slightly different failure mechanism for geometries with different head sizes, as discussed before. If the head size is small, and especially if it does not increase proportionally with the embedment depth, there is a strong inhomogeneity of the strain field (localisation of damage due to the cutting effect of the head) which does not decrease significantly with increase of the embedment depth. Consequently, the damage is localised in the volume whose size is small relative to the specimen size and the response close to the LEFM prediction, *i.e.,* the damaged zone compared to the embedment depth is relatively small. On the contrary, when the head size increases, the localisation of damage is less pronounced and the damaged volume at peak load is larger relative to the embedment depth. Due to this, the non-linear effect on the crack growth is stronger and the size effect deviates from the LEFM prediction.

3.4.3. *Scaling of the concrete specimen size*

Figure 17 shows that the test as well as the calculated data for the first series of calculations deviate from the LEFM prediction as well as from most of the tests. This may be caused by the fact that in the experiments [ELI 92] and numerical investigations, the specimen size was assumed as shown in Figure 2, that means the diameter d of the test member was not proportional to the embedment depth but $d = h_{ef}$+200 mm. To investigate the influence of the non-proportional member scaling, the analysis for the member geometry with proportional scaling of the specimen was carried out. The basic geometry and the material properties were the same as in the experiment except that the total specimen height was taken as $h = 3h_{ef}$ (in the test $h = 2h_{ef}$).

In Figure 18, the nominal strength is plotted as a function of the embedment depth for the proportional and not proportional scaling of the member geometry. Furthermore, the test data (non proportional scaling) as well as the prediction according to the LEFM design formula are also shown. As can be seen, for smaller embedment depths (h_{ef} = 50 and 150 mm), the member geometry does not significantly influence the failure load. On the contrary, for a large embedment depth (h_{ef} = 450 mm), the nominal strength is approximately 20% higher for the specimen with non proportional scaled geometry. The size effect is stronger and practically identical to the prediction according to LEFM (β = 1.51, see Figure 18).

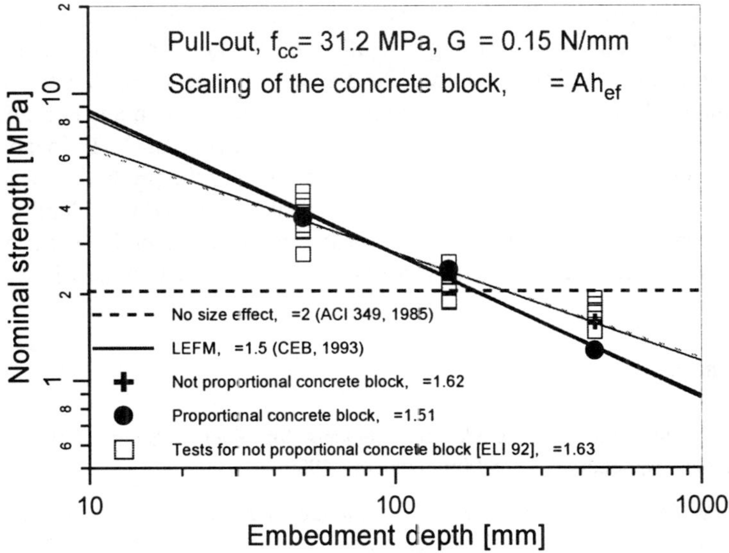

Figure 18. *Size effect on the nominal pull-out strength as a function of the size scaling) of the concrete member*

4. Conclusions

– The comparison between experimental and numerical results shows that a finite element analysis based on the smeared crack approach can realistically predict the concrete cone failure mode and ultimate resistance.

– The experimental results as well as the numerical simulations confirm that a headed stud embedded into a plain concrete block is able to transfer a pull-out force into the concrete utilising only the tensile resistance of the concrete with no need for reinforcement. The main reason for this is a stable crack growth which assures insensitivity of the structural response on the variation of the concrete tensile strength.

– Calculated and measured L-D curves for headed anchors with relatively small head sizes exhibit a rather ductile response. When the head size increases, the ductility of the response decreases, however, the pull-out resistance increases.

– The failure mode depends on the size of the head. For anchors with relatively small heads, before the typical mode-I crack opens, a strong shearing under the head can be observed. This reduces the embedment depth and therefore the resistance as well. For anchors with relatively large heads, the crack growth is for the whole load history close to the mode-I failure type.

– As a consequence of the stable crack growth in the pre-peak regime, the test and numerical results show a strong size effect on the nominal pull-out strength for a broad range of embedment depths. The size effect depends on the scaling of the head of the stud as well as on the scaling of the concrete member. For small heads and proportional scaling of the whole geometry (embedment depth, head size and member geometry), the size effect is strong and close to the prediction according to the LEFM. On the contrary, when the ratio head size to embedment depth is large or when the concrete member is not scaled proportional to the embedment depth, the size effect is weaker and it deviates from the prediction based on the LEFM.

5. References

[ACI 80] ACI Standard 349, "Code Requirements for Nuclear Safety Related Concrete Structures", Appendix B - Steel Embedments, 1980.

[BAL 86] BALLARINI R., SHAH S., KEER L., "Failure Characteristics of Short Anchor Bolts Embedded in a Brittle Material", *Proc. R. Soc. London*, A 404, p. 35-54, 1986.

[BAZ 83] BAZANT Z.P., OH, "Crack Band Theory for Fracture of Concrete", *RILEM*, 93(16), p. 155-177, 1983.

[BAZ 84] BAZANT Z.P., "Size Effect in Blunt Fracture: Concrete, Rock, Metal", *Journal of Engineering Mechanics, ASCE*, 110(4), p. 518-535, 1984.

[BAZ 86] BAZANT Z.P., "Mechanics of Distributed Cracking", *Applied Mechanics Reviews, ASME*, Vol. 39, p. 675-705, 1986.

[BOD 85] BODE H., HANENKAMP W., "Zur Tragfähigkeit von Kopfbolzen bei Zugbeanspruchung", Bauingenieur, p. 361-367, 1985.

[BOR 89] DE BORST R., ROTS J., "Occurrence of Spurious Mechanisms in Computations of Strain Softening Solids", *Eng. Comp.*, 6, p. 272-280, 1989.

[BOR 92] DE BORST R., MÜHLHAUS H.B., "Gradient-dependent Plasticity: Formulation and Algoritmic Aspects", *IJNME*, 35, p. 521-539, 1992.

[CEB 90] Committee Euro-International du Béton, "CEB-FIP Model Code - Final Draft", Paris, 1990.

[ELI 85] ELIGEHAUSEN R., SAWADE G., "Behaviour of Concrete in Tension", *Betonwerk + Fertigteil Technik*, 5 and 6, 1985.

[ELI 88] ELIGEHAUSEN R., FUCHS W., MAYER B., "Load-bearing Behaviour of Anchor Fastenings in Tension", *Betonwerk + Fertigteil Technik*, Vol. 12/87(1), p. 826-832 and 1/88(2), p. 29-355, 1988.

[ELI 89] ELIGEHAUSEN R., SAWADE G., "Analysis of Anchorage Behaviour (Literature Review)", editor L. Elfgren, *RILEM Report*, Chapman and Hall, London, p. 263-280, 1989.

[ELI 89] ELIGEHAUSEN R., SAWADE G., "A Fracture Mechanics Based Description of the Pull-out Behaviour of Headed Studs Embedded in Concrete Structures", editor L. Elfgren, *RILEM Report*, Chapman and Hall, London, p. 281-299, 1989.

[ELI 90] ELIGEHAUSEN R., OZBOLT J., "Size Effect in Anchorage Behaviour", *Proceedings of the 8th European Conference on Fracture - Fracture Behaviour and Design of Materials and Structures*, Torino, Italy, p. 2671-2677, 1990.

[ELI 92] ELIGEHAUSEN R., OZBOLT J., "Size Effect in Concrete Structures", editor A. Carpinteri, *Application of Fracture Mechanics to Reinforced Concrete*, Elsevier Applied Science, Torino, Italy, p. 17-44, 1992.

[ELI 92] ELIGEHAUSEN R., BOUŠKA P., CERVENKA V., PUKL R., "Size Effect of the Concrete Cone Failure Load of Anchor Bolts", In Bazant, Z.P., editor, *Fracture Mechanics of Concrete Structures (FRAMCOS 1)*, p. 517-525, Breckenridge, Elsevier Applied Science, London, New York, 1992.

[ELI 92] ELIGEHAUSEN R., FUCHS W., ICK W., MALLEE R., SCHIMMELPFENNING K., SCHMAL B., REUTER M. "Tragverhalten von Kopfbolzenverankerungen bei zentrischer Zugbeanspruchung", *Bauingenieur*, 67, p. 183-196, 1992.

[ELI 97] ELIGEHAUSEN R., MALLEE R., REHM G., *Befestigungstechnik*, Ernst & Sohn, Berlin, Germany, 1997.

[FUR 94] FURCH J. "Zum Trag- und Verschiebungsverhalten von Kopfbolzen bei zentrischem Zug", Dissertation, University of Stuttgart, 1994.

[OTT 81] OTTOSEN, N.S., "Nonlinear Finite Element Analysis of Pull-out Tests", *Journal of Structural Division*, 107(4), p. 591-603, 1988.

[OZB 95] OZBOLT J., "Maßstabseffekt in Beton und Stahlbetonkonstruktionen", Postdoctoral Thesis, Stuttgart, 1995.

[OZB 96] OZBOLT J., BAZANT Z.P., "Numerical Smeared Fracture Analysis: Nonlocal Microcrack Interaction Approach", *IJNME*, 39(4), p. 635-661, 1996.

[OZB 97] OZBOLT, J., LI, Y.-J., KOZAR, I., "Microplane Model for Concrete – Mixed Approach", Submitted for publication in *IJSS*, 1997.

[PIJ 87] PIJAUDIER-CABOT G., BAZANT Z.P., "Nonlocal Damage Theory", *Journal of Engineering Mechanics, ASCE*, 113(10), p. 1512-1533, 1987.

[RIL 90] RILEM TC 90 – FMA, "Round Robin Analysis of Anchor Bolts - Invitation", *Materials and Structures*, 23, p. 78, 1990.

[SAW 94] SAWADE G., "Ein energetisches Materialmodell zur Berechnung des Tragverhaltens von zugbeanspruchtem Beton", Dissertation, University of Stuttgart, 1994.

Chapter 2

Durability Mechanics

Shrinkage and Weight Loss Studies in Normal and High Strength Concrete

B. Barr — A.S. El-Baden

Division of Civil Engineering
Cardiff School of Engineering
University of Wales Cardiff, UK

ABSTRACT. *The paper reports on an extensive study of shrinkage and weight loss behaviour of a range of normal strength (NSC) and high strength concrete (HSC) ranging from 40 MPa to 120 MPa at 28 days. Three test specimen sizes were used in the study ; two being cylinders with diameters of 100 and 150 mm and a height/diameter ratio of two and the third being prisms with dimensions of 76x76x254 mm. All specimens were subjected to standard air drying (RH 65 ± 5%, T = 20 ± 2°C) for 100 days. Ultimate weight loss and shrinkage were also recorded after oven drying at 105 ± 5°C following the standard drying period. In general the results show that HSC has similar variation of shrinkage and weight loss with time to that observed for NSC. The main difference is that HSC has a higher early rate of shrinkage and lower weight loss development compared to NSC. For all concrete Grades and specimen sizes used in this study, the shrinkage displays a linear relationship with weight loss. The test results are compared to the ACI Model and the more recent RILEM B3 Model. The authors are pleased to report that the experimental work was initiated following a discussion between Professor Zdenek Bazant and the first author during a conference held at Barcelona. It is therefore most appropriate that the paper is included as part of the celebration of the 60th birthday of Professor Bazant.*

KEY WORDS: *High Strength Concrete, Shrinkage, Weight Loss, Shrinkage Models, Shrinkage Prediction.*

1. Introduction

During the last twenty years, there has been a growing interest in the use of HSC with its enhanced mechanical properties and improved durability. The most common means of producing such concrete is by the incorporation of silica fume (SF) and superplasticizer (SP) in the mix [1]. Most studies relating to HSC have been based on the evaluation of its enhanced mechanical properties, its economic advantages in the construction processes and its significantly improved durability, [1-3]. However, its time-dependent properties have not been studied in the same detail.

Numerous studies have addressed the time-dependent strains due to drying shrinkage of NSC and code expressions for prediction values are now readily available. Unlike NSC, HSC is a relatively new material and therefore data on its shrinkage behaviour are very limited [4]. It is well known that shrinkage of concrete is related to moisture loss, due to drying. Whereas the relationship between shrinkage and moisture loss has been well established for NSC [5-7] the same has not yet been established for HSC. A limited amount of research has been published [2,4] in this area which shows that the drying shrinkage-weight loss relationships are almost linear as in the case of NSC. The significant weight loss observed in the case of shrinkage of HSC (with SF addition) has also been recorded by other workers, [8,9].

Recently it has been proposed that it is possible to update the shrinkage prediction models by using short term data of both shrinkage and weight loss [10-12]. The main aim of this study was to extend this area of research by investigating both shrinkage and weight loss for a range of NSC and HSC mixes. The test results have been compared to two existing Models i.e. the ACI Model and the B3 Model.

According to ACI Model [13], the shrinkage $\varepsilon sh(t,t_0)$ at time t (days), measured from the start of drying at t_0 (days), is expressed as follows:

$$\varepsilon sh(t,t_0) = \frac{(t-t_0)^{\alpha}}{a+(t-t_0)^{\alpha}}\ \varepsilon sh_{\infty} \qquad [1]$$

in which εsh_{∞} is the ultimate shrinkage and a and α are constants which can be determined experimentally for any particular concrete. If this is not possible average values for the two constants can be taken from ACI recommendation tables [13] in which a =35 and 55 for moist and steam curing respectively, α=1 and $\varepsilon sh_{\infty} = -780 \times 10^{-6}$. For any conditions which are different from the standard conditions, the ultimate shrinkage strain can be evaluated by the following equation:

$$\varepsilon sh_{\infty} = -780 \times 10^{-6}\, K_{cp}\, K_h\, K_s\, K_f K_c\, K_A\, K_H \qquad [2]$$

where K_{cp} , K_h , K_s , K_f , K_c , K_A , and K_H are correction factors for curing period, member size in terms of volume/surface, slump value, fine aggregate/total aggregate ratio, cement content, air content and ambient relative humidity respectively.

The more recent B3 Model complies with the general guidelines recently formulated by RILEM Committee TC 107, 1995 [14]. In this Model the average shrinkage of the cross - section due to drying $\varepsilon sh(t,t_0)$ can be calculated at any time t (days), measured from the start of drying at t_0 (days), by the following equations:

$$\varepsilon sh(t,t_0) = - K_h \, \varepsilon sh_\infty \tanh[(t-t_0)/\tau_{sh}]^{1/2} \qquad [3]$$

$$\tau_{sh} = k_t \, (k_s D)^2 \qquad [4]$$

where K_h is the humidity factor, εsh_∞ is the ultimate shrinkage, τ_{sh} is the shrinkage half time, ks is the shape factor (k_s=1 considered for simplified analysis), D is the effective cross section thickness = 2V/S and k_t is a correction factor . Both εsh_∞ and k_t are calculated from the concrete composition. (For further details of the two models the reader is referred to references 13 and 14).

Both the ACI Model and B3 Model have been used in this study to provide comparison with the experimental results. The Models are not discussed in detail and have been reported briefly only to make the paper complete.

2. Experimental Details

2.1 Mix proportions and materials used

Five concrete mixes with medium to high workability were used in the investigation. These are realistic, practical mixes which can be used for construction of full scale structures and have developed as much for their rheological properties as for their 28-day compressive strength. The mix details, shown in Table 1, were selected from previous work carried out in the same laboratory on mix proportions for NSC and HSC [15]. The mixes were designed to have a 28 - day nominal cube compressive strength in the range 40 to 120 MPa. The materials used were normal Portland cement, sea dredged sand with a maximum particle size of 5 mm and crushed limestone coarse aggregate. All mixes were prepared using a horizontal pan mixer. A standardised mixing procedure was used in which the cement and sand were mixed first and then the water and coarse aggregate added. The complete mixing time was about 5 minutes and some of the fresh concrete properties (slump and fresh density) were also measured.

Table 1. *Mix proportions for concrete mixes*

Conc. Grade	Mix proportions by mass					w/c	cement content (kg/m^3)	SP (ml /kg of cement)	Slump (mm)
	Cement	SF	FA	CA	Water				
40	1	0	2.00	2.50	0.56	0.56	400	0	150
60	1	0	1.81	2.81	0.50	0.50	400	0	100
80	1	0.11	2.12	3.50	0.45	0.50	340	13.5	160
100	1	0.11	1.77	2.97	0.32	0.35	400	23.0	140
120	1	0.11	1.28	2.13	0.22	0.24	510	35.9	180

SF = Silica Fume; FA = Fine Agg.; CA = Coarse Agg.

The specimens for shrinkage and mechanical tests were cast in steel moulds, covered with moist hessian and polythene sheets, and demoulded after 24 hours. Immediately after demoulding they were weighed and transferred to the curing room (RH 65 ± 5%, T= 20 ± 2^0 C). Three 76x76x254 mm prisms, two 100x200 mm cylinders and two 150x300mm cylinders, were used for evaluating shrinkage and weight loss. Compressive, flexural and splitting tensile strength were also determined at 28 days.

A demec strain gauge, having a gauge length of 100 mm and an accuracy of 0.002 mm per division, was used for strain measurements. Demec discs (pips) were fixed to each specimen, using " plastic padding " adhesive (which has shown good ability to stick to concrete in about 10 minutes and retain its adhesive properties even under water). The pips were located at 120° intervals around the cylinder specimens and at the centre of the upper face of the prism specimens, as shown in Fig. 1 .

A balance with an accuracy of 0.01 gm was used for the weight loss measurements for the prisms and 100 mm diameter cylinders and another balance with an accuracy of 10 gm was used for the 150 mm diameter cylinder.

2.2 Testing procedure

The shrinkage and weight loss readings were taken daily immediately after demoulding. The frequency of the readings was gradually reduced as the specimens matured. The readings were taken according to the following schedule: once a day for the first 4 days, at the end of the first week, at 10 days, at 2 weeks, at 20 days, and then once every 10 days up to 100 days. Two or three readings were taken on each pair of pips, and the average values recorded, while for weight loss the results were expressed as a percentage of the original weight of the samples.

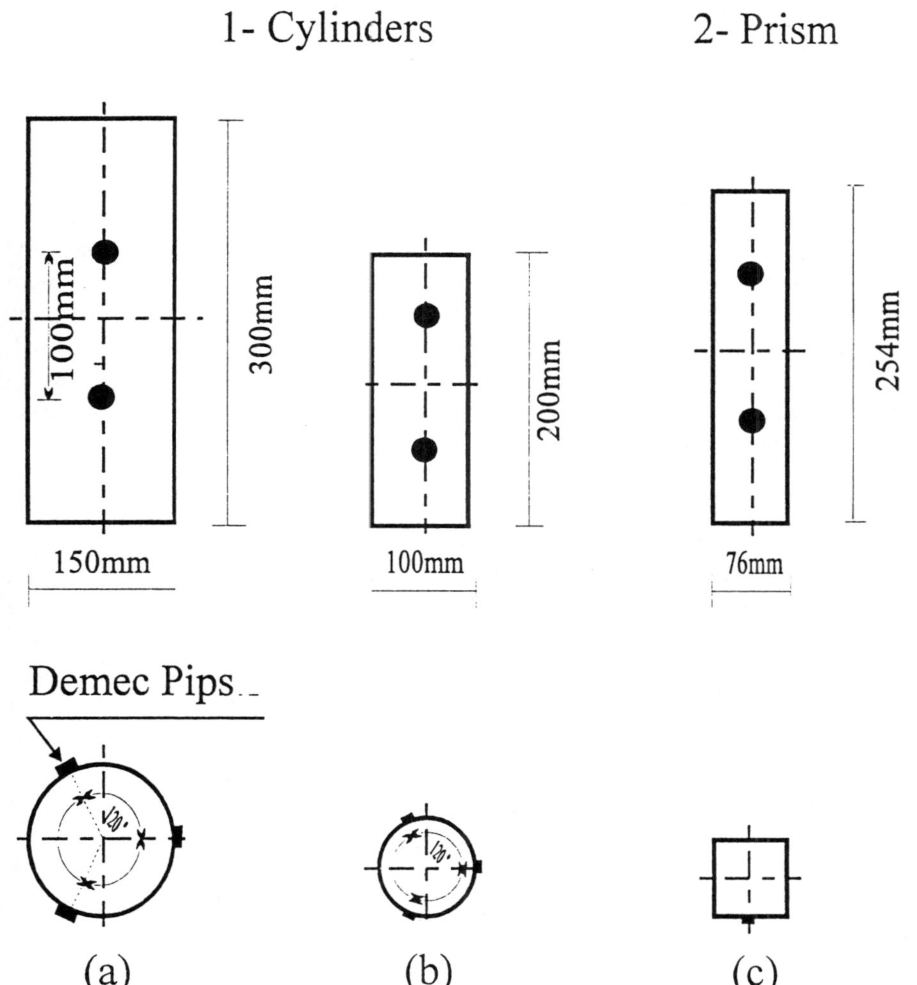

Figure 1. *Schematic representation of test samples and demec point locations*

At the completion of 100 days of drying the shrinkage samples were placed in an oven at 105 ± 5°C, until an approximate weight loss equilibrium was attained. The ultimate weight loss was recorded and thereafter the samples were transferred to air tight cooling desiccators for 24 hours before taking the final shrinkage measurements.

3. Results and Discussion

The time dependent deformations of concrete are complex and potentially involve several variables. However, water migration seems to be the most important factor. Therefore, the following discussion deals primarily with this factor, although other factors may play some role.[8,21,22].

3.1 Shrinkage development

The shrinkage strains measured over 100 days are shown in Fig. 2 for the five concrete grades and three specimen sizes used in the study. It is observed that HSC has similar shrinkage curves as those for NSC. However, the HSC shows a more rapid increase in the early shrinkage during the first 30 days. The results also show that the rate and magnitude of shrinkage reduced as the size of specimen increased. The variation in shrinkage strains between the grades investigated was not so significant with comparable shrinkage being observed for all grades. These observations are in agreement with other investigations, [9,23].

The development of shrinkage depends primarily on the pore structure of the matrix together with the influence of the coarse aggregate which restrains shrinkage. From the pore structure point of view, the pores control the overall shrinkage strain in terms of the amount and rate of water lost. The pore structure is affected by the use of silica fume which produces a finer pore structure with a less interconnected pore system. This slows down considerably the rate of water loss [8,24,25]. The increase in the early shrinkage of HSC may be related to the considerable quantity of cement used in HSC compared to that for NSC, as reported in other investigations [23, 24]. The increased relative amount of coarse aggregate used in the Grade 80 and 100 mixes may also have contributed to a reduction in the shrinkage values observed for these mixes.

3.2 Weight loss development

The variation of weight loss (expressed as a percentage of the initial weight) against drying time (in days) is shown in Fig.3. It is observed that the weight loss reduced considerably as the grade and specimen size increased. The explanation of these results can be related to two contributions: The first contributing factor is the amount of the original water content. In the case of NSC, a larger amount of adsorbed and evaporable water is available which tends to increase the drying response. However, in the case of HSC, the thickness of the adsorbed water layers is smaller, contributing to a lower water diffusion. In addition the expulsion of moisture from the gel pores becomes more difficult as the porosity and water content are decreased. In the case of HSC incorporating SF, a second contributing factor is the microstructure. The pore refinement caused by SF addition leads to low diffusibility [4,8,24]. In addition, size of samples affect the weight loss values by varying the drying path which, in turn, affects the time required and the quantity of water migrating from the concrete sample.

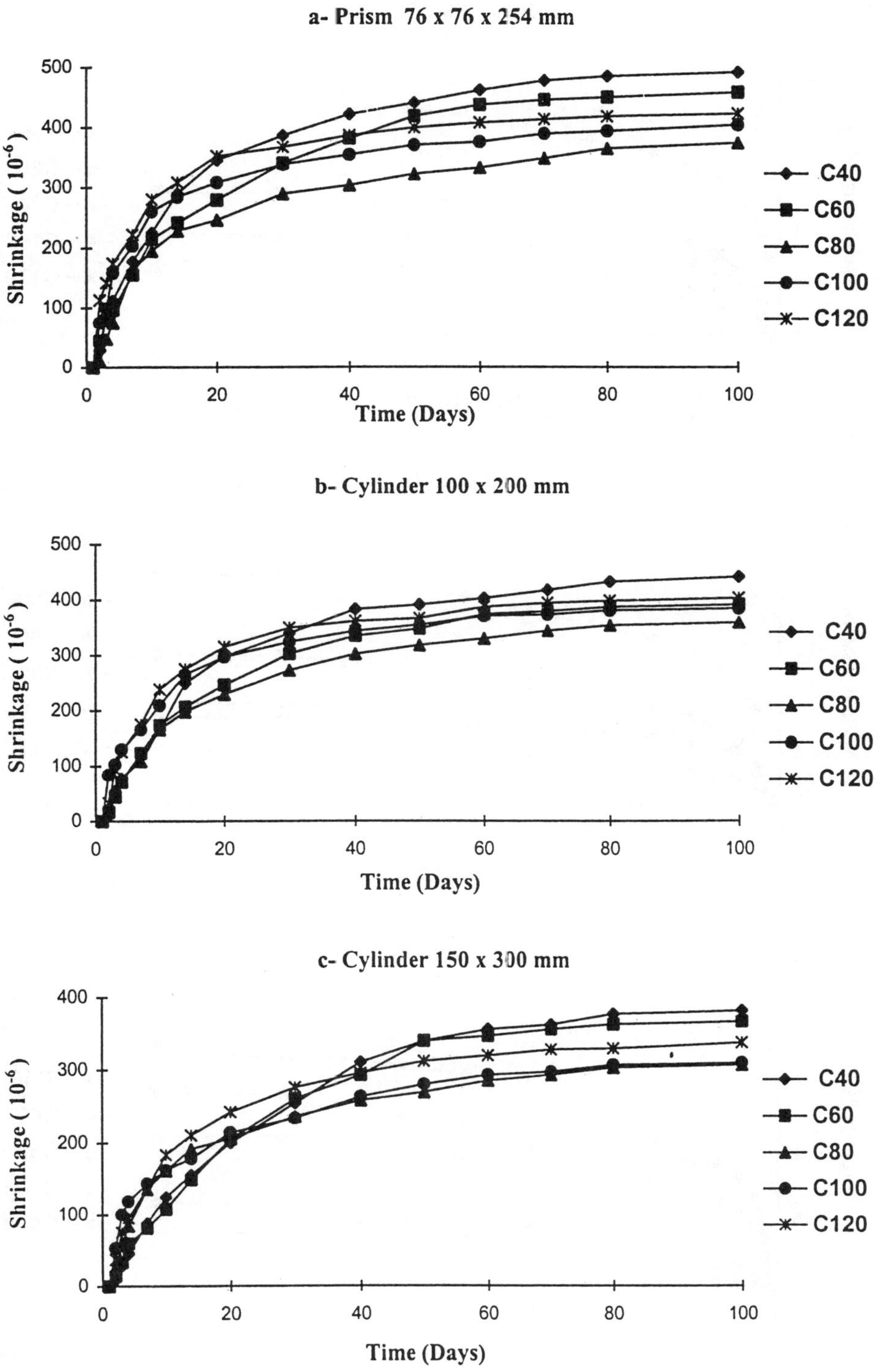

a- Prism 76 x 76 x 254 mm
Shrinkage (10⁻⁶)
500
400
300
200
100
0
0
20
40
60
80
100
Time (Days)
C40
C60
C80
C100
C120
b- Cylinder 100 x 200 mm
Shrinkage (10⁻⁶)
500
400
300
200
100
0
0
20
40
60
80
100
Time (Days)
C40
C60
C80
C100
C120
c- Cylinder 150 x 300 mm
Shrinkage (10⁻⁶)
400
300
200
100
0
0
20
40
60
80
100
Time (Days)
C40
C60
C80
C100
C120

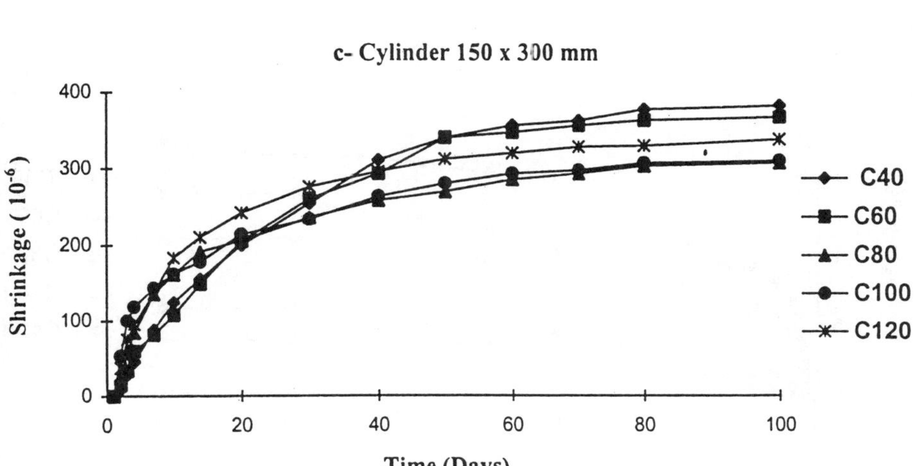

Figure 2. *Shrinkage - time relationships for five concrete Grades*

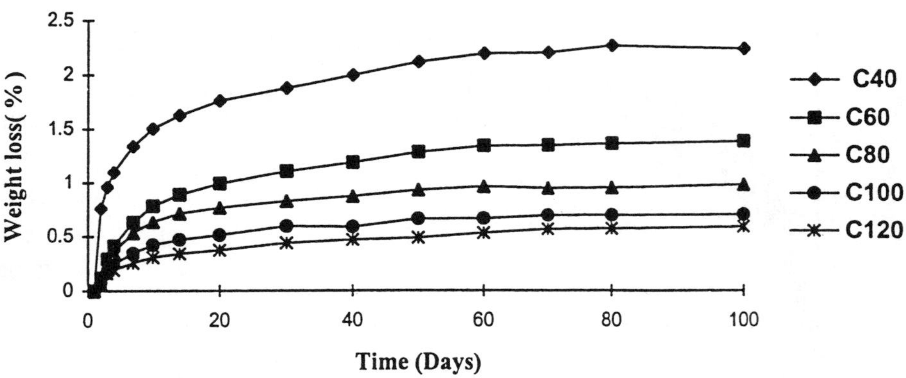

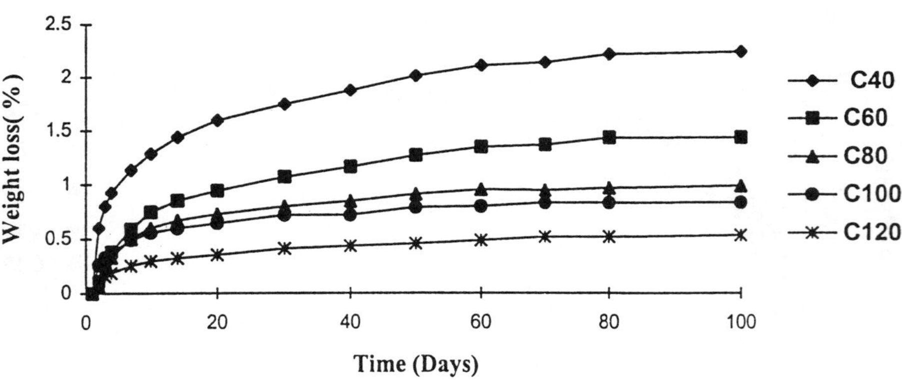

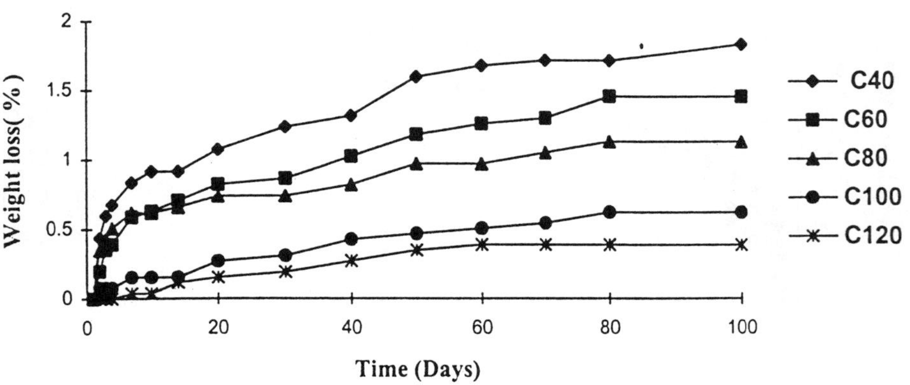

Figure 3. *Weight loss - time relationships for five concrete Grades*

3.3 Shrinkage and weight loss relationships

Fig. 4 shows the shrinkage values plotted against the corresponding weight loss for the different sample sizes and concrete Grades. It is observed that a linear relationship (for the period considered) exists between shrinkage and weight loss. This is in agreement with some previous observations for both NSC and HSC [2,4,11,12]. The best fit for this linear relation can be expressed in the following form:

$$y = ax + b \qquad [5]$$

where y and x are shrinkage and weight loss as defined previously and a and b are constants. Values for constants a and b for the current results together with the correlation coefficient (R^2) are given in Table 2.

Table 2. *Regression analysis results for shrinkage - weight loss relationship*

R.E.	Prism (76x76x254 mm)					Cylinder (100x200 mm)					Cylinder (150x300 mm)				
	C 40	C 60	C 80	C 100	C120	C 40	C 60	C 80	C 100	C120	C 40	C 60	C 80	C 100	C120
a	319	346	404	529	656	277	303	396	551	857	296	323	365	407	460
b	-229	-35	-47	28	64	-168	-38	-46	-75	-27	-134	-68	-78	79	157
R^2	0.993	0.980	0.981	0.994	0.959	0.985	0.991	0.970	0.983	0.973	0.975	0.961	0.936	0.925	0.931

R.E. = Regression Equation, y=ax+b

R = Correlation coefficient

From the shrinkage weight loss relationships it may be observed that the slope of the graphs increase with increasing concrete Grade and reduce with increasing specimen size. These observations are compatible with earlier observations [2].

From the results shown in Fig. 4, it is observed that for an equal amount of weight loss, the highest concrete Grade results in the maximum shrinkage. This is in agreement with earlier observations [2,8] and shows that for equal amounts of water evaporated during drying, the shrinkage strains are higher in the case of pastes with a low w/c ratio. This may be as a result of greater capillary stresses which are induced as water is evaporated from smaller pores [26]. From the linear relationship obtained between shrinkage and weight loss it may be possible (to some extent) that shrinkage can be predicted from such a relationship by using the weight loss results to update

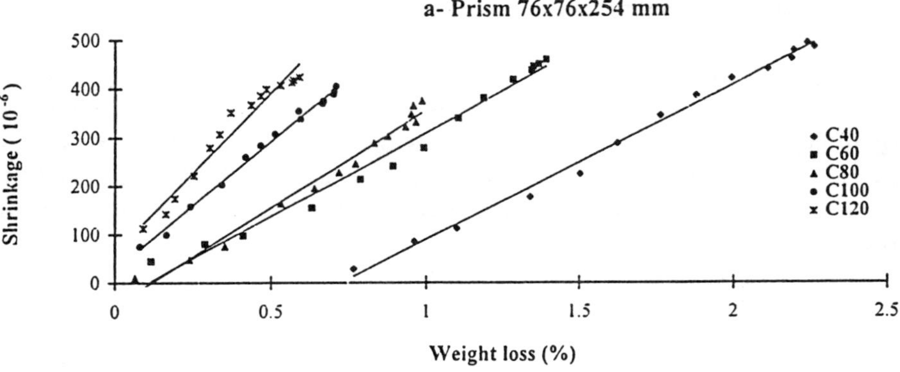

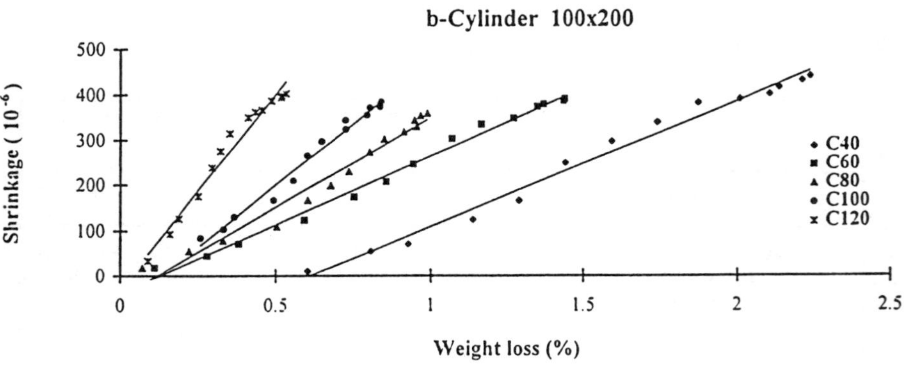

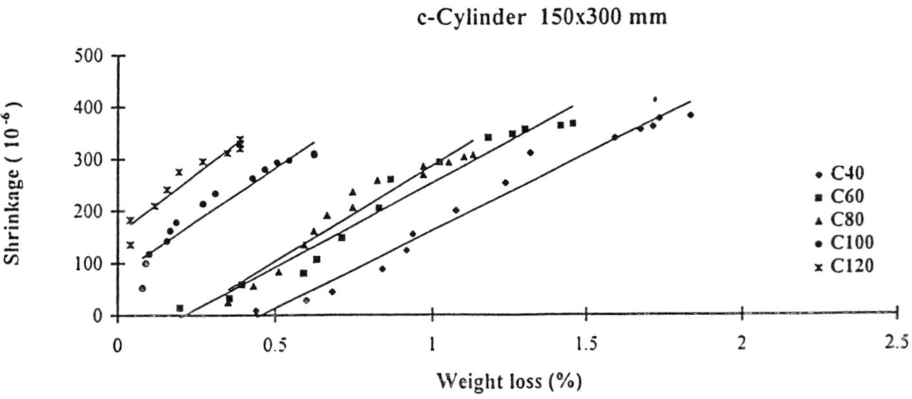

Figure 4. *Shrinkage - weight loss relationships*

the current prediction models as suggested recently [10-12]. However, it is not clear that such a linear relationship will be valid for drying periods in excess of 100 days.

3.4 Shrinkage development after oven drying

The ultimate shrinkage values after 100 days of standard air drying and subsequent oven drying for prism samples are shown in Fig. 5 as a function of water/binder ratio (w/b) for all Grades considered. It is observed that the NSC mixes have the largest shrinkage values and also that the oven dried specimens show shrinkage values which follow the same trend as those subjected to only air drying. The air drying shrinkage represents approximately 90% of the oven drying shrinkage values.

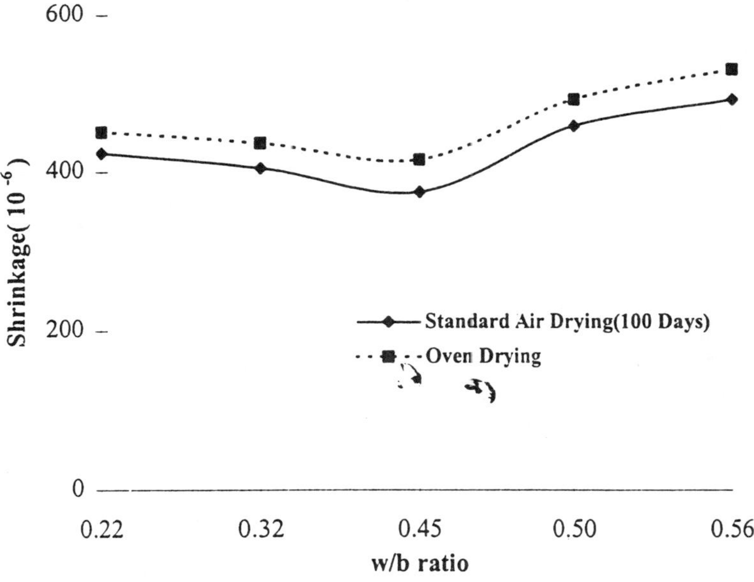

Figure 5. *Effect of oven drying on ultimate shrinkage (Prism 76x76x254 mm)*

4. Comparison of the predicted and experimental results of shrinkage

The predicted results of shrinkage, using both ACI and B3 models, together with the measured values are presented in Figs. 6 to 8. The humidity was considered to be constant over the 100 days of drying with an average value of 65%.

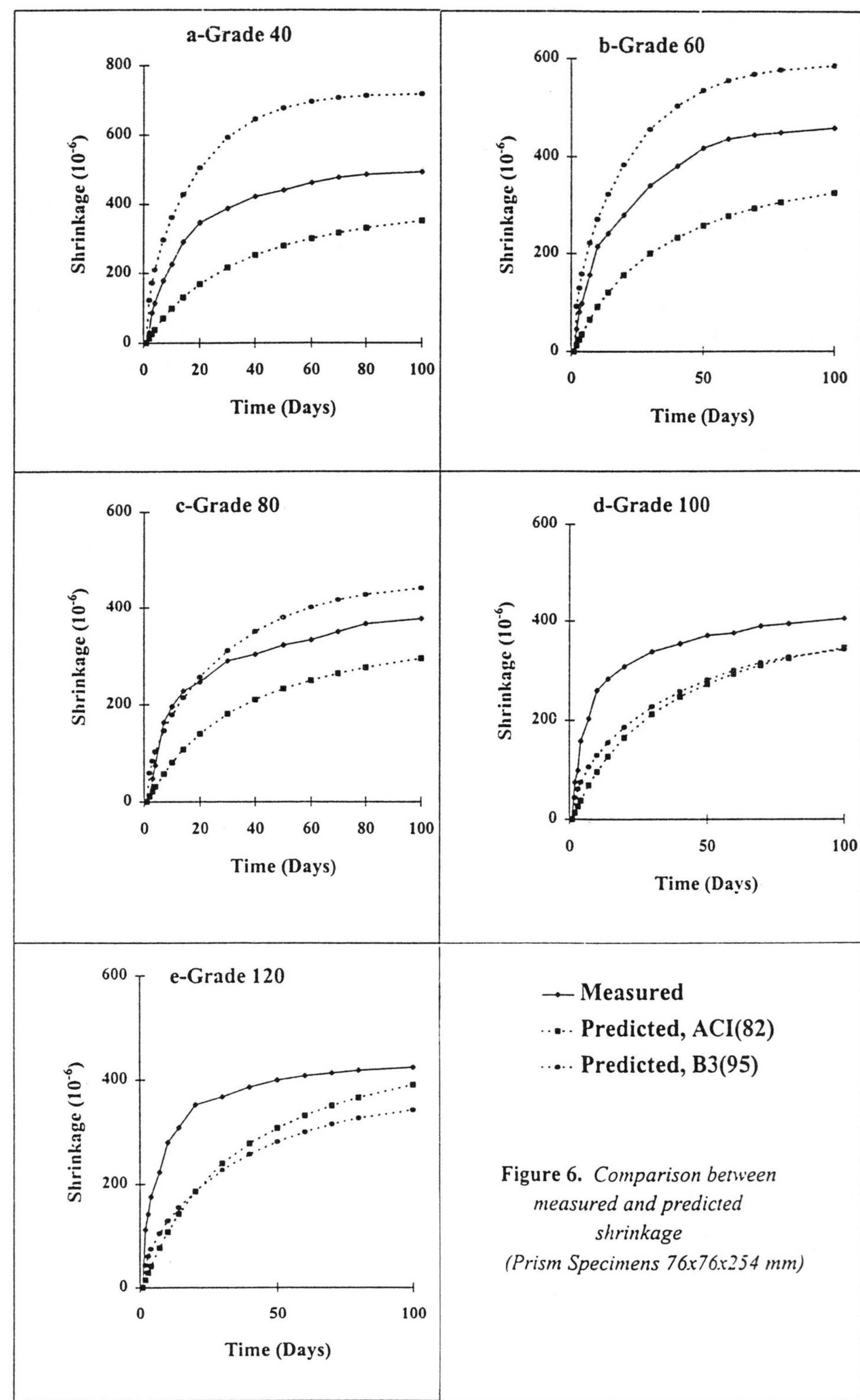

Figure 6. *Comparison between measured and predicted shrinkage (Prism Specimens 76x76x254 mm)*

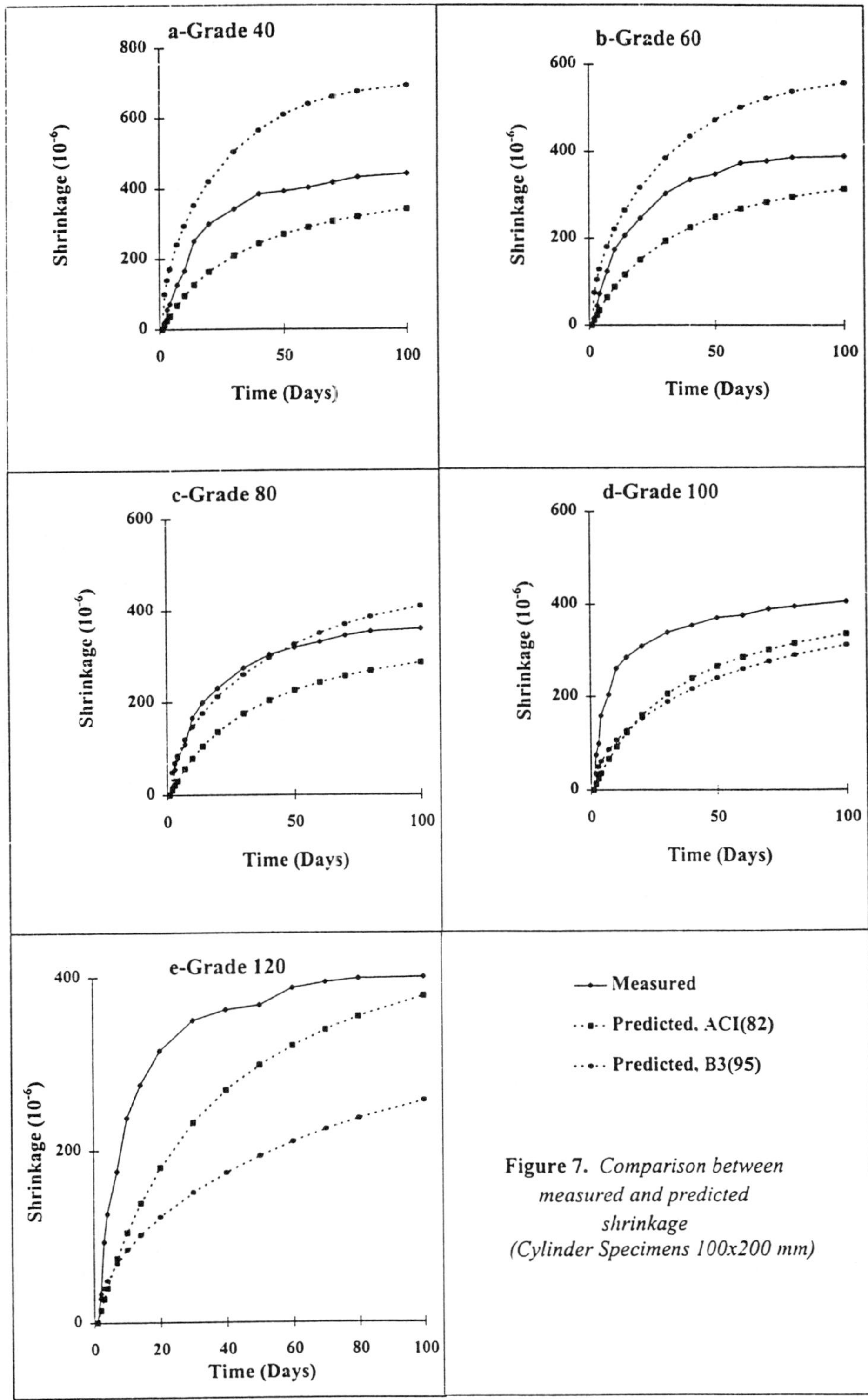

Figure 7. *Comparison between measured and predicted shrinkage (Cylinder Specimens 100x200 mm)*

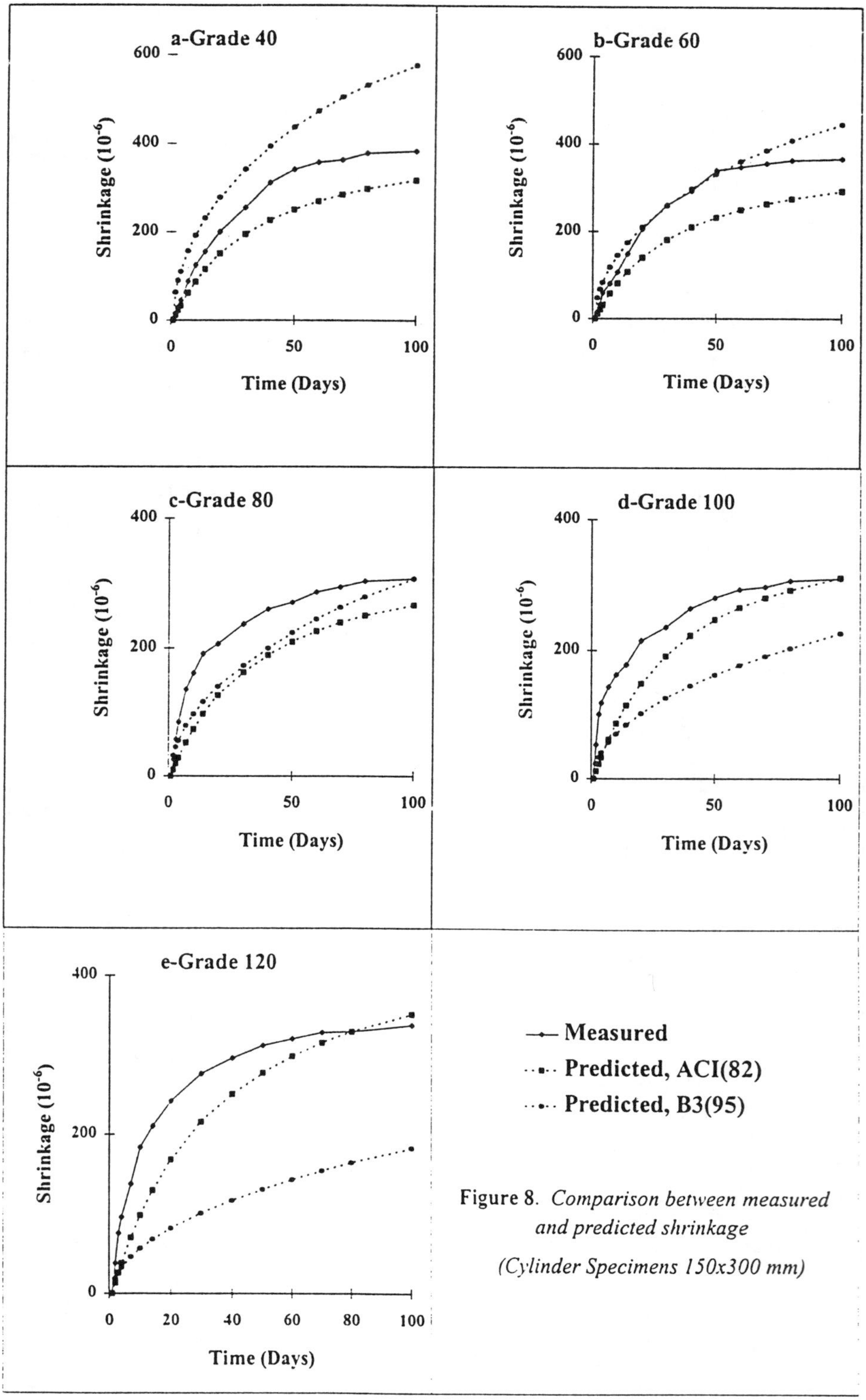

Figure 8. *Comparison between measured and predicted shrinkage*

(Cylinder Specimens 150x300 mm)

The ACI model tends to under estimate the values of shrinkage for all specimen sizes and concrete Grades investigated in this study. However, the difference between the measured and the predicted values reduced as the specimen size and the concrete Grade increased. In general, the ACI model is good for predicting the shrinkage in HSC mixes.

The B3 model over estimates shrinkage in the case of NSC (40 to 60 MPa), and shows good agreement for Grade 80. However, in the case of HSC, the B3 model tends to give an under estimation of shrinkage for all sizes considered.

Both the ACI and the B3 models fail to give a good estimation of the shrinkage during the first month of drying. Further work is in progress to determine how the accuracy of prediction can be improved by considering short-term data for both shrinkage and weight loss.

5. Conclusions and Future Work

The main purpose of the work reported in this paper was to study the shrinkage weight loss relationships for both NSC and HSC. The data presented should provide a base for future work to update shrinkage predictions using short term data of both shrinkage and weight loss. The present results concerning the shrinkage weight loss relationships lead to the following conclusions:

(a) HSC was observed to have a high rate of drying shrinkage during the first month of air drying. Beyond one month the rate of shrinkage was considerably reduced.
(b) HSCs have only a small weight loss during standard air drying compared to other Grades of concrete. This can be attributed to the effect of SF on the microstructure of HSCs.
(c) Shrinkage of both NSC and HSC showed a linear relationship with corresponding weight loss for the 100 day period considered.
(d) The air drying shrinkage during the first 100 days of drying represents approximately 90% of the ultimate shrinkage recorded following prolonged oven drying. This was observed for all concrete Grades.
(e) Both the ACI and the B3 models have limitations in predicting short-term shrinkage. The models must be used with care when applied to determine early shrinkage values.

The experimental results reported here are being extended to include the shrinkage-weight loss relationships in fibre-reinforced high strength concrete. The same basic concrete Grades have been used together with three concentrations of fibre reinforcement (1, 2 and 3% by weight). The shrinkage-weight loss relationships will be used to investigate the possible limitations of existing shrinkage models and thereafter used in conjunction with a neural network study. These studies will be reported later when the work is completed.

6. Acknowledgements

The second author wishes to acknowledge the financial support provided by Al-Fateh University, Tripoli, Libya, to carry out the work described in the paper.

7. References

[1] FIP/CEB, *State of the art report, high strength concrete*, 1st Edn. Thomas Telford, London, 1990.

[2] BENTUR, A. and GOLDMAN, A., "Curing effects strength and physical properties of high strength silica fume concretes", *ASCE Journal of Materials in Civil Engineering*, 1 (1), pp.46-59, 1989.

[3] SHAH, S.P. and AHMAD, S.H., *High performance concrete and applications*, Edward Arnold, London, 1994.

[4] ELIE, E., BUGUAN, M., OMAR, C. and AITCIN, P.C., "Drying shrinkage of ready mixed high performance concrete", *ACI Materials Journal*, 91(3), pp.300-305, 1994.

[5] HOBBS, D.W., "Influence of specimen geometry upon weight change and shrinkage of air-dried concrete specimens", *Magazine of Concrete Research*, 29(99), pp.70-80, 1977.

[6] HOBBS, D.W. and MEARS, A.R., "The influence of specimen geometry upon weight change and shrinkage of air-dried mortar specimens", *Magazine of Concrete Research*, 23 (75), pp.89-98, 1971.

[7] SAKATA, K., "A study on moisture diffusion in drying and drying shrinkage of concrete", *Cement and Concrete Research*, 13 (2), pp.216-224, 1983.

[8] TAZAWA, E. and YONEKURA, "Drying shrinkage and creep of concrete with condensed silica fume", *American Concrete Institute* (SP-91) Detroit, pp.903-921, 1986.

[9] CARRETTE ,G.G. and MALHOTRA, V.M., "Mechanical properties, durability and drying shrinkage of Portland cement concrete incorporating silica fume"., *Cement, Concrete and Aggregates*, 5 (1), pp.3-13, 1983.

[10] BAZANT , Z. P. and BAWEJA, S., "Justification and refinements of model B3 for concrete creep and shrinkage : Part 2-Updating and theoretical basis", *Materials and Structures*, 28(181), pp.488-495, 1995.

[11] GRANGER, L., TORRENTI, J.M. and ACKER, P., "Thoughts about drying shrinkage : Scale effect and modelling", *Materials and Structures*, 30(196), pp.96-105, 1997.

[12] GRANGER, L., TORRENTI, J.M. and ACKER,P., "Thoughts about drying shrinkage: Experimental results and quantities of structural drying creep", *Materials and Structures*, 30(204), pp.588-598, 1997.

[13] ACI Committee 209, "Prediction of creep, shrinkage and temperature effects of concrete structures, designing for the effect of creep, shrinkage , temperature in concrete structures", *American Concrete Institute* (SP-76), Detroit,.pp.255-301, 1982.

[14] RILEM Draft Recommendation (107-GCS), "Creep and shrinkage prediction model for analysis and structures , Model B3",.*Materials and Structures*, 28(180), pp.357-365, 1995.

[15] TAYLOR, M.R, LYDON, F.D and BARR, B.I.G., "Mix proportions for high strength concrete", *Construction and Building Materials*, 10(6), pp.445-450, 1996.
[16] British Standard Institute, *Specification for Portland cement*, BSI, London, 1991.
[17] British Standard Institute, *Aggregate from natural sources for concrete*, BS 882, BSI, London, 1983.
[18] British Standard Institute, *Recommendations for determination of strain in concrete*, BS 1881: Part 206 ' , BSI, London, 1986.
[19] MORICE, P.B. and BASE, G. D., "The design and use of a demountable mechanical stain gauge for concrete structures", *Magazine of Concrete Research*, 5(60), pp.37-42, 1953.
[20] British Standard Institute, *Methods of testing hardened concrete for other than strength*, BS 1881, Part 5, BSI, London, 1970.
[21] SMADI, M M., SLATE, F.O and NILSON, A.H., "Shrinkage and creep of high, medium and low strength concretes, including overloads", *ACI Materials Journal*, 84(3), pp.224-234, 1987.
[22] NGAB, A.S., NILSON, A.H. and Slate, F.O., "Shrinkage and creep of high strength concrete", *ACI Journal*, 78 (4), pp.255-261, 1981.
[23] SWAMY, R.N. and ANAND, K.L, "Shrinkage and creep properties of high strength structural concrete", *Civil Engineering Public Works Review*, 65 (10), pp.859-868, 1973.
[24] GHOSH, S. and NASSER, K.W., "Creep, shrinkage, frost and sulphate resistance of high strength concrete", *Canadian Journal of Civil Engineering*, 22(3), pp.621-636, 1995.
[25] LARRARD, F., "Creep and shrinkage of high-strength field concretes", *High-Strength Concrete*, American Concrete Institute, (SP-121), Detroit, pp.577-598, 1990.
[26] HANSEN, W, "Drying shrinkage mechanisms in Portland cement paste", *Journal of American Ceramic Society* , 70 (5), pp.323-328, 1987.

Time-Dependent Behaviour of Cracked and Ageing Concrete

B.L. Karihaloo * — S. Santhikumar **

* *Cardiff School of Engineering, University of Wales Cardiff, UK*
** *Department of Civil Engineering, University of Sydney, Australia*

A shorter version of the paper was presented at EURO-C 1998, Badgastein, Austria.

ABSTRACT. *For the slow or static fracturing (i.e., cracking) of concrete structures, it is important to include the interaction between cracking and time-dependent behaviour of concrete both inside and outside the cracked zone This paper gives an overview of the constitutive modelling techniques that are able to describe at least some aspects of the time-dependent behaviour of cracked and ageing concrete and are easily incorporated into finite element programs using fictitious or blunt crack models. For completeness, the constitutive models describing the creep of uncracked concrete under sustained compressive loads are also included. The time-dependent behaviour of concrete has been successfully described by many constitutive models under sustained compressive stresses. The creep rates under compressive and tensile stresses have been found to be nearly the same for uncracked concrete. However, once the concrete has cracked, the situation alters dramatically.*

KEY WORDS: *Creep, Relaxation, Cracked Concrete, Ageing Concrete, Tension Softening, Micromechanical Modelling, Kelvin Chain Rheological Model, Solidification of Hydrated Cement.*

1. Introduction

The long term performance of concrete structures is fundamentally affected by the behaviour of concrete after it has cracked. This is due to the fact that the fracture of concrete is preceded by micro-cracking so that there is no well-defined crack tip, but rather a diffuse damage zone, within which cracking increases and stresses decrease as the overall deformation increases. This results in the softening of the material in the fracture process zone.

There is little information on the development of cracks with time and how this affects the macroscopic stress in concrete. The problem is further complicated by the ageing of concrete. It should however be mentioned that the time-dependent behaviour of concrete has been explored to some extent under dynamic conditions when this behaviour is dominated by inertial effects and wave propagation

[MIN 87], [DU 89], [TAK 91], [YOU 92]. But we shall not be concerned with the dynamic behaviour in the present paper.

We will first give overview of the popular creep constitutive models for uncracked concrete. We will then focus on how the behaviour of cracked concrete is influenced by its ageing and creep under quasi-static conditions, when inertial effects can be disregarded. For this, we shall consider three approaches. The first is based on the concept of activation energy and rate-dependent softening, and has been developed in a series of papers by Bazant and co-workers [BAZ 90], [BAZ 92b, 93b], [BAZ 92a], [BAZ 93a], [WU 93].

The second approach is based on the inclusion of a standard rheological model for creep and relaxation into the fictitious crack model for concrete in order to accommodate the time dependency of crack opening, the latter in some instances being established by fitting stress relaxation test results [HAN 90, 91], [ZHO 92], [ZHA 92a, b], [CAR 95].

The third approach combines a Kelvin chain rheological model associated with solidification theory for studying the time-dependent behaviour of ageing concrete with a micro-mechanical model for the static softening behaviour of cracked concrete in the fracture process zone [SAN 96, 98a, b].

2. Creep of Uncracked Concrete under Tension or Compression

There are numerous models for explaining the creep mechanism in concrete under compression. None of them can explain all the observed phenomena, but all of them stress the importance of the movement of water both in the capillary (meso-) and gel (micro-) pores that increases the capillary stress by surface tension. Another important mechanism involves the physical rearrangement of solid particles at the micro-level as a result of the induced (capillary) internal or externally-applied stresses [BAZ 88]. The amount of creep and its rate of development depend on the applied stress level and the age or the degree of hydration at first loading. Creep increases at a faster rate as the sustained stress level is raised. On the other hand, it decreases as the age at first loading increases.

There are many models available in the literature and in the codes for the design of concrete structures for estimating the rate of growth of creep strains with time. These range from simple power laws to exponential and hyperbolic expressions which contain many constants that are established from uniaxial creep tests in compression. By far the most versatile model is due to Bazant – the so-called B3 model which is an updated version of his earlier models. According to this model, the total creep strain due to an applied sustained compressive stress σ_0 is

$$\varepsilon(t) = \Phi(t,\tau)\sigma_0 + \varepsilon_0 \qquad [2.1]$$

where ε_0 consists of the shrinkage and other thermal strains. The creep strain is divided into basic and drying components, as is the creep compliance function

$$\Phi(t,\tau) = q_1 + C_b(t,\tau) + C_d(t,\tau,t_0) \qquad [2.2]$$

Here, q_1 is the instantaneous strain due to a unit applied stress, $C_b(t,\tau)$ is the compliance function for basic creep, and $C_d(t,\tau,t_0)$ that for drying creep. t is the age of concrete at the time of loading, $(t-\tau)$ the duration of sustained load, and t_0 the age at which drying begins.

The basic creep compliance function is defined as

$$C_b(t,\tau) = q_2 Q(t,\tau) + q_3 \ln\left[1 + (t-\tau)^n\right] + q_4 \ln\left(\frac{t}{\tau}\right) \qquad [2.3]$$

where

$$Q(t,\tau) = Q_f(\tau)\left[1 + \left\{\frac{Q_f(\tau)}{Z(t,\tau)}\right\}^{r(\tau)}\right]^{-1/r(\tau)} \qquad [2.4]$$

with

$$\begin{aligned} r(\tau) &= 1.7\tau^{0.12} + 8, \\ Z(t,\tau) &= \tau^{-m} \ln\left[1 + (t-\tau)^n\right] \\ Q_f(\tau) &= \left[0.086\tau^{2/9} - 1.21\tau^{4/9}\right]^{-1} \end{aligned} \qquad [2.5]$$

m and n are constants to be determined from a uniaxial creep test.

The additional drying creep compliance is defined as

$$C_d(t,\tau,t_0) = q_5\left[\exp\{-8H(t)\} - \exp\{-8H(\tau)\}\right]^{1/2} \qquad [2.6]$$

in which $\tau \geq t_0$ and $H(t) = 1 - (1-h)S(t)$. h is the relative humidity, and $S(t)$ the time function for shrinkage.

The coefficients $q_1,\ldots, q_5$ appearing in formulae [2.2] - [2.6] are obtained from the best fit of available test data.

$$q_1 = 0.6 \times 10^6 / E_{28}, \quad E_{28} = 57000(f_c')^{1/2}$$

$$q_2 = 451.1c^{0.5}(f_c')^{-0.9}, \quad q_3 = 0.29(w/c)^4 q_2 \qquad [2.7]$$

$$q_4 = 0.14(a/c)^{-0.7}, \quad q_5 = 7.57\times10^5 (f_c')^{-1}\varepsilon_{sh\infty}^{-0.6}$$

In [2.7], f_c' should be in *psi*. *w/c* denotes water to cement ratio, and *a/c* aggregate to cement ratio.

Even though, the mechanisms of creep in concrete under tension are different from those in compression, although it is observed that the magnitude and rate are of the same order as those in compression at a similar stress level. Many tests reported in the literature confirm this similarity in behaviour. Therefore, the models for prediction of creep rate under compression can be used with minor adjustments for the prediction of creep rate under tension.

The creep, and the related phenomenon of stress relaxation, in concrete can also be studied by using rheological models of varying degrees of sophistication. The predictive capability of these models depends upon the accuracy of the material constants assigned to the various elements that constitute a given rheological model. We shall consider several rheological models later on in the paper.

We now turn to the question of how the behaviour of cracked concrete is influenced by its ageing and creep under tension.

3. Crack Length and Crack Opening Growth Rates

Bazant and co-workers [BAZ 90], [BAZ 92b, 93b], [BAZ 92a] approached the time-dependency of crack growth from the point of view of a thermally activated process and used an equivalent linear elastic fracture model and the R-curve concept to arrive at:

$$\dot{a} = \frac{da}{dt} = A\left(\frac{K}{K_R}\right)^m \qquad [3.1]$$

where a is the crack size, A and m are constants to be found experimentally, K is the local stress intensity factor and K_R the plateau value in the R-curve.

Replacing the elastic modulus E in the expression for crack mouth opening displacement w by the corresponding creep compliance gave the time variation

$$w(\tau) = \frac{1}{b}\int J(\tau, t')d[P(t')\delta(t')] \qquad [3.2]$$

where τ is the current time, t the time at first loading, and $J(\tau, t')$ the compliance function for creep in the bulk of the specimen. The geometric compliance δ is time dependent because it varies with the relative crack length that increases as the crack propagates.

Moreover, the effective fracture process zone length c_f in the size effect model of Bazant was shown to vary as follows:

$$c_f = c_{f0}\left(\frac{\dot{a}}{\dot{a}_0}\right)^{\frac{1}{n}} \qquad [3.3]$$

where $\dot{a}_0$ is a normalising factor, n is an empirical constant and c_{fo} the value of c_f at time $t = 0$.

However, as this approach was found not to predict test data with any reasonable accuracy, it was argued that although the classical linear elastic fracture mechanics does not involve time, the breakage of bonds required for crack extension happens at a certain finite rate. This rate of breakage of bonds was calculated by considering the activation energy Q. In this manner the rate of opening of a crack due to thermally activated bond breakages was calculated to be:

$$\dot{w} = C_0 \sinh\{\psi(\sigma, w)\} e^{\left(\frac{Q}{RT_0} - \frac{Q}{RT}\right)} \qquad [3.4]$$

with

$$\psi(\sigma, w) = \frac{T_0}{T} k_0 \{\sigma - \phi(w)\} \qquad [3.5]$$

where T, T_0 are actual and reference temperatures, respectively, R is the universal gas constant, k_0 and C_0 are empirical constants and σ is the applied stress.

The stress σ_b on the bonds that undergo fracturing was assumed to be proportional to the net crack bridging stress $\sigma - \phi(w)$, where $\phi(w)$ described the static tension-softening diagram of concrete. In [3.4] and [3.5] the constants have to be determined experimentally, for instance from relaxation tests.

The function $\psi(\sigma, w) = 0$ describes approximately the stress-displacement diagram of the fictitious crack model for extremely slow rates of crack opening, i.e., $\dot{w} \to 0$.

4. Visco-Elastic Fictitious Crack Models

Although equations [3.4] and [3.5] can be directly used in the fictitious crack model, the creep deformation in concrete used at normal temperatures is essentially stress driven. There is, thus, no need to consider the thermally activated contribution to the creep of concrete. For this reason, it is often sufficient to supplement the FCM with a rheological model for concrete in order to follow the evolution of the static tension-softening diagram with time.

4.1. *Linear Visco-Elastic FCM*

Hansen [HAN 90] was the first to study the evolution of the linear tension-softening relation:

$$\sigma(w) = f_t\left(1 - \frac{w}{w_c}\right) \quad [4.1]$$

with time using the rheological model shown in Figure 1. Here, f_t is the tensile strength of concrete, w the crack opening, and w_c its critical value.

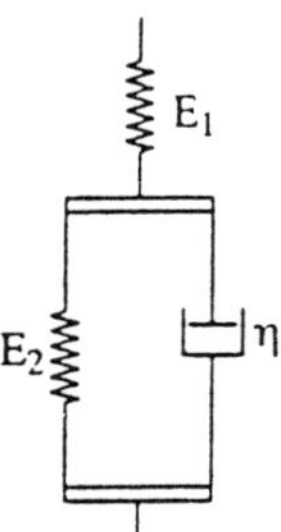

Figure 1. *A rheological model for concrete*

The stress-strain relationship for the rheological model of Figure 1 is:

$$E_1\eta\dot{\varepsilon} + E_1E_2\varepsilon = \eta\dot{\sigma} + (E_1 + E_2)\sigma \quad [4.2]$$

from which one can construct the creep function for a constant stress $\sigma(0) = \sigma_0$ applied at $t = 0$.

$$\varepsilon(t)=\sigma_0\left[\frac{1}{E_1}+\frac{1}{E_2}\left(1-e^{-t/T_1}\right)\right] \tag{4.3}$$

and the relaxation function for a constant strain $\varepsilon(0)=\varepsilon_0$ applied at $t = 0$.

$$\sigma(t)=\varepsilon_0\left[E_\infty+\left(E_1-E_\infty\right)e^{-t/T_2}\right] \tag{4.4}$$

In [4.3] and [4.4], $T_1 = \eta / E_2$ is the retardation time, $T_2 = \eta / (E_1 + E_2)$ is the relaxation time and $E_\infty = [E_1 E_2 / (E_1 + E_2)]$.

Hansen [HAN 90] implemented the creep [4.3] and relaxation [4.4] functions into the FCM after making suitable adjustments to simulate the evolution of $\sigma(w)$ relation [4.1], and replacing ε by w via an internal length scale to give:

$$w(t)=-\sigma_0\left[\frac{1}{M_1}+\frac{1}{M_2}\left(1-e^{-t/T_1}\right)\right] \tag{4.5}$$

$$\sigma(t)=-w_0\left[M_\infty+\left(M_1-M_\infty\right)e^{-t/T_2}\right] \tag{4.6}$$

where $M_1 = E_1, M_2 = -E_2$ and $T_1 = -\eta / M_2$. However, in order to retain a descending $\sigma(w)$ curve, not only has the stiffness of the springs to be negative, but also the dashpot viscosity. This was necessary to retain a positive retardation time. Thus, the model loses its physical significance, although it is useful for gaining an insight into the evolution of $\sigma(w)$ with time, as shown in Figure 2.

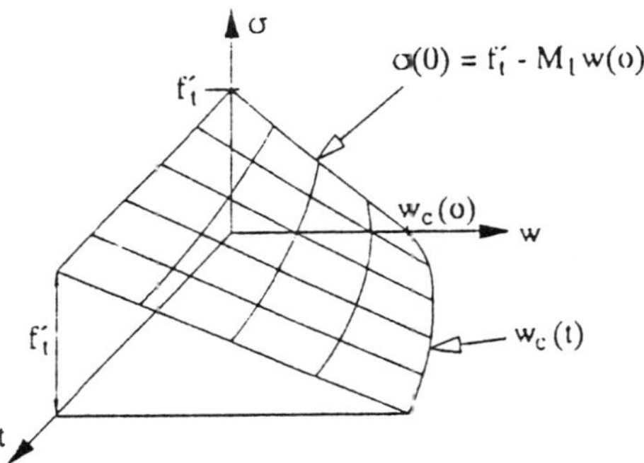

Figure 2. *Development of the linear $\sigma(w)$ approximation to tension softening with time according to the linear visco-elastic model of Figure 1 (after [HAN 90])*

Another direct approach to the inclusion of the visco-elastic behaviour of concrete into the FCM was proposed by Zhou & Hillerborg [ZHO 92]. The total stress decrement $d\sigma$ in the time interval dt was decomposed into the decrement $d\sigma^R$ caused by relaxation and the decrement $d\sigma^I$ due to the increase in crack opening by dw. By using a modified Maxwell rheological model for concrete, the two stress decrement components can be written as:

$$d\sigma^I = F\,dw\frac{e^{-dt/\tau}+1}{2} \qquad [4.7]$$

$$d\sigma^R = (\sigma - \alpha\sigma_0)\left(e^{-\frac{dt}{\tau}} - 1\right) \qquad [4.8]$$

where σ is the stress at the time t, σ_0 is the stress corresponding to deformation w in the static softening relation $\sigma(w)$, α and τ are material constants and F is the unloading slope in the tension softening regime.

This approach was further developed by Carpinteri et al. [CAR 95] to enhance the range of its applicability. They made modifications to the stress decrement $d\sigma^R$ based on stress relaxation tests and numerical simulations and proposed to replace [4.8] with:

$$d\sigma^R = -F\,dw^c = -F\,\dot{w}^c dt \qquad [4.9]$$

where the creep rate $\dot{w}^c$ was derived by fitting uniaxial tensile creep data:

$$\frac{\dot{w}^c}{w_{ult}^c} = 0.0096\left(\frac{\sigma_s}{\sigma_u}\right)^{27}\left\{0.556 + 10.2\left(\frac{w^c}{w_{ult}^c}\right)^5 + 4.45e^{-50\frac{w^c}{w_{ult}^c}}\right\} \qquad [4.10]$$

Here σ_s is the sustained stress level and σ_u the stress level at unloading.

It should be mentioned that the approach of Zhou & Hillerborg [ZHO 92] and Carpinteri et al. [CAR 95] is not concerned with the evolution of the tension softening with time. Rather, it is concerned with stress levels σ_s and σ_u that are below the levels corresponding to the static tension softening. It determines the time it would take for the cracked material to creep from these lower stress levels to attain the deformation corresponding to that on the static tension softening curve. In that respect, it will predict a local creep rupture time in the sense that the deformation may still be well below the critical static value w_c at which a fictitious crack becomes a real crack.

4.2. *Non-Linear Hereditary Visco-Elastic FCM*

Zhang & Karihaloo [ZHA 92a, 92b] combined the Rabotnov fractional exponential creep kernel for concrete with the static FCM in order to include the history-dependent time response of cracked concrete into the FCM. The hereditary visco-elastic response of a material can be expressed as:

$$\varepsilon(t)=C(0)\sigma(t)+\int_0^t \sigma(\tau)\phi(t-\tau)d\tau \qquad [4.11]$$

where $\phi(t)=\partial C/\partial t$ is the so-called uniaxial creep kernel at $\sigma(t)$ = constant, and $C(0)=1/E(0)$ is the compliance at $t = 0$.

For the description of uncracked concrete, the Rabotnov fractional exponential kernel has been found suitable (see, e.g. [KAM 80]):

$$\phi(t)=\frac{\lambda}{E}t^{-\alpha}\sum_{n=0}^{\infty}\frac{-\beta^{n}t^{n(1-\alpha)}}{\Gamma[(n+1)(1-\alpha)]}=\frac{\lambda}{E}E(\alpha,-\beta,t) \qquad [4.12]$$

where $\lambda>0$, $\beta>0$, $0\leq\alpha<1$ are material constants to be determined from uniaxial creep tests, and $\Gamma(\ )$ is the Gamma function. From the definition of creep compliance, it follows from [4.11] and [4.12] that:

$$C(t)=C(0)\left[1+\lambda\int_0^t E(\alpha,-\beta,\tau)d\tau\right] \qquad [4.13]$$

In particular, the long term creep compliance $C(\infty)$ is given by:

$$C(\infty)=C(0)(1+\lambda/\beta) \qquad [4.14]$$

which corresponds to $1/E_{\infty}$ appearing in [4.4].

Zhang & Karihaloo chose the following relation in the fracture process zone:

$$\frac{\sigma(\varsigma)}{f_t}=\frac{e^{-c\varsigma^2}-e^{-c}}{1-e^{-c}} \qquad [4.15]$$

in order to study its evolution with time. Here, c is a material constant, $\varsigma=\xi/\ell_p$ is the distance from the fictitious crack tip (Figure 3). The choice of the rather complicated approximation to the stress decay in the fracture process zone (FPZ) in fact facilitates the evaluation of integrals, such as [4.13], in an analytical form. It is assumed here that ℓ_p or f_t, but not both, is independent of time. This limits the application of the model to large concrete structures, such as dams, in which ℓ_p is indeed small in comparison with the structural size, so that it can be regarded as a static parameter.

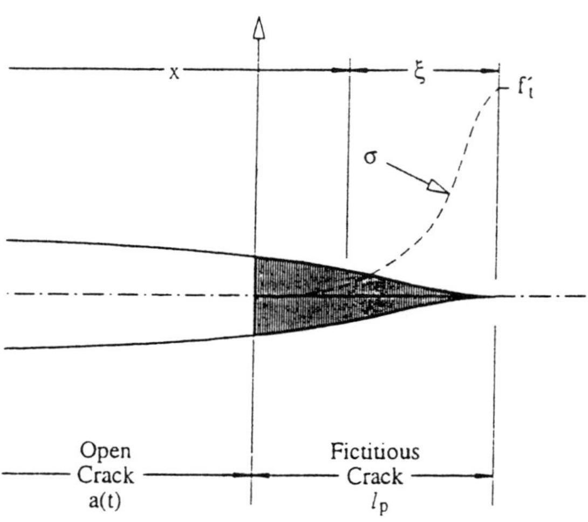

Figure 3. *The fictitious crack model (FCM), showing the co-ordinate system (after [KAR 95])*

For the assumed stress distribution [4.15] in the FPZ, Zhang & Karihaloo used the Green function formalism [KAR 95] to write a formal expression of the elastic crack opening displacement at any time *t*. However, to make further progress they discussed two limiting cases by assuming that either ℓ_p or f_t' does not exhibit time dependence. In the sequel, this assumption is highlighted by assigning the subscript ℓ_p or f_t' to the corresponding derived quantity. Without loss of generality, it is further assumed that the external stress σ does not vary with time. The elastic solution with a parameter *t* has the form

$$w_e[x, a(t)] = C'f[x, a(t)] \qquad [4.16]$$

where, following [SCH 75], it can be shown that

$$f_{ft}[x, a(t)] = \frac{4I_2}{\sigma}\left(\frac{K_I(t)}{I_1}\right)^2 F[s(t)]$$

$$f_{\ell p}[x, a(t)] = \frac{8I_2K_I(t)}{I_1}\sqrt{\frac{\ell p}{2\pi}}\; F[s(t)] \qquad [4.17]$$

with

$$F(s) = \sqrt{1-s} + \frac{s}{2}\ln\frac{1-\sqrt{1-s}}{1+\sqrt{1+s}}, \quad s(t) = 1-\varsigma \qquad [4.18]$$

Here, $C' = 1/E'$ is the elastic compliance for plane stress or plane strain, and

$$I_1 = \int_0^1 \frac{\sigma(\varsigma)}{\sqrt{\varsigma}}d\varsigma, \quad I_2 = -\int_0^1\left(\frac{d\sigma}{d\varsigma}\right)\varsigma^{-1/2}d\varsigma \qquad [4.19]$$

The classical viscoelastic-elastic correspondence principle is no longer valid, once the crack begins to grow. However, under the condition that $a(t)$ and/or $\ell p(t)$ is a monotonic, non-decreasing function of *t*, the opening displacement of a growing crack can be calculated using the Volterra integral operator of the second kind to give

$$w[x,t] = w_e[x,t] + E\int_0^t \phi(t-\tau)w_e[x,\tau]d\tau \qquad [4.20]$$

Finally, the condition for crack propagation is:

$$w[x,t]_{x=a(t)} = w_e[a(t)] + E\int_{t'}^t \phi(t-\tau)w_e[a(t),\tau]d\tau = w_c \qquad [4.21]$$

where t' is the time when the crack tip first reaches the point *x*.

Zhang & Karihaloo also calculated the creep rupture time. They recognised that the time taken by a crack to become unstable consists of the incubation and initial-growth periods, a prolonged period of slow subcritical crack growth, and finally, a brief period of rapid growth before rupture, as shown in Figure 4. However, they argued that for pre-existing cracks in large concrete structures, the principal subcritical growth period Δt is considerably longer than Δt_i or t_d and for all practical purposes determines the rupture lifetime *T*. Based on this approximation, they calculated the subcritical crack growth velocity for the visco-elastic model [4.12] and the $\sigma(\varsigma)$ relation [4.15] to be:

$$\left(\frac{da}{dt}\right)_{\ell p} = \frac{a}{\kappa}\left[\frac{\gamma\,\dfrac{K_I(t)}{K_{Ic}} - 1}{1 - \dfrac{K_I(t)}{K_{Ic}}}\right]^{\frac{1}{1-\alpha}} \quad [4.22]$$

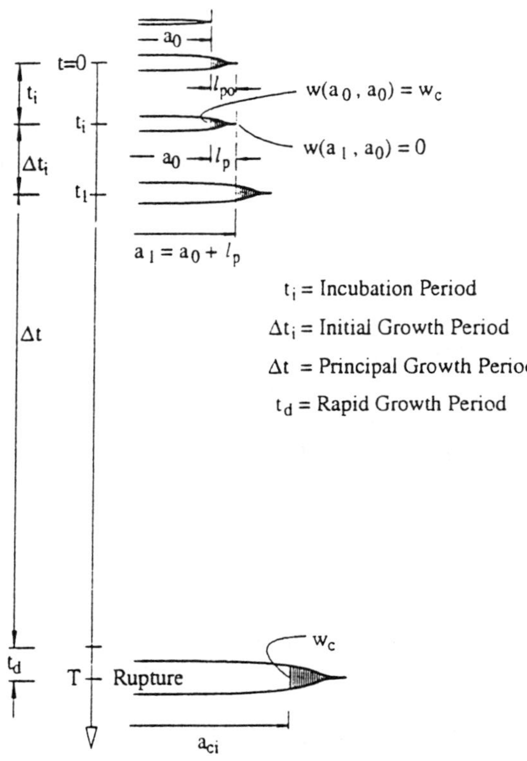

Figure 4. *A schematic illustration of the various periods in the growth of a crack from its inception to rupture (after [KAR 95])*

$$\left(\frac{da}{dt}\right)_{ft} = \frac{\pi K_I^2(t)}{2I_1^2 f_t'^2}\,\frac{\ell p}{a+\ell p}\,\kappa\left[\frac{\gamma\left(\dfrac{K_I(t)}{K_{Ic}}\right)^2 - 1}{1 - \left(\dfrac{K_I(t)}{K_{Ic}}\right)^2}\right]^{\frac{1}{1-\alpha}} \quad [4.23]$$

where K_{Ic} is the critical stress intensity factor for instantaneous crack growth and:

$$\kappa = \frac{a+\ell_p}{\ell_p}\left[\frac{\Gamma(2-\alpha)}{\beta}\right]^{\frac{1}{1-\alpha}} \qquad [4.24]$$

$$\gamma = 1+\frac{\lambda}{\beta}k(\alpha) \qquad [4.25]$$

$$k(\alpha)=\frac{\sqrt{\pi}\,\Gamma(2-\alpha)}{2(2-\alpha)\Gamma(2.5-\alpha)} \qquad [4.26]$$

When ℓ_p is assumed to be independent of time it is given in terms of $K_I(t)$ by

$$K_I(t)=\sqrt{\frac{2\ell p}{\pi}}\,f_t'I_1 \qquad [4.27]$$

The above approach was quite successful in predicting delayed propagation of an existing crack in a large concrete dam [KAR 95].

5. A Model for Ageing Visco-Elastic Tension Softening Material

A general constitutive modelling approach for the time-dependent fracture process was proposed by Santhikumar & Karihaloo [SAN 96, 98a, b] based on a combination of micromechanical approach to tension softening, Kelvin chain representation of the visco-elastic behaviour and solidification theory for ageing. Let us consider each element of this general approach separately.

5.1. *Micromechanical Model for Tension Softening*

The micromechanical model of Li & Huang [LI 90] assumes that the tensile strength of concrete is limited by the branching of the largest interfacial crack into the mortar matrix which then becomes the dominant discontinuous macroflaw. This dominant crack is assumed to grow in a "homogenised" medium with fracture toughness K_{Ic}^{hom}, although the real material is highly nonhomogeneous and disordered. Li & Huang show that K_{Ic}^{hom} can be related to the K_{Ic}^{m} of the matrix and the volume fraction of aggregate V_f:

$$K_{Ic}^{\mathrm{hom}} = K_{Ic}^{m}\left[\frac{1+0.87V_f}{1-\frac{\pi^2}{16}V_f\left(1-\nu^2\right)}\right]^{\frac{1}{2}} \qquad [5.1]$$

They approximated the dominant crack by a flat crack in an infinite medium and calculated the net inelastic deformation in the medium of the crack plane. To determine the post-peak material separation w at each stress level σ, the crack opening area was smeared over the total crack length. In this manner, they showed that:

$$w = \frac{\left(K_{Ic}^{\mathrm{hom}}\right)^2}{E\left(1-V_f\right)f_t\left(\frac{\sigma}{f_t}\right)}\left[1-\left(\frac{\sigma}{f_t}\right)^3\right] \qquad [5.2]$$

5.2. *Kelvin Chain Rheological Model*

A single Kelvin unit is too simple to allow a good representation of the time-dependent behaviour of uncracked concrete. A combination of a large number of Kelvin units and one Hookean elastic element H_0 in parallel, as shown in Figure 5 to form the so-called Kelvin chain model, is called for.

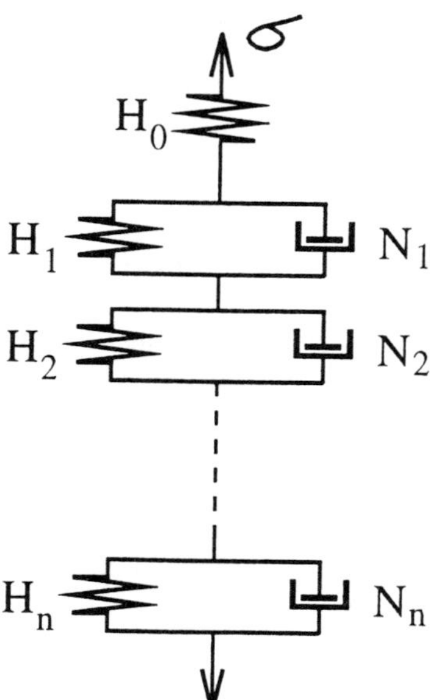

Figure 5. *Kelvin chain rheological model*

The stress in all Kelvin units is the same:

$$\sigma = E_0\varepsilon_0 = E_1\varepsilon_1 + \lambda_1 \frac{d\varepsilon_1}{dt} = E_2\varepsilon_2 + \lambda_2 \frac{d\varepsilon_2}{dt} = \cdots = E_n\varepsilon_n + \lambda_n \frac{d\varepsilon_n}{dt} \qquad [5.3]$$

while the strains are additive to give the following stress-strain relation:

$$\varepsilon = \frac{\sigma}{E_0} + \sum_{k=1}^{n} e^{-\frac{t}{\theta_k}} \left[\frac{1}{\lambda_k} \int_0^t \sigma(\tau) e^{\frac{\tau}{\theta_k}} d\tau + \varepsilon_{0,k} e^{\frac{t_0}{\theta_k}} \right] \qquad [5.4]$$

Here, $\varepsilon_{0,k}$ are the initial strains in the individual Kelvin units at time t_0, and $\theta_k = \lambda_k / E_k$ are the respective retardation times [SOB 1984].

Equation [5.4] can be solved for a constant applied stress σ_0 to study the creep deformation:

$$\varepsilon(t) = \sigma_0 \left[\frac{1}{E_0} + \sum_{k=1}^{n} \frac{1}{E_k} \left(1 - e^{\frac{-(t-t_0)}{\theta_k}} \right) \right] + \sum_{k=1}^{n} \varepsilon_{0,k} e^{\frac{-(t-t_0)}{\theta_k}} \qquad [5.5]$$

On the other hand, the solution of equation [5.4] for a constant applied strain ε_0 gives the relaxation function:

$$\sigma(t) = \left[E_0\varepsilon_0 - E_0 \sum_{k=1}^{n} \varepsilon_{0,k} e^{\frac{-(t-t_0)}{\theta_k}} \right] - \left[E_0 \sum_{k=1}^{n} e^{\frac{-t}{\theta_k}} \frac{1}{\lambda_k} \int_0^t \sigma(\tau) e^{\frac{\tau}{\theta_k}} d\tau \right] \qquad [5.6]$$

which can be rewritten in the following form:

$$\sigma(t) = \left\{ E_0\varepsilon_0 - E_0 \sum_{k=1}^{n} \varepsilon_{0,k} e^{\frac{-(t-t_0)}{\theta_k}} \right\} - E_0 \int_0^t K(t-\tau)\sigma(\tau) d\tau \qquad [5.7]$$

where $K(t-\tau) = \sum_{k=1}^{n} \frac{1}{\lambda_k} e^{-\frac{t-\tau}{\theta_k}}$

Using the Krylov-Kantorovich method of successive approximations, this integral equation can be numerically solved for a fixed ε_0.

5.3. *Ageing Effect in Visco-Elasticity*

Since Young's modulus E and viscosity λ of concrete change with age as a result of cement hydration or development of cracks in concrete, it is important to include also the time-dependent behaviour of these material properties. The time-dependent $E(t)$ and $\lambda(t)$ are usually determined by fixing the number of chains $\underline{n}$ and their retardation times θ_k in the Kelvin chain model and by matching the predicted behaviour with creep data.

However, small variations in the creep data are known to cause large variations in the resulting functions $E_k(t)$, in that, disparate functions $E_k(t)$ closely approximate the same creep data. Due to this ambiguity, some of the functions $E_k(t)$ determined in this manner can take on negative values for certain short time periods, although the overall modulus of the chain is always positive [BAZ 89], [BAZ 88]. This can still give acceptable numerical results in the overall sense, but does not seem satisfactory from the theoretical point of view.

For this reason, it is expedient to regard ageing as the result of a progressive solidification of a basic constituent with intrinsically non-ageing behaviour [BAZ 89], [CAR 93]. Non-ageing constitutive models (e.g., the Kelvin chain model) are used for modelling the visco-elastic behaviour of the constituent, with all the ageing effects restricted to the description of the evolution of the solidification process itself. Thus, the ageing modulus $E_k(t)$ of a Kelvin unit can be written as:

$$E_k(t) = \upsilon(t) E_k \qquad [5.8]$$

where $\upsilon(t)$ is the effective load-bearing volume fraction of hydrated cement which changes with age. It is assumed that all the moduli of the Kelvin chain vary in proportion to a single ageing function, $\upsilon(t)$. The evolution of the moduli $E_k(t)$ according to [5.8] eliminates the mathematical difficulties due to ill-posedness of the problem of moduli identification mentioned above. It also ensures positiveness of all moduli and non-divergence of the creep curves. Similarly, ageing dashpot viscosities $\lambda_k(t) = \theta_k E_k(t)$ can be written as:

$$\lambda_k(t) = \theta_k \upsilon(t) E_k \qquad [5.9]$$

Denoting by $\theta_k E_k$ the non-ageing dashpot viscosities (i.e. $\lambda_k = \theta_k E_k$), the ageing dashpot viscosities [5.9] become

$$\lambda_k(t) = \upsilon(t) \lambda_k \qquad [5.10]$$

It will be noted that in the above description the effect of ageing upon $E_k(t)$ and $\lambda_k(t)$ is represented by a single ageing function $\upsilon(t)$.

5.4. *Time-Dependent Tension-Softening of Ageing Concrete*

It is notable from the tension softening relation [5.2] that the residual tensile stress carrying capacity of cracked concrete reduces progressively with increasing crack opening displacement until it vanishes at a critical crack opening. But the tension softening relation [5.2] is not of much help to the problem at hand in its current form. We therefore invoke the concept of "effective" spring [HUA 95], and assume that the damage along the eventual fracture plane is tantamount to the faces of this plane being connected by springs whose stiffness is related to the crack opening displacement (Figure 6).

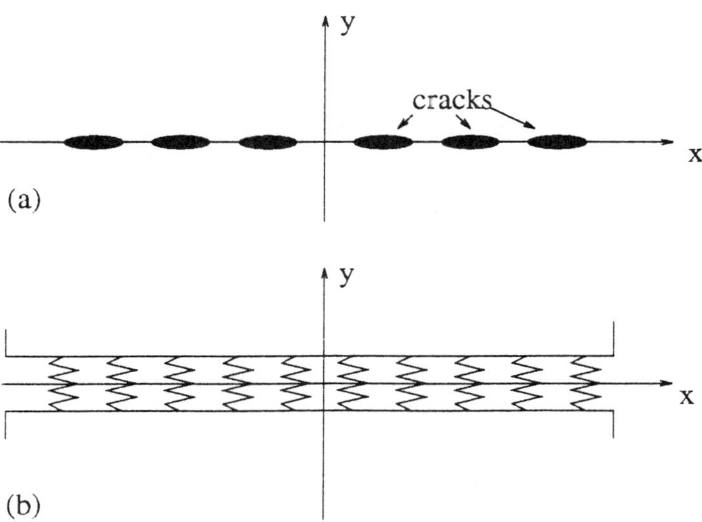

Figure 6. *Unbroken ligaments between two half planes y<0 and y>0 transmitting a remote normal stress, showing (a) the complementary view in terms of coplanar cracks on y = 0, and (b), the representation by distributed springs*

The constitutive relation between the stress transmitted by the "effective" springs, and the vertical displacement of the upper half-plane which is related to the total spring stretch w, is given by:

$$\sigma = kw \qquad [5.11]$$

where k is the stiffness of the ageing tension softening spring.

Though ageing plays an important role in the creep of normal concrete, its effect on stiffness of cracked concrete is not so pronounced. The latter decreases with time mainly because of the progressive development of cracks. This effect upon k can be represented through the solidification concept via a relation similar to [5.8]

$$k(t)=\upsilon(t)k_0 \tag{5.12}$$

where k_0 is a constant which is numerically equated to Young's modulus of the spring used in the rheological model at time t_0. From [5.2], [5.11] and [5.12] it follows that

$$\upsilon_k(t)=\frac{k_{0,k}}{k_{0,0}}F(\sigma,w) \tag{5.13}$$

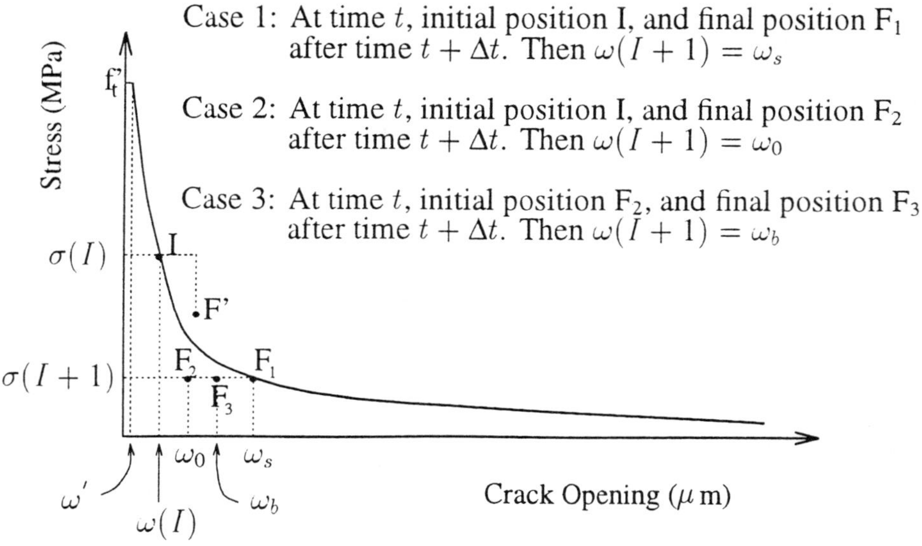

Figure 7. *Time-dependent tension softening diagram: I - initial position at $T = t$, F - possible final position at $T = t + \Delta t$*

where (refer to Figure 7)

$$F(\sigma,w)=[1.0-Q(x)]\frac{\sigma(I+1)}{w_0+w'}+Q(x)\frac{\sigma(I+1)}{w(I+1)+w'} \tag{5.14}$$

with $Q(x) = \frac{w(I+1) - w_0}{w_s - w_0}$, and w' is an infinitesimally small value of crack opening displacement which is introduced to avoid the singularity at the very beginning of tension softening.

5.5. *Computational Algorithm*

To predict the behaviour of time-dependent ageing tension softening materials using the above model, an incremental procedure is necessary. The time increment has to be small to ensure the desired degree of accuracy.

- The non-ageing Young modulus E_k and dashpot viscosity λ_k are determined by judiciously selecting the number of chains $\underline{n}$ and their retardation times θ_k in the model and by matching the predicted behaviour with creep data of uncracked concrete. (See Appendix.)
- The non-ageing (initial) stiffness $k_{0,k}$ of each spring is assumed to be numerically equal to its Young's modulus E_k.
- The time-dependent stiffness of the spring $k_k(t)$ is computed at each time step using [5.12], with the time-dependent function $\upsilon_k(t)$ calculated from [5.13] and [5.14] (see also Figure 7). To avoid the singularity at the onset of tension softening (at $\sigma = f_t'$), it is expedient to assume an infinitesimally small value of crack opening displacement w'.
- In the creep [5.5] and relaxation functions [5.7] E_0 and E_k are replaced with k_0 and $k_k(t)$ to take account of their time-dependency.
- Before the creep and relaxation functions [5.5] and [5.7] are evaluated at the current time step $t + \Delta t$ using [5.12]-[5.14], the time-dependent stress and crack opening displacement at the end of the previous time step t are calculated to check whether or not they lie on the instantaneous tension softening curve $\sigma(t) - w(t)$ (point F_1 in Figure 7). If not, then the following procedure is adopted (refer to Figure 7).

  If, for example, the point (σ, w) is located above the instantaneous $\sigma(t) - w(t)$ curve (point F' in Figure 7), then it is clear that relaxation alone must take place until the stress drops to the level on the curve. If, on the other hand, the point (σ, w) is located below the instantaneous curve (point F_2 or F_3 in Figure 7), then the creep alone must take place until the crack opening displacement increases to the level on the curve.

- The time is again incremented by Δt and the procedure repeated until the crack opening displacement reaches its critical value w_c which is assumed to be independent of time.

The above computational scheme was used to predict the time-dependent tension softening response of a real concrete mix [HUA 89] whose mechanical properties are listed in Table 1.

The generic rheological properties of this mix, calculated by using the Kelvin chain model [BAZ 74], are given in Table 2 for two sets of retardation time θ_k differing by a factor of five.

The static tension-softening response of this mix, calculated by using [5.2], is shown in Figure 8 and its evolution with time in Figure 9.

Parameter	Value
V_f	0.72
E_c	30 GPa
K_{Ic}^{m}	0.3 MPa $\sqrt{\text{m}}$
f_t	3 MPa
w_c	350 μm

Table 1. *Mix properties*

Unit	E(GPa)	θ_k (hours)	
Single spring	30	-	-
Kelvin unit 1	950	0.024	0.12
Kelvin unit 2	350	0.240	1.20
Kelvin unit 3	270	2.400	12.00
Kelvin unit 4	510	24.000	120.00
Kelvin unit 5	420	240.000	1200.00

Table 2. *Rheological properties*

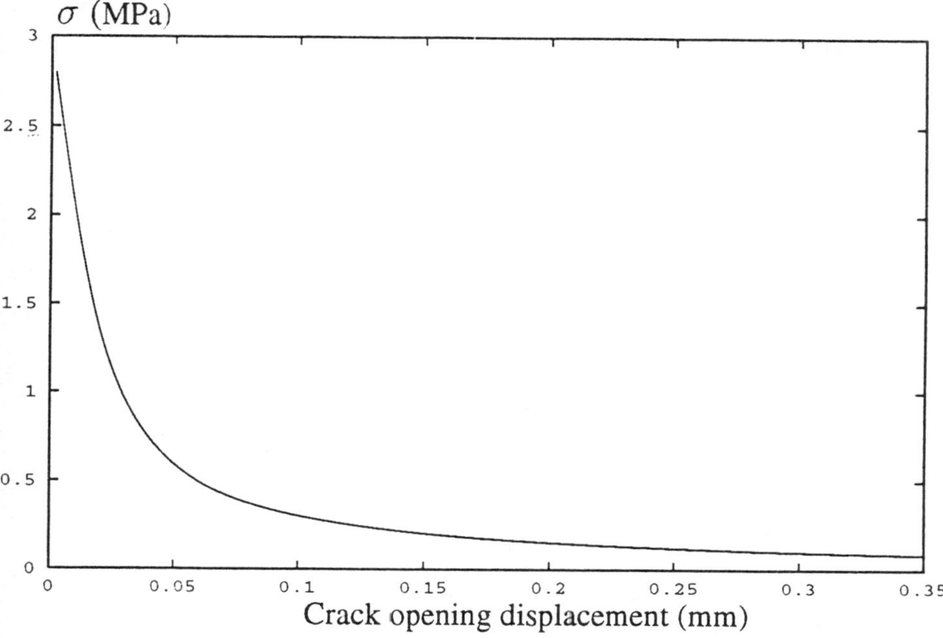

Figure 8. *Static tension-softening relation for the material properties of Table 1*

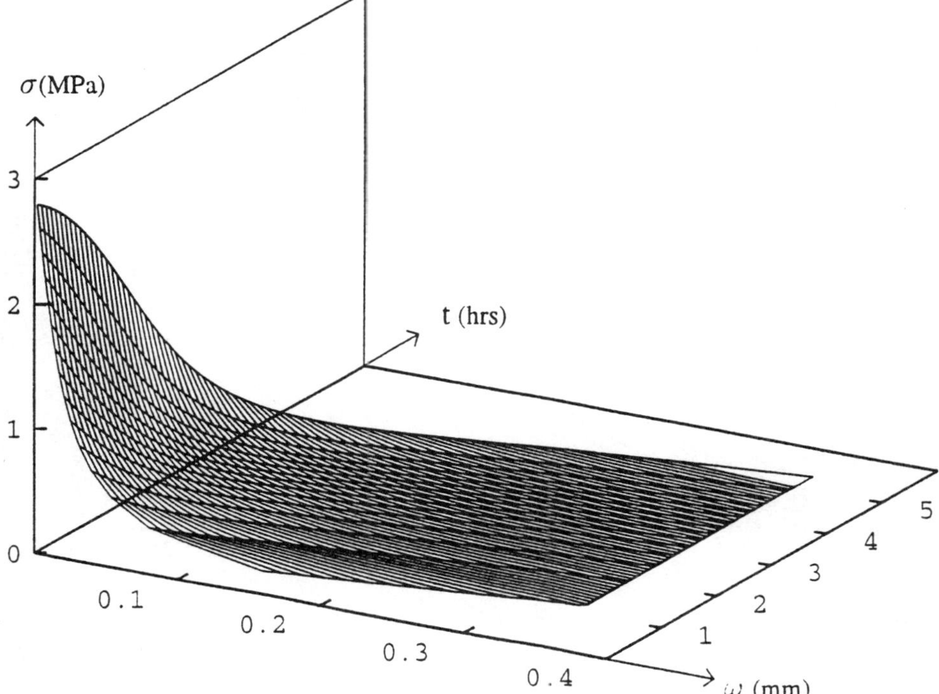

Figure 9. *Variation in residual tensile carrying capacity and crack opening displacement with time as predicted by the model for the mix of Table 1 and Table 2 (lower θ_k values)*

Figure 10 shows the reduction in the residual tensile carrying capacity and the increase in the crack opening displacement with time for different starting values at $t = 0$ on the static softening curve.

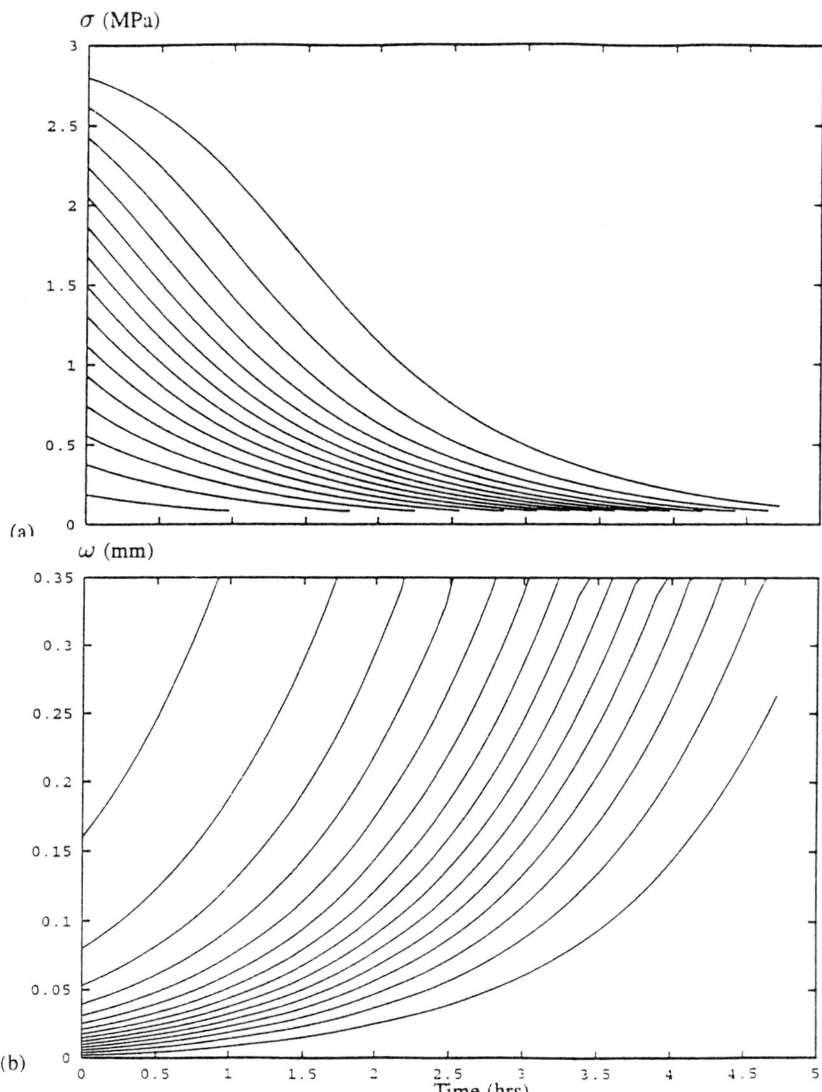

Figure 10. *(a) Relaxation of residual tensile carrying capacity and (b) increase of crack opening displacement with time as predicted by the model for the mix with lower retardation times of Table 2*

The creep and relaxation curves corresponding to the two sets of retardation time in Table 2 are shown in Figures 11 and 12 respectively, in order to highlight the important role played by the retardation time in these phenomena.

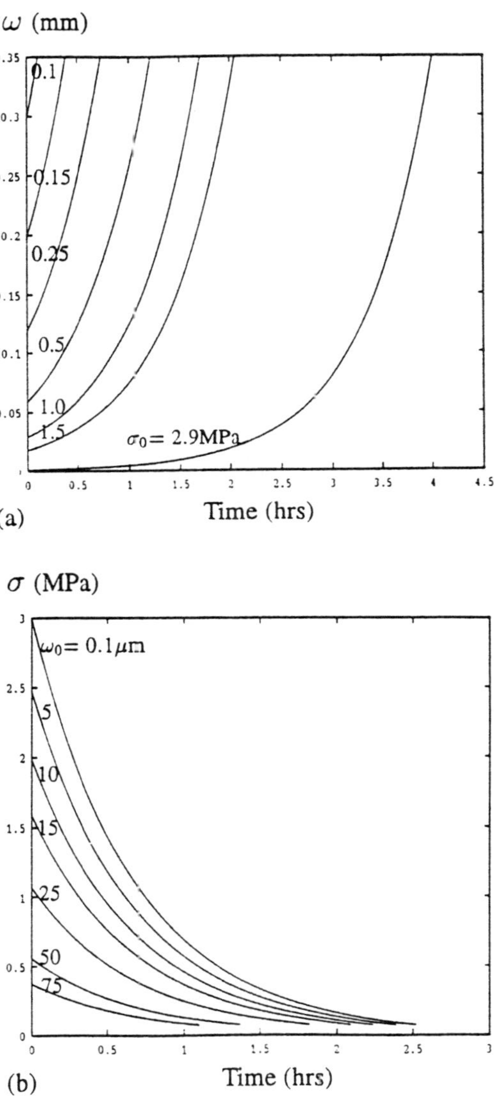

Figure 11. *Predicted (a) creep curves from different constant stress levels, (b) relaxation curves at different constant crack openings for lower retardation times of Table 2*

The above constitutive model based on a judicious combination of two-dimensional micromechanical approach to tension softening, the Kelvin chain rheological model and the solidification theory for ageing of concrete can be easily implemented in a finite element analysis using the FCM. The time-dependent Young's modulus and retardation time of each Kelvin unit in the chain are calibrated from available test data for uncracked concrete.

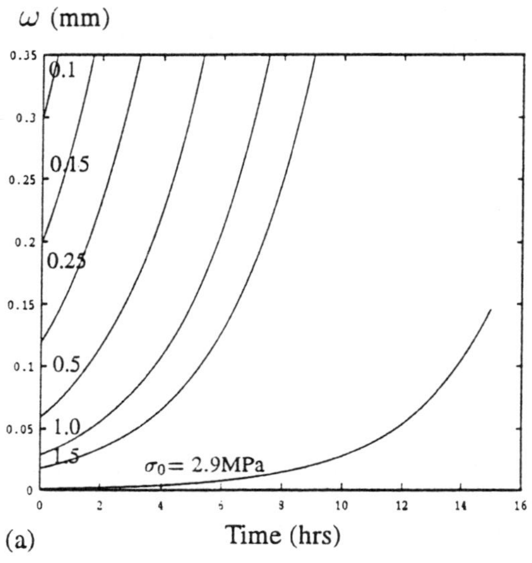

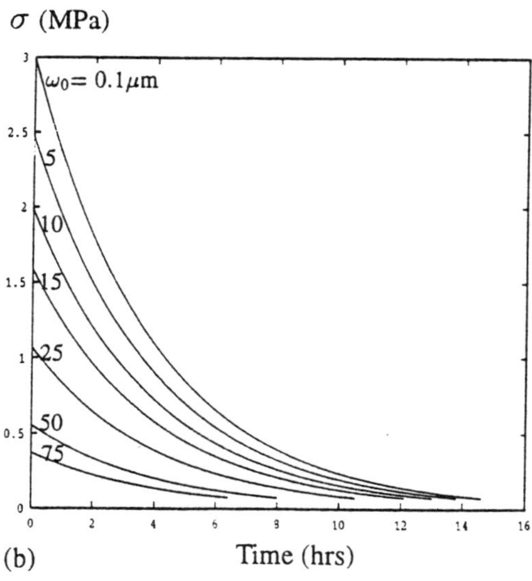

Figure 12. *Predicted (a) creep curves from different constant stress levels, (b) relaxation curves at different constant crack openings for higher retardation times of Table 2*

6. Concluding remarks

The computational algorithm outlined in Section 5 assumes implicitly that the static tension softening diagram [5.14] is continuous to the first order. In practice, this diagram is often approximated by simple curves, some of which are discontinuous. Among the most popular discontinuous approximation is the bi-linear diagram of Figure 13.

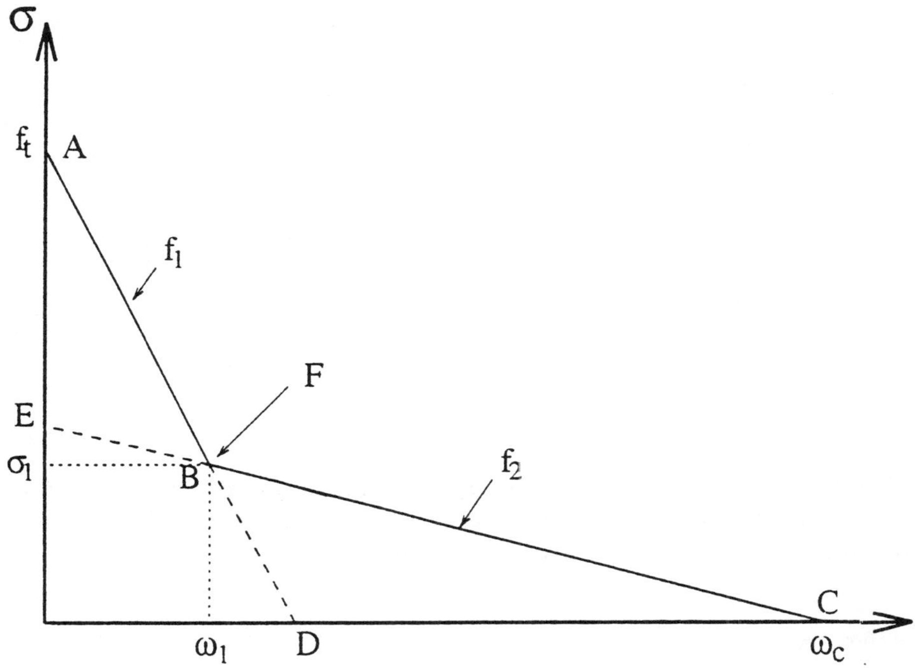

Figure 13. *Bi-linear tension softening diagram, showing the co-ordinates of the knee*

The computational algorithm can be applied to the bi-linear tension softening diagram after approximating it as closely as possible by a continuous function with a continuous first derivative. A smoothing procedure has been outlined by Santhikumar & Karihaloo [SAN 98b] which ensures that the continuous approximation is practically indistinguishable from the original bi-linear diagram. They also revealed that the bi-linear tension softening diagram predicts the wrong response in the creep rupture region. Figure 14 shows a typical response.

They argued that the severe inadequacy of the bi-linear tension softening diagram is due to the fact that the linear tail segment inhibits the rate of increase in crack opening, especially when $w(t)$ approaches the critical value w_c. They also found that this wrong trend in the rate of increase of $w(t)$ near creep rupture is unaffected by the mix rheological properties, the location of the knee or the slope of the linear tail segment.

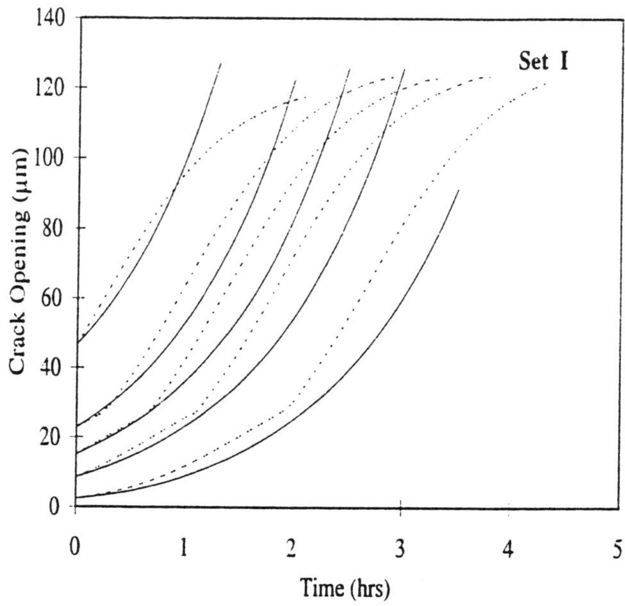

Figure 14. *Evolution of w(t) with time from the initial value at $t_0 = 0$ as predicted with the continuous softening curve of Figure 8 (solid lines) and a continuous approximation to the bi-linear diagram (broken lines)*

References

[BAZ 88] BAZANT, Z.P., (ed.). *Mathematical Modeling of Creep and Shrinkage of Concrete*. London: John Wiley, 1988.

[BAZ 90] BAZANT, Z.P., Rate effect, size effect and nonlocal concepts for fracture of concrete and other quasi-brittle materials, in S.P. Shah (ed.), *Mechanisms of Quasi-brittle Materials*, 131-153. Dordrecht: Kluwer Academic, 1990.

[BAZ 93a] BAZANT, Z.P., Current status and advances in the theory of creep and interaction with fracture, in Z.P. Bazant & I. Carol (eds), *Creep and Shrinkage of Concrete*, 291-307, London: E. & F.N. Spon, 1993a.

[BAZ 74] BAZANT, Z.P., ASGHARI, A., Computation of Kelvin chain retardation spectra of aging concrete, *Cem. & Conc. Res.*, 14:797-806, 1974.

[BAZ 92a] BAZANT, Z.P., GETTU, R., Rate effects and load relaxation in static fracture of concrete, *ACI Mater.J.*, 89:456-467, 1992.

[BAZ 92b] BAZANT, Z.P., JIRASEK, M., R-curve modeling of rate effect in static fracture and its interference with size effect, in Z.P. Bazant (ed), *Fracture Mechanics of Concrete Structures*, 918-923. London: Elsevier, 1992.

[BAZ 93b] BAZANT, Z.P., JIRASEK, M., R-curve modeling of rate and size effects in quasibrittle fracture, *Int.J.Fracture*, 62:355-373, 1993.

[BAZ 89] BAZANT, Z.P., PRASANNAN, S., Solidification theory for concrete creep, I:Formulation, *ASCE J.Eng.Mech.*, 115:1691-1703, 1989.

[CAR 93] CAROL, I., BAZANT Z.P., Viscoelasticity with aging caused by solidification of nonaging constituent, *ASCE J.Eng.Mech.*, 119:2252-2269, 1993.

[CAR 95] CARPINTERI, A., VALENTE, S., ZHOU, Z.P., Crack propagation in concrete specimens subjected to sustained loads, in F.H. Wittmann (ed), *Fracture Mechanics of Concrete Structures*, 1315-1328, Frieburg: Aedificatio Publishers, 1995.

[DU 89] DU, J., KOBAYASHI, A.S., Hawkins, N.M., FEM dynamic fracture analysis of concrete beams, *ASCE J.Eng.Mech.,* 115:2136-2149, 1989.

[HAN 90] HANSEN, E.A., A visco-elastic fictitious crack model, in S.P. Shah, S.E. Swartz & M.L. Wang (eds), *Micromechanics of Quasi-brittle Materials,* 156-165, London: Elsevier, 1990.

[HAN 91] HANSEN, E.A., Influence of sustained load on the fracture energy and the fracture zone of concrete, in J.G.M. van Mier, J.G. Rots & A. Baker (eds), *Fracture Processes in Concrete, Rock & Ceramics*, 829-838, London: E. & F.N. Spon., 1991.

[HUA 89] HUANG, J., LI, V.C., A meso-mechanical model of the tensile behaviour of concrete, Part II: Modeling of post-peak tension softening, *Composites*, 20:370-378, 1989.

[HUA 95] HUANG, X., KARIHALOO, B.L., ROSE, L.R.F., Effective spring constant for planar arrays of circular cracks, *Int. J. Damage Mech.*, 4:103-116, 1995.

[KAM 80] KAMINSKII, A.A., *Fracture Mechanics of Visco-elastic Bodies* (in Russian), Kiev: Naukova Dumka, 1980.

[KAR 95] KARIHALOO, B.L., *Fracture Mechanics and Structural Concrete*, Harlow, England: Addison Wesley Longman, 1995.

[LI 90] LI, V.C., HUANG, J., Relation of concrete fracture toughness to its internal structure, in H.P. Rossmanith (ed.), *Fracture and Damage of Concrete and Rock*, 39-46, Oxford: Pergamon, 1990.

[MIN 87] MINDESS, S., BANTHIA, N., YANG, C., The fracture toughness of concrete under impact loading, *Cement & Conc. Res.*, 17:231-241, 1987.

[SAN 96] SANTHIKUMAR, S., KARIHALOO, B.L., Time-dependent tension softening, *Mech. of Cohesive-Frictional Mater.*, 1:295-304, 1996.

[SAN 98a] SANTHIKUMAR, S., KARIHALOO, B.L., A model for ageing visco-elastic tension softening materials, *Mech of Cohesive-Frictional Mater.*, 3:27-40, 1998.

[SAN 98b] SANTHIKUMAR, S., KARIHALOO, B.L., Inadequacy of the bi-linear approximation to ageing visco-elastic tension softening materials, *Mech. of Cohesive-Frictional Mater.*, (submitted), 1998.

[SCH 75] SCHAPERY, R.A., Theory of crack initiation and growth in viscoelastic media I - III, *Int. J. Frac.*, 11:141-159, 369-388, 549-562, 1975.

[SOB 84] SOBOTKA, Z., *Rheology of Materials and Engineering Structures*, Amsterdam: North-Holland, 1984.

[TAK 91] TAKEDA, J., KOMOTO, H., TANIKAWA, T., Favourable and unfavourable rate effects on fracture process of concrete, in J.G.M. van Mier, J.G. Rots & A. Baker (eds), *Fracture Processes in Concrete, Rock and Ceramics,* 849-859, London: E. & F.N. Spon., 1991.

[WU 93] WU, Z.S., BAZANT, Z.P., Finite element modeling of rate effect in concrete with influence of creep, in Z.P. Bazant & I. Carol (eds), *Creep and Shrinkage of Concrete*: 427-432, London: E. & F.N. Spon., 1993.

[YOU 92] YOU, J.H., HAWKINS, N.M., KOBAYASHI, A.S., Strain-rate sensitivity of concrete mechanical properties, *ACI Mater. J.*, 89:146-153, 1992.

[ZHA 92a] ZHANG, C., KARIHALOO, B.L., Stability of a crack in a linear visco-elastic tension softening material, in Z.P. Bazant (ed.), *Fracture Mechanics of Concrete Structures,* 75-81, London: Elsevier, 1992.

[ZHA 92b] Zhang, C. & Karihaloo, B.L., Stability of a crack in a large concrete dam, *C.E. Trans. Inst. Engrs. Aust.*, CE34:369-375, 1992b.

[ZHO 92] Zhou, Z.P. & Hillerborg, A., Time-dependent fracture of concrete: testing and modeling, in Z.P. Bazant (ed.), *Fracture Mechanics of Concrete Structures*, 906-911, London: Elsevier, 1992.

Appendix

Once the number of chains in the Kelvin model has been judiciously selected, the non-ageing Young moduli E_k are determined as follows:

- The retardation times θ_k are chosen arbitrarily. However, they must not be spaced too sparsely on the $\ln(t-t_0)$ scale and should cover the entire time period of interest. Thus, $\theta_k' s$ may be chosen in intervals spaced a decade apart, with the smallest θ_k less than the age of concrete at the time of first loading (say, $\theta_1 \leq 0.1t_0$).

- The Kelvin chain relation [5.5] for a constant applied stress σ_0 is differentiated with respect to t to reduce it a Dirichlet series

$$\frac{d\varepsilon}{dt} = \sum_{k=1}^{n} \frac{\sigma_0}{\theta_k E_k} e^{-(t-t_0)/\theta_k} \qquad \text{[A.1]}$$

- Next, the creep strain rate $\dot{\varepsilon}$ curve is constructed from the known creep rate data as a function of $(t-t_0)$ and equation [A.1] solved by the method of least squares to determine E_k

On the Residual Tensile Properties of High Performance Siliceous Concrete Exposed to High Temperature

Roberto Felicetti — Pietro G. Gambarova

Milan University of Technology
Piazza Leonardo da Vinci 32
20133 Milano, Italy

ABSTRACT. *High-Performance Concrete containing silica fume (f_c = 60-120 MPa) is definitely more sensitive to high temperature (T = 100-800°C) than ordinary concrete, due to its denser cementitious matrix, with lower diffusivity, and to the greater role of the aggregate, whose thermal decay is much more harmful in HPC. Here, two highly-siliceous concretes (f_c = 72 and 95 MPa) are tested in direct and indirect tension (3-point bending), to investigate their residual mechanical properties after a single thermal cycle at high temperature (T = 105-500°C), and more specifically: (1) to measure the complete stress-CMOD response in notched cylinders; (2) to evaluate the tensile strength in direct and indirect tension; and (3) to assess to what extent fracture energy, toughness and characteristic length are affected by high temperature.*

KEY WORDS: *Characteristic Length, Fracture Energy, High-Performance Concrete, High Temperature, Residual Properties (after a thermal cycle), Tensile Behavior, Toughness.*

1. Introduction and objectives

The mechanical properties of ordinary concrete ($f_c \leq 50$ MPa) at high temperature (up to 800-1000°C) are well known thanks to the many tests carried out since the Sixties and Seventies [RIL 85] and even very recently [PAP 91; TAC 93; THI 93]. However, certain aspects related to concrete composition, multiaxial loading and steel-concrete interaction are still under investigation, as confirmed by the research activity which is in progress on these topics. Furthermore, in the case of High-Performance Concrete (f_c = 60-120 MPa) the test data are scanty indeed and research work is badly needed [DIE 88, 89], [FEL 95], [NIS 97], [PHA 97].

Within this framework, the objective of this project is to investigate the residual mechanical properties of two high-performance highly siliceous concretes, after being exposed to a high-temperature cycle, as should be expected of the sustained temperatures at the inner surface of a R/C or P/C secondary containment-shell, due to a major accident in the reactor core of a nuclear power plant (200°C and above, loss of coolant and core melting). In this regard, both the behavior *under* a sustained temperature and that *after* one cycle at a high sustained temperature are still open to investigation, not to speak of the thermal shocks (including hot spots) with their hygrothermal problems, which are characterized by a severe transient behavior and are not considered here [MAR 78], [BAZ 79, 81, 96].

As regards test modalities at high temperature, many test results and documents (see Refs. in [RIL 85]) show that the damage induced by exposure to high temperature is mostly irreversible and remains "frozen" in the concrete, even after cooling down to room temperature. For this reason the mechanical properties measured directly under sustained high temperature or after cooling down to room temperature are very close: for instance, when normal-strength concrete is tested 1-2 months after the thermal cycle, the residual compressive strength at room temperature is close to 85-90% of that at high temperature; then a certain recovery occurs. Since the tests at room temperature are far simpler (a test in direct tension at high temperature is difficult to perform, not to mention structural tests), the research program presented here is based on specimens subjected to a high-temperature cycle, but tested *after* cooling down to room temperature (105, 250, 400 and 500°C, cylinders and prisms, Figure 1).

A series of tests on cylinders in pure compression, and on cylinders and prisms subject to direct tension and to bending (3-point bending tests, notched and unnotched specimens) was carried out, to evaluate Young's modulus E_c, the strength both in tension and in compression (f_t and f_c), and – more important – the entire stress-strain (or displacement) curves in uniaxial compression (tension). Here reference is made only to the tests in direct and indirect tension, but the main results in compression are also recalled [FEL 95, 96, 98].

Testing in tension makes it possible to evaluate the specific fracture energy G_f, which is one of the fundamental parameters (together with E_c and f_t) required by most mathematical models based on concrete fracture mechanics or concrete damage mechanics.

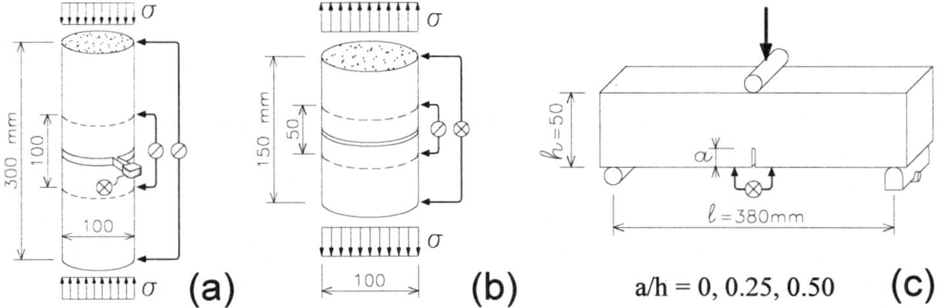

Figure 1. *Specimens for concrete characterization in: (a) compression [FEL 95,96,98]; (b) direct tension; and (c) indirect tension;* ⊗ *= control parameter*

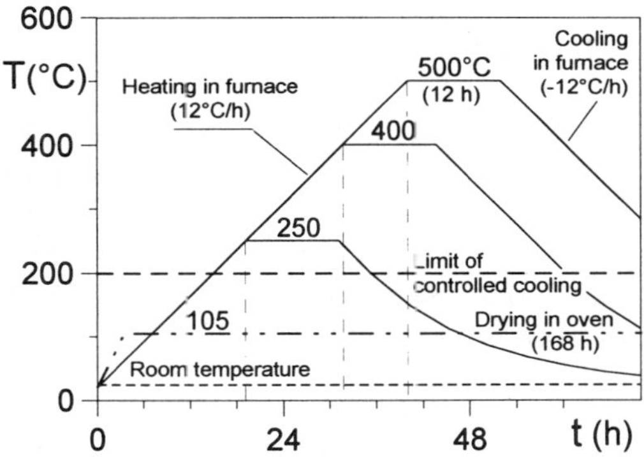

Figure 2. *Dehydration process and thermal cycles*

Components (kg/m³)	f_c = 72 MPa	f_c = 95 MPa
Portland cement	290	415
Silica fume	30 (9.4%)	30 (6.7%)
Calcareous filler	105	0
Aggregate: crushed flint (sand /gravel /pebbles)	831 / 287 / 752	
Water (water/binder ratio)	138 (43%)	133 (30%)
Superplasticizer: (melammino-sulfonate)	10.6 (3.3%)	14.8 (3.3%)
Retarder	1.7 (0.5%)	2.4 (0.5%)
Slump (mm)	230	210

Table 1.. *Concrete mix-design*

The mix-design of the concrete, and the dehydration and heating processes are shown in Table 1 and Figure 2 respectively. Two similar concretes were investigated (f_c = 72 and 95 MPa - Table 1, the percentages in brackets being the ratios with respect to the total binder content = cement + silica fume). The main difference between the two mix-designs regards the cement content, since the calcareous filler of the former concrete was replaced with more cement in the latter. The water content also includes the amount contained in the superplasticizer and retarder. The highly siliceous aggregate (mostly flint consisting of quartz, opal and calcedonium) is commonly used in certain regions of Central Europe for its good mechanical properties. However, the relatively high content of zeolitically-bound water makes flint-based concrete very sensitive to high temperature, since the water is slowly expelled between 100 and 600°C, with relevant volume changes and subsequent splitting. The aggregate consisted of round and crushed pebbles.

2. Previous Results in Uniaxial Compression

With reference to compression, a previous study on cylindrical specimens (∅ = 100 mm; h/∅ = 3 [FEL 95, 96, 98]) clearly shows the astonishing decrease in the residual mechanical properties after a high-temperature cycle (Figures. 3a, b) and the remarkable flattening-off of the falling branch of the stress-strain response (Figure 3a). In the case of 500°C the residual strength is so small (< 10% of the virgin-material strength) that for practical purposes the material cannot bear any load at all! The trend is the same for Young's modulus (Figure 3b).

Unexpectedly, the two types of concrete behave in virtually the same way beyond 250°C, but an explanation may be found in the presence of identical water content, and the same aggregate type and grading curve, which equalize the two types of concrete at high temperature when the more homogeneous matrix of the higher-grade concrete has been impaired by the thermally-induced microcracking.

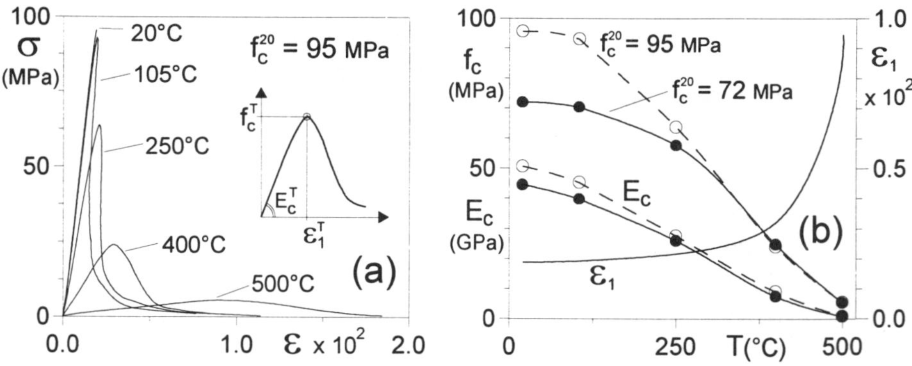

Figure 3. *Summary of test results in compression [FEL 95, 96, 98]: (a) stress-strain curves at room temperature and after one cycle at high temperature and (b) plots of the Young's modulus, compressive strength and associated strain, as a function of the temperature*

The reduction of compressive strength with the temperature is more severe than that which occurred in tests by other authors (see also Figure 14a, T ≥ 250°C). Two possible reasons may be given: firstly, flint-based aggregates are very sensitive to high temperature, as already mentioned, and, secondly, the exposure to the nominal temperature was longer than in the vast majority of the tests reported in the literature (12 hours compared to 1-3 hours, see for instance [PAP 91]).

3. Tensile Behavior in Uniaxial Tension and in Bending

Testing in direct tension is, in principle, the most suitable way [HOR 91] to characterize the behavior of concrete in tension and to work out the values of such fundamental parameters as Young's modulus, peak strength, fracture energy, critical crack width and slope of the falling branch. However, this is not so, since the lack of homogeneity in the concrete at the micro- and meso-level (different components, microcracks, voids, pores and sand pockets) unavoidably introduces some sort of "structural" effect into the behavior of the specimen (e.g., bending), and these phenomena can hardly be separated from the "constitutive" behavior of the material.

For instance, bending is likely to occur even if the specimen ends are perfectly glued to the platens of the loading machine and the specimens are perfectly centered along the loading axis. Even the loading modalities are critical, since fixed platens favor test stability and induce more homogeneous behavior in the test specimen, while hinged platens cause more structural effects, which may overshadow concrete constitutive behavior.

In order to prevent or limit the undesired structural effects, the displacements astride the notched section should be controlled and kept uniform [GIU 87; CAR 94]. To this end, a set of extensometers is generally fixed to the specimen, and their baselength should be small enough to prevent the loss of control ensuing from the elastic-energy release caused by crack formation and propagation.

Since direct-tension tests are highly demanding in terms of test procedure, 3-point bending tests are often preferred because of their greater simplicity and stability. However, in the latter case the structural effects, which are evident and intentional, require careful modeling in order to work-out the stress-strain relationship of the material [ROS 91].

Beyond the stress-strain and stress-displacement relationships, there is another aspect which plays a fundamental role in the characterization of the tensile behavior of concrete. The various inhomogeneities and flaws inside the concrete mass (such as coarse aggregates, microcracks and pores) are involved in crack formation and propagation, whose early phase is characterized by a distributed process within a finite volume of material (fracture process zone). The size of this volume is related to the material texture and is only marginally affected by the structural size, giving rise to the so-called "size effect". This effect is represented by a "scaling law", which describes the apparent concrete strength at failure, as a decreasing function of the size [BAZ 90].

As a result, there is one material parameter, related to the depth of the fracture process zone (characteristic length), which can be assessed only by comparing the response of geometrically similar specimens (but of highly different sizes) or by performing different types of tests, in such a way that the boundary conditions and the specimen shapes involve different structural effects and brittleness [DIP 97].

Summing up, both the strain-controlled tests in uniaxial tension and the tests in bending (on notched or unnotched specimens) are fundamental for the identification of the material parameters in tension, provided that the experimental results are interpreted by means of suitable theories [ROS 94].

For the above-mentioned reasons, two series of tests in tension were planned and carried out. In the former tests the specimens were notched cylinders glued to fixed platens, while in the latter tests concrete prisms were adopted, with or without a notch along the bottom of the mid-span section.

4. Tests in Direct Tension

Twenty cylinders (Figure 4a) were successfully tested in uniaxial tension by controlling the elongation of the entire specimen (baselength = specimen length = 150 mm). Since the specimens heated up to 500°C were completely cracked (both the aggregate and the mortar) and did not show any appreciable residual strength, the results reported here regard only 4 temperature levels (20, 105, 250 and 400°C).

Because of the greater scatter to be expected of brittle behavior, the specimens not cycled or cycled at 105°C (more brittle), were tested in batches of 3, while the specimens cycled at 250 and 400°C were tested in batches of 2.

Some details on technology, instrumentation and loading procedure are given below:

– The end sections of the cylindrical specimens (h = 150 mm, ∅ = 100 mm) were ground and polished in order to guarantee both parallelism and planarity (ASTM C39-81 Standard).

– The circumferential notch (depth 8 mm, width = 2.5 mm) was cut by means of a diamond disk, *before* the dehydration and heating processes (Figure 4a).

– Both press platens were fitted with specially-built heads provided with 4 movable pointers at 90°, with a sensitivity of 0.1 mm, in order to guarantee the perfect alignment of the specimen with the loading axis (Figure 4d).

– Before fixing the specimen between the press platens, the end sections were smeared with a thin layer of epoxy glue; then, after the specimen had been aligned, a compressive force of 0.5 kN was applied during the hardening of the glue (12 hours).

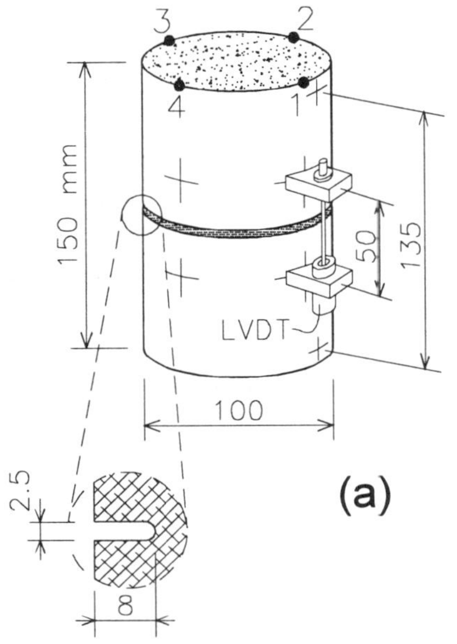

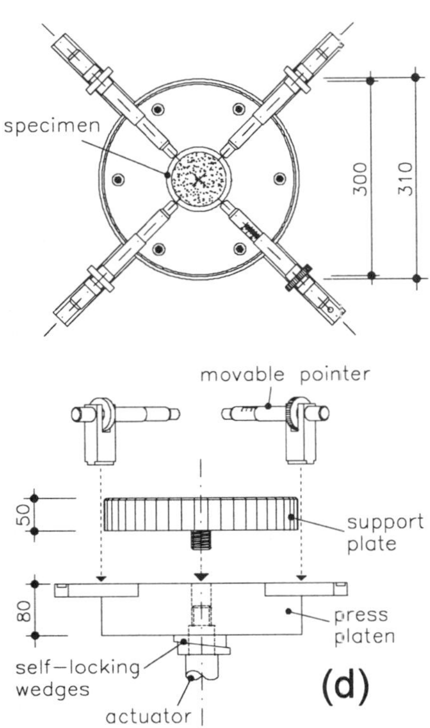

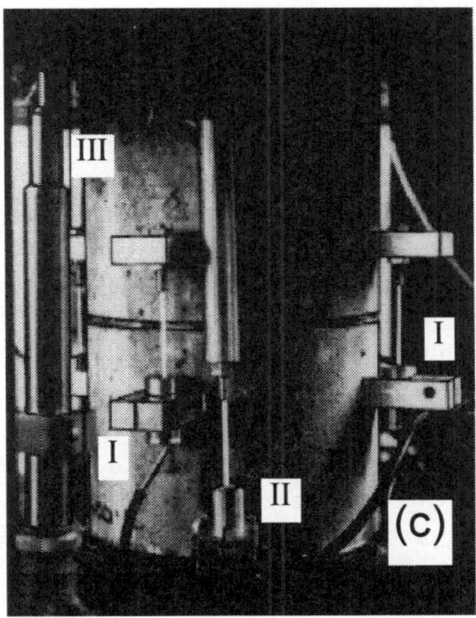

Figure 4. *Tests in direct tension: (a) specimen geometry and instrumentation; (b) Instron electromechanical press; (c) instrumented specimen; and (d) movable head of the press (lower head) with centering pointers*

– The tests were carried out with an Instron electro-mechanical press (capacity 100 kN, Figure 4b), which was controlled by the average signal of 2 LVD Transformers placed between the platens (at 180°); in this way, the elastic energy release during cracking may have affected the sensitivity of the control – which did not happen – in relation to the solution based on the same number of LVDTs placed astride the notch; however, the set-up adopted here allows the press to be controlled even if the crack does not form in the notched zone.

– The movable head of the press (lower head, Figure 4b) was fitted up with 4 adjustable rods, which where blocked to the cross-head (Figure 4b, top); these stiffening rods bypass the load cell (which measures the actual load applied to the specimen) and allow the end rotations of the specimen to be controlled in order to keep the crack opening in the notched section as uniform as possible.

– 4 LVDTs (at 90°) were assembled astride the notch (baselength of 50 mm), in order to monitor the deformation and crack opening in the notched zone (I in Figure 4c).

– 2 LVDTs (at 180°) were placed astride the notch, with a baselength of 135 mm, in order to measure the elongation of the specimen *with no* disturbances (if any) from the glue layers at the specimen ends (II in Figure 4c).

– 4 LVDTs (at 90°) were placed between the press platens (baselength 150 mm, III in Figure 4c), in order to measure the total elongation of the specimen, end disturbances and end rotations (if any) included.

– The tests were carried out at a constant displacement-rate (0.1 mm/s up to 1/3 of the softening branch, and 0.2-0.4 μm/s during the last part of the load-displacement response).

The results are shown in Figure 5 (stress-displacement curves at 20, 250 and 400°C) and in Figure 6 (average stress-displacement curves). In Figure 7a, the residual tensile strength is plotted against the temperature of the thermal cycle. Contrary to the tests in compression, there is a limited but not negligible increase between 20 and 105°C, while the decrease between 105 and 400°C is even greater than in compression, as shown by the ever-increasing ratio between the compressive strength f_c and the tensile strength f_t (Figure 7b, f_c/f_t).

Also, the ratio between the tensile strength in bending f_t^* and the direct tensile strength f_t increases with the temperature as a result of the concrete's greater deformability and non-linearity, with more stress redistribution in bending (Figure 7b, f_t^*/f_t).

Finally, the plots of the fracture energy G_f are shown in Figure 12 (see also Figure 13a) for both concretes: there is no clear dependency on the temperature of the thermal cycle, to the point that G_f may even be considered as practically unaffected by the thermal damage ($G_f = 200 \div 205 \pm 45 \div 60$ J/m^2).

On the contrary, toughness K_{IC} decreases and characteristic length l_{ch} increases (Figure 13b). This means that the damaged material is more prone to cracking.

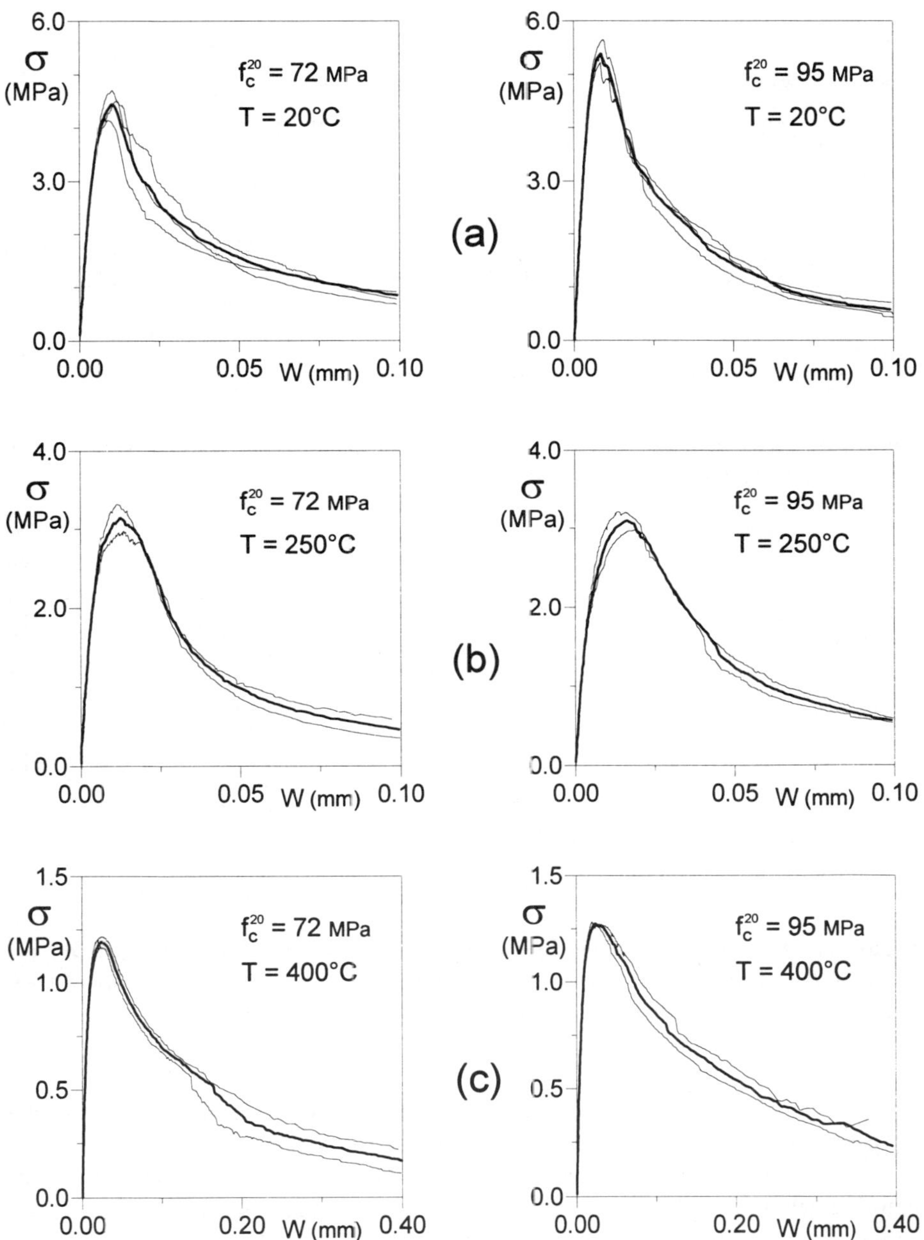

Figure 5. *Direct tension, notched cylinders: stress-CMOD curves at 20°C (a); 250°C (b); and 400°C (c). Thin line for individual curves; heavy line for mean curves*

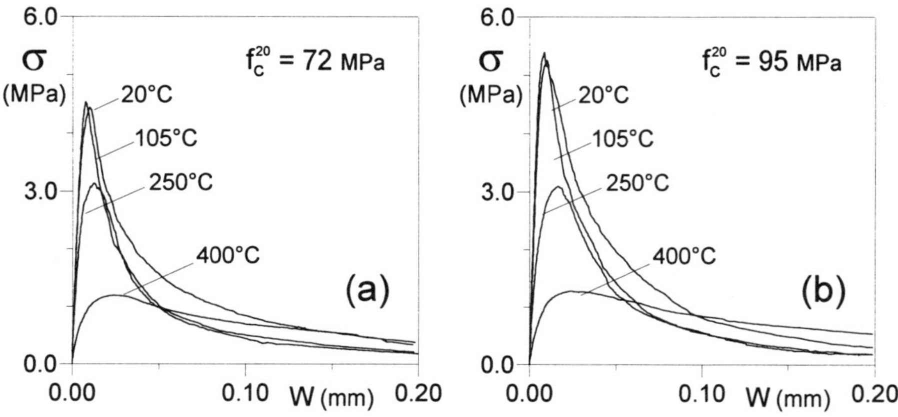

Figure 6. *Direct tension, notched cylinders: average stress/CMOD curves at room temperature and after a high-temperature cycle*

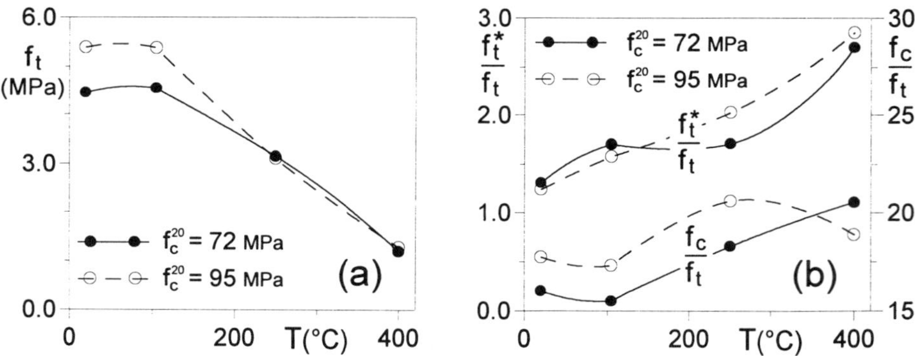

Figure 7. *Direct tension, notched cylinders: (a) plots of the residual strength; and (b) plots of two strength ratios (f_c/f_t = compressive-to-tensile strength ratio; f_t^*/f_t = indirect-to-direct strength ratio; indirect strength = 3-point bending)*

However, the structural ductility may be increased, because the dissipation involves a greater volume, as a result of the greater characteristic length.

The test data require a deeper analysis in order to work out the stress-strain law before cracking and the stress-opening law after cracking. As a matter of fact, the stress-displacement curves shown in Figures 5 and 6 are nothing more than the load divided by the net area of the notched section. Consequently, the stress is an "average stress" which disregards the stress concentration close to the notch. Nevertheless, the stress-CMOD curves are indicative of the softening effects of the thermal cycles, as well as of the astonishing effect that the temperature has on the tensile strength. Last but not least, the test results show a very limited scattering.

5. Tests in Bending (Indirect Tension)

Sixty prisms (Figure 8) were successfully tested in bending (indirect tension, 3-point bending tests) by controlling the elongation of the bottom fibers (mid-span section, unnotched specimens), cracking included, or the opening of the notch mouth (notched specimens).

For each temperature level (20, 105, 250, 400 and 500°C), two nominally-identical specimens were tested, and besides the unnotched specimens (20), other specimens with two different notch depths (c/a = 0.25, 0.50) were tested (40 specimens).

The following provides some details on technology, instrumentation and loading procedure:

– The specimens (415x100x100 mm, span 380 mm) were cast in a steel formwork and the notch was cut *before* the dehydration process, by means of a diamond disk.

– The specimens could not have the dimensions suggested by RILEM (1190x200x100 mm span 1130 mm, for max. aggregate size between 16 and 32 mm) because of the limited dimensions of the furnace chamber (500x310x660 mm, the largest general-purpose off-the-shelf furnace in 1993-94, with T_{max} = 1000-1100°C). On the other hand RILEM-suggested dimensions refer to specimens to be tested in displacement-controlled processes with relatively deformable machine frames. This is not the case of the Instron press used in this research project, which is very stiff. Furthermore, the control adopted here (clip-gage at the intrados of the specimen) makes the test very stable up to the complete loss of strength.

– A specially-built loading bridge was designed and prepared (Figure 8a). It consisted of (a) a stiff horizontal beam (HE200B) fixed to the lower movable platen of the press; (b) one roller-type support and one double-hinge support along the intrados of the specimen; (c) one cylindrical loading device coupled to a spherical seat, along the extrados of the specimen; (d) two systems of levers and counterweights to eliminate the effects of the self-weight of the specimen, both in the uncracked and in the cracked phases.

– A specially-built swan-neck light-alloy double frame was designed and adopted (Figure 8a) in order to measure the mid-span deflection at the intrados, with no spurious effects due to the elastic and inelastic settlements underneath the supports. The double frame was provided with a fixed upright which was kept in contact with the intrados by two sets of springs placed between the extrados of the specimen and the cross-ties of the frame at the extremities of the specimen (Figure 8c).

– Two pairs of LVDTs were attached to the cross-ties of the frame, in contact with the extrados of the specimen, in order to measure the mid-span deflection (Figure 8c).

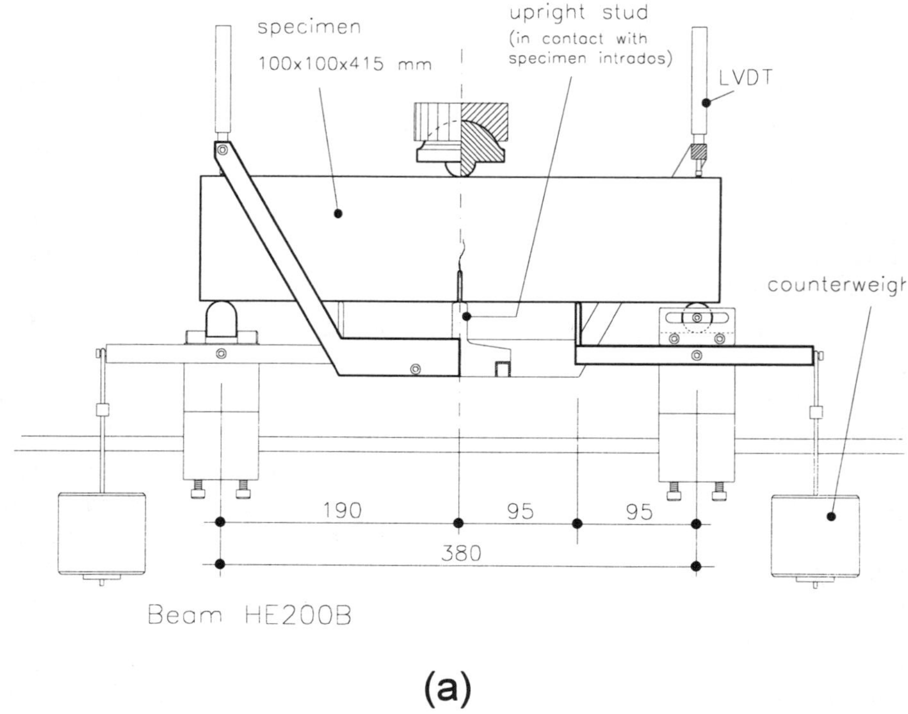

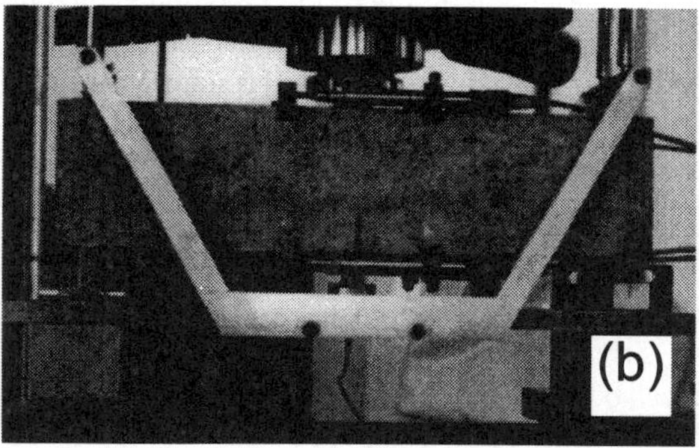

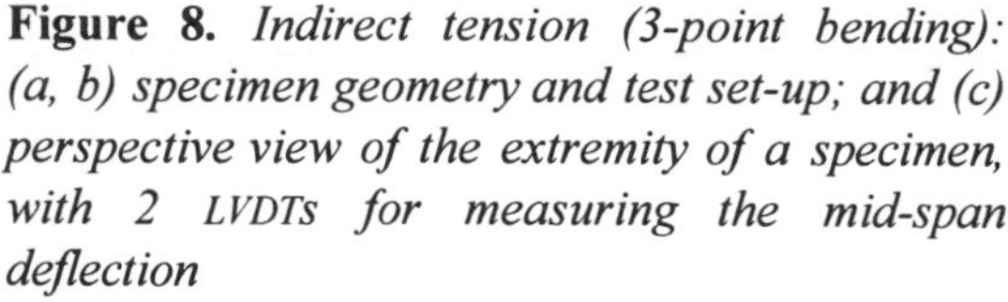

Figure 8. *Indirect tension (3-point bending): (a, b) specimen geometry and test set-up; and (c) perspective view of the extremity of a specimen, with 2 LVDTs for measuring the mid-span deflection*

– One clip-gage, placed astride the mid-span section, kept the press under control, and was automatically disengaged in the case of a sudden collapse of the specimen due to blunt crack propagation. The baselength of the clip-gage was 15 mm in the notched specimens and 100 mm in the unnotched specimens.

– 4 LVDTs (baselength 100 mm) were placed astride the mid-span section, along the 4 edges of the specimen, in order to measure the relative rotation between two undisturbed sections close to the notched or cracked zone (these data are instrumental in refining the numerical analysis of strain localization, Figure 8b).

– The tests were carried out at a constant displacement rate (0.2 μm/s in the elastic stage; 0.5 μm/s close to the load-peak and beyond, in the softening-stage).

The results are shown in Figure 9 (load-displacement curves of all specimens), in Figure 10 (average load-displacement curves for each notch depth) and in Figure 11 (peak stress – or residual strength after the thermal cycle – as a function of the temperature).

As might have been expected, the scattering of the test results is more pronounced in unnotched specimens, where the regularizing effects of the notch are missing.

At the highest temperatures, the curves for different notch depths tend to mix. A possible explanation is to be found in the notch-like effects that are produced by aggregate and mortar cracking during the thermal cycle. These effects tend to become more important than the effects produced by the geometrical notch.

As in the tests in direct tension, the peak stress (= tensile strength) increases slightly between 20 and 105°C (Figure 11), while it falls rapidly between 105 and 500°C: at 500°C. The role of the notch is very limited, since other "notches" appear, such as the cracks due to the thermal process.

Comparing the tensile strengths in direct and indirect tension (Figure 7b), it appears that their ratio f_t^*/f_t increases with the temperature of the thermal cycle. Since f_t^* is related to concrete non-linear behavior (softening included), the tendency of (f_t^*/f_t) to grow, beyond 250°C, is indicative of the greater deformability of the concrete, even if the peak strength decreases dramatically. The decay of the tensile strength is shown also by the plot of the ratio f_c/f_t (= compressive strength/tensile strength, Figure 7b), which increases with the temperature of the thermal cycle (from 16 at 20°C to more than 20 after a cycle at 400°C). Summing up, the tensile strength tends to decrease more than the compressive strength after a cycle at high temperature, and this fact should be taken into account in R/C structures, when considering the resistant mechanisms based on the tensile strength, such as bond, bar tension-stiffening, web shear and punching shear.

The average values of the fracture energy (notched cylinders in direct tension and notched/unnotched prisms in 3-point bending) are reported in Figure 12. As already observed, there is no clear dependency, neither on the temperature nor on concrete grade. The definition of the fracture energy G_f is shown in Figure 13a, while the characteristic length ℓ_{ch} and the toughness K_{IC} are plotted in Figure 13b as mentioned at the end of Section 4.

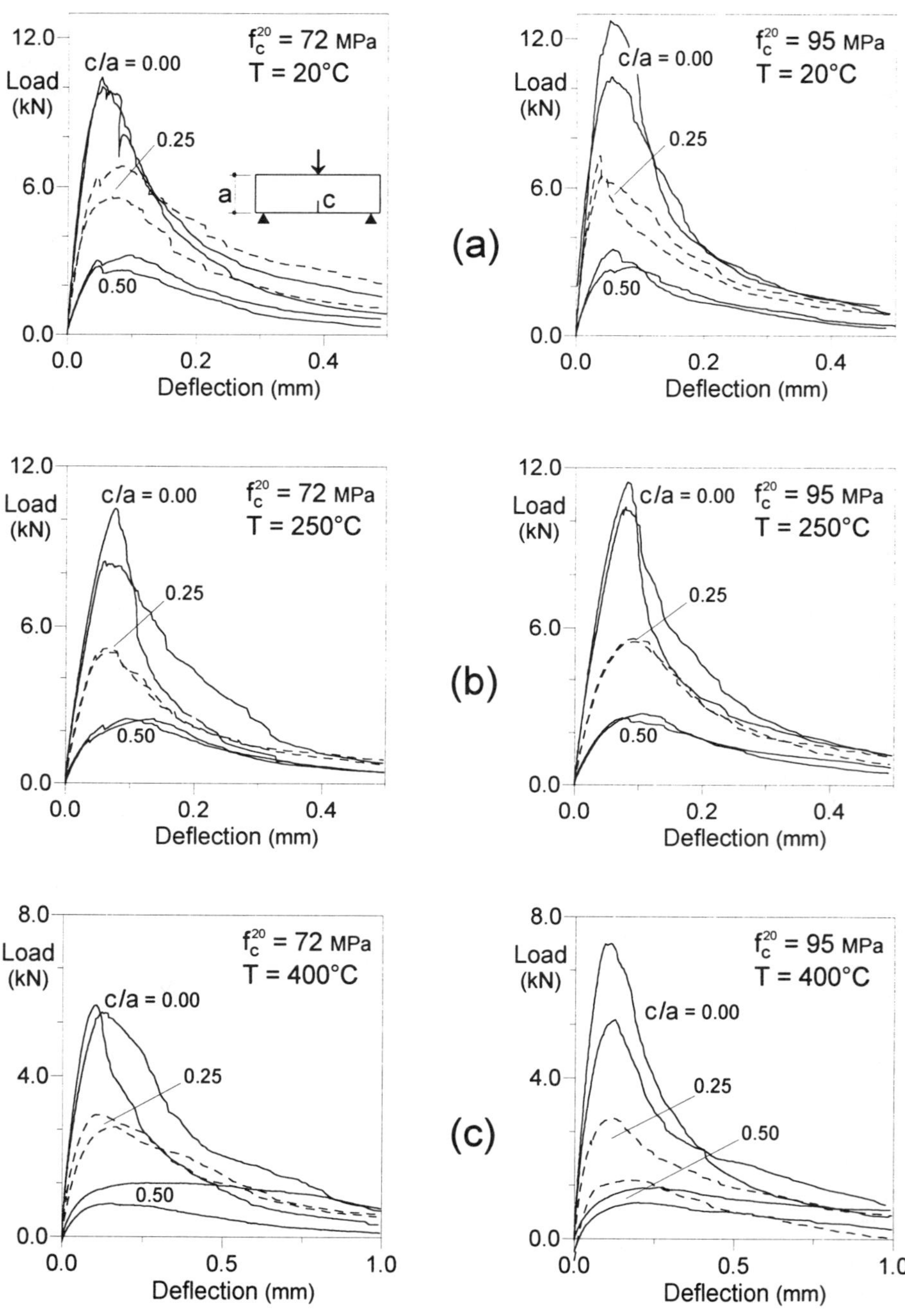

Figure 9. *Indirect tension, notched and unnotched prisms: load-deflection curves at 20°C (a); 250°C (b); and 400°C (c); in each case 2 specimens were tested for repeatability*

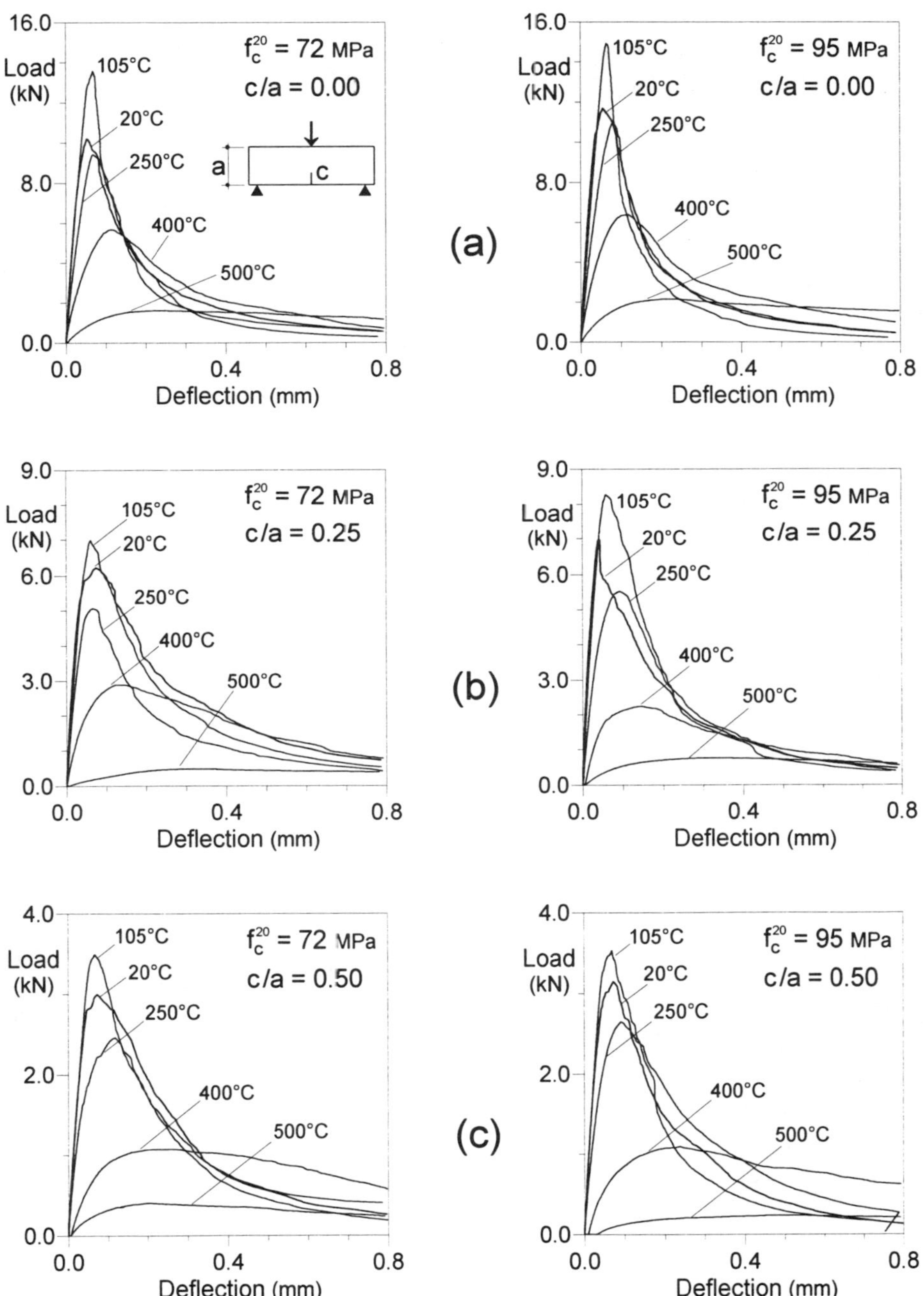

Figure 10. *Indirect tension, notched and unnotched prisms: load-displacement curves (mean curves) for various notch depths, c/a = 0.0 (a); 0.25 (b); and 0.50 (c)*

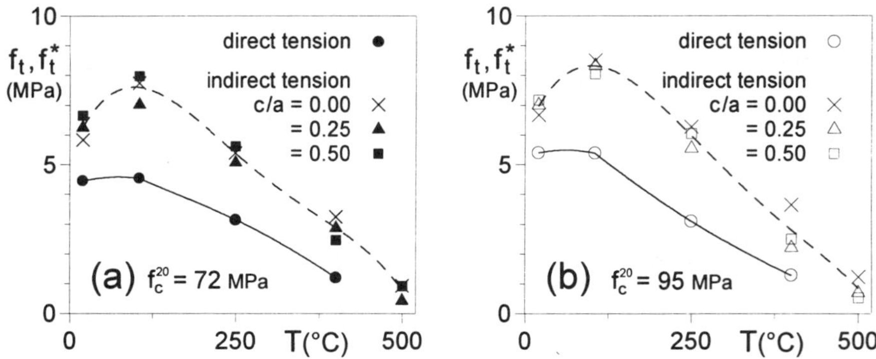

Figure 11. *Direct and indirect tension: plots of the nominal strength in tension*

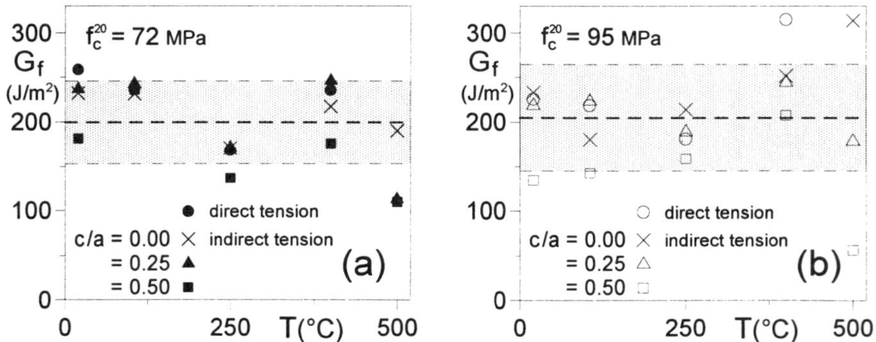

Figure 12. *Direct and indirect tension: plots of the specific fracture energy G_f, as a function of the temperature: (a) G_f^{AV} = 200±45 J/m²; and (b) G_f^{AV} = 205±60 J/m²*

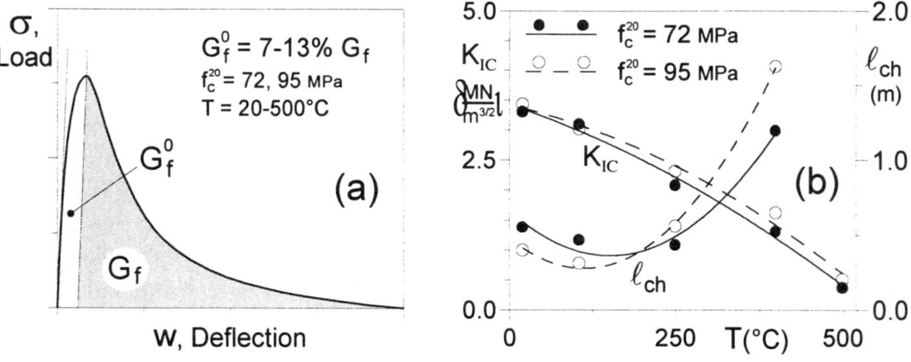

Figure 13. *(a) Energy dissipated in direct-tension tests and in 3-point bending tests; and (b) plots of the toughness K_{Ic} and characteristic length l_{ch}, as a function of the temperature;* $K_{IC} = \sqrt{G_f \cdot E_C}$; $\ell_{ch} = G_f \cdot E_C / f_t^2$

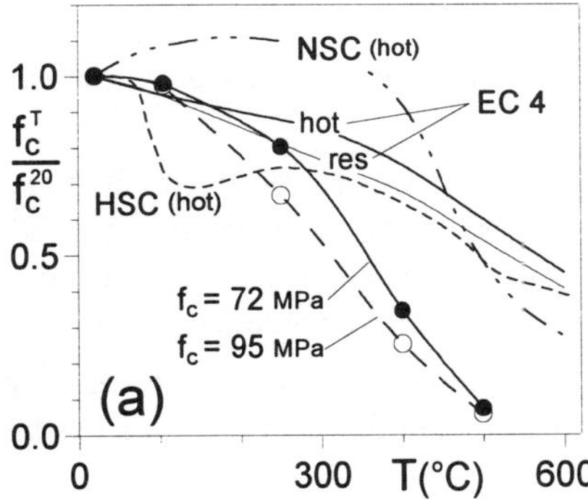

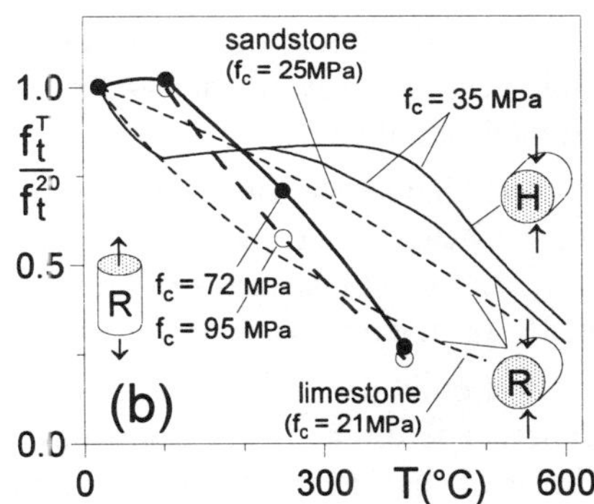

Figure 14. *Test results at high temperature (hot, H) and after cooling down to room temperature (res, R) in compression (a) [DIE 89]; and in direct/indirect tension (b), see Thelandersson and Harada [RIL 85]*

6. Concluding Remarks

The systematic series of tests in compression and tension, carried out on two mostly quartzitic concretes (f_c = 72 and 95 MPa at 20°C), after being exposed to high temperature (105-500°C), leads to the following remarks:

– The mostly quartzitic aggregates, the high-grade cement and the silica fume used in the two mixes give the concretes good mechanical properties at room temperature, even with relatively-high water/binder ratios (= 0.42, f_c = 72 MPa).

– Both concretes are very brittle when tested at room temperature, with no previous exposure to high temperature: the fracturing process regards mostly the aggregates, which appear neatly split at the end of the test. In compression, if the tests are force-controlled, the specimens blow up and project concrete chips in all directions.

– Both concretes are very sensitive to high temperature. Exposure to temperatures higher than 250°C weakens the concretes, in terms of both strength and stiffness, but makes the material softer. However, after a cycle at 400°C the residual strength in compression is only 35% to 25% of the virgin-concrete strength, and after a cycle at 250°C 1/5 to 1/3 of strength is gone (f_c = 72 and 95 MPa). In tension, the situation is even worse, since after a cycle at 400°C (250°C) the residual strength is 0.27 and 0.24 (0.70 and 0.60) times the tensile strength at 20°C (f_c = 72 and 95 MPa respectively). A comparison with EC4 provisions and with two typical diagrams quoted in the literature (NSC and HSC) is shown in Figure 14a. As for the Young's modulus, it is also adversely affected by high temperature: after a cycle at 250/400°C, the residual value is equal to 70/20% with respect to the virgin-concrete (f_c = 72 and 95 MPa).

– The strength in direct tension is more sensitive to temperature than the compressive strength; as a result, the strength ratio f_c/f_t increases from 16-18 (virgin specimens) to 20 and more after a cycle at 400°C.

– In direct tension, the hardly noticeable increase in strength up to 105°C is followed by a roughly linear loss with the temperature, with hardly any residual strength left at 500°C.

– In indirect tension, the three-point bending tests confirm the improvement of the tensile behavior up to 105°C (+33%, unnotched specimens), and the pronounced decrease beyond this temperature: after a cycle at 500°C the residual capacity is from 7 to 15% of the original capacity (virgin material, notched and unnotched specimens).

– Both the direct-tension and 3-point bending tests show a remarkable increase in concrete deformability, as shown by the extended and smooth softening branches of the stress-displacement and load-displacement curves.

– As might have been expected, the behavior in bending is less affected by the thermal cycles: after a cycle at 250°C (400°C) the indirect tensile strength (which is a linear function of the peak load) is still equal to 93-84% (55-34%) of the original strength, with the largest values for the unnotched specimens (f_c = 72-95 MPa).

– For both concretes, toughness decreases and characteristic length increases after a cycle at high temperature, while fracture energy is little affected by the temperature. Such behavior may even lead to more ductile structures, owing to the greater damaged volume, which may offset the weakening of the mechanical links within the material.

– The aggregates seem to be the main culprit for the disappointing behavior at high temperatures: at 500°C most of the quartzitic pebbles appear cracked or split, with a different color compared to the virgin material. This behavior may be due mostly to the expulsion of water from the aggregate (adsorbed and zeolitically-bound water). As a matter of fact, the water content of opal and calcedonium (which make up a sizable part of the flint pebbles) may be up to 20% of the aggregate weight.

Unfortunately these results cannot be compared with previous results regarding the tensile behavior of high-strength concrete subjected to high temperatures, since – to authors' knowledge – no tests of this type have ever been carried out before. However, a comparison is possible with the indirect-tension tests carried out on NSC, as shown in the last figure (Figure 14b, splitting tests).

Acknowledgement

The authors wish to thank the Italian Agency for Energy, Technological Innovation and Environment - ENEA (Project 1794/20-3-95), and the Italian National Council for Research - CNR (Project "Special Materials for Better Structures", 1994-97) for providing the financial support for this investigation. The understanding and cooperation of MS Engineers Franco Corsi and Giuseppe Giannuzzi of ENEA are gratefully acknowledged.

The authors are very honored to dedicate this paper to Professor Zdenek P. Bazant, whose studies and innovative thinking have given – and are still giving – so many stimuli to theoretical, numerical and experimental research in the many-faceted domain of concrete nonlinear and fracturing behavior.

References

[BAZ 79] BAZANT Z.P., THONGUTHAI W., "Pore pressure in heated concrete walls: theoretical prediction", *Magazine of Concrete Research*, V. 31, No.107, pp. 67-76, 1979.

[BAZ 81] BAZANT Z.P., CHERN J.C., THONGUTHAI W., "Finite Element Program for Moisture and Heat Transfer in Heated Concrete", *Nuclear Engineering and Design*, V. 68, pp. 61-70, 1981.

[BAZ 90] BAZANT Z.P., KAZEMI M.T., "Determination of fracture energy, process-zone length and brittleness number from size effect, with application to rock and concrete", *Int. Journal of Fracture*, V. 44, pp. 111-131, 1990.

[BAZ 96] BAZANT, Z.P., KAPLAN, M.F., *Concrete at High Temperature: Material Properties and Mathematical Models*, Concrete Design and Construction Series, ed. by F.K. Kong and R.H. Evans, Longman Group Limited, Essex (UK), 412 p., 1996.

[CAR 94] CARPINTERI A., FERRO G., "Size effect on tensile fracture properties: a unified explanation based on disorder and fractality of concrete microstructure", *Materials and Structures*, V.27, pp. 563-571, 1994.

[DIE 88] DIEDERICHS U., JUMPPANEN U.M., PENTTALA V., "Material properties of high-strength concrete at elevated temperatures", IABSE 13th Congress, Helsinki (Finland), pp. 489-494, 1988.

[DIE 89] DIEDERICHS U., JUMPPANEN U.M., PENTTALA V., "Behaviour of high-strength concrete at high temperatures", Report 92, Helsinki University of Technology, Helsinki (Finland), 76 p., 1989.

[DIP 97] DI PRISCO M., FELICETTI R., GAMBAROVA P.G., "On the evaluation of the characteristic length in high-strength-concrete", Proc. Int. Conf. on High Strength Concrete, ASCE - Engrg. Foundation, Kona (Hawaii, USA), 1997 (in press).

[FEL 95] FELICETTI R., GAMBAROVA P.G., Volpe M., "On the thermo-mechanical behavior of siliceous high-strength concretes exposed to high temperature" (in Italian), Studi e Ricerche, V.16, School for the Design of R/C Structures, Milan University of Technology, Milan (Italy), pp. 59-92, 1995.

[FEL 96] FELICETTI R., GAMBAROVA P.G., ROSATI G., CORSI F., GIANNUZZI G., "Residual mechanical properties of high-strength concretes subjected to high-temperature cycles", *Trans. 4th Int. Symposium on Utilization of High-Strength/High-Performance Concrete*, Paris (France), pp. 579-588, 1996.

[FEL 98] FELICETTI R., GAMBAROVA P.G., "The effects of high temperature on the residual compressive strength of high-strength siliceous concretes", *ACI - Materials Journal* (in press), 1998.

[GIU 87] GIURIANI E., ROSATI G.P., "On the behavior of concrete elements subjected to tension after cracking" (in Italian), Studi e Ricerche, School for the Design of R/C Structures, Milan University of Technology, V.8-86, pp. 65-82, 1987.

[HOR 91] HORDIJK D.A., "Local approach to fatigue of concrete", Doctoral Thesis, Delft University of Technology, Delft (The Netherlands), 210 p., 1991.

[MAR 78] MARCHERTAS A.H., FISTEDIS S.H., BAZANT Z.P., BELYTSCHKO T.B., "Analysis and application of prestressed concrete reactor vessels for Liquid Metal Fast Breeder Reactor Containment", *Nuclear Engineering and Design*, V. 49, pp. 155-173, 1978.

[NIS 97] NIST - Nat. Inst. of Standards and Technology, Proc. Int. Workshop on Fire Performance of High-Strength Concrete, S.P. 919, ed. by L.T. Phan, N.J. Carino, D. Duthinh and E. Garboczi, Gaithersburg (Md, USA), 164 pp., 1997

[PAP 91] PAPAYIANNI J., VALIASIS T., "Residual mechanical properties of heated concrete incorporating different pozzolanic materials", *Materials and Structures*, RILEM, V. 24, pp. 115-121, 1991.

[PHA 97] PHAN L.T., CARINO N.J., "Review of mechanical properties of HSC at elevated temperature", *ASCE, Journal of Materials in Civil Engineering*, V.10, No 1, pp. 58-64, 1997.

[RIL 85] RILEM-COMMITTEE 44-PHT, Behaviour of Concrete at High Temperatures. Ed. by U. Schneider, Dept. of Civil Engineering, Gesamthochschule, Kassel Universität, Kassel (Germany), 122 p., 1985.

[ROS 91] ROSATI G.P., SCHUMM C.E., "An identification procedure of fracture energy in concrete: mathematical modeling and experimental verification", *Proc. Of Int. Conf. on Fracture Processes in Brittle Disordered Materials: Concrete, Rock and Ceramics*, ed. by J.G.M.van Mier, J.G. Rots and A. Bakker, publ. by E and FN Spon, Noorwijk (The Netherlands), pp. 533-542, 1991.

[ROS 94] ROSATI G.P., SCHUMM C.E., FERRARA G., "Evaluation of an objective stress-COD relationship for cracked concrete under tension", *Proc. Of the Europe-U.S. Workshop on Fracture and Damage in Quasi-Brittle Structures: Experiment, Modeling and Computer Analysis,* ed. by Z.P. Bazant, Z. Bittnar, M. Jirasek and J. Mazars, Prague (Czech Republic), F and FN Spon, pp. 183-190, 1994.

[TAC 93] TACHEUCHI M., HIRAMOTO M., KUMAGAI N., "Material properties of concrete and steel bars at elevated temperatures", Trans. of 12th SMiRT Int.Conf., Vol. H, Stuttgart (Germany), 1993.

[THI 93] THIENEL K.C., ROSTASY F.S., "Behaviour of biaxially restrained concretes under high temperature", Trans. of 12th SMiRT Int. Conf., Vol. H, Stuttgart (Germany), 1993.

Numerical Evaluation of the Mechanical Contribution of Pore Pressure in Spalling of Concrete at Elevated Temperatures

Gregory Heinfling — Jean-Marie Reynouard

URGC, Structures, INSA de Lyon
Bât. 304
F-69621 Villeurbanne cedex

ABSTRACT. *A computational method is proposed to account for the mechanical contribution of pore pressure in the analysis of the behaviour of concrete structures submitted to thermo-mechanical loadings at high temperatures. Considering concrete as an homogenous isotropic material, the temperature and pore pressure fields are derived from a coupled heat and moisture transport calculation. A thermo-plasticity based model is used for the analysis of the behaviour of the skeleton through the effective stress. The evolution of the hydro-mechanical scalar coupling coefficient with temperature is identified with help of mercury porosimetry tests performed at different elevated temperatures. Experiments performed on an axisymmetrical high performance concrete specimen are simulated. The effect of pore pressures on the possible spalling failure mechanism is emphasised.*

KEY WORDS: *Concrete, High Temperature, Numerical Modelling, Thermoplasticity, Cracking, Thermomechanical Interaction, Pore Pressure, Spalling.*

Introduction

In severe accidental situations such as fire or hypothetical core disruptive nuclear accident, reinforced concrete structures can be submitted to extremely high transient temperatures. High temperatures induce strong chemical and physical changes of the micro-structure of concrete which affects its mechanical behaviour. The major phenomena which have been identified include the release and evaporation of significant amount of water that induce pressure gradients under which water is transported towards the surface through pores.

Experimental and theoretical analyses tend to consider that pore pressure built up could contribute to the explosive spalling of high performance concrete structures at elevated temperatures [HAR 65], [MEY 75], [ZHU 77], [SER 77] [BAZ 81], [NOU 96], [AND 97]. These spalling phenomena are a major problem in the evaluation of the safety of concrete structures under such conditions.

The issues of the numerical analysis of moisture migration and prediction of pore pressures under different conditions have been investigated by several authors [ENG 73], [BAZ 81], [DAY 82], [SCH 87], [KON 95], [JOU 97]. In order to analyse spalling phenomena, most of the previously mentioned works have focused on the possible local mechanisms that could lead to spalling and are based on qualitative fracture mechanics considerations. Very few studies deal with the numerical modelling of the mechanical effects of pore pressures on the behaviour of concrete structures at elevated temperatures. Within the framework of finite elements, a numerical method has been proposed by [MAJ 95]. However, in this study, the concrete was assumed to behave elastically and pore pressure was considered as a direct internal stress, regardless of the actual nature of concrete porous micro-structure.

Following the same general approach, we present here the first step of a numerical study which aims at providing a general numerical tool allowing for the non-linear analysis of the possible mechanical contribution of pore pressure in the spalling of concrete at high temperatures. This paper focuses on an extension of a thermo-plasticity based model for concrete at elevated temperatures to account for pore pressures within the framework of the mechanics of porous media.

1. Constitutive Equations

In the proposed approach, the concrete is considered as a porous medium which is a superposition of two interacting continuum media: the solid skeleton and a fluid phase inside the connected pores. This material is considered homogeneous and isotropic, and it is assumed that all the physical properties can be expressed as effective properties. An important assumption is that the transfer properties of concrete are considered to be only dependent on temperature, moisture and pore pressure. These are of course rather crude assumptions since damage can strongly affect the permeability and the porosity of concrete and can induce anisotropy of these properties [GER 96]. The problem is then solved in two sequences. Temperature and pore pressures are first derived from a heat and moisture transfer analysis and are then used as input for the stress analysis.

1.1. *Heat and Moisture Transport Calculations*

For the applications presented further in this study, and within the framework of the assumptions mentioned above, the temperature and pore pressure fields have been evaluated by [NOU 96] in a previous study. Heat and moisture transports in concrete at elevated temperatures are highly coupled mechanisms. The mathematical model used by these authors for the coupled heat and moisture transport calculation of the tested specimens is based on the model of [BAZ 81].

This approach uses the complementary assumptions that the Dufour's flux is negligible and that normal Darcy's law applies. The set of governing equations for coupled heat and moisture transport are given as follows.

The conservation of mass is represented by:

$$\frac{\partial w}{\partial t} = -\operatorname{div} J + \frac{\partial w_d}{\partial t} \qquad [1]$$

where t is the time, w is the free water content and w_d is the total mass of free water that has been released into the pores by dehydration of the cement matrix. The mass flux of moisture J is given by Darcy's law:

$$J = -a \operatorname{grad} p \qquad [2]$$

where a is the water permeability of concrete, and p is the pore water pressure. The conservation of energy is represented by:

$$\rho C \frac{\partial T}{\partial t} = C_a \frac{\partial w}{\partial t} + C_w J \operatorname{grad} T - \operatorname{div} q \qquad [3]$$

where ρ is the unit mass of concrete, and C, C_a, C_w respectively are the isobaric heat capacities of concrete, adsorbed water and free capillary water. Finally, the heat transfer rate q is given by :

$$q = -k \operatorname{grad} T \qquad [4]$$

where q is the heat flux and k is the heat conductivity of concrete.

The introduced material properties are dependent on pore pressure and temperature. These governing equations are complemented by semi-empirical sorption isotherms, relating the free water content w, pressure p and temperature T. The finite element scheme is based on theGalerkin method and a step by step solution with iterations is used for time integration of the nonlinear set of variational equations.

1.2. *Definition of the Effective Stress*

Within the framework of the mechanics of porous continua [COU 95], and considering a thermo-poro-plastic behaviour for concrete, the effective stress vector σ' responsible for the deformation of the skeleton can be expressed in the following general form as a function of the total stress vector σ and pore pressure p:

$$\sigma' = \sigma + B_0\left[p + \varphi\left(\varepsilon, \varepsilon^P, p, T\right)\right] \qquad [5]$$

where B_0 is the initial Biot's tensor of the hydraulics to mechanics coupling. φ is a function describing its evolution with the deformation of the skeleton, pore pressure and temperature. ε and ε^p respectively are the total strain vector and the plastic strain vector of the skeleton. Considering the assumption of isotropy of the material and neglecting as a first approximation the effects of damage of the skeleton on the Biot's tensor, the effective stress can be expressed in the following simple form:

$$\sigma' = \sigma + b_0\left[1 + \varphi(p, T)\right]pi \qquad [6]$$

where i is the identity vector and b_0 is the initial Biot's coefficient. For general porous materials, this coefficient is bounded below by the value of the porosity of the material Φ, and above by 1:

$$\Phi \leq b \leq 1 \qquad [7]$$

For concrete, the Biot's coefficient can be assumed to be given by the porosity Φ which has been shown to be highly temperature dependent [NOU 96]. The effective stress vector is then finally expressed by:

$$\sigma' = \sigma + \Phi_0\left[1 + \psi(p, T)\right]pi \qquad [8]$$

where Φ_0 is the initial porosity of concrete and ψ is a function representing its evolution with pore pressure and temperature. Empirical laws implemented in TEMPOR 2 provides the evolution of porosity with pore pressure. The evolution of this parameter with temperature can be obtained from mercury porosimetry tests performed after cooling of concrete specimens heated at different elevated temperatures [NOU 96].

1.3. *Multi-Surface Thermo-Plasticity Based Model for the Behaviour of the Skeleton*

The characteristic points of the thermo-plastic behaviour identified for concrete [HEI 97] are supposed to hold true for the skeleton despite the fact that pore pressure can contribute to some of the experimental observations. The model proposed and implemented in the finite element code CASTEM2000 by these authors for concrete at elevated temperatures, is then used here for the analysis of the behaviour of the solid skeleton through the effective stress.

1.3.1. *Incremental Formulation*

In this model, assuming small strains, the total strain rate of concrete $\dot{\varepsilon}$ is decomposed into the sum of an elastic strain rate $\dot{\varepsilon}^{e}$, a plastic strain rate $\dot{\varepsilon}^{p}$, a thermal expansion strain rate $\dot{\varepsilon}^{\theta}$ and a thermo-mechanical interaction strain rate $\dot{\varepsilon}^{tr}$:

$$\dot{\varepsilon}=\dot{\varepsilon}^{e}+\dot{\varepsilon}^{p}+\dot{\varepsilon}^{\theta}+\dot{\varepsilon}^{tr} \qquad [9]$$

Elastic as well as inelastic properties are temperature dependent. Their variations with temperature are irreversible. The elastic strain rate determines the stress rate through the temperature dependent elastic stiffness ratio matrix D(T) which is assumed to be isotropic throughout this paper:

$$\dot{\sigma}=D\dot{\varepsilon}^{e}+\dot{D}\varepsilon^{e} \qquad [10]$$

The temperature dependency of the mechanical properties correspond to a phenomenological description of the micro-structural and chemical changes that take place in heated concrete. In particular, drying shrinkage is implicitly taken into account through the variations of the coefficient of thermal expansion of concrete $\alpha(T)$. The thermal strain rate is given as follows, where i is the identity vector:

$$\dot{\varepsilon}^{\theta}=\alpha\dot{T}i \qquad [11]$$

Investigation tests on plain concrete have shown that the thermal deformation of concrete is strongly dependent on the stress applied during heating up [AND 76]. It is not entirely clear which mechanisms cause this phenomenon usually called transient creep, but it seems that between 100°C and 250°C, drying of concrete is an important factor. Froml a phenomenological point of view, in order to describe the response of concrete under combined thermal and mechanical action, it is then necessary to abandon the usual assumption that thermal strain and mechanical strain can be treated as mutually independent components. Thermo-mechanical interaction strains have then to be taken into account.

In one dimensional context, a simple formula has been proposed by [AND 76]. This formula has been generalised to a multiaxial state of stress by [BOR 87] and has been successfully incorporated by [KHE 92] in a thermo-plasticity model. According to these authors, the thermo-mechanical interaction strain rate can be written as :

$$\dot{\varepsilon}^{tr}=H\sigma'\dot{T} \qquad [12]$$

where σ' is the effective stress vector , $\dot{T}$ is the rate of heating, and H, for an isotropic material is given as:

$$H_{ijkl} = \frac{\alpha k}{f'_c}\left[-\gamma\delta_{ij}\delta_{kl} + \frac{1}{2}(1+\gamma)\left(\delta_{ik}\delta_{jl} + \delta_{il}\delta_{jk}\right)\right] \qquad [13]$$

whith δ_{ij}, the Kroenecker symbol and $f'_c(T)$, the temperature dependent uniaxial compression strength of concrete. k and γ are coefficients of thermo-mechanical interaction that can be evaluated from transient creep tests. For usual concrete, k varies from 1.8 and 2.35 and γ has been found to be equal to 0.285.

The temperature driven phenomenological descriptions of the strains and changes of physical properties, induced by the hygral modifications of the cement matrix, determine the limits of this kind of model. Heating rate sensitivity of the modifications of the mechanical properties have, for example, to be prescribed explicitly by the variation laws introduced in data. This is not straightforward for non-uniformly heated complex structures. However, simulations of tests performed on plain concrete specimens as well as reinforced concrete structures give satisfying results as soon as the parameters and their variations are identified under thermal and hygral conditions corresponding as closely as possible to the structure analysed [HEI 97].

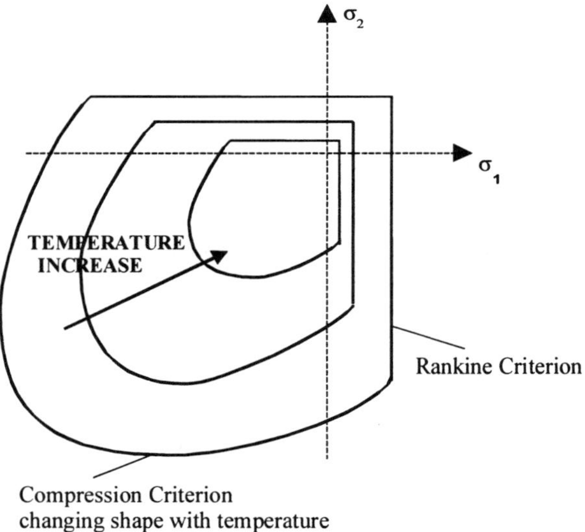

Figure 1. *Failure criterion in two dimensional space with varying temperature*

As shown in Figure 1, a temperature dependent non smooth multi-surface criterion is used to describe the non symmetrical failure envelope of concrete. The maximum tensile stress criterion of Rankine is used to bound the tensile strength:

$$\sigma_d \leq f_t'(T) \qquad [14]$$

where σ_d is the principal tensile stress in the principal direction d and $f_t'(T)$ is the temperature dependent uniaxial tensile strength of concrete.

Biaxial compression tests performed at elevated temperatures show a significant increase in the sensitivity of compression strength to hydrostatic pressure with temperature [EHM 85], [KOR 85]. This makes the ultimate strength envelopes changing shape with temperature. In order to account for this effect, a Drucker-Prager type criterion changing shape with temperature is used to describe the failure surfaces of concrete under compression. This criterion is given:

$$\sqrt{3J_2} + \eta I_1 - \mu f_c'(T) \leq 0 \qquad [15]$$

where I_1 and J_2 respectively are the first invariant of the stress tensor and the second invariant of the stress deviator and:

$$\eta = \frac{\beta - 1}{2\beta - 1} \quad \text{and} \quad \mu = \frac{\beta}{2\beta - 1} \qquad [16]$$

where $\beta(T)$ is the ratio of the biaxial compression strength to the uniaxial compression strength.

The variations of β with temperature can be obtained from biaxial compression tests performed at different elevated temperatures. An example of variation law is presented in figure 2 for the case of the concrete tested by [KOR 85].

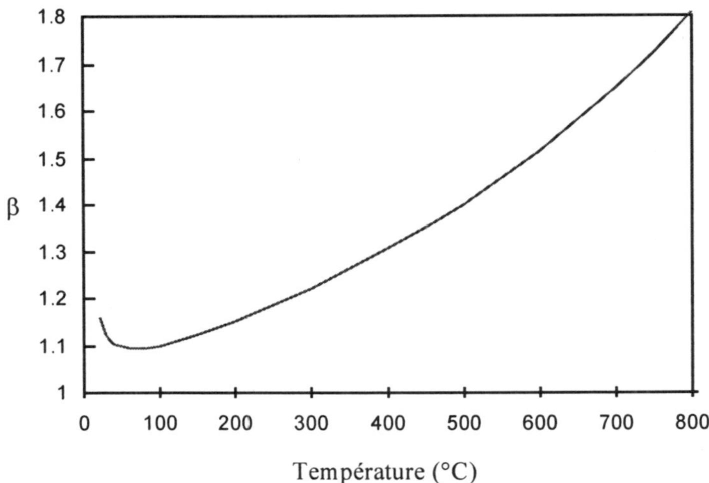

Figure 2. *Variations of* β *with temperature*

The experimental strength envelopes obtained by [KOR 85] are shown in Figure 3 together with the model curves. The change of shape of the failure surfaces is captured and the agreement between the predicted and the experimental strength envelopes is acceptable.

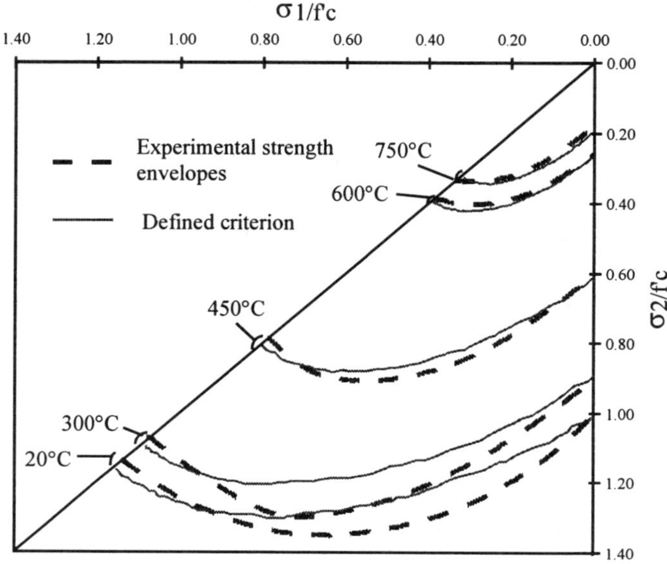

Figure 3. *Biaxial compression strength envelopes obtained by[KOR 85] together with the proposed criterion*

Within the framework of multi-surface thermo-plasticity theory, the existence of yield surfaces f_i, function of stress σ, hardening parameter κ_i, and temperature T is assumed:

$$f_i(\sigma, \kappa_i, T) = 0 \quad [17]$$

Mode I cracking is described within this thermo-plasticity framework through tensile plastic flow. The isotropic Rankine flow theory proposed by [FEE 95] is used. This approach corresponds to an isotropic smeared rotating description of cracking. The yield surfaces corresponding to the Rankine criterion are given as:

$$f_1(\sigma', \kappa_1, T) = \sigma'_1 - \tau_1(\kappa_1, T) \quad [18]$$

where σ'_1 is the major principal tensile stress and $\tau_1(\kappa_1, T)$ is an equivalent stress which is given by a softening function of the internal parameter κ_1. This softening function is identified by the tensile strength f'_t and the temperature dependent tensile fracture energy $G_f = G_f(T)$ of concrete.

The yield surfaces corresponding to the compression criterion are given as:

$$f_2(\sigma', \kappa_2, T) = \frac{1}{\mu(T)}\left(\sqrt{3J_2} + \eta(T)I_1\right) - \tau_2(\kappa_2, T) \qquad [19]$$

where $\tau_2(\kappa_2, T)$ is an equivalent stress which is given by a hardening/softening function of the internal parameter κ_2. This hardening/softening function is identified by the compressive strength $f'_{c(T)}$ and the temperature dependent compressive fracture energy $G_c(T)$ of concrete.

The plasticity conditions are imposed on the two surfaces during the plastic flow:

$$\begin{cases} f_1(\sigma, \kappa_1, T) = 0 \\ f_2(\sigma, \kappa_2, T) = 0 \end{cases} \qquad [20]$$

Isotropic hardening and associated plasticity are assumed for compression as well as tension plastic flow. The evolution of the plastic strain rate is given by the associated flow rule. The ambiguity of the plastic flow direction at the corner is removed according to [MAI 69], considering the contribution of each individual loading surface separately. In the general case, where two loading surfaces are activated, the plastic strain rate is then given as:

$$\dot{\varepsilon}^p = \dot{\lambda}_1 \frac{\partial f_1}{\partial \sigma} + \dot{\lambda}_2 \frac{\partial f_2}{\partial \sigma} \qquad [21]$$

where $\dot{\lambda}_i$ are plastic multipliers that have to comply with the Kuhn-Tucker conditions:

$$\dot{\lambda}_i \geq 0, \ f_i \leq 0, \ \dot{\lambda}_i f_i = 0 \qquad [22]$$

The Hillerborg method is employed in order to solve partially the pathological mesh dependency induced by the softening behaviour. An equivalent length related to the mesh size is then introduced in the definition of the $\tau(\kappa, T)$ laws.

It is assumed that the internal mechanical damage in the material as reflected in the internal parameter κ is governed by a work-hardening hypothesis. The internal variable is determined by the inelastic work rate $\dot{W}^p$ defined by:

$$\dot{W}^P = \sigma^T \dot{\varepsilon}^p = \tau(\kappa, T)\dot{\kappa} \qquad [23]$$

Using Euler's theorem, we obtain the evolutionary equation for the two criteria:

$$\dot{\kappa}_i = \dot{\lambda}_i \qquad [24]$$

1.3.2. *Thermo-Plastic Return Mapping Algorithm*

A typical trapezoidal Euler backward scheme is used for the integration of the thermo-plastic constitutive equations. The updated effective stress vector σ'_{n+1} is obtained by:

$$\sigma'_{n+1} = \sigma'_e - D_{n+1}\left\{\Delta\lambda_1 \frac{\partial f_1}{\partial \sigma'} + \Delta\lambda_2 \frac{\partial f_2}{\partial \sigma'}\right\}_{n+1} \qquad [25]$$

The subscript n+1 refers to the time step. The thermo-elastic predictor σ'_e is obtained by freezing inelastic flow during the time step :

$$\sigma'_e = D_{n+1}\left(\Delta\varepsilon_{n+1} - \Delta\varepsilon^{th}{}_{n+1} - \Delta\varepsilon^{tr}{}_{n+1}\right) + \Delta D \varepsilon^e{}_n + \sigma'_n \qquad [26]$$

ΔD captures the change of elastic properties with the temperature increase on the loading step.

Considering the evolutionary equation [22] and assuming uncoupling of tensile and compressive hardening mechanisms, the problem finally consists of the determination of the inelastic incremental multipliers which enforce the plasticity conditions at the temperature T_{n+1}. In the general case where two loading functions are active, this reads:

$$\begin{cases} f_1(\Delta\lambda_1, \Delta\lambda_2, T_{n+1}) = 0 \\ f_2(\Delta\lambda_1, \Delta\lambda_2, T_{n+1}) = 0 \end{cases} \qquad [27]$$

A local Newton-Raphson method is used to solve this set of nonlinear equations. The updated inelastic incremental multiplier is calculated with help of Broyden method:

$$\begin{Bmatrix} \Delta\lambda_1 \\ \Delta\lambda_2 \end{Bmatrix}^{(i+1)} = \begin{Bmatrix} \Delta\lambda_1 \\ \Delta\lambda_2 \end{Bmatrix}^{(i)} - J_i^{-1} \begin{Bmatrix} f_1 \\ f_2 \end{Bmatrix}^{(i)} \qquad [28]$$

The superscript (i) refers to the internal iterations during the solving process. The Jacobian J_i is comprised of the partial derivatives. In the case of a single criterion, the Broyden method corresponds to the secant method. Figure 4 shows the iterative return mapping process corresponding to this scheme for the general case where the two loading functions are active.

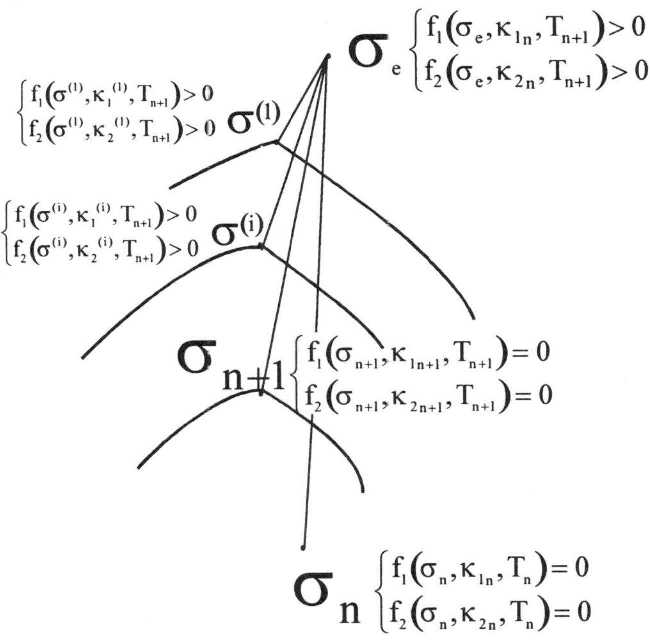

Figure 4. *Iterative return mapping process*

1.3.3. *Equilibrium Equations*

The equilibrium equations resulting from the finite element discretisation are given as:

$$K_{n+1}^{(i)} \Delta a_{n+1}^{(i+1)} = f_{ex\,n+1} + f_{p_{n+1}} - f_{in\,n+1}^{(i)} \qquad [29]$$

where $K_{n+1}^{(i)}$ is the tangent stiffness matrix and $\Delta a_{n+1}^{(i+1)}$ is the nodal displacement vector increment and:

$$f_{p_{n+1}} = \int_V B^T b_{n+1} p_{n+1} i dV \quad [30]$$

is the pore pressure induced load vector, where B is the strain-nodal displacement matrix of the elements employed, b is the Biot's coefficient and i is the identity vector,

$$f_{in\,n+1}^{(i)} = \int_V B^T \sigma'^{(i)}_{n+1} dV \quad [31]$$

is the internal load vector,

$$f_{ex\,n+1} = \int_V N^T F_{v\,n+1} dV + \int_S N^T F_{s\,n+1} dS \quad [32]$$

is the generalised external load vector where N is the interpolation matrix of the elements employed and F_V and F_S respectively are the body and traction forces vectors.

In the above expressions, the subscript n+1 refers to the time step and superscript (i) refers to the external iteration during the solving process.

This set of non linear equilibrium equations is solved using the Newton-Raphson method. An explicit expression is derived for a consistently linearised tangent stiffness matrix associated with the Euler backward scheme used for the integration of the constitutive equations.

2. Application to the Analysis of the Behaviour of High Strength Concrete Specimens

Tests performed by [NOU 96] on high strength concrete specimens have been simulated. The results presented here don't pretend to be of any definitive character. The main objectives of this application are to analyse the capability of the proposed model to provide a realistic prediction of the behaviour of concrete structures at elevated temperatures, and to quantify the consequences of the assumptions introduced in its formulation. The tests have been carried out on axisymmetrical (16cm x 32cm) specimens heated up at the controlled rate of 1°C/min. The mix proportions of the high strength concrete employed are detailed in [NOU 96]. It had a low cement content with additions of silica fume and fillers. Water reducer and retarding admixture were also added.

Axisymmetrical calculations of temperature and pore pressure have been carried out by [NOU 96] with help of TEMPOR2. Material parameters used in computations are given in table 1.

Age of concrete	30 days
Relative humidity of concrete	95 %
Initial temperature	25 °C
Saturation water content at 25°C	171 kg/m^3
Cement content	363 kg/m^3
Unit weight of concrete	2449 kg/m^3
Thermal conductivity	1.6 J/ms°C
Permeability	10^{-12} m/s
Water/cement ratio	0.44

Table 1. *Material parameters used in TEMPOR2 computations. After [NOU 96]*

The boundary conditions applied uniformly on all the faces of the specimens were Dirichlet's type for temperature and perfect moisture transfer. Zero thermal and moisture fluxes were imposed on the symmetry axis. The temperatures obtained were in agreement with the experimental measurements. The thermal gradient increased with the temperature rising, up to a maximum of about 325°C then decreased. It was very high near the heating surface (about 39°C/cm) and lower in the central part (about 2°C/cm). Figure 5 presents the temperature difference between the central part of the specimens and the surface as a function of the surface temperature.

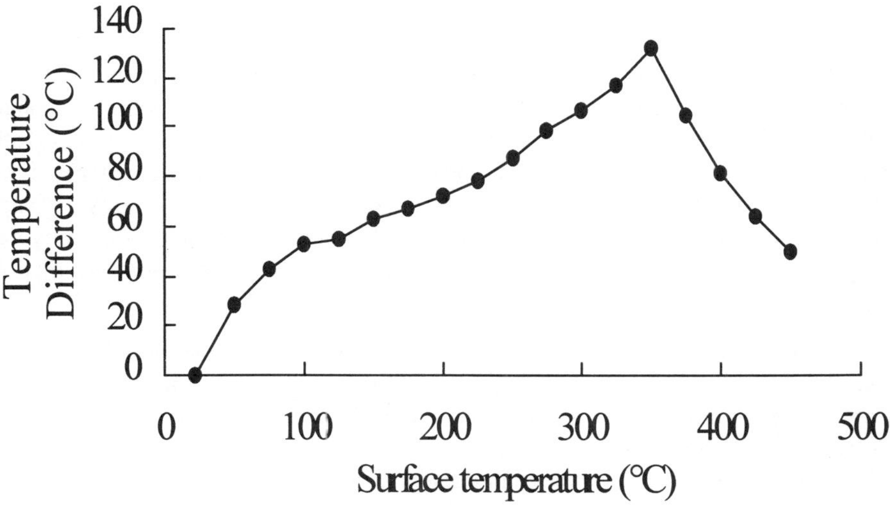

Figure 5. *Temperature difference between the central part of the specimens and the surface. After [NOU 96]*

Figure 6 presents the distribution of the water vapour pressure on the radius of the specimens at middle height. The maximum pressure was found to increase with the temperature up to a maximum and then decreased. It was observed that during heating, the pressure peak moved towards the centre of the specimen.

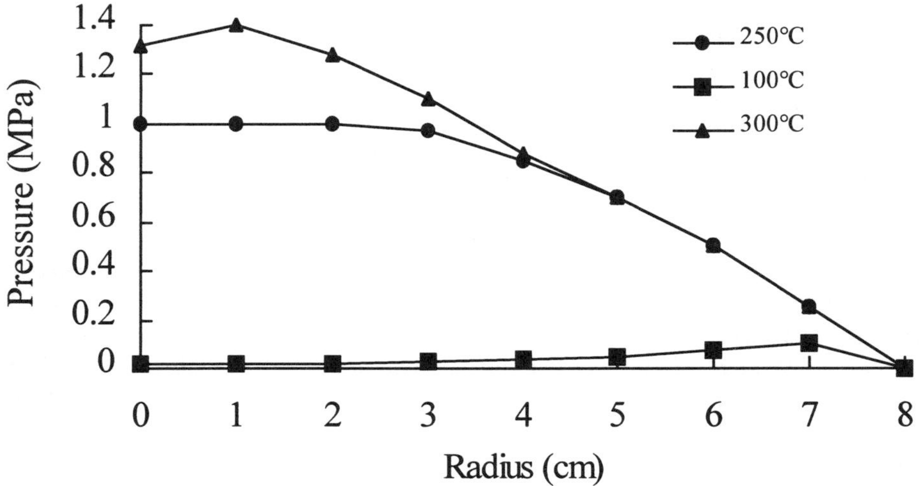

Figure 6. *Distribution of the water vapour pressure on the radius of the specimens at middle height. After [NOU 96]*

Figure 7 presents the evolution of the calculated pore pressure at the centre of the specimen as a function of the temperature at the surface. Results are compared with saturating vapour pressure values obtained from ASTM tables and with pressures obtained by [JOU 97] with a model formulated within the framework of the mechanics of open reactive porous media. It can be seen that the peak of pore pressure is predicted at the same temperature for the two coupled thermo-hygral analysis and that saturated vapour pressure is unable to provide a good estimation after this temperature.

Experimental observations indicate that about one third of the cylinders spalled explosively during heating. Spalling took place between 275 and 350°C, in the rising phase of the thermal gradient. It is interesting to note that no spalling or explosion was observed when heating ordinary concrete specimens which presented almost the same experimental thermal gradients during heating up [NOU 96].

The temperature and pore pressure fields previously calculated have been projected on the finite element mesh employed for the mechanical analysis presented in figure 8 together with the applied boundary conditions. Massive eight nodes isoparametric elements have been employed. The initial mechanical properties as well as their variations with temperature introduced in the computations are based on experimental measurements by [NOU 96] namely

Young's modulus, coefficient of thermal expansion, uniaxial compressive strength, uniaxial tensile strength and porosity. Commonly employed values and variation laws have been used for the other properties.

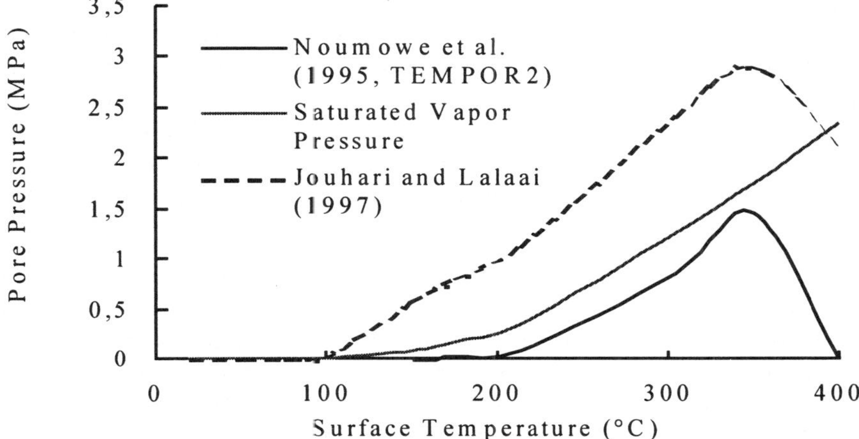

Figure 7. *Calculated pore pressure in the centre of the specimens*

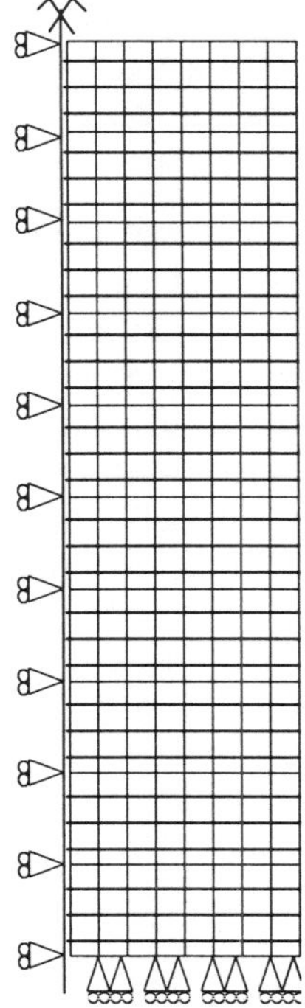

Figure 8. *Axisymmetrical finite element mesh and boundary conditions for the mechanical analysis*

The computations have been performed for the case where only thermal loading is considered and for the case where thermal loading and pore pressures are taken into account. The results of the calculations indicate the same global behaviour for both calculations. They show a triaxial state of tensile stresses in the central part of the specimen, while, towards the surface, compressive stresses appear in the axial and circumferential directions. Figure 9 presents the isovalues of the hardening parameter in tension obtained at different temperatures considering thermal loading and pore pressures. Cracking occurs in the central part at 70°C, much before significant pore pressure is built up in that part. The cracked zone is extending during heating up and reaches the top surface at 350°C. These results are consistent with the experimental observations of spalled specimens that describe a completely pulverised central zone [NOU 96].

Figure 10 presents the hardening parameter in tension as a function of surface temperature for both calculations. It can be observed that after 200°C on the surface, the hardening parameter, that we can somehow relate to crack opening, does not increase anymore. This can be explained by stress release due to the decrease of Young's modulus. For the case where pore pressures are taken into account, this parameter increases up to a maximum value of 350°C which can be considered as the predicted failure temperature.

Hence, we can consider that in these experiments, the spalling failure mode is initiated by the thermal gradient only. However, it can be deduced from the compared crack propagation analysis that pore pressures can play a significant role in the kinetics of this failure mechanism since, as shown in figure 10, that thermal loading only seems not to predict failure at 350°C.

The results presented here are of course strongly influenced by the assumptions introduced for simplifying the problem. One has in particular to consider the effects of the local volume change induced by cracking on the evolution of pore pressure during heating. This volume increase could make pore pressure drop after cracking if diffusion cannot supply rapidly enough the mass of further steam to fill the dynamically expanding volume of the crack [BAZ 97]. One may also consider the possible increase of the Biot's tensor coefficients that would increase the effect of pore pressure on the effective stress. The issue of these possible cross effects should be scrutinised in the future.

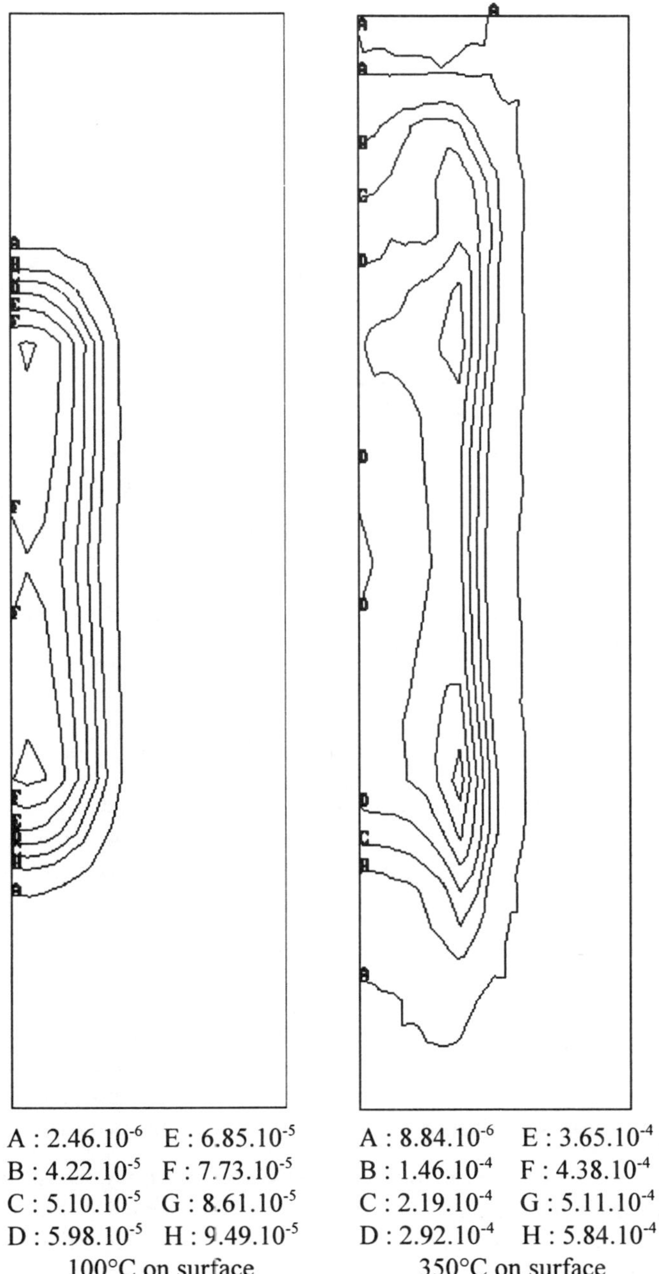

Figure 9. *Isovalues of the hardening parameter in tension at different surface temperatures*

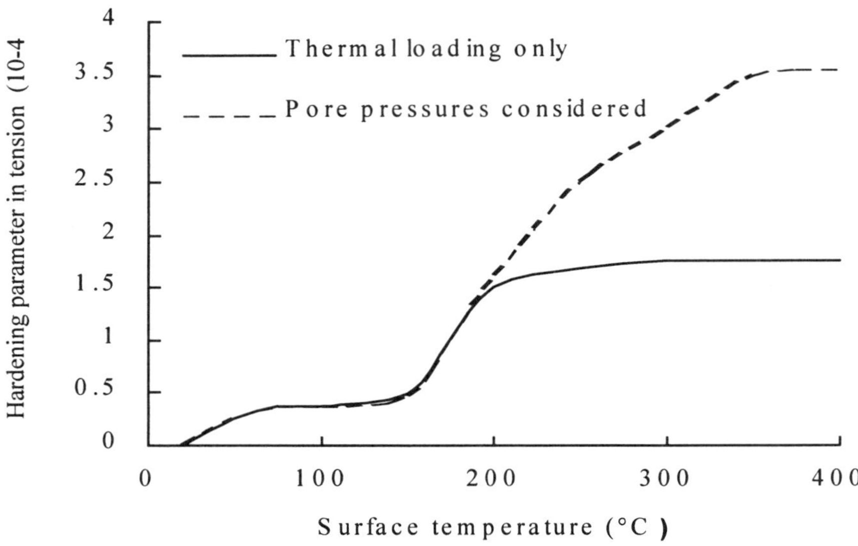

Figure 10. *Hardening parameter in tension as a function of surface temperature for a cracked point in the central part of the specimen*

4. Conclusions

Within the framework of the mechanics of porous continua, an extension of a thermo-plasticity based model to account for pore pressures has been proposed for the analysis of the behaviour of concrete structures at elevated temperatures.

This approach, based on simple assumptions, is a first tool developed for the evaluation of the possible mechanical contribution of pore pressures to the spalling of concrete and to identify the main parameters involved in this contribution.

This model has been applied to the analysis of the behaviour of high strength concrete specimens submitted to high temperatures. Results emphasise that the predicted failure mechanism is in agreement with the experimental observations and that pore pressures can significantly affect the kinetics of this mechanism.

These results are of course influenced by the assumptions introduced for simplifying the problem. Further developments are thus needed in order to account for the effects of cracking on the heat and moisture transfer properties as well as on the hydraulics to mechanics coupling parameters such as the Biot's tensor.

These developments associated with an improvement of the thermo-plastic model allowing to account explicitly for dehydration effects would allow an accurate analysis of spalling mechanisms of concrete structures.

Finally, a closer examination of strain localisation in the presence of coupled diffusive and mechanical phenomena involving softening behaviour should be undertaken in order to evaluate the objectivity of the numerical solution provided by this kind of analysis.

Acknowledgements

This work has been done within the framework of a cooperation with the Structural Mechanics division of the Design Department for thermal and nuclear projects of Electricité de France. The financial support is fully acknowledged by the authors.

5. References

[AND 97] ANDERBERG Y., "Spalling phenomena of HPC", Proceedings of the Int. Workshop on fire performance of high-strength concrete, NIST, Gaithersburg, Maryland, 1997, NIST special publication, No 919.

[AND 76] ANDERBERG Y., THELANDERSSON S., Stress and deformation characteristics of concrete at high temperature. Part 2. Experimental investigation and material behaviour model. Lund (Suède), Lund Institute of technology, 1976, 84 p., Bulletin n° 54.

[BAZ 97] BAZANT Z.P., "Analysis of pore pressure, thermal stresses and fracture in rapidly heated concrete", Proceedings of the Int. Workshop on fire performance of high-strength concrete, NIST, Gaithersburg, Maryland, 1997, NIST special publication, No 919.

[BAZ 81] BAZANT Z.P., CHERN J.C., THONGUTHAI W., "Finite element program for moisture and heat transfer in heated concrete", *Nuclear Engineering and Design*, 1981, No 68, pp. 61-70.

[COU 95] COUSSY O., *Mechanics of porous continua,* John Wiley & Sons 1993.

[DAY 82] DAYAN A., GLUEKLER E.L., "Heat and mass transfer within an intensely heated concrete slab", *Int. J. of heat and mass transfer*, 1982, Vol. 25, pp. 1461-1467.

[BOR 89] DE BORST R., PEETERS P.P.J.M., "Analysis of concrete structures under thermal loading", *Comp. Meth. Appl. Mechs. Engng.,* 1989, Vol. 77, pp. 293-310.

[EHM 85] EHM C., SCHNEIDER U., "The high temperature behavior of concrete under biaxial conditions", *Cement Concrete Res.*, 1985, Vol. 15, pp. 27-34.

[ENG 73] ENGLAND G.L., SKIPPER M.E., "On the prediction of moisture movement in heated concrete up to 550°C", Trans. 2nd SMIRT, BAM, Berlin, 1973, Vol.3, Paper H6/2.

[FEE 95] FEENSTRA P.H., DE BORST R., "A plasticity model and algorithm for mode-I cracking in concrete", *Int. Journal Num. Meth. Engng*, 1995, Vol 38, pp. 2509-2529.

[GER 96] GERARD B., BREYSSE D., AMMOUCHE A., HOUDUSSE, O., DIDRY 0., "Cracking and permeability of concrete under tension", *Materials and Structures*, 1996, Vol. 29, pp.141-151.

[HAR 65] HARMATHY T.Z., "Effect of moisture on the fire endurance of building materials.", ASTM, Philadelphia, 1965, ASTM Special Technical Publication No. 385, pp. 74-95.

[HEI 97] HEINFLING G., REYNOUARD J.M., MERABET O., DUVAL C., "A thermo-elastic-plastic model for concrete at elevated temperatures including cracking and thermo-mechanical interaction strains", Proceedings of the 5th COMPLAS, Barcelona, 1997, pp. 1493-1498.

[JOU 97] JOUHARI L., LALAAI I., "A constitutive model for thermo-hydro-chemo-mechanical response of decomposing high performance concrete under high temperature", Proceedings of the 2nd International Conference on Thermal Stresses and Related Topics, Rochester, 1997, pp. 221-224.

[KON 95] KONTANI O., SHAH S.P., "Pore pressure in sealed concrete at sustained high temperatures" Concrete Under Severe Conditions: Environment and loading (Volume Two) Ed.by K. Sakai, N. Banthia and O.E. Gjorv, E & FN Spon, 1995, pp. 1151-1162.

[KHE 92] KHENNANE A., BAKER G., "Thermoplasticity model for concrete under transient temperature and biaxial stress" Proceedings of the R. Soc. Lond., 1992, A, No 439, pp. 59-80.

[KOR 95] KORDINA K., EHM C., SCHNEIDER U., "Effect of biaxial loading on the high temperature behavior of concrete", Proceedings of the 1st Symp. of fire safety Science, Gaithersburg, 1985, p. 281-290.

[MAJ 95] MAJUMDAR P., GUPTA A., MARCHERTAS A., "Moisture propagation and resulting stress in heated concrete walls", *Nuclear Engineering and Design*, 1995, No 156, pp 147-158.

[MAI 69] MAIER G., "Linear flow-laws of elastoplasticity : a unified general approach", *Lincei-Rend. Sci. Fis. Mat. Nat.*, 1969, Vol. 47, pp. 266-276.

[MEY 75] MEYER-OTTENS C., "Spalling of structural concrete under fire", *Deutscher Ausschuss für Stahlbeton*, Berlin (in German): W. Ernst & Sohn, 1975, Heft 248.

[NOU 96] NOUMOWE A., CLASTRES P., DEBICKI G., COSTAZ J.L., "Thermal stresses and water vapour pressure of high performance concrete at high temperature", Proceedings of the 4th International Symposium on Utilisation of High-Strength/High-Performance concrete, Paris, 1996, pp. 561-570.

[SCH 87] SCHNEIDER U., HERBST H.J., "Transport processes in thick concrete structures at high temperatures", Trans. Of the 9th SMIRT, 1987, Vol. H, pp. 167-172.

[SER 77] SERTMEHMETOGLU Y., A mechanism of spalling of concrete under fire conditions, PhD Thesis. Kings College, Univ. of London, 1977.

[ZHU 77] ZHUKOV V.V., "Reasons of explosive spalling of concrete by fire", *Concrete and Reinforced Concrete*, 1977, No. 3.

Organic Fluids Penetrating into Cracked Concrete

Hans W. Reinhardt

Institute of Construction Materials, University of Stuttgart
and FMPA (Otto-Graf-Institute), Pfaffenwaldring 4
D-70550 Stuttgart, Germany

ABSTRACT. *Penetration tests with organic fluids were carried out on precracked concrete specimens. Testing procedure and results are described and discussed. Conclusions are drawn which are valuable for the assessment of concrete structures with respect to the tightness against organic fluids.*

KEY WORDS : *Penetration, Concrete, Organic Fluids, Bending Cracks, Wedge-Splitting Test.*

1. Introduction

Concrete is able to retain fluids permanently or at least for a certain time. Tunnels, pipes and pools are made of concrete and can resist high hydraulic pressure. Chemical plants, catching basins, and fuel stations have often a concrete floor which retains organic fluids temporarily. When cracks occur, one has to check whether these are tensile (or through) cracks, bending cracks or surface cracks. It has been shown in experiments and through experience that tensile cracks heal again if the crack width is less than 0.15 mm and if water permeates through it. The mechanism has been explained in [EDV 96]. On the other hand, organic fluids permeate also through cracks having only 0.10 mm width [BIC 97] and drying permeability and diffusivity are considerably increased by 0.10 mm wide cracks [BAZ 87].

Surface cracks are mainly caused by non-linear temperature and moisture distribution in a cross-section. Eigenstresses develop and may cause map-cracking. The depth of these cracks is typically 0.1 to 0.2 times the thickness of the concrete wall or slab [REI 89].

Every reinforced concrete structure has bending cracks and the question is how detrimental are these cracks and how do they impair the tightness of a concrete structure. A series of experiments was started to show the effect of bending cracks on the penetration of organic liquids which penetrate either from the cracked side or from the opposite side into the concrete.

2. Test geometry and crack generation

Two types of specimens were prepared which are designated as a wedge-splitting (WS) and a notched beam (NB).

2.1 Wedge splitting

The wedge splitting test is nowadays standard in fracture mechanics of concrete investigations [BRÜ 90, HIL 77, TSC 91]. The specimen used is a prism 200 x 200 x 100 mm which was cut from a 200 mm cube after 7 days moist curing. The specimens were stored at least 3 months at 20°C and 65% RH before testing. A 3 mm wide notch was cut to a depth of 40 to 120 mm.

Fig. 1 shows the specimen and loading arrangement to generate a crack. At the top of the specimen, two steel plates are glued to the concrete with two edges cut at an inclination angle of 8.84°. The wedge consists of a hardened steel plate with inclined edges which fit to the glued plates. PTFE sheets are placed in contact as to reduce friction. The vertical force P_v generates a horizontal force P_H which is about 3.2 times P_v.

During the test the crack mouth opening displacement (CMOD) is measured by two LVDTs, indicated as W2 in Fig. 1.

When the load increases, CMOD increases almost linearly up to about 70% of the maximum load. The stiffness decreases thereafter until the maximum load is reached. Then, the load decreases in a stable manner with increasing CMOD. The length of the crack was recorded using a magnifying glass or a measuring microscope. A micro-crack is defined as a crack with a width of 10 μm which can be measured optically [RIL 94].

In the linear range, crack length and width develop proportional to each other. The horizontal load is resisted mainly in an elastic way. When the peak load has been reached, the CMOD increases strongly due to the softening of the concrete while the length of the visible crack increases less than proportionally.

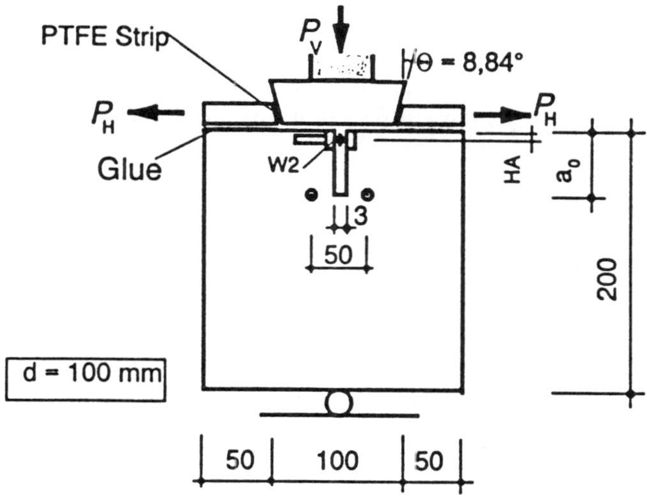

Figure 1. *Loading arrangement of wedge splitting test, dimensions in mm*

2.2 Notched beam

The wedge splitting test uses a specimen economically made but it was argued whether the crack was the same as in a real structure subject to a bending moment. Therefore, a beam was designed according to Fig. 2 with a small reinforcement which enabled transport and handling without damage.

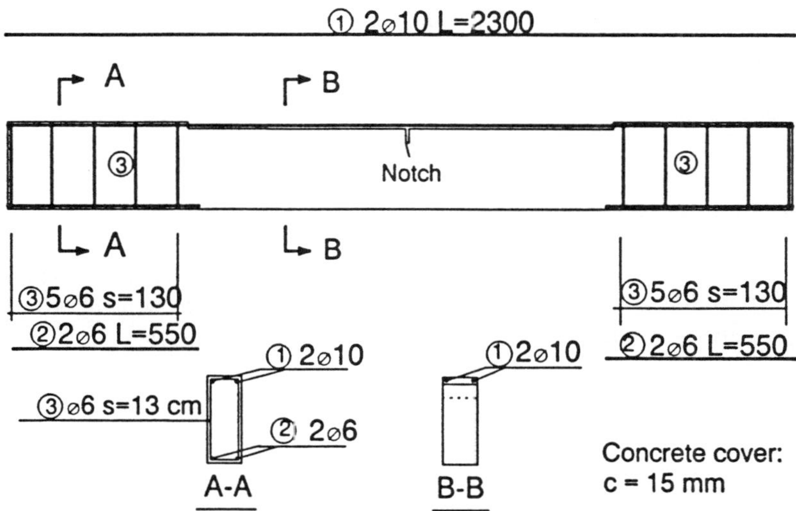

Figure 2. *Notched beam*

In the middle portion, the steel reinforcement is placed outside the concrete in order to make sure that no reinforcement can interfere with the penetrating liquid. The loading arrangement allowed a four point bending test with continuous displacement control. Due to the low stiffness of the unboned steel reinforcement, only one crack could develop starting from the notch.

3. Testing equipment for fluid penetration

3.1 Wedge splitting specimens

After the specimens were cracked to a certain depth, the lateral surfaces were coated with a transparent epoxy resin which allowed the visual inspection of the penetration of the liquid into the concrete. Then, the specimen was placed in a steel frame and the crack was fixed at a certain width. Details of the arrangement can be found in [REI 98].

The testing fluid was held at a constant head of 1.4 m. The absorbed amount of fluid and the penetration front were recorded continuously.

In a real structure, the fluid can also penetrate from the compression zone of the cross-section. To show the influence of such a crack on the fluid transport, another arrangement has been developed.

The specimens were coated by a transparent epoxy resin which allowed optical monitoring of the penetration front. However, in order to measure the front in the interior of the specimen, it is split after the test perpendicular to the crack. An infra-red-camera was used in order to determine the penetration depth on the split surface since acetone is very volatile and would have been detected only unprecisely by optical observation [SOS 94].

3.2 Notched beam specimens

After the notched beam showed a certain crack length, the beam was unloaded and a prism of 200 x 230 x 100 mm^3 which contained the crack was cut. The infiltration procedure was the same as with the wedge splitting specimen.

4. Materials used

One type of concrete was used throughout the tests. Composition and standard properties are given in Table 1. The composition is typical for a concrete which follows the German Guideline for concrete structures for ecotoxic liquids [DEU 96].

Table 1. *Concrete composition and standard properties*

Cement type	Portland cement, CEM I 32.5 R
Cement content, kg/m^3	320
Aggregates	Quartzitic, max. 16 mm
Water-cement ratio	0.50
Admixture	Plasticizer, 2 ml/kg cement
Compressive strength [1)], MPa	43.1 (min. 38.5, max. 51.1)
Density, kg/m^3	2340

[1)] Mean of 6 cubes 200/200/200 mm^3 for 28 days

The test fluids are given in Table 2 and compared with water. Acetone is water soluble; n-Decane and n-Heptane are not.

Table 2. *Test fluids and physical properties at 20°C [LID 95]*

Test fluid	Viscosity η mPa s	Surface tension σ mN/m	$(\sigma/\eta)^{0.5}$ $(m/s)^{0.5}$	Density kg/dm^3
Acetone	0.324	23.70	8.53	0.79
n-Decane	0.907	23.90	5.10	0.75
n-Heptane	0.409	20.00	7.06	0.69
Water	1.002	72.85	8.52	1.00

5. Experimental results

5.1 Wedge splitting specimens

The fluid penetration could be observed through the transparent epoxy coating. Fig. 3 shows a schematic of the penetration during several time steps. Three areas can be distinguished: the upper part where the fluid penetration occurs almost without interaction with the crack, the area along the crack where the fluid from the crack is absorbed, and finally the area around the crack tip.

The depths of the fluid front are designated with e_1, e_2, e_3. In this particular case, the experiment yields $e_3 < e_1 < e_2$, i.e., the penetration in front of the crack tip is slower than in the undisturbed zone. This result was not obvious in advance because the visible crack is preceded by a cluster of sub-micro cracks (so-called crack process zone) which could have been more pervious than the bulk concrete.

To assess the tightness of a structure, the total penetration e which is crack length plus e_3 has to be known. Fig. 4 shows the penetration depth as function of square root of time.

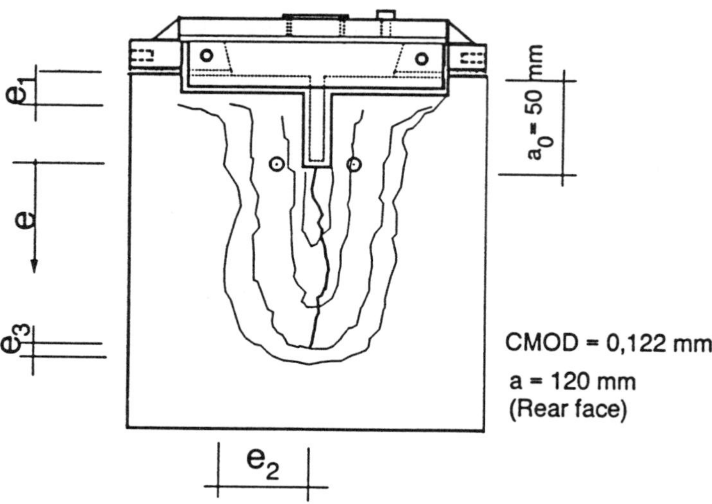

Figure 3. *Schematic of fluid penetration in the vicinity of a crack*

The depths of the fluid front are designated with e_1, e_2, e_3. In this particular case, the experiment yields $e_3 < e_1 < e_2$, i.e., the penetration in front of the crack tip is slower than in the undisturbed zone. This result was not obvious in advance because the visible crack is preceded by a cluster of sub-micro cracks (so-called crack process zone) which could have been more pervious than the bulk concrete.

To assess the tightness of a structure, the total penetration e which is crack length plus e_3 has to be known. Fig. 4 shows the penetration depth as function of square root of time.

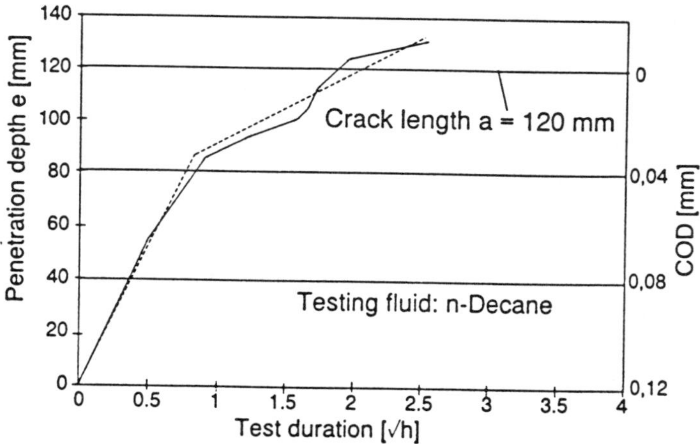

Figure 4. *Penetration depth as function of square root of time, Crack opening displacement (COD) is for illustration*

The square root function is shown since capillary absorption and the permeation into a crack follow such a time function [REI 97]. The diagramme can be divided into three parts: the first part shows a rapid increase of penetration depth until about 90 mm, a reduced speed until 120 mm, and even more delayed penetration thereafter. On the right ordinate, the crack opening displacement (COD) is indicated. Correlating the three speeds of penetration, one can see that the fluid penetrates within one hour till a crack width of 0.04 mm. It took about four hours to reach the crack tip which is equivalent to 0.01 mm. Finally, the penetration speed is retarded after the total crack length has been crossed by the fluid. The penetration rate of the last part will be discussed later on.

Penetration of a fluid from the opposite side of the crack has been tested with acetone. Fig. 5 shows the absorbed volume vs. square root of time.

There is a straight line until about 24 hours and, after that, a slight increase compared to the square root relation. The more important result is that the penetration front which has been observed optically has reached the crack tip after four hours but that the penetration behaviour has not changed, i.e., the crack has not acted as a sink for the fluid.

The specimen has been split normal to the crack plane after 65 hours. Since acetone evaporates quickly, a thermo-image has been taken, which demonstrated that the fluid concentration was uniform in uncracked concrete and in concrete containing a crack width ≤ 0.04 mm width.

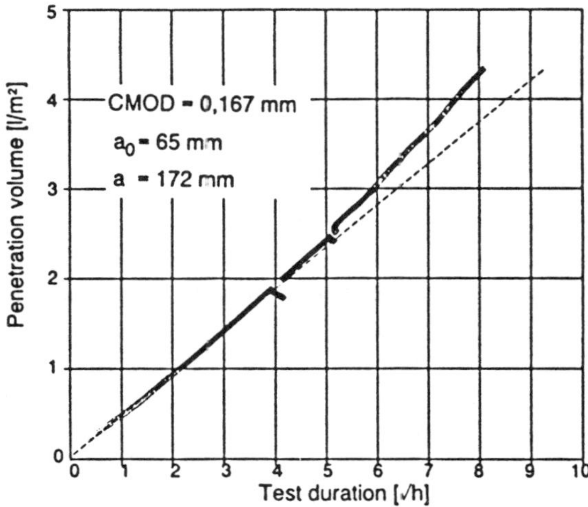

Figure 5. *Absorbed fluid volume vs. square root of time, penetrating into the compression zone*

5.2 Notched beam specimen

A prism has been cut from a beam according to Fig. 2 which contained a crack of 0.1 mm at the notch. Acetone was the test fluid which penetrated from the notched side into the beam. The crack length was 75 mm. Fig. 6 shows the penetration depth versus square root of time. Similar to Fig. 4, there is a fast penetration in the first minutes and a deceleration until the crack tip is reached.

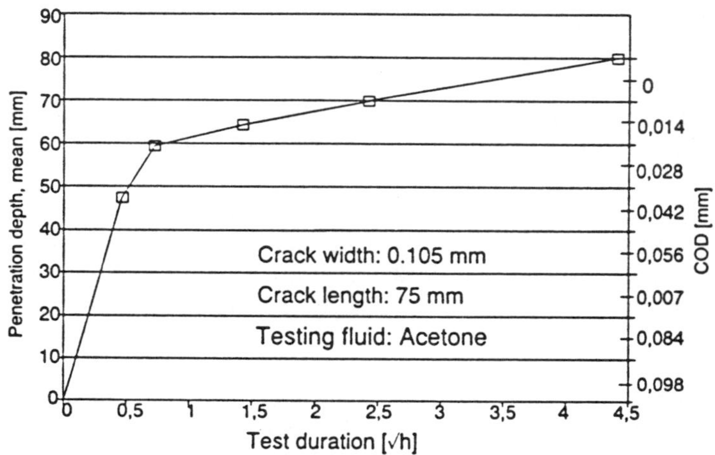

Figure 6. *Penetration depth vs. square root of time*

Also in this figure, a crack width (COD) of about 0.04 mm leads to the transition of the penetration speed. It can be stated that the penetration behaviour of a wedge splitting and a notched beam specimen is alike.

6. Discussion of results

One result is that visible bending cracks are filled with liquid very quickly when they are in contact with the fluid. On the other hand, when the fluid penetrates into concrete from the opposite side (compression zone), the crack does not act as a sink. This is an important result in the view of judging the tightness of a slab on the ground. If a bending crack is visible, one has to assume that it will fill up with fluid. If there is no visible crack, the concrete behaves as uncracked concrete although there might be a bending crack on the lower surface.

Another result concerns the penetration into the process zone of a crack. The tests have shown that, at a crack opening displacement of about 0.04 mm, concrete seems to behave like uncracked concrete. However, one has to keep in mind that the penetration into the tip of a crack is retarded by geometrical reasons. [WILL 94] gave an analytical solution for the slit with a cylindrical end. It comes out that the penetration is retarded the more compared to the one-dimensional case the longer the duration of the fluid transport. This means that a portion of the tensile zone in a bent cross-section contributes to the tightness of a structure. For a practical judgement the following steps can be taken: First, the neutral axis of a cracked beam under service load is calculated; Second, the crack width at the surface of the tensile zone is analysed; Third, assuming a wedge shaped crack, the position of the 0.03 mm crack width is calculated. The measure 0.03 mm implies a certain safety margin against the 0.04 mm which was observed in the tests. The contributon of the tensile zone to the tightness of concrete is very appreciated, because there are many practical cases where the compression zone alone would not fulfill the requirements of [DEU 96].

Permeability tests on cracked concrete have shown that a tensile strain of 10^{-4} reduces the water permeability by two orders of magnitude [BRE 97], i. e., microcracks increase the transport velocity considerably. Although this may contradict the findings of the tests on bendng cracks, one has to take into account that the process zone ahead of a visible crack leads to a compressed concrete area, whereas the concrete in a uniform tensile strain field does not. Further investigations on this subject are encouraged.

7. Conclusion

Organic liquids permeate through cracks very quickly. Bending cracks are filled up quickly when they are in contact with the liquid. If the liquid front approaches the crack from the compressive zone the crack does not act as a sink. The process zone ahead of a bending crack behaves like uncracked concrete as long as the crack opening displacement is less than 0.04 mm.

The results are very valuable for assessing the tightness of structures. More research is encouraged to compare and model penetration and permeation of (mirco-) cracked concrete.

Acknowledgement

At this special occasion, the author would like to thank Zdenek Bazant for his steady interest in our work. Many stimulating discussions promoted new ideas.

The valuable contribution to the test series by Dr. X.-f. Zhu is gratefully acknowledged.

8. References

[BAZ 87] BAZANT, Z.P., SENER, S., JIN-KEUN KIM, Effect of cracking on drying permeability and diffusivity of concrete. ACI Mat. J. 84 (1987), No. 5, pp 351-357

[BIC 97] BICK, D., CORDES, H., TROST, H., *Eindring- und Durchströmungsvorgänge umweltgefährdender Stoffe an feinen Trennrissen in Beton.* Deutscher Ausschuss für Stahlbeton, Bulletin No. 475, Berlin 1997, pp 101-171

[BRE 97] BREYSSE, D., GÉRARD, B., *Transport of fluids in cracked media.* Chapter 4 in "Penetration and Permeability of Concrete. Barriers to organic and contaminating liquids", ed. by H.W. Reinhardt, E & FN Spon, London 1997, pp 123-153

[BRÜ 90] BRÜHWILER, E., WITTMANN, F.H., *The wedge splitting test, a method of performing stable fracture mechanics tests.* Eng. Fracture Mech. 35 (1990), No. 1-3, pp 117-126

[DEU 96] DEUTSCHER AUSSCHUSS FÜR STAHLBETON. Richtlinie Betonbau beim Umgang mit wassergefährdenden Stoffen, Berlin Sept. 1996, Part 1-6

[EDV 96]EDVARDSEN, C.K., *Wasserdurchlässigkeit und Selbstheilung von Trennrissen in Beton.* Deutscher Ausschuss für Stahlbeton, Bulletin No. 455, Berlin 1996.

[HIL 77] HILLEMEIER, B., HILSDORF, H.K., *Fracture mechanics studies on concrete compounds.* Cement and Concrete Res. 7 (1977), pp 523-536

[LID 95] LIDE, D.R. (ED.), *CRC Handbook of Chemistry and Physics.* CRC Press, 76th Ed., Boca Raton 1995

[REI 89] REINHARDT, H.W., *Beurteilung von Rissen hinsichtlich der Durchlässigkeit von Betonbauteilen*, Darmstädter Massivbauseminar, Band 2, Darmstadt 1989, pp VII /1-16

[REI 97] REINHARDT, H.W. (ED.), *Penetration and Permeability of Concrete. Barriers to organic and contaminating liquids.* E & FN Spon, London 1997

[REI 98] REINHARDT, H.W., SOSORO, M., ZHU, X.-F., *Cracked and repaired concrete subject to fluid penetration.* Materials and Structures 31 (1998), No. 206, pp 74-83

[RIL 94] RILEM TC 122-MLC. State-of-the-art report on microcracking and lifetime of concrete. Part 1, 1994

[SOS 94] SOSORO, M., REINHARDT, H.W., *Eindringverhalten von Flüssigkeiten in Beton in Abhängigkeit von der Feuchte der Probekörper und der Temperatur.* Deutscher Ausschuss für Stahlbeton, Bulletin No. 445, Berlin 1994, pp 87-108

[TSC 91] TSCHEGG, E.K., *New equipments for fracture tests on concrete.* Materialprüfung 33 (1991), pp 338-343

[WILL 94]WILSON, M.A., HOFF, W.D., *Water movement in porous building materials - XII. Absorption from a drilled hole with a hemispherical end.* Building and Environment 29 (1994), No. 4, pp 537-544

Testing and Modeling Alkali-Silica Reaction and the Associated Expansion of Concrete

Y. Xi * — A. Suwito * — X. Wen ** — C. Meyer *
W. Jin *****

** Department of Civil, Environmental, and Architectural Engineering
University of Colorado, Boulder, CO 80309, USA*

*** Department of Civil and Architectural Engineering
Drexel University Philadelphia, PA 19104, USA*

**** Department of Civil Engrg. and Engrg. Mechanics
Columbia University New York, NY 10027, USA*

ABSTRACT. *The first part of this paper introduces the most recent ASR test results which examined the ASR problem of concrete in a systematic way, including the effects of aggregate size, types of aggregate (natural aggregate and crushed mix-color glass), and chemical composition of cementitious binders (regular Portland cement and a new cementitious material, called ashcrete). The second part presents a recently developed material model which can characterize the effects of various influential parameters on ASR behaviors and predict the development of ASR expansion. The model is based on a modified version of the generalized self-consistent theory. It also takes into account the chemomechanical coupling of the ASR expansion process by using two opposing diffusion processes. One is the diffusion of chemical ions from pore solution into aggregate, and the other one is the permeation of ASR gel from the surface of aggregate out to the surrounding porous cement matrix. The material parameters involved in the expansion model are determined, based on ASR kinetics and cement chemistry.*

1. Introduction

Alkali-Silica Reaction (ASR) is a reaction which involves the hydroxide ions of the pore solution in cement paste and reactive silica in aggregates. ASR can be deleterious to the concrete due to the expansion and possible cracking of the concrete associated with the reaction. An aggregate can be reactive or unreactive depending on the crystal structure of the silica in the aggregate. An unreactive aggregate has an ordered silicon-oxygen tetrahedra, while a reactive aggregate (i.e., opal) has a disordered network of silicon-oxygen tetrahedra which has a higher internal surface area and exists in a higher state of energy and thus is unstable.

Various testing methods have been developed which are designed to reveal whether an aggregate is potentially reactive and whether it could cause abnormal expansion and cracking of concrete. Direct measurement of the long term expansion has been considered to be the most dependable way to evaluate aggregate reactivity, although it is quite time consuming. A number of accelerated testing procedures have been devised, and are currently being used with some confidence. The most recently developed accelerated testing method is ASTM C1260 (accelerated mortar bar test) which requires only 14 days of testing. This method is used in the present study.

The ASR mechanisms have also been studied extensively. Research on ASR of regular concrete showed that maximum expansion (refered to as pessimum expansion) due to ASR depends on many parameters, such as the volume fraction, the type, and the size of reactive aggregate, the composition and alkali content of the cement, the rate of strength development, and the mixture proportion of the concrete including water-cement ratio. Based on research results on ASR mechanisms, many methods have been developed to prevent the damages induced by ASR. Comparing with the developments of testing methods and studies on ASR mechanisms, development of theoretical models for predicting ASR is still in a very preliminary stage.

In a recent joint research project, the authors performed a systematic experimental and theoretical study on reducing expansion of ASR. In the project, cushed mixed-color waste glass was used as aggregate in concrete. Given the condition that the glass will definitely cause ASR problem, various new additives have been examined to reduce the ASR expansion. It was found that reactive metakaolin, lithium and chromium based chemicals are very effective in reducing ASR expansions even if 100% of natural aggregate was replaced by glass. Moreover, when the same additive was used, different processing techniques resulted in different expansion reductions [Jin 98].

There are two purposes of this paper. One is to present some test results on reducing ASR by using a new concrete, which have not been reported in previous publications. The other one is to present a new theoretical model for predicting ASR expansion, which has been briefly introduced in a recent conference [Jin et al. 98].

2. ASR Test of Regular Concrete with Glass Aggregates

Waste glass has similar mechanical properties (e.g. strength) as regular sand and mineral rock aggregates. However, waste glass as aggregates reacts with cement paste and causes severe ASR expansion. In the first set of test, only 10% of natural aggregate was replaced by the glass. The so-called spectrum analysis was conducted to examine the reactivity of glass aggregate of different sizes. The cement used in this project was Type I Portland cement with equivalent sodium content lower than 0.06% (low alkali cement defined by ASTM C150); the glass aggregates were made from waste beverage bottles, which were crushed by a crusher, thoroughly washed to get rid of the attached paper and sugar, and carefully sieved with #8, #16, #30, #50 and #100 sieves. In this test, there were 6 groups of specimens with 2 specimens in each group. The two reference specimens were made in accordance with ASTM C 1260, with no glass aggregates. The specimens marked with #8 had 10% #8 size natural aggregates replaced by the glass aggregates of the same size. The specimens marked with #16, #30, #50, #100 were made following the same rule.

The test results are shown in Fig. 1. One can see that the specimens of #16 glass aggregate give the highest expansion (0.40%), the specimens of #8 give the second highest expansion (0.35%) and the reference specimens give the least expansion (0.06%). Generally speaking, the smaller particles have a higher surface area which leads to a faster rate of reaction, and thus to a larger ASR expansion. This is not true, however, in the test results with waste glass as aggregate. This phenomena will be further investigated and discussed in later sections.

3. ASR Test of Ashcrete with Glass Aggregates

To reduce ASR expansion in concrete, many different methods have been developed. In the present study, we examined the effectiveness of a new concrete in terms of reducing ASR expansion. First, the properties of the new concrete will be introduced briefly, and then, the effectiveness of reducing ASR expansion will be discussed in detail.

The new material is called Ashcrete [Sam et al. 95][Sam 96][Sil 97]. Ashcrete is a low cost and environmentally friendly new cementitious material. It has a great potential for various applications in the construction industry. Ashcrete has been made of activation chemicals, Class-F fly ash, coarse and fine aggregates without any Portland cement. It is completely different from conventional utilization of fly ash in concrete. The controlling parameters of ashcrete have been identified and optimized in terms of strength development and cost.

Ashcrete develops strengths up to 62 MPa (9,000 psi) in 24 hours. The activation chemicals we used are sodium hydroxide and sodium silicate solution. There is no high calcium lime involved, and thus the nature of the chemical reactions is not pozzolanic, which results in a very fast strength development. The controlling parameters that are particularly important for ashcrete have been identified, namely, the ratio of total chemicals to fly ash, the molar ratio of silica dioxide to sodium oxide (SiO_2/Na_2O), and the water to binder ratio (binder is the sum of sodium hydroxide, sodium silicate solution, and fly ash). Other controlling parameters are similar to those for Portland cement concrete, such as binder to aggregate ratio and coarse to fine aggregate ratio.

The ratio of the total chemical to fly ash has an effect on both ultimate strength and strength development. The basic trend is that the higher the ratio, the higher the resulting strength of the ashcrete. For economical reasons, the ratio should be kept below 20%. The molar ratio SiO_2/Na_2O has a critical effect on both ultimate strength and strength development. Our extensive test results showed that SiO_2/Na_2O varies upon specific chemical composition of the ash to be used, particularly on Al_2O_3 and SiO_2 contents of the ash. The water to binder ratio has a similar effect as water to cement ratio in regular concrete. Different water to binder ratios have been tested and the basic trend is that the lower the ratio, the higher the resulting strength. The lowest water to binder ratio tested was 0.3.

Ashcrete samples were made according to ASTM C1260. The sand, the coarse aggregates and the glass aggregate used for the ashcrete samples were the same as those used for the regular concrete samples. Test results are shown in the Fig. 2(a) and Fig. 2(b). One can see that the ASR expansions are very small in all the specimens. The maximum expansion is about 0.02%, which is much lower than the expansion of regular concrete shown in Fig. 1. The expansion of the reference sample (natural aggregate only without glass aggregate) is about 0.017%. The same aggregate was used in the specimens of Portland cement concrete (see Fig. 1), and the resulting expansion was about 0.05%. All curves except the curve with a #50 glass particle showed shrinkage after a certain testing period. The shrinkage started about 10 days after the test for most of the specimens. For specimens with #100 glass particles, the shrinkage started only after three days. The maximum amount of shrinkage measured from the specimens was about 0.015%, which is very small comparing with the drying shrinkage of regular concrete. It is very important to note that ASR is not a problem for the ashcrete.

4. Effect of the Size of the Natural Aggregate on ASR

In order to understand the effect of the aggregate size on ASR expansion, we performed two parallel spectrum analyses on Portland cement concrete as well as on ashcrete with uniformed aggregates, which means that each specimen only contained the natural aggregates of one size. The test results are shown in Fig. 3. The expected result of the test is that the expansion due to ASR increases as the size of the aggregates decreases. In Fig. 3, however, the expansion reaches its maximum at the aggregates size of #50 (300 μm). The minimum expansion occurs at the size of #16 (1.19 mm). This spectrum analysis explained the previous test results (see Fig. 1) in which 10% glass aggregate was used to substitute natural aggregates. When #16 glass was used to replace #16 natural aggregates, a maximum expansion resulted. Because #16 natural aggregates was the most non-reactive aggregates among all sizes of the aggregates, and when this non-reactive aggregates was replaced by glass aggregates, a maximum expansion occured. On the other hand, the #50 and #100 natural aggregates themselves cause the highest expansion compared with aggregates of other sizes. When they were replaced by the glass aggregates, the ASR expansion did not change very significantly.

A similar spectrum analysis was performed for ashcrete. The test results are shown in Fig. 4. Basically, the specimens length increased in the first 5 - 7 days and then decreased, which corresponds to shrinkage. The samples with natural aggregates of size #8 and #100 had a low expansion on the 3rd and 4th days of the test (about 0.008%) and a relatively large shrinkage at the end of the test (around 0.01 %). The samples with natural aggregates of size of #16, #30 and #50 showed a higher expansion (0.01 ~ 0.02%) and a lower shrinkage (> -0.005%). From these results, we can see that the pessimum size of the natural aggregates in ashcrete was in the range of #16 and #30, which was different from the results of the Portland cement concrete. More importantly, the pessimum expansion was very small, much lower than the critical value specified by the ASTM C1260 (0.05%).

Now that the behavior of each size of the aggregates is known, the data from each sample (with one size aggregate) can be multiplied by a percentage of the weight according to ASTM C1260 (10% for #8, 25% for #16, 25% for #30, 25% for #50, and 15% for #100), then a curve of ASR expansion can be obtained by a weighted average as shown in Fig. 6. Comparing the weighted average with the test result of the reference samples (from Fig. 2), the two curves agreed quite well. This confirmed the correctness and repeatability of our test results.

5. The Composite Model for ASR Expansion

The expansion due to ASR has been modeled by several researchers. The dilatation theory [Gro 90] was developed based on the observation of the microstructure of an expanded mortar composed of silica glass particles in ordinary Portland cement mortar. The absorption theory [Fur 94] was used to develop a model for predicting the concrete expansion behavior due to ASR. The expansion was assumed to occur because of the absorption of water by gel. In this model, concrete is considered as a heterogeneous porous material and reactive aggregates as spherical inclusions surrounded by a layer which exert an isotropic pressure on the matrix. The concrete expansion process is divided into two stages, i.e.: (1) The diffusion of the hydroxide and the alkali ions into aggregate, followed by the reaction of these ions with the reactive silica in the aggregate, (2) The development of the expansion induced by ASR. It is further assumed that there exists a porous zone around the aggregates and that the expansion is initiated only when the volume of the reaction product exceeds the volume of the porous zone. The double-layer theory [Pre 97] was proposed to explain the behavior of ASR gel. The model was developed based on the fundamental surface-chemistry principles, the expansion characteristics of mortar bars, the reaction product gels in the samples or the volume change behavior of a colloidal system. The expansion of reaction-product gels is attributed to swelling caused by electrical double layer repulsion. In other words, the volume expansion behavior is expected to be determined by the surface phenomena. A quantitative estimate of the expansive pressure in concrete induced by ASR is made using the Gouy-Chapman model of the diffusing part of the electric double layer.

The model we developed in this study is an extension of the absorption theory [Fur 94]. The model is divided into two parts. The first part is the application of the composite theory to simulate ASR expansion. ASR expansion may be simulated as thermal expansion, and thus, the expansive strain of ASR can be determined by using similar theoretical models developed for thermal expansion of composites. One of the models for evaluating effective expansion of composites is based on the generalized self-consistent method [Xi 97], in which the basic element is a composite spherical system with one constituent phase associated with another. The center sphere is the aggregate and the outside layer is the cement paste matrix. With this type of model, the effective expansion of the concrete is

$$\alpha = \frac{K_a \alpha_a V_a (3K_m + 4G_m) + K_m \alpha_m V_m (4G_m + 3K_a)}{K_m (3K_a + 4G_m) - 4V_a G_m (K_m - K_a)} \tag{1}$$

in which the subscripts m and a represent the matrix and the aggregate respectively; V, K, and G are the volume fractions, the bulk moduli and the shear moduli of the phases respectively; and α is the expansive strain of the phases. In the case of concrete, the aggregate is the constituent phase that expands during ASR, and the cement paste matrix does not expand. Therefore, Eq. (1) can be simplified by taking a_m as zero.

The composite model used in the present study is based on an assumption that the concrete can be treated as an elastic material, even though experimental results have shown that ASR induced expansion can result in inelastic damage and fracture to concrete. This means that the model developed in this paper is applicable only for predicting the concrete behaviors with initial and moderate ranges of ASR expansions. When severe crack maps due to ASR are observed on concrete, more sophisticated models will have to be developed to account for the inelastic and fracture damages around each individual aggregate as well as the resulting macroscopic crack map on the surface of concrete.

Another important point is that, although this model is capable of introducing as many layers as possible in the basic composite spherical element, the ASR product surrounding the aggregate cannot be chosen as one distinct phase (layer). This is because one of the basic assumptions of the model is that the volume ratios of the spheres in each basic element are fixed for all basic elements with different sizes. This fixed volume ratio requires that the thickness of the ASR product layer surrounding an aggregate must be proportional to the size of the aggregate, which, however, is not true in the case of ASR. Since the ASR layer cannot be treated as an independent phase, one must keep in mind that, when the ASR occurs, the aggregate phase in the model is a composite itself including the ASR product and the unreacted aggregate. The outside layer is the bulk cement paste in the concrete.

The unknown parameter in Eq. (1) is the expansion of aggregate due to ASR, α_a. As discussed earlier, the aggregate of the same mineralogical feature but of a different size reacts in different ways. As a result, the overall ASR expansion of all aggregates (as one phase) should be determined by volumetric average of the ASR expansion of the aggregates with different sizes

$$\alpha_a = \sum \phi_i \alpha_i \tag{2}$$

in which ϕ_i and α_i are the volume fraction and ASR expansion of the aggregate with size R_i. ϕ_i is usually known from the concrete mixing design. In ASTM C1260, five different sizes of aggregates are specified with the range from 4.75 mm to 0.15 mm (#4 to #100 seizes).

6. The Ion Diffusion Model of ASR

As stated earlier, there are two transport processes involved in the ASR. One is the penetration of ions in the pore solution into the aggregate, and the other is the permeation of the ASR gel formed in the aggregate moving out to the interface region of the aggregate. The ASR expansion is actually the combined effect of these two opposing transport processes. The ionic penetration can be described by Fick's law

$$B_{ion}\frac{\partial C_{ion}}{\partial t}=\nabla\left(D_{ion}\nabla C_{ion}\right) \tag{3}$$

in which C_{ion} is the free ion concentration of the pore solution inside the aggregate, which could be hydroxide ions, calcium ions, sodium ions, etc. B_{ion} and D_{ion} are the binding capacity and ion permeability of the aggregate respectively. B_{ion} and D_{ion} may not be simple constants, and they depend on the microstructure of the cement paste as well as the type of ions under consideration. When the angularity of aggregate is low (round shaped aggregate such as river sand), the aggregate may be simulated as a sphere and the same diffusion equation may be formulated in a polar system and solved analytically. The initial condition of the equation is $C_{ion} = 0$ for $t = 0$ in the aggregate. The boundary condition on the edge of the aggregate is $C_b = C_0$ (a constant ion concentration), and the boundary condition at the center of the aggregate is $C_{ion} = 0$. The solution of Eq. (3) is

$$C_{ion}(r,t)=C_0+\frac{2R_iC_0}{\pi r}\sum_{n=1}^{\infty}\frac{(-1)^n}{n}\exp\left(\frac{-\kappa n^2\pi^2 t}{R_i^2}\right)\sin\frac{n\pi r}{R_i} \tag{4}$$

in which $\kappa = D_{ion} / B_{ion}$.

It should be pointed out that at the initial stage of the mortar bar test and also for concrete members used in practice, C_b is usually not a constant, but a function of time. With a time dependent boundary condition, $C_b = C_0(t)$, Eq. (3) can also be solved analytically, the solution will not be listed here. $C_0(t)$ should be obtained from macroscopic diffusion analysis of the whole concrete member.

The ASR occurs in the region of the aggregate where C_{ion} reaches a certain concentration level. Eq. (4) may be solved inversely for the depth of ASR, x, in an aggregate with size R_i. After the ASR depth r is determined, the volume of the reacted aggregate with size R_i during ASR, V_{Ri}, can be calculated

$$V_{Ri} = \left[\frac{R_i^3 - (R_i - x)^3}{R_i^3} \right] \left(\frac{4}{3} \pi R_i^3 \right) = \left[1 - \left(1 - \frac{x}{R_i} \right)^3 \right] \left(\frac{4}{3} \pi R_i^3 \right) \tag{5}$$

Eq. (5) is the reacted volume of one aggregate with size R_i, and the total reacted volume of all aggregates with size R_i is $V_R = V_a \phi_i V_{Ri}$. This volume can then be converted into the volume of the ASR gel created, V_{gel}, $V_{gel} = \eta V_R$, in which η is the volume ratio of the ASR gel to the reacted aggregate, and $\eta > 1$ for regular Portland cement concrete due to the ASR expansion. η can be determined approximately from the reaction kinetics of ASR [Gro 90][Fur 94]. For ashcrete, η varies with time during the accelerated mortar bar test, as shown in Fig. 2 and Fig. 4. η is larger than one during the first few days of the test, and then drops to one in about 10 days, and becomes smaller than one thereafter, which corresponds to shrinkage.

7. The Gel Permeation Model for ASR

The ASR gel accumulates in the interface zone and generates interfacial pressure, which push the gel to permeate through the porous cement paste. The ASR gel permeation can be characterized by Darcy's law for a viscous flow as follows

$$\frac{\partial C_{gel}}{\partial t} = \nabla \left(\frac{K_{gel}}{\eta_{gel}} \nabla P \right) \tag{6}$$

in which C_{gel} and η_{gel} are the concentration and viscosity of the gel respectively; K_{gel} is the gel permeability of the porous cement paste; and P is the interface pressure distribution due to ASR gel. One can see that Eq. (6) is a chemo-mechanical coupled equation, because the interfacial pressure P needs to be evaluated simultaneously from the equilibrium of the composite system, the diffusion of ions and the resulting ASR reaction. Also, Eq. (6) cannot be solved unless a relationship between the pressure P and the gel concentration C_{gel} is established.

In order to solve this coupled equation, a state equation relating C_{gel}, P, and temperature T was introduced first. In the case of ASTM C1260 test, where T can be treated as a constant (80 °C), the state equation may be written as $C_{gel} = \beta P$, where β is the state function for the cement paste. When the pores in the cement paste are saturated by ASR gel, $C_{gel} = C_p$, where C_p is the porosity of interface zone. At the same time, the pressure P reaches the saturation pressure, which can be taken approximately as the tensile strength of the cement paste, f_t. Therefore, $\beta = C_p/f_t$. With the state equation, Eq. (6) can then be expressed in terms of C_{gel} (or P). The initial condition is $C_{gel}(r,0) = 0$. The boundary condition at the interface is $C_{gel}(R_i,t) = \beta P(t)$; and the boundary condition at the far field is $C_{gel}(R_{if},t) = 0$; where R_{if} is the far field distance, which can be taken as half of the average spacing between two aggregates. Eq. (6) can be solved analytically

$$C_{gel}(r,t)=\frac{2\beta}{r\left(R_{if}-R_i\right)}\sum_{n=1}^{\infty}\exp\left[\frac{-\nu n^2\pi^2 t}{\left(R_i-R_{if}\right)^2}\right]\sin\left[\frac{n\pi\left(r-R_i\right)}{\left(R_{if}-R_i\right)}\right]$$
$$\left\{\frac{-n\nu\pi R_i}{\left(R_i-R_{if}\right)}\int_0^t\exp\left[\frac{\nu n^2\pi^2\lambda}{\left(R_i-R_{if}\right)^2}\right]P(\lambda)d\lambda\right\} \tag{7}$$

in which $\nu = K_{gel} / \eta_{gel}$.

One can see that Eq. (7) is not a closed-form solution because $P(t)$ as the boundary condition at the interface is an unknown and is a function of time. A step by step numerical procedure was used to solve the coupled equations. The details of the numerical procedure can be found elsewhere [Suw et al. 98].

The gel volume in the porous cement paste surrounding an aggregate at time t, V_{gpi}, may be evaluated by integrating over the surrounding cement paste around the aggregate of size R_i,

$$V_{pgi}=\int_{R_i}^{R_{if}}4\pi r^2 C_{gel}\,dr \tag{8}$$

The total ASR gel volume absorbed in the pores of all aggregates with size R_i is $V_{pg}=V_a\phi_i V_{pgi}$. The total volume change due to ASR of all aggregates with size Ri, that is, the volume of the ASR gel beyond the accommodating capacity of the interface zone of the aggregates, is $\Delta V_i=V_{gel}-V_{pg}$. Then by assuming the isotropic behavior for concrete, the coefficient of expansion for the aggregate with size R_i due to ASR is

$$\alpha_i=\frac{\Delta V_i}{V_{ai}}=\frac{V_{gel}-V_{pg}}{\phi_i V_a} \tag{9}$$

The volume change ΔV_i generates a pressure between the aggregate and the matrix, which is $P(t)$. With given ΔV_i, $P(t)$ can also be solved from the equilibrium of the composite system. The $P(t)$ thus solved must be equal to the $P(t)$ in the state equation, which was used to solve Eq. (6). Substituting Eq. (9) into Eq. (2) gives the final expression for ASR expansion for all aggregates together. Then using Eq. (1), the effective ASR expansion of concrete can be determined.

As one can see, there are many material parameters involved in the two transport models and the composite model which need to be determined by considering physical, chemical, and microstructural features of cement paste and aggregate. Determination of these parameters will not be discussed further, and can be seen elsewhere [Xi et al. 98].

Eqs. (1) - (9) form a complete model for ASR expansion. Some preliminary results from the prediction model are shown in Figs. 6-9. Figure 6 shows the profile of ion concentration in an aggregate; Fig. 7 shows the profile of ASR gel concentration in the interface zone of the aggregate. The cut-off zone near the interface shows that ASR gel concentration reaches the interface porosity, which means the pores are filled up with ASR gel. Fig. 8 shows the development of the interface pressure with time; and Fig. 9 shows the increase of ASR expansion with time. The basic trends predicted by the model are all correct.

8. Conclusions

The new cementitious material, ashcrete, is very effective in reducing ASR expansion. The maximum ASR expansion of ashcrete with natural aggregate as well as with glass aggregate is much lower than the specified critical value from ASTM C1260.

The expansion vs. time curves of ashcrete are completely different from those of regular concrete. The maximum ASR expansion of ashcrete appears at about 10 days of accelerated testing, and then shrinkage occurs. The shrinkage is very low, and does not create any shrinkage cracking.

The Pessimum effect is different between regular concrete and ashcrete. The maximum expansion of regular concrete is induced by #50 aggregate, while the maximum expansion of ashcrete with the same aggregate is caused by #16 aggregate.

A theoretical model is developed for predicting expansion of ASR. The model takes into account the transport-mechanical coupling of ASR expansion. The mechanical part is characterized by a modified generalized self-consistent model. The transport behaviors are described by two opposing transport processes: ion penetration from the pore solution into aggregate; ASR gel permeation from the aggregate-cement paste interface to the far field. The overall ASR expansion is the combined effect of the two transport processes and the interface pressure generated by the ASR gel.

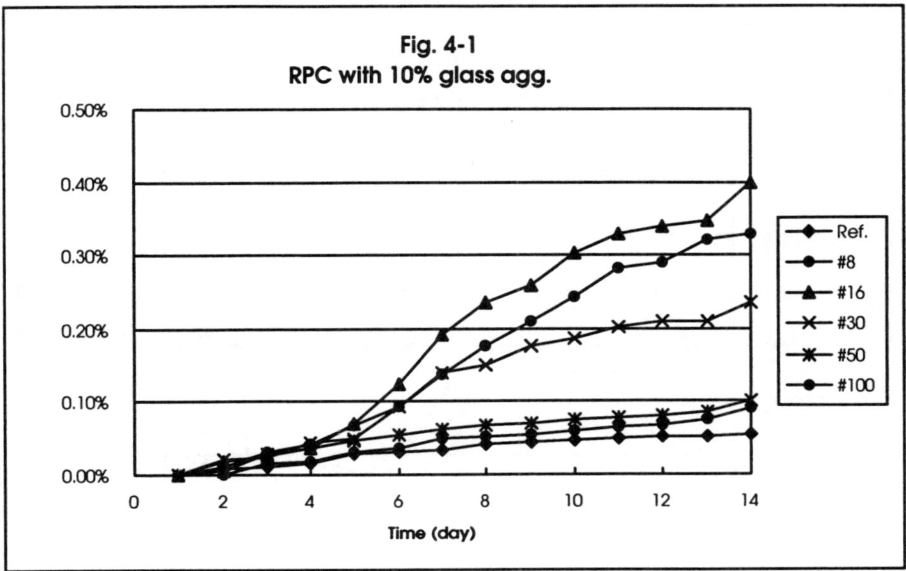

Figure 1. *ASR test of regular Portland cement (RPC) concrete samples with 10% glass aggregate of different sizes*

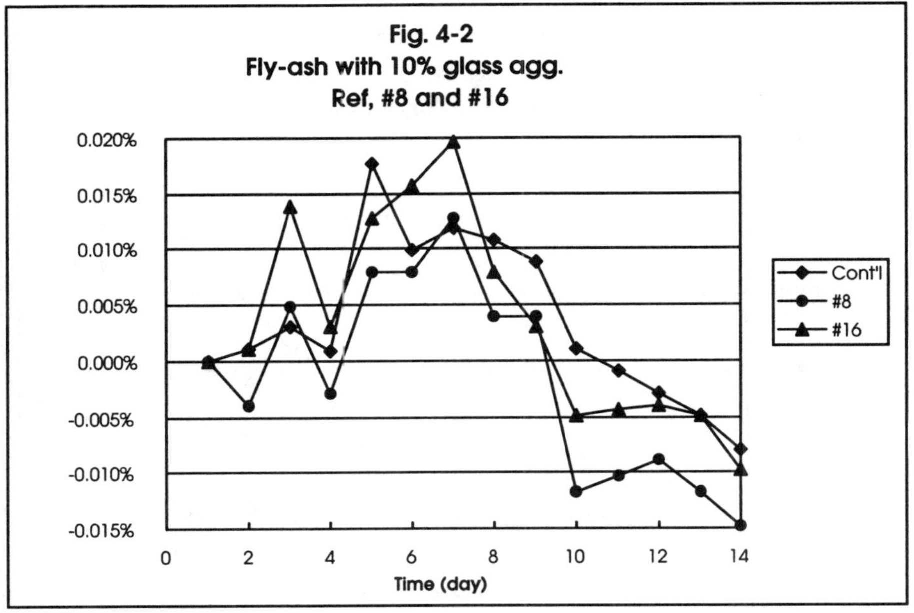

Figure 2(a). *ASR test of ashcrete samples with 10% glass aggregate of different sizes*

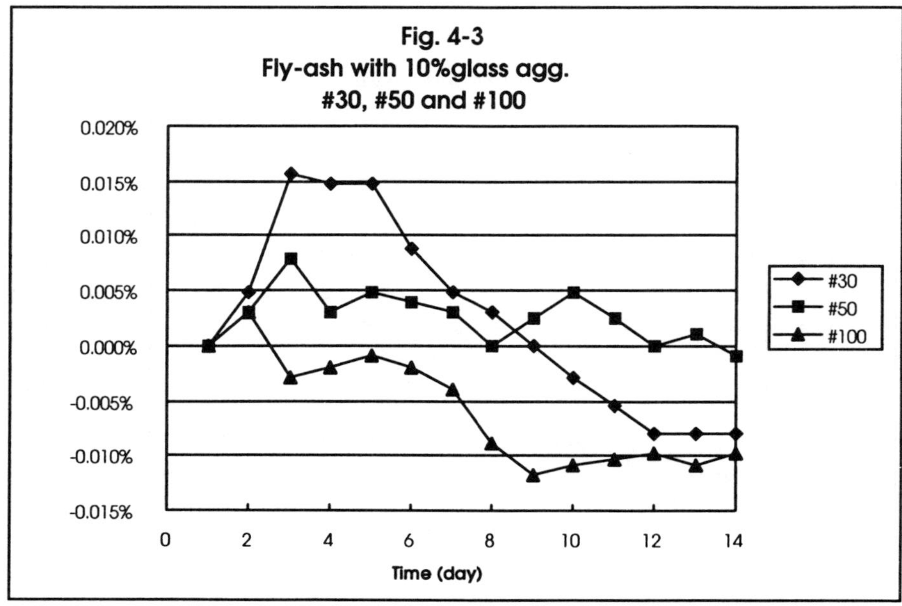

Figure 2(b). *ASR test of ashcrete samples with 10% glass aggregate of different sizes*

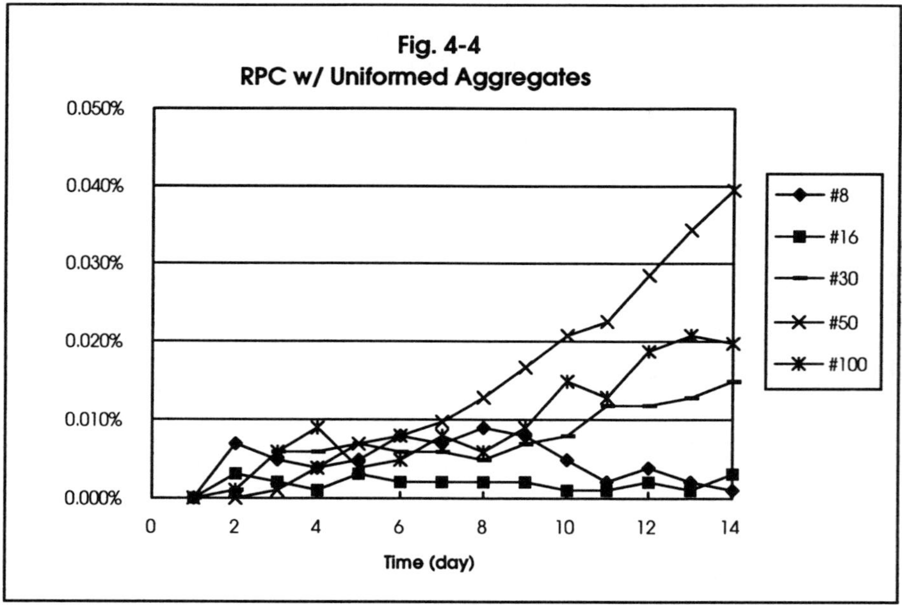

Figure 3. *ASR test of regular concrete samples with natural aggregate of one size*

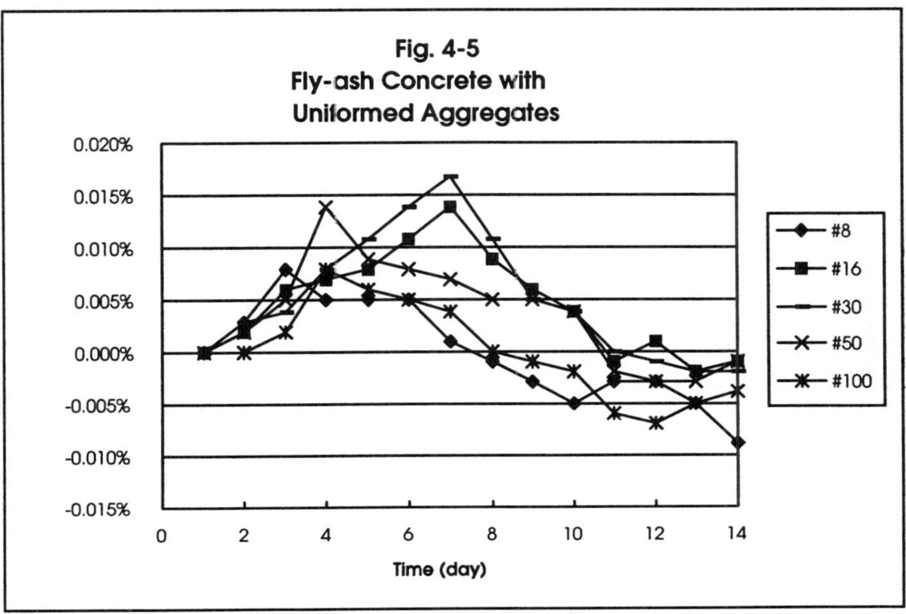

Figure 4. *ASR test of ashcrete samples with natural aggregate of one size*

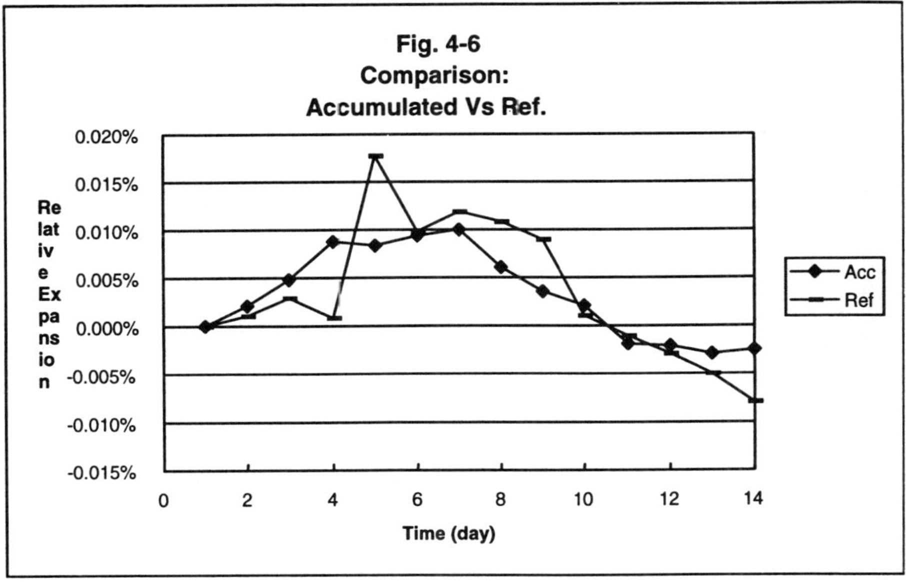

Figure 5. *Comparison of the weighted average ASR cuve (accumulated) with the Ref. curve shown in Fig. 2*

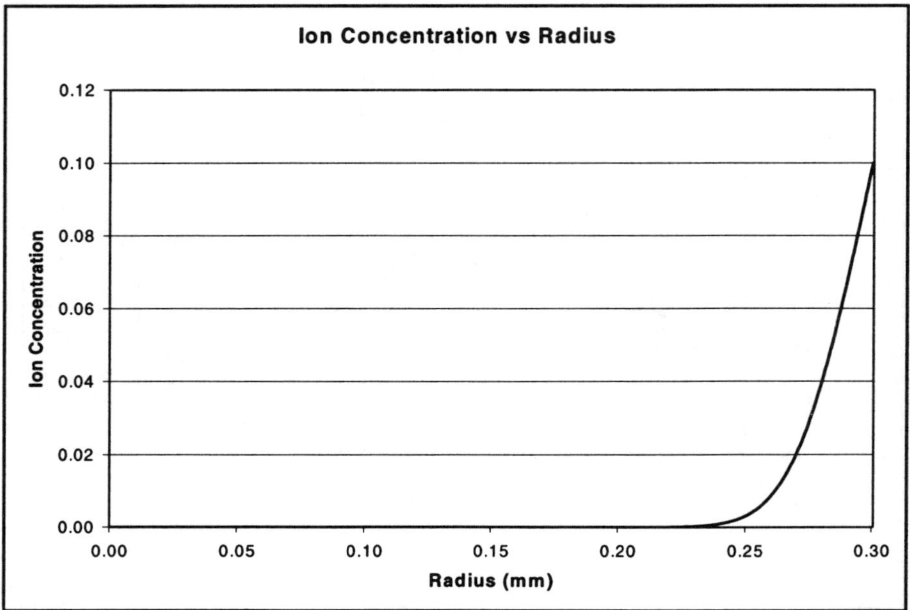

Figure 6. *The Profile of ion concentration in the aggregate*

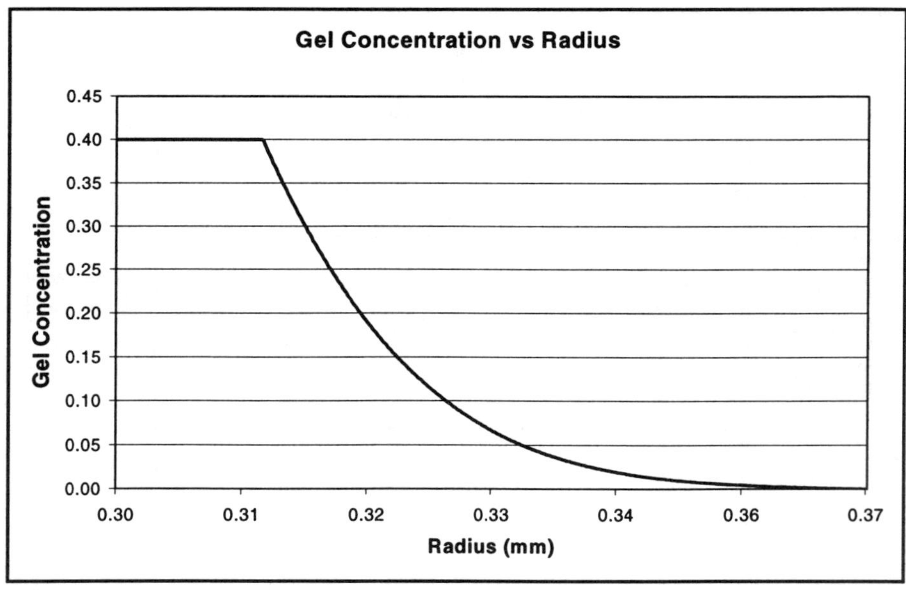

Figure 7. *The Profile of ASR gel concentration in the interface zone of the aggregate*

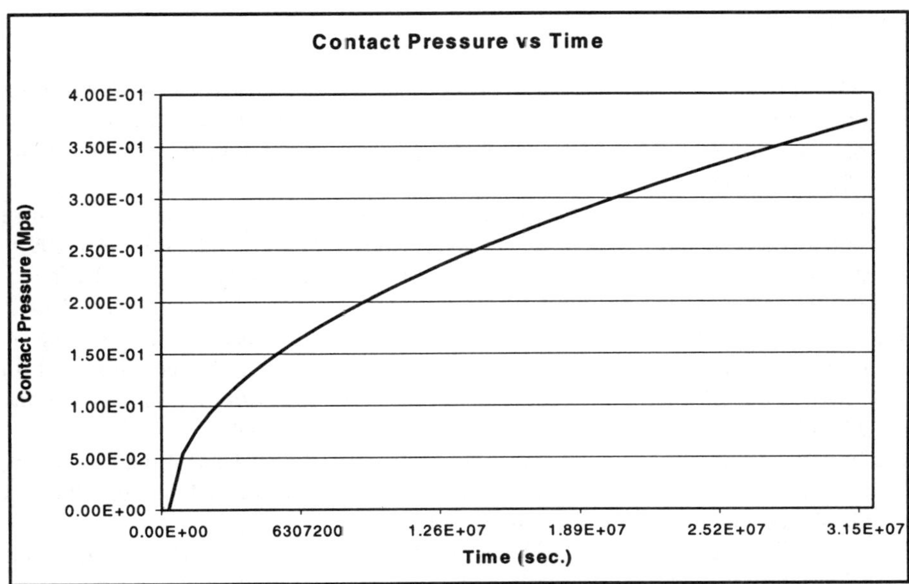

Figure 8. *The development of the interface pressure with time*

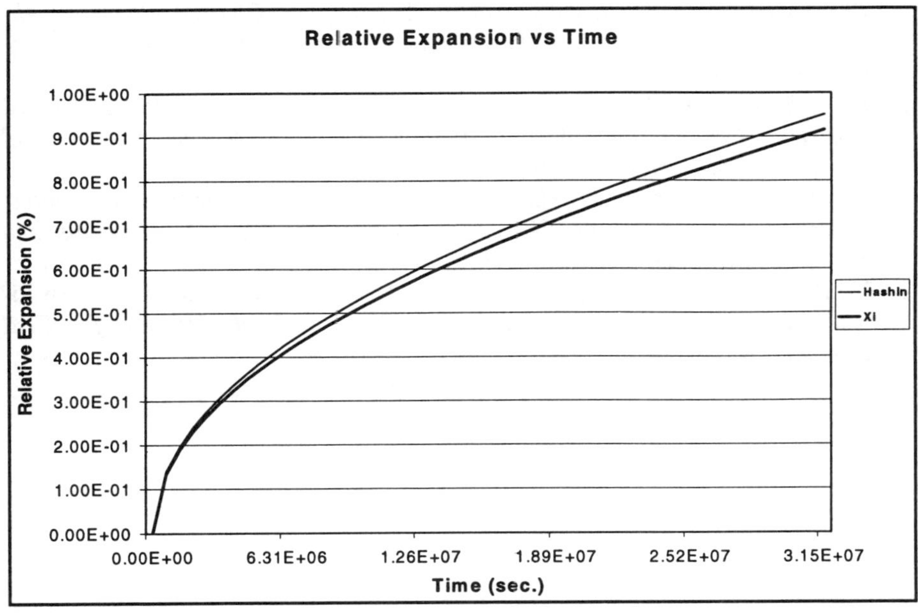

Figure 9. *The increase of ASR expansion with time*

References

[Fur 94] Furusawa, Y., Ohga, H., and Uomoto, T. "An Analytical Study Converning Prediction of Concrete Expansion Due to Alkali-Silica Reaction", *Proc. of 3rd Int. Conf. on Durability of Concrete*, Nice, France, Editor Malhotra, SP 145-40, 757-780, 1994.

[Gro 90] Groves, G.W., and Zhang, X. "A Dilatation Model for the Expansion of Silica Glass/OPC Mortars", *Cement and Concrete Research*, 20, 453-460, 1990.

[Jin 98] Jin, W. *Alkali-Silica Reaction in Concrete with Glass Aggregate - A Chemo-Physico-Mechanical Approach*, Ph.D. Dissertation, Columbia University, 1998.

[Jin et al. 98] Jin, W., Suwito, A., Meyer, C., and Xi, Y. "Theoretical Modeling on Expansion and Damage due to Alkali-Silica Reaction", Proceedings of 12th Engineering Mechanics Conference: *Engineering Mechanics: A Force for the 21st Century*, May 18-20, San Diego, CA., 1175-1178, 1998.

[Pre 97] Prezzi, M., Monteiro, P.J.M., and Sposito, G. "The Alkali-Silica Reaction, Part I: Use Double-Layer Theory to Explain the Behavior of Reaction -Product Gels", *ACI materials journal*, 10-17, 1997.

[Sam 96] Samadi, A. *Treatment of Fly Ash to Increase its Cementitious Characteristics*, Ph.D. Dissertation, Drexel University, Philadelphia, PA., 1996.

[Sam et al. 95] Samadi, A., Xi, Y., Martin, J.P., and Cheng, J. "A Unique Concrete Made with Fly Ash and Sodium Silicate Solution", *Proc. of April Meeting, Ame. Cer. Soc.*, Cincinnati, OH., 1995.

[Sil 97] Silverstrim, T., Rostami, H., Xi, Y., and Martin, J. "High Performance Characteristics of Chemically Activated Fly Ash (CAFA)", *Proc. of the PCI/FHWA Int. Sym. On High Performance Concrete*, New Orleans, Louisiana, Oct. 20-22, 135-147, 1997.

[Suw 98] Suwito, A., Xi, Y., Jin, W., and Meyer, C. "Solution for a Coupled Transport-Mechanical Problem for Porous Media", to be submitted to *J. of Engineering Mechanics*, ASCE.

[Xi 97] Xi, Y., and Jennings, H.M. "Shrinkage of Cement Paste and Concrete Modelled by a Multiscale Effective Homogeneous Theory", *Materials and Structures (RILEM)*, 30, July, 329-339, 1997.

[Xi 98] Xi, Y., Suwito, A., Jin, W., and Meyer, C. "Theoretical Modeling of Alkali-Silica Reaction in Concrete", to be submitted to *J. of Materials for Civil Engineering*, ASCE.

Measurement of Pore Water Pressure in Concrete and Fracturing Concrete by Pore Pressure

Hideki Oshita * — Tada-aki Tanabe **

* *Department of Civil Engineering of Chuo University*
1-13-27, Kasuga, Bunkyou-ku
Tokyo, Japan

** *Department of Civil Engineering of Nagoya University*
Hurouchо, Chikusa-ku, Nagoya
Aichi, Japan

ABSTRACT. *Water migration characteristics in concrete is one of the main components to be considered in designing marine structures, underground structures and water containing structures. The radioactive waste disposal facility may be another important structure for which the same consideration is necessary. In the first part of the study, the behaviour is examined measuring the pore pressure and pore pressure gradient inside of concrete as well as developing the mathematical model for analysing the phenomena. In the second part of the study, experiments of fracturing concrete by pore pressure alone are performed. A fracture energy model based on the effective stress is developed and concrete fracture in which pore water pressure is existing is discussed.*

KEY WORDS: *Water Migration, Pore Water Pressure, Total Stress, Effective Stress.*

Introduction

There exist quite numbers of concrete structures for which water proofing is the imperative. For those structures, water migration inside of concrete should be studied in detail since reasonable water proofing technique may be established only after the true mechanism of water migration is clarified. Beside these, the mechanical characteristics of concrete affected sometimes beyond the extent we usually assume by the existence of water in the pore. For example, it is noted that the larger the water content of concrete specimen, the smaller the bending and shear strength. The experiment of shear strength of reinforced concrete beams carried out at Tottori University [Ino 90] showed that the shear strength decreases for the specimen which contains more water compared with the one with less water. М и х а й л о в [М и х 74] showed that the decrease of the uni-axial compressive strength of concrete occurs exponentially due to the increase of water content. For the creep rate of concrete at early ages, there exist several reports that tell the degree of water migration effect is much larger than the other factors, e.g. [Mor 93], [Osh 93], [Tan 93].

Although there are many phenomenon concerned with the water migration in concrete, study in this direction has been relatively scarce. Moreover, a study has hardly been carried out in which pore water pressure occurring is measured.

In understanding the water migration in concrete, measurement of pore water pressure, its gradient and the quantitative estimation of water migration in concrete are indispensable.

In this study, the experimental apparatus to measure the pore water pressure occurring in concrete is developed and the pore pressure for various states of concrete are measured in various loading conditions. Moreover, the experiment in which concrete specimens are subjected to high pore pressure and failure is generated by pore pressure, were performed.

Finally, the permeable and deformational theory of concrete proposed by authors is evaluated and a new analytical theory to identify the fracture energy for concrete with pore water is presented.

Part I. Measurement of Pore Water Pressure in Concrete and the Analytical Model

1. Experimental Set-up

1.1. *Development of Pressure Cell which allows External Loading*

A new pressure cell which contains a concrete specimen and is capable of pressurizing a specimen in 3D space with the uni-axially applicable external loading apparatus was developed. The standard non-saturated tri-axial cell proposed by [JGS 89] is improved so as to be applicable for concrete material. The apparatus of developed cell is shown in figure1.1. The cell is consisting of steel chamber having the diameter of 250mm and the height of 600mm. The concrete specimen is a cylinder having the diameter of 100mm and the height of 200mm and the pore water in concrete is undrained during testing due to the coating of all the surface of a specimen by special rubber sleeve. The pressure meters measuring pore water pressure and pore air pressure, are set up at the top and bottom of the specimen as shown in figure 1.2 and the glass fiber filter and ceramic plate are set at each interfaces. The glass fiber filter is made of glass and passes air only, therefore the pore air pressure alone can be measured. The ceramic plate is made of pottery and passes water and air, therefore the pore water pressure and pore air pressure, i.e. pore

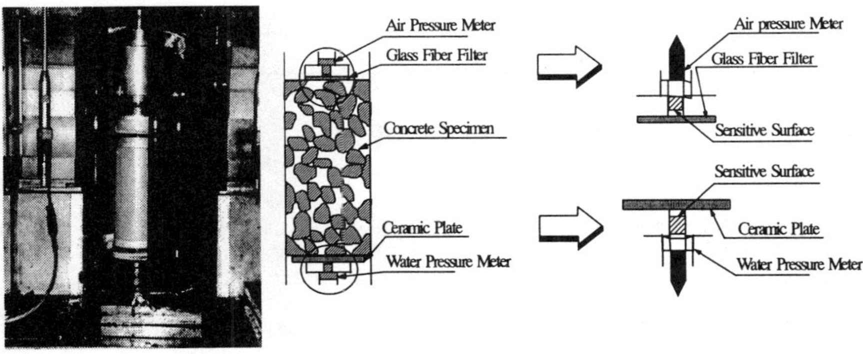

Figure 1.1. *Apparatus Pressure*

Figure 1.2. *Water Pressure Meter and Air Meter of Developed Cell*

mixed pressure, can be measured. The pore water pressure is measurable indirectly by the differences of the pore mixed pressure and pore air pressure measured at the top and the bottom of a specimen. The pressure resistance of steel chamber is 10MPa and the maximum capacity of pressure meter set up at the top and bottom of the specimen is 2MPa and 10MPa, respectively.

1.2. *Factors adopted in the Experiment*

Since the extent of pore pressure generation by lateral loading and external compressive loading is not known in the past, various preparatory experiments have been performed and finally the following factors are adopted to clarify the pore pressure generation mechanism. Namely, the first factor to be considered is rigidity of concrete skeleton, since pore pressure generation may be affected by its rigidity. The second factor to be considered is lateral confining effect on the generation of pore pressure. The third factor adopted is the extent of water content in concrete. To realize these objectives, the experimental parameters shown in table 1.1 are adopted. They are loading age, lateral pressure and curing method.

Loading Age	Examination of the effect of concrete skeleton strength on pore water pressure
Lateral Pressure	Rxamination of the effect of confined effects on pore water pressure
Curing Method	Examination of the effect of water content on pore water pressure

Table 1.1. *Experimental Parameters and Objectives*

Each specimen bears the identification label in which each letter signify certain meanings which is demonstrated in the following taking the example of (Ⅲ-0.5-W)

Ⅲ : loading age (day) (=3days)

0.5 : lateral pressure (MPa) (=0.5MPa)

W : cured in water (A : cured in air)

Specimen No.	Disposal Method of Concrete Specimen	Pressur (MPa)
Ⅲ-0.5-W	A B,D Air Curing C; Age (days 0 1 2 3	0.5
Ⅲ-1.0-W		1.0
Ⅲ-2.0-W		2.0
Ⅶ-0.5-W	A B Water Curing C; Age (days 0 1 2 3 4 5 6 7	0.5
Ⅶ-1.0-W		1.0
Ⅶ-2.0-W		2.0
Ⅶ-0.5-A	A B Air Curing C; Age (days 0 1 2 3 4 5 6 7	0.5
Ⅶ-1.0-A		1.0
Ⅶ-2.0-A		2.0

A: PlacingB: Removal of Forms

Table 1.2. *Curing Method of Concrete Specimen*

Total numbers of specimens are made as listed in table 1.2 and have been tested according to the schedule. It's should be noted that the adoption of these parameters is not directly related to young age concrete.

In the initial stage, the 3D pressure applied to the specimens is raised to the designated pressure value, then the uni-axial load was applied at an end of the cylindrical spcemen.

The loading machine is a universal testing machine of which capacity is 100ton. The displacement control loading with the loading speed of $0.05mm/\sec$, i.e., strain speed of $250\mu/\sec$ is adopted. The loading stress initially increase and attain peak, then decrease to almost zero gradually. During the experiments, the lateral pressure has been kept constant.

2. Pore Water Pressure Generation due to Deformation

2.1. *General Tendency of Pore Water Pressure Generation*

In this section, the effects of rigidity of skeleton, lateral pressure and water content on pore water pressure in the pre-peak region and the post-peak region will be discussed. Before entering the discussion, the sign convention of stress and strain

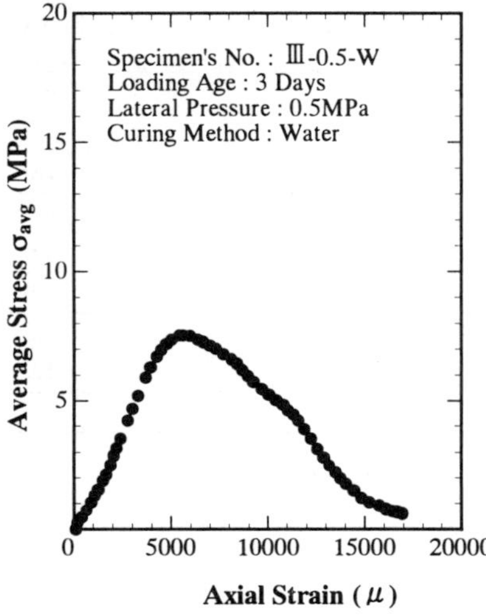

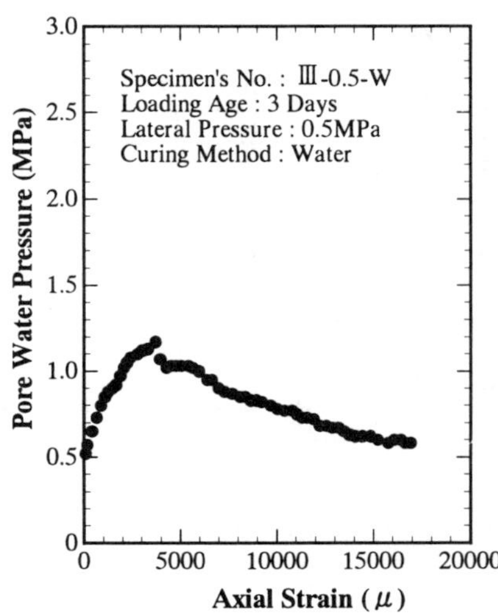

(a) *Characteristics of Average Stress* (b) *Characteristics of Pore Water Pressure*

Figure 1.3. *General Tendency of Pore Water Pressure*

used should be mentioned. As the conventional usage of positive sign for the compressive pore pressure, every strain and stress in the compressive direction are denoted as positive while the tensile strain and stresses are taken as negative. In the beginning, general trend of pore pressure generation is explained taking Ⅲ-0.5-W specimen as an example. The experimental results of Ⅲ-0.5-W specimen, in which the loading age is 3 days, and the lateral pressure is 0.5MPa and which is cured in water after removal of forms, were shown in figure 1.3. Figure's 1.3(a), 1.3(b) show the relationship between the average stress and total strain, and the pore water pressure and total strain, respectively. Here, the average stress σ_{ave} is defined as follows.

$$\sigma_{ave} = \frac{I_1}{3} = \sigma_1 + \sigma_2 + \sigma_3 \qquad [1.1]$$

σ_{ave} : average stress

I_1 : first order invariants of stress tensor

σ_α : principal stress (α =1, 2, 3)

As shown in figure 1.3(b), the positive pore water pressure occur with the external loading, reaches the peak and decreases. The occurrence of maximum pore water pressure looks to correspond to the occurrence of yield strain of concrete or the maximum volume strain of concrete which is not necessarily be the maximum loading point. The yielding strain is corresponding to the strain at which concrete becomes plastic from elastic. The maximum value of pore water pressure is almost 1.2MPa, which is 15% of the maximum average stress. There exists no doubt for that these measured value is the value of pore water pressure which is occurring in concrete judging from the calibrated function of developed apparatus.
Therefore, the experiment has, for the first time, confirmed the existence of pore water pressure in concrete with the order of MPa.

2.2. *Characteristics of Pore Water Pressure due to Deformation*

The value of pore water pressure at the application of lateral pressure is almost same as the value of given lateral pressure as shown in figure 1.3. These tendencies appeared for all specimens except for the specimens cured in air. It may be said that the concrete specimen are nearly fully saturated since equilibrium of pore pressure and lateral confining pressure can only be assured by the full saturation of concrete pore under the condition of strict hindrance of water migration from outside.
The pore water pressure occurring due to the compressive load after applying the lateral pressure is caused by the deviatoric stress shown in the following equation.

$$P(\sigma_1-\sigma_3) = \frac{2}{3}(p-\sigma_3) \qquad [1.2]$$

Therefore, the deviatoric stresses and pore water pressure caused by the deviatoric stresses were used in the following examination,
Figure 1.4 and figure 1.5 show the experimental results for the specimen Ⅲ-0.5,1.0,

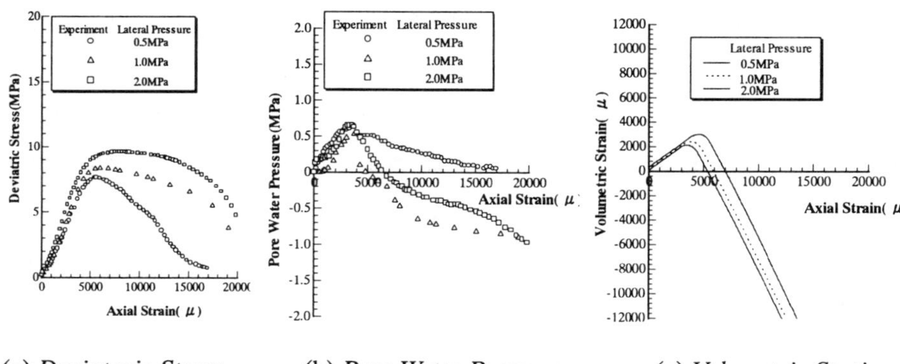

(a) *Deviatoric Stress* (b) *Pore Water Pressure* (c) *Volumetric Strain*

Figure 1.4. *Comparison of Lateral Pressure (Loading Age: 3 Days)*

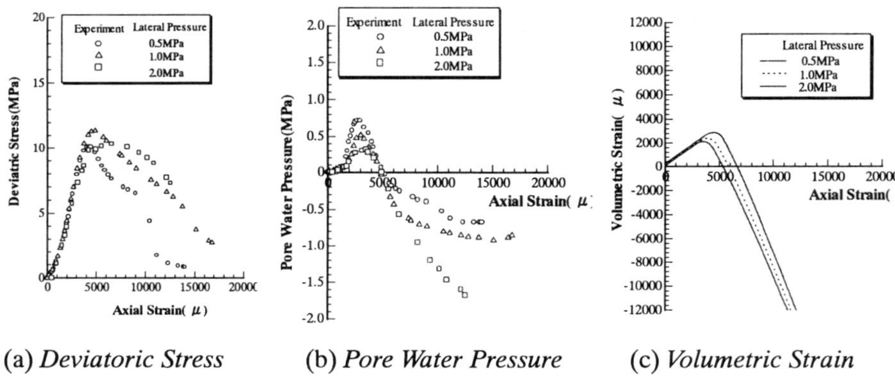

(a) *Deviatoric Stress* (b) *Pore Water Pressure* (c) *Volumetric Strain*

Figure 1.5. *Comparison of Lateral Pressure (Loading Age: 7 Days)*

2.0-W and Ⅶ-0.5, 1.0, 2.0-W of which loading age is 3 and 7 days ,(i.e. week skeletal concrete versus strong skeletal concrete)and which are cured in water after removal of forms, respectively. Figure's 1.4(a).(b),(c), and figure's 1.5(a), (b), (c) show the relationship between the deviatoric stress and axial strain, between the pore water pressure and axial strain and between the volumetric strain and axial strain, respectively. In each figures, the experimental results were shown with circle, triangle and square for the cases of lateral pressure of 0.5, 1.0, 2.0MPa, respectively. In the pre-peak region, the positive (compression) pore water pressure gradient occur due to the increase of deviatoric stress, while in the post-peak region, the negative (tension) pore water pressure gradient occur and finally the value of pore water pressure becomes negative, as shown in figure 1.4(b) and figure 1.5(b). Comparing such tendency of pore water pressure with the deviatoric stress, it is recognized that the axial strain at the maximum value of positive pore water pressure corresponds to the yield strain or the strain at the maximum value of deviatoric stress, as shown in figure's 1.4(a), 1.5(a). Moreover, the strain also corresponds to the maximum value of volumetric strain, as shown in figure's 1.4(c), 1.5(c), though it should be mentioned that figure 1.4(c) and figure 1.5(c) are analytical values for which the detailed description is made in the following section. Namely, in the pre-peak region, the positive pore water pressure gradient occur due to the volumetric contraction of concrete specimen, while in the post-peak region, the negative pore water pressure gradient occur due to the volumetric expansion. Therefore, it should be noted that the main effective factor on the pore water pressure generation is the volume changes of concrete skeleton.

2.2.1. *Effect of Lateral Pressure on Pore Water Pressure*

The first observation of the test results is that the maximum value of pore water pressure is constant of 0.6MPa in the pre-peak region with no relation to the magnitude of lateral pressure in case of 3 days of loading age (weaker skeleton) as shown in figure 1.4(b) while its value becomes smaller with the increase of lateral pressure in case of 7 days of loading age (stronger skeleton) as shown in figure 1.5(b). These difference is discussed with analytical results in the next section, however, it is related with relative rigidity of skeletal system and the volume contraction. The less pronounced generation of pore pressure may be interpreted as that the increase of lateral confinement resulted in less volume contraction.

The second observation of the test results is that in the post-peak region, the rapid decrease of negative pore water pressure occur in the range of axial strain of 5000 ~7000 μ and then it becomes more gradual beyond the range of axial strain of 7000 μ with no relation to the lateral pressure and comes close to the atmospheric pressure.

2.2.2. *Effect of Skeletal Strength on Pore Water Pressure*

Figure 1.6 to figure 1.8 show the experimental results with the lateral pressure of 0.5, 1.0 and 2.0MPa, respectively and show the comparison of the loading age for the specimen Ⅲ-0.5-W and Ⅶ-0.5-W, Ⅲ-1.0-W and Ⅶ-1.0-W, Ⅲ-2.0-W and Ⅶ-2.0-W, respectively. Figure's (a), (b) and (c) show the relationships between the deviatoric stress and axial strain, between the pore water pressure and axial strain and between the volumetric strain and axial strain, respectively and then the experimental results were shown with circle and triangle as the loading age of 3 and 7 days, respectively.

As shown in figure's 1.6～1.8, the maximum value of deviatoric stress at the loading age of 3 days is naturally smaller than that of 7days, however, the maximum value of pore water pressure is similar with no relation to the loading age i.e. skeletal strength variation. From the volumetric strain shown in figure's 1.6(c)～1.8(c), this reason may be seen that the volumetric contraction stain in the pre-peak region has almost same value on each loading ages. The remarkably different effect of loading age(i.e. skeletal rigidity) is that the sudden reduction of deviatoric stress in the post-peak region occur at the loading age of 7 days rather than that of 3 days and more brittle failure are seen for the specimen of the loading age of 7 days, as shown in figure's 1.6(a) ～ 1.8(a). These brittle behavior influences the characteristics of pore water pressure i.e, the sudden decrease of pore water pressure as shown in figure's 1.6(b)～1.8(b). From the volumetric strain shown in figure's 1.6(c)～1.8(c), this reason may be attributed to that the volumetric expansion strain in the post-peak region at the loading age of 7 days is larger than that of 3 days and indicates that cracks formed develop rather rapidly. The above phenomenon is confirmed by the experimental results shown in figure 1.9 in which crack width is very large at the loading age of 7 days rather than that of 3 days.

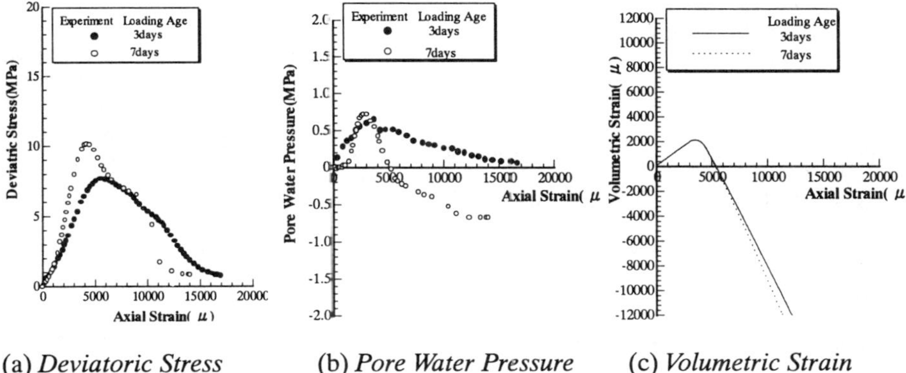

(a) *Deviatoric Stress* (b) *Pore Water Pressure* (c) *Volumetric Strain*

Figure 1.6. *Comparison of Loading Age (Lateral Pressure: 0.5 MPa)*

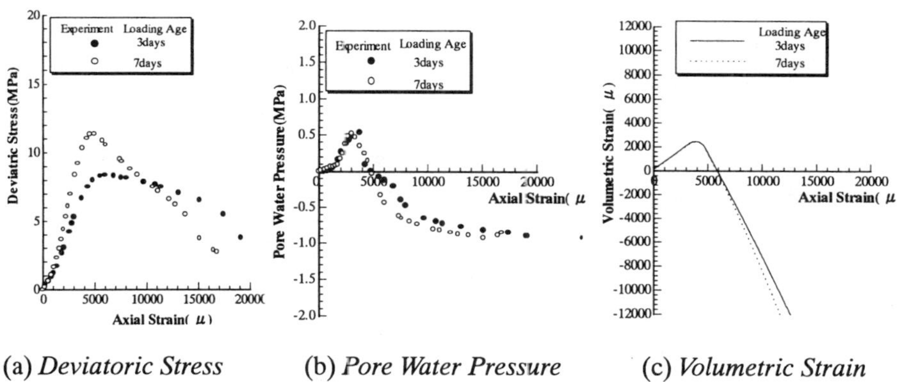

(a) *Deviatoric Stress* (b) *Pore Water Pressure* (c) *Volumetric Strain*

Figure 1.7. *Comparison of Loading Age (Lateral Pressure: 1.0 MPa)*

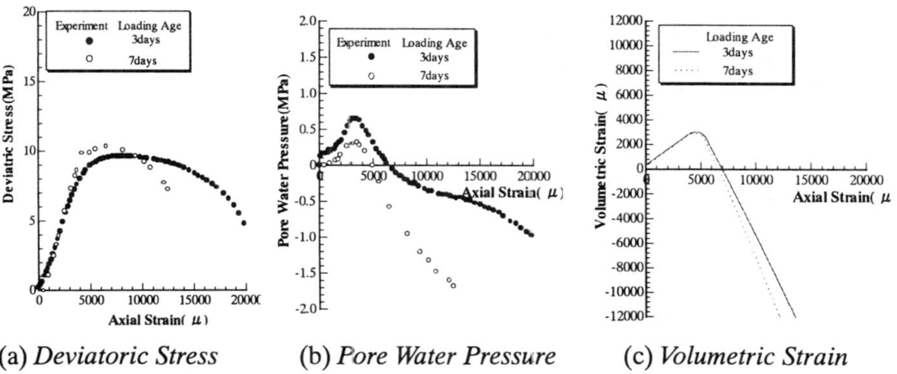

(a) *Deviatoric Stress* (b) *Pore Water Pressure* (c) *Volumetric Strain*

Figure 1.8. *Comparison of Loading Age (Lateral Pressure: 2.0 MPa)*

2.2.3. *Effect of Water Containment on Pore Water Pressure Generation*

Figure's 1.10 ~ 1.12 show the experimental results for the lateral pressure of 0.5, 1.0, 2.0MPa, respectively and show the comparison of the curing method for the specimenⅦ-0.5-W and Ⅶ-0.5-A, Ⅶ-1.0-W and Ⅶ-1.0-A, Ⅶ-2.0-W and Ⅶ-2.0-A, respectively. Figure's(a),

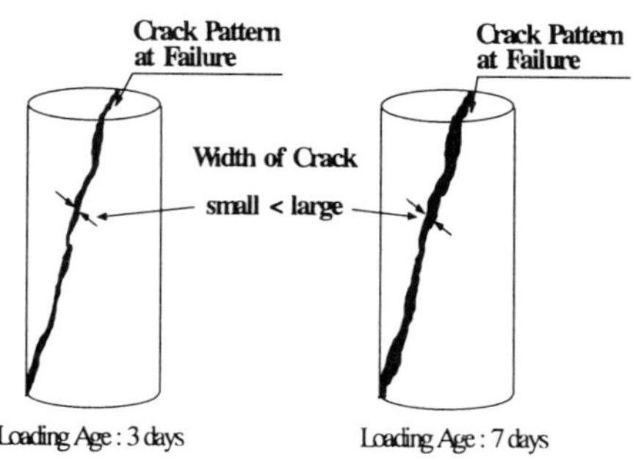

Figure 1.9. *Crack Pattern*

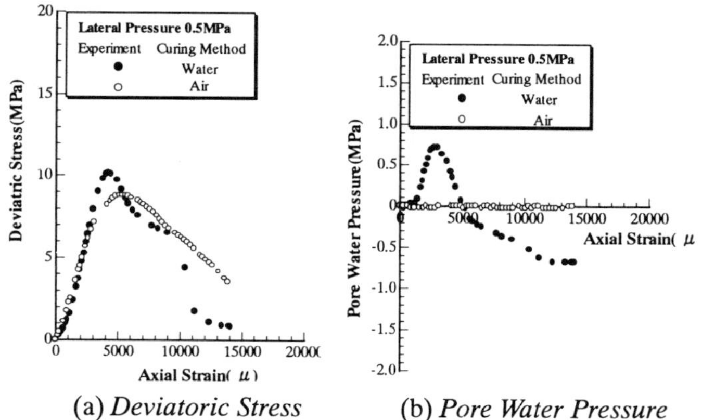

(a) *Deviatoric Stress* (b) *Pore Water Pressure*

Figure 1.10. *Comparison of Curing Method (Lateral Pressure: 0.5 MPa)*

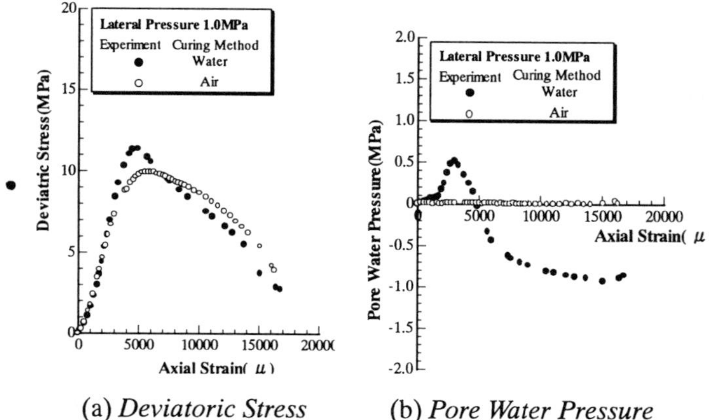

(a) *Deviatoric Stress* (b) *Pore Water Pressure*

Figure 1.11. *Comparison of Curing Method (Lateral Pressure: 1.0 MPa)*

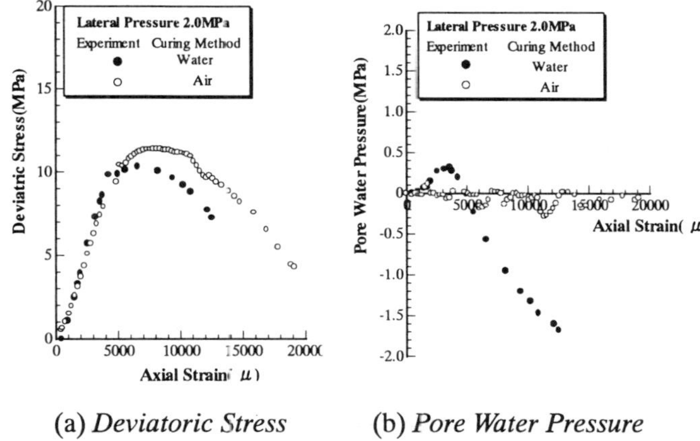

(a) *Deviatoric Stress* (b) *Pore Water Pressure*

Figure 1.12. *Comparison of Curing Method (Lateral Pressure: 2.0 MPa)*

(b) show the relationship between deviatoric stress and axial strain, between pore water pressure and axial strain, respectively. The experimental results were shown with the black circle for water cured specimen, and white circle for air cured specimen.

The pore water pressure of the specimen cured in air does not occur at all due to low water content , as shown in figure's 1.10(b)~1.12(b). Therefore, the total stress for the specimen cured in air is same as the effective stress which occur in the skeleton phase.

As shown in figure's 1.10(a)~1.12(a), the maximum value of the deviatoric stress for the specimen cured in water with the lateral pressure of 0.5 and 1.0MPa is larger than that of cured in air, while for the specimen cured in water with the lateral pressure of 2.0MPa ,the tendency has changed to the contrary and the peak value is smaller than that of cured in air. We consider that it is simply due to the mismatching of the transition point of volume contraction and skeletal peak point affected by the stronger lateral pressure. Namely, the occurrence of negative pore pressure at the peak skeletal strength point has happened. These phenomena is dully possible as indicated in figure 1.13.

So far, we have discussed pore pressure generation in concrete mainly from experimental investigation. In the following chapter, the analytical model we

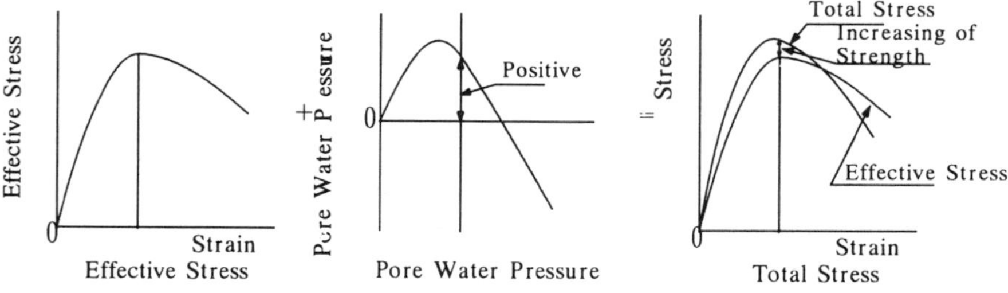

(a) *Positive pore Water Pressure (Compression)*

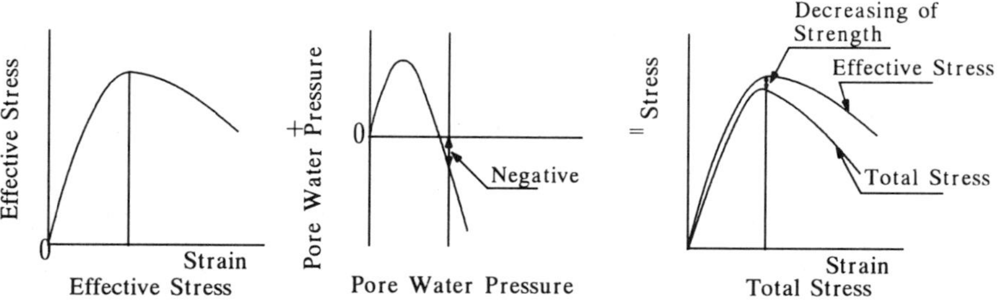

(b) *Negative Pore Water Pressure (Tension)*

Figure 1.13. *Characteristics of Compressive Strength*

adopted to analyze the phenomena, will briefly introduced and the experimental results will be reviewed back again in more detail.

3. Analytical Model for Water Migration

Concrete as non homogeneous material is assumed to be porous composite material which consist of aggregate, cement paste, water, interfacial cracks occurring at interface of aggregate and cement paste and relatively large crack band forming the fracture surface. The crack categories are due to flow characteristics of water. The modeling of concrete is shown in figure 1.14. The symbols which denote the aggregate, cement paste, water, interfacial cracks and crack band are A, C, W, IC, and CB, respectively in the modeling. The symbols $V^{(A)}$,$V^{(C)}$,$V^{(CB)}$ as shown in figure 1.14 denote the volumes of aggregate, cement paste, crack band and the

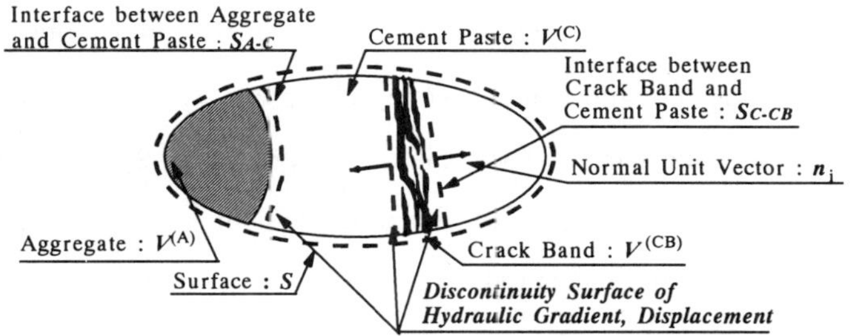

Figure 1.14. *Modeling of Concrete as Non-Homogeneous Material*

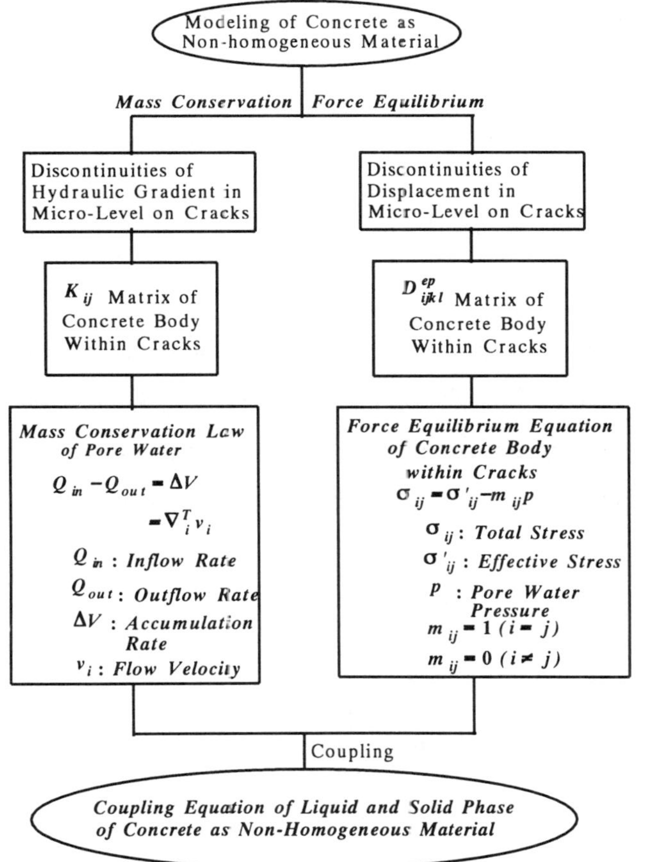

Figure 1.15. *Flowchart of Modeling of Concrete*

symbols S, S_{A-C}, S_{C-CB} denote the surface area of concrete body, interfacial surface area between aggregate and cement paste, of cement paste and crack band. The permeability matrix and elasto-plastic matrix of concrete body having cracks were formulated assuming discontinuity of hydraulic gradient and displacement on crack surface in micro level and the analytical theory for water migration in concrete as a non homogeneous material was developed through the coupling of mass conservation law and force equilibrium equation, as shown in figure 1.15.

Though the details of the theory can be referred to [Osh 93], the equilibrium equation which takes the pore water pressure effect into account is written as equation [1.3] as a differential forms using the principle of virtual work and using the appropriate shape functions.

$$K_{T_{ij}}\frac{d\bar{u}_j}{dt} - L_j\frac{d\bar{p}}{dt} - A_j\frac{d\bar{T}}{dt} - \frac{d\bar{f}_j}{dt} = 0 \qquad [1.3]$$

where the vector $\bar{u}_i$ and $\bar{p}$ are the nodal displacement vector and nodal pore water pressure, respectively. The matrices $K_{T_{ij}}$, L_i and A_i are the tangential stiffness matrix, the effect of pore water pressure on the compressibility of solid phase and the effect of temperature, respectively. The vector $d\bar{f}_i$ is the incremental external force.

The mass conservation law of pore water is written as

$$-H \cdot \bar{p} - L_i^T \frac{d\bar{u}_i}{dt} - S\frac{d\bar{p}}{dt} - W\frac{d\bar{T}}{dt} - \frac{dg_p}{dt} - \frac{df_p^{cr}}{dt} + f_p^{ext} = 0 \qquad [1.4]$$

where the matrices H, L, S and W are the effect of pore water pressure on the compressibility of liquid phase, the effect of displacement on the compressibility of solid phase, the effect of pore water pressure on the compressibility of solid phase, and the effect of the temperature and vectors g_p, f_p^{cr}, and f_p^{ext} are the volumetric contraction by hydration of liquid phase, the effect of creep, and the flow rate.

Finally, the governing equation for water migration in concrete which satisfy the mass conservation law and equilibrium equation simultaneously, can be written as the matrix form as follows.

$$\begin{bmatrix} 0 & 0 \\ 0 & -H \end{bmatrix}\begin{Bmatrix} \{\bar{u}\} \\ \{\bar{p}\} \end{Bmatrix}+\begin{bmatrix} K_T & -L \\ -L^T & -S \end{bmatrix}\begin{Bmatrix} \dfrac{d\{\bar{u}\}}{dt} \\ \dfrac{d\{\bar{p}\}}{dt} \end{Bmatrix}=\begin{Bmatrix} \dfrac{d\{\bar{f}\}}{dt}+A\dfrac{d\{\bar{T}\}}{dt} \\ W\dfrac{d\{\bar{T}\}}{dt}+\dfrac{d\{f_p^{creep}\}}{dt}-\{f_p^{ext}\}+\dfrac{d\{g_P\}}{dt} \end{Bmatrix} \quad [1.5]$$

Solving equation [1.5] in which the nodal displacement and pore water pressure are unknown, the characteristics of water migration in concrete which varies from homogeneous to non homogeneous material can be obtained substituting the boundary condition and initial condition.

Figure's 1.16 and 1.17 show the analytical and experimental results for the specimen Ⅲ-0.5, 1.0, 2.0-W and Ⅶ-0.5, 1.0, 2.0-W of which loading age is 3 and 7 days and the specimen is cured in water after removal of forms, respectively. Figure's 1.16(a).(b),(c),and figure's 17(a), (b), (c) show the relationship between the deviatric stress and axial strain, between the pore water pressure and axial strain and between the volumetric strain and axial strain obtained by analysis, respectively. The parameters adopted in the analysis are listed in table 1.3. In each figures, the experimental results were shown with circle, triangle and square and the analytical results with solid, dotted and broken line for the cases of the lateral pressure of 0.5, 1.0 and 2.0MPa.

In the pre-peak region, the analytical results for pore water pressure show a good agreement with the experimental results as shown in figure's 1.16(b), 1.17(b). The maximum value of pore water pressure in the analytical results becomes higher with the increase of lateral pressure while its value is constant of 0.6MPa in the experimental results with no relation to the lateral pressure in case of 3 days of loading age as shown in figure 1.16(b). The values in the experiment become even

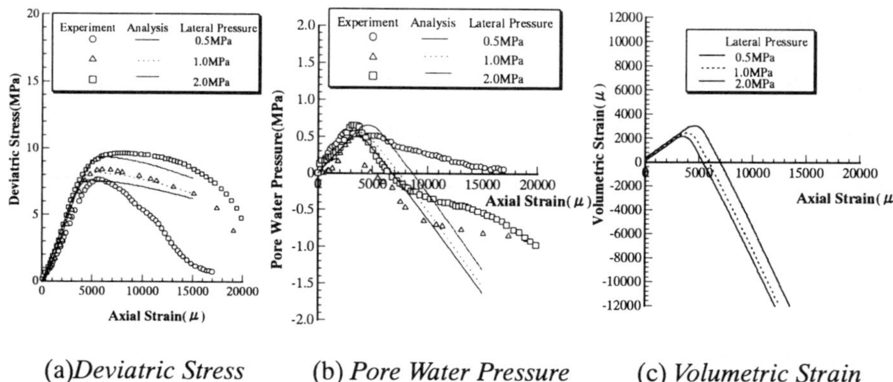

(a)*Deviatric Stress* (b) *Pore Water Pressure* (c) *Volumetric Strain*

Figure 1.16. *Analytical Estimation (Loading Age: 3 Days)*

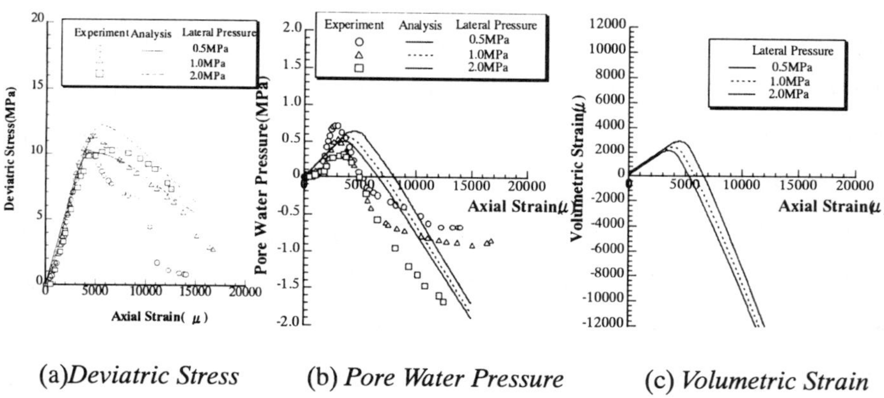

(a)*Deviatric Stress* (b) *Pore Water Pressure* (c) *Volumetric Strain*

Figure 1.17. *Analytical Estimation (Loading Age: 7 Days)*

smaller with the increase of lateral pressure in case of 7 days of loading age as shown in figure 1.17(b). Though the reasons for this difference is not clear at this moment, the disagreement of boundary conditions in post-peak region may be one of the reasons. Namely, in the present experiments, the surface of concrete specimen was coated with the rubber sleeve to assume the undrained condition for pore water and this rubber sleeve which is well used in soil testing, is elastic material. In the process of volumetric contraction of concrete specimen shown in figure 1.18, when the pore water pressure at the surface of concrete specimen becomes higher than the lateral pressure, the out flow of pore water from specimen may occur and water will gather in the interface of specimen and rubber sleeve. If the above phenomenon

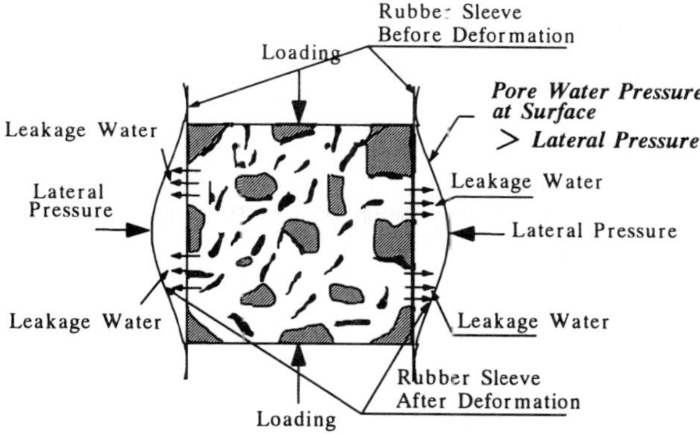

Figure 1.18. *Water Behavior between Concrete Surface and Rubber Sleeve*

occur, the experimental conditions do not match with the analytical boundary conditions due to the imperfect undrained condition for pore water.

In the post-peak region, the analytical results for pore water pressure show a good agreement with the experimental results as a whole, as shown in figure's 1.16(b), 1.17(b). However, in the range beyond the axial strain of 8000μ, the simple extension of calculation assuming same material parameters result in negative pore water pressure beyond *-1.0Mpa* in which the existence of water as liquid phase should be questioned. The experiment show the sudden decrease of negative pore water pressure gradient around here and suggests the necessity of more detailed analysis beyond the region.

4. Conclusion

The experimental measurement of pore water pressure occurring in concrete due to external load has been performed and generation of non-negligible pore water pressure in concrete by external load is identified. The pore water pressure will surely affect the mechanical characteristics of concrete structures which contain pore water and should carefully analyzed The amount of the pressure will be determined by load itself as well by boundary conditions. The analytical model to predict those pore pressure and pore water migration was presented and shown to be capable of estimating the pressure in practical application region.

Part II. Fracture of Concrete by Pore Water Pressure and the Analysis of Water inside of Concrete

In the part I of the paper, it was recognized that the substantial amount of pore pressure is generated in concrete by external load and its mechanism are discussed with analytical results. In this investigation, the study is extended to find out the fracturing of concrete by pore pressure alone. The investigation will contribute to clarify the fracture characteristics of saturated or partially saturated concrete and some of the flow mechanism of water inside of concrete.

1. Experimental Apparatus

1.1. *Development of Testing Cell for Lateral Pressurization by Water*

The cell in which water exists and is pressurized, is composed of the cylindrical container made of steel as shown in figure 2.1. Concrete specimen, the diameter of which is 15cm and the height is 30cm is set at the center of the cell and the O-ring and the back up ring are situated at the both end of concrete specimen to prevent water leakage. Pressurization is performed by injection of water between a concrete specimen and the cell. The injecting velocity of water is constant at 0.04MPa/sec. The applied lateral pressure is measured by the pressure transducer shown in figure 2.1. Pore water pressure in concrete is measured at the center, a half diameter location and surface in the center section of a concrete specimen by the pressure meter fixed at the end of 1mm diameter stainless steel pipe embedded in a concrete specimen as shown in figure 2.2. Moreover, the ceramic disk plate which allows liquid and gas to pass is set between the stainless steel pipe and pressure meter to measure the pore water pressure in detail. Concrete strains in the longitudinal and

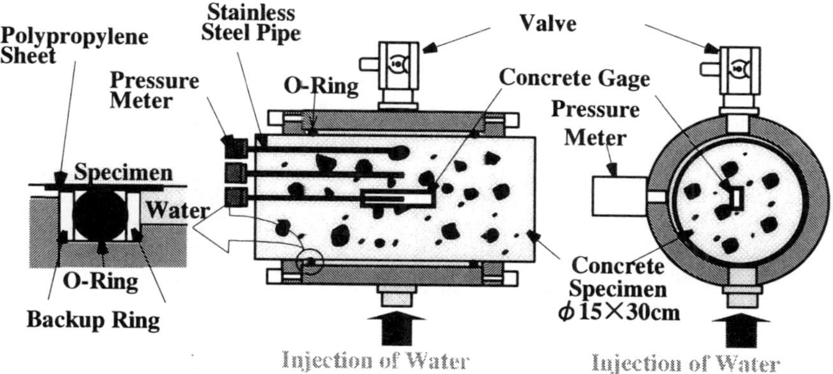

Figure 2.1. *Experimental Apparatus*

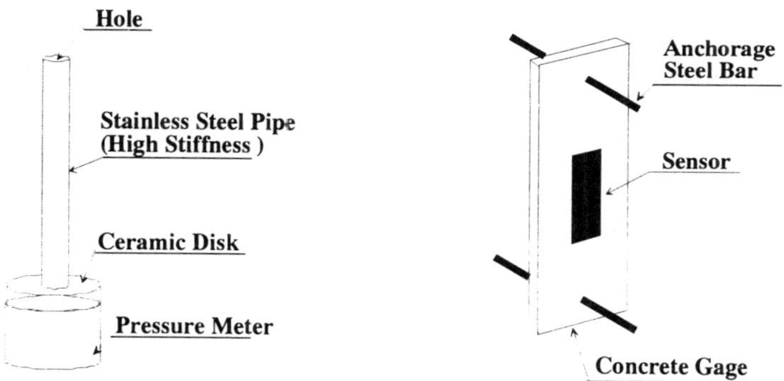

Figure 2.2. *Pressure Meter* **Figure 2.3.** *Mold Gage*

radius direction are measured by concrete gage embedded in a concrete specimen with the anchorage bar shown in figure 2.3.

Specimens tested are exactly same to the one adopted in the research part I in terms of mix proportion, size, and testing age. In other words, experimental variable is only skeletal strength. However, the deformational behavior are essentially similar between the two, only the results obtained at the age of 3 days will be described in the following sections.

1.2. *Loading History for the Lateral Pressure*

The lateral pressure is applied up to the fracture of concrete as shown in figure 2.4.

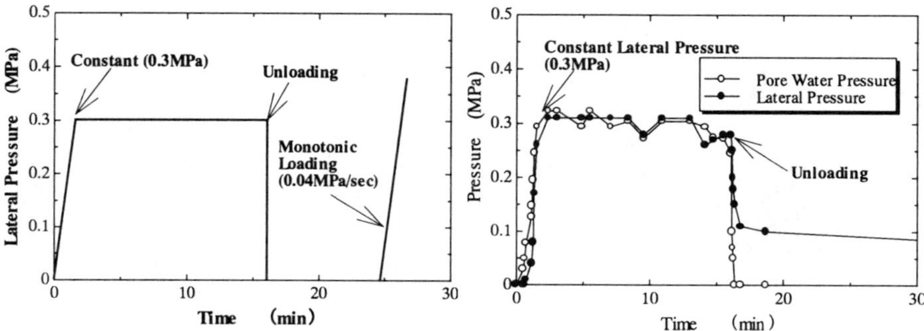

Figure 2.4. *Lateral Pressure History* **Figure 2.5.** *Water Pressure History*

The loading history of lateral pressure is that the lateral pressure is applied up to the one third of the uniaxial tensile strength of concrete specimen, i.e., 0.3 MPa. After, the lateral pressure is held at 0.3 MPa for fifteen minutes (Stage 1), the lateral pressure is unloaded to zero. The lateral pressure, then is reapplied monotonically up to the fracture of concrete specimen at the loading speed of 0.04 MPa/sec (Stage 2). The holding time of lateral pressure in Stage 1, is determined such that the pore water pressure occurring at the centre, one half of the radii location and surface of concrete specimen at the centre section, become the same as the applied lateral pressure of 0.3 MPa. At this moment, the concrete specimens are considered to be nearly saturated with water.

2. Deformational Behavior of Concrete Subject to Constant Lateral Pressure (Stage 1)

2.1. *Characteristic of Pore Water Pressure*

The pore water pressure occurring in concrete specimen subjected to the lateral pressure history is shown in figure 2.5. In this figure, the symbol ○ and ● show the lateral pressure and pore water pressure at the center of the concrete specimen.
It is noted that the pore water pressure is generated close to the given lateral pressure

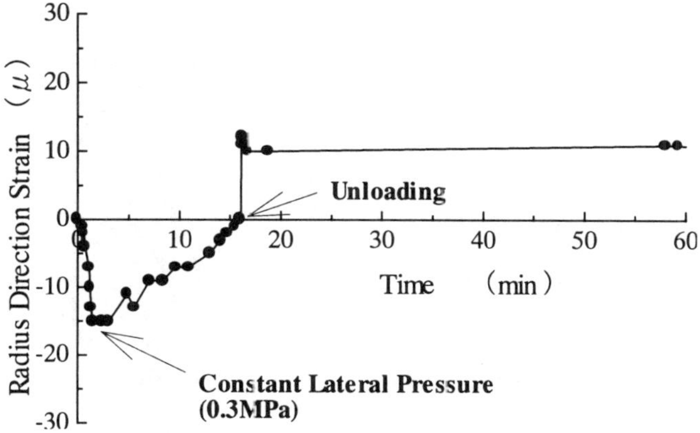

Figure 2.6. *Radius Direction Strain History*

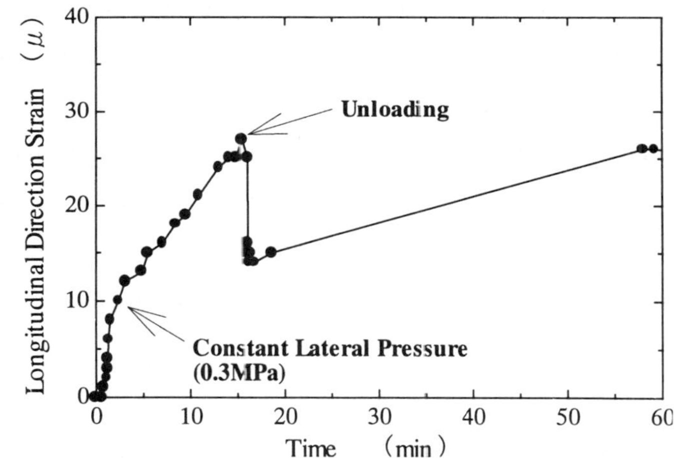

Figure 2.7. *Longitudinal Direction Strain History*

during the period of sixteen minutes in the Stage 1. After unloading of lateral pressure, the sudden decrease of pore water pressure occur and residual pore water pressure with 0.1MPa remains. The good agreement of pore water pressure with the lateral pressure indicates that the concrete specimen is nearly saturated with water.

2.2. *Characteristic of Deformational Behavior*

Concrete strain history in the longitudinal and radial direction is shown in figure's 2.6 and 2.7. At the loading of lateral pressure, it is noted that the radial strain is in compression and the longitudinal strain is in tension due to the Poisson's effect. Then, at the stage of constant lateral pressure at 0.3MPa, the radial compressive strain gradually decreases and longitudinal strain keeps on increasing at constant at 0.3MPa lateral pressure. These deformational behaviors are considered to be due to the extension of skeletal phase by pore water pressure. In particular, it should be noted that the magnitude of effective stress in the longitudinal direction occurring in solid phase is equal but with opposite sign to the pore water pressure due to the equilibrium since no external force is applied in longitudinal direction. The experimental results indicate that the creep has occurred in the very rapid rate due to the water migration and due to the decrease of combining force between solid phase. The water migrate from center to the both edge surface of which pressure is equal to the atmospheric pressure as shown in figure 2.8.

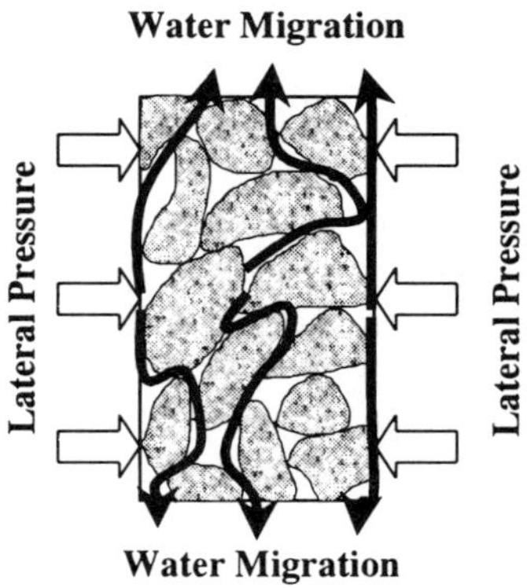

Figure 2.8. *Water Migration Phenomenon*

3. Deformational Behavior of Concrete Specimen Subjected to Lateral Pressure (Stage 2)

3.1. *Characteristic of Pore Water Pressure*

The relationship between pore water pressure and lateral pressure in Stage 2 loading, is shown in figure 2.9. In this figure, the symbols ○, △ and □are shown at the center, a half radii location and surface of concrete specimen at the center section, respectively. The relationship between the pore water pressure and the longitudinal strain at the center and the surface of concrete specimen is shown with the solid and the dotted line in figure 2.10.

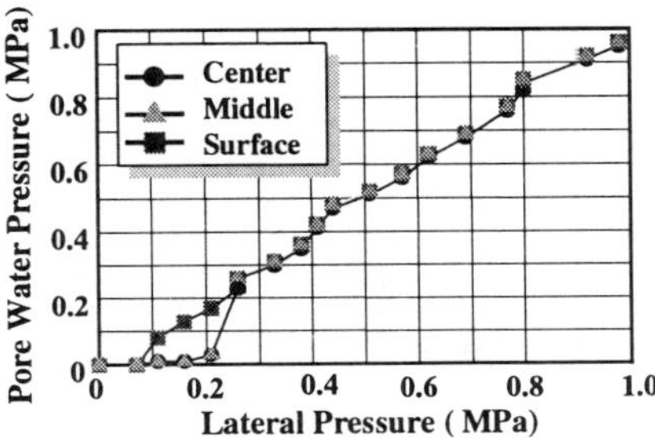

Figure 2.9. *Pore Water Pressure and Lateral Pressure Relation on Each Point*

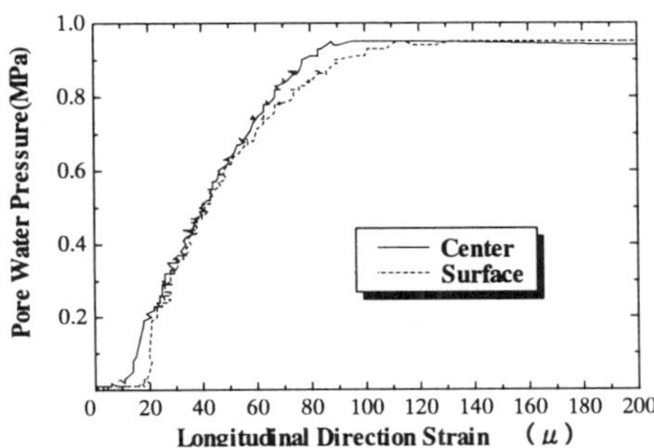

Figure 2.10. *Pore Water Pressure and Longitudinal Direction Strain Relation*

It should be noted as mentioned in the previous section that the concrete specimen is nearly saturated from the initial stage of reloading as can be judged from nearly equal pore water pressures to the lateral pressure as shown in figure 2.9. Moreover, it should be noted that the influence of pore water pressure on the deformational behavior of concrete specimen is similar at any location within the section judging from nearly equal reading of the pore water pressure at the surface as well as at the center of concrete specimen as shown in figure 2.10.

3.2. *Characteristic of Deformational Behavior*

Figure 2.11 shows the relationship between the pore water pressure and the strain at the center of the concrete specimen. In this figure, the symbols ● and ○ show the longitudinal and the radial strains, respectively.

The linear increase of tensile strain in the longitudinal direction occurs with the increase of pore water pressure and then the splitting fracture occurs at the 0.98MPa of lateral pressure and the final sudden drop of pore water pressure is observed. This phenomenon is similar to that of uniaxial tensile test. On the other hand, the scarce radial strain is observed in the final state and its value can be ignored compared with the longitudinal strain. These are the process that the lateral loading pressure is

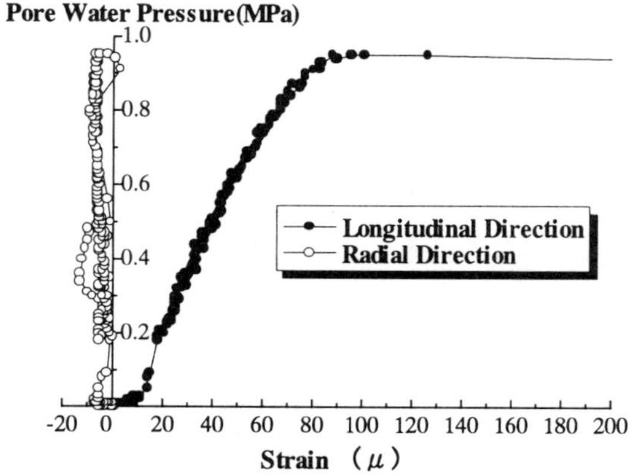

Figure 2.11. *Pore Water Pressure and Strain Relation*

transferred gradually to the pore water pressure in the saturated skeleton which induces recovery of compressive radial strain, while longitudinal deformation keeps on increasing due to persistent skeletal stress.

More precisely, the effective stress in the longitudinal direction, the effective stress σ' for the porous permeable material is defined simply in the following equation.

$$\sigma = \sigma' - p \qquad [2.1]$$

in which the sign is positive for tension for stresses and positive for compression for pressure. With the application of boundary total stress is zero,

$$\sigma' = p \qquad [2.2]$$

Hence it is clear that the above mentioned experimental fracture of concrete specimen has occurred due to the effective stress caused by pore water pressure.

In the Part I of the paper, it is noted that the pore water pressure occurs with the order of MPa in concrete. Then, it is no wonder that this phenomenon will cause the reduction of shear strength of concrete under water environment as well as the reduction of uni-axial compressive strength in the same condition.

4. Effective Stress Based Fracture Energy and Modeling of Porous Material

4.1. *Effective Stress Based Fracture Energy*

From foregoing discussion, concrete can be regarded as porous permeable elasto-plastic material composed of aggregate and cement paste whose porosity is partly saturated with water. External force to such material induces effective stress in solid phase, while the pore pressure is induced in liquid and gas phase. To express such phenomenon analytically, both of the force equilibrium equation and the mass conservation law need to be satisfied.

In this study, aggregate is assumed to be elastic body and cement paste is assumed to be elasto-plastic body. Moreover, yield function of cement paste is a function of damage parameter μ and fracture energy G_f as the authors have proposed in [Den 91].

In general, the fracture energy G_f is written using strain energy per unit volume W_f in the following

$$G_f = w_c W_f \qquad [2.3]$$

where, w_c is width of fracture process zone or crack band width.

The fracture energy is defined by the effective stress σ' which occurs in the solid phase instead of total stress which is usually done, and denoted as $G(\mu)$. The incremental form is read as

$$dG(\mu) = w_c \{\sigma'\}^T d\{\varepsilon^p\} = w_c \sigma_e d\varepsilon_p \qquad [2.4]$$

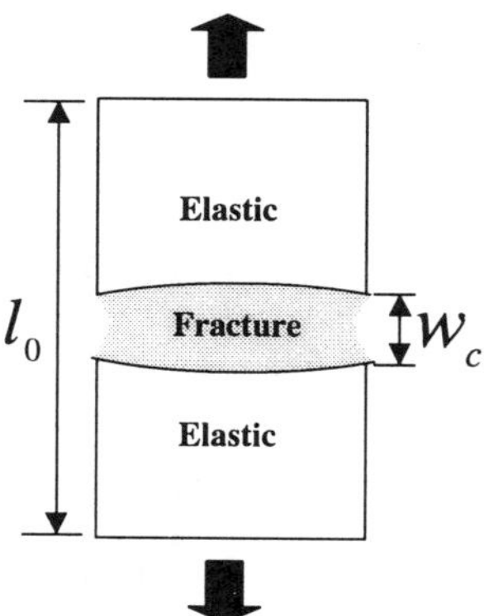

Figure 2.12. *Smeared Crack Model*

where, μ is damage parameter and represented as

$$\mu = \frac{\sigma_e d\varepsilon_p}{f_t \varepsilon_0} = \frac{dG(\mu)}{w_c f_t \varepsilon_0} \qquad [2.5]$$

Referring to figure 2.12, the incremental effective strain is written as

$$d\{\varepsilon_t'\} = d\{\varepsilon_u^e\} + \frac{w_c}{l_0} d\{\varepsilon_c^p\} = d\{\varepsilon_u^e\} + \beta\, d\{\varepsilon_c^p\} \qquad [2.6]$$

where β is defined as

$$[\beta] = \frac{\omega_c}{l_0} \begin{bmatrix} 1 & 0 & 0 \\ & \eta & 0 \\ sym. & & \xi \end{bmatrix} \qquad [2.7]$$

and $d\{\varepsilon_u^e\}$ and $d\{\varepsilon_c^p\}$ are incremental elastic and plastic strain, respectively, l_0 is a scale of concrete element. β is a ratio for scale l_0 and width of fracture zone ω_c in uni-axial situation.

4.1.1. *Elasto-Plastic Stiffness Matrix of Cement Paste*

A yield function f of cement paste can be represented as a function of effective stress and fracture energy, that is

$$f = f(\ \{\sigma'\},\ G(\mu)\ ) = 0 \qquad [2.8]$$

Introducing fracture energy to plasticity model proposed by Denzil and Tanabe[Den 91], elasto-plastic matrix is written in terms of effective stress as

$$\left[D_C^{ep}\right] = \left[D_C^e\right] - \frac{\left[D_C^e\right]\left[\beta\right]\left\{\frac{\partial f}{\partial\{\sigma'\}}\right\}\left\{\frac{\partial f}{\partial\{\sigma'\}}\right\}^T\left[D_C^e\right]}{\left\{\frac{\partial f}{\partial\{\sigma'\}}\right\}^T\left[D_C^e\right]\left[\beta\right]\left\{\frac{\partial f}{\partial\{\sigma'\}}\right\} - w_C\frac{\partial f}{\partial G(\mu)}\{\sigma'\}^T\left\{\frac{\partial f}{\partial\{\sigma'\}}\right\}} \qquad [2.9]$$

The focal point of the above equation is that the elasto-plastic stiffness matrix is defined with the effective stress $\{\sigma'\}$ instead of the total stress $\{\sigma\}$.

4.1.2. *Elasto-Plastic Stiffness Matrix of Concrete as Porous Media*

The incremental effective strain of concrete is defined as a summation of weighted incremental effective strains of aggregate and cement paste and is written as

$$d\{\varepsilon'\} = \frac{V_A}{V}d\{\varepsilon'_A\} + \frac{V_C}{V}d\{\varepsilon'_C\} \qquad [2.10]$$

Where, $d\{\varepsilon'\}$, $d\{\varepsilon'_A\}$ and $d\{\varepsilon'_C\}$ are incremental effective strains of concrete, aggregate and cement paste, respectively and V, V_A and V_C are volume of concrete, aggregate and cement paste, respectively. Adopting series model, the elasto-plastic stiffness matrix of concrete from equation [2.10] is written as

$$\left[D_S^{ep}\right] = \left[\frac{V_A}{V}\left[D_A^e\right]^{-1} + \frac{V_C}{V}\left[D_C^{ep}\right]^{-1}\right]^{-1} \tag{2.11}$$

4.2. *Modeling a Porous Material*

4.2.1. *Force Equilibrium Equation*

The relationship between total stress $\{\sigma\}$, effective stress $\{\sigma'\}$ and pore water pressure in porous media p is simply written as

$$\{\sigma\} = \{\sigma'\} - \{m\}p \tag{2.12}$$

in which tension is taken for positive in the stress field while compression is taken for positive in the pore water pressure field, and $\{m\}$ is Kronecker's delta. Applying equation. [2.12] to the principle of virtual work and reducing to the discrete form, the following differential equation is obtained as

$$K_T \frac{d\{\overline{u}\}}{dt} - L\frac{d\{\overline{p}\}}{dt} - A\frac{d\{\overline{T}\}}{dt} - \frac{d\{f\}}{dt} = 0$$

$$K_T = \int_\Omega (1-\xi) B^T D_S^{ep} B d\Omega, \quad L = \int_\Omega \xi B^T \{m\}\overline{N} d\Omega \tag{2.13}$$

$$A = \int_\Omega (1-\xi) B^T D_S^{ep} \{m\}\alpha \overline{N} d\Omega, \quad f = \int_\Omega N^T \{b\} d\Omega + \int_\Gamma N^T \{t\} d\Gamma$$

where, $\{\overline{u}\}$ and $\{\overline{p}\}$ are the nodal displacement and nodal pore water pressure, respectively and $\{T\}$, $\{f\}$ and dt are nodal temperature, nodal force and time increment, respectively.

4.2.2. *Mass Conservation Law of Pore Water*

Assuming Darcy's law for flow of pore water in pore system, flow velocity $\{v\}$ is represented as

$$\{v\} = -k\nabla h = -k\nabla(z + p/\gamma) \qquad [2.14]$$

in which k is a permeability, γ is a unit weight of water, z is the coordinate. The mass conservation law is written as

$$\Delta Q = Q_{out} - Q_{in} = -\nabla^T \{v\} \qquad [2.15]$$

Accumulation expressed in equation [2.15] may include the following factors:

a) A change in total strain

b) A change in the particle volume due to pore water pressure

c) a change in the volume of fluid

d) a change in fluid volume due to temperature

e) a change in fluid volume due to hydration

f) a change in the particle volume due to the change in effective stress

Reducing the mass conservation law into the discrete form using those factors, the following equation can be obtained as

$$-H\{\overline{p}\} - L^T \frac{d\{\overline{u}\}}{dt} - S\frac{d\{\overline{p}\}}{dt} - W\frac{d\{\overline{T}\}}{dt} - \frac{d\{g_p\}}{dt} + \{f_p\} = 0 \qquad [2.16]$$

where,

$$H = \int_\Omega (\nabla\overline{N})^T k' \nabla\overline{N} d\Omega, \quad L^T = \int_\Omega \xi\overline{N}^T \{m\}^T B d\Omega, \quad S = \int_\Omega \overline{N}^T \frac{\xi}{k_f} \overline{N} d\Omega$$

$$W = \int_\Omega \overline{N}^T \{3(1-\xi)\alpha - 3\xi\mu\} \overline{N} d\Omega, \quad g_p = \int_\Omega \overline{N}^T \frac{\eta\gamma_p}{\rho_w} C_H d\Omega$$

$$fp = \int_{\Omega} \overline{N}^T q d\Omega - \int_{\Omega} \left(\nabla \overline{N}\right)^T \frac{k}{\gamma} \nabla \gamma z d\Omega + \int_{\Gamma} \overline{N}^T \left(\{v\}^T \cdot \boldsymbol{n}\right) d\Gamma$$

4.2.3. *Governing Equations*

A coupling equation for the force equilibrium equation and mass conservation of pore water are now expressed in matrix form as

$$\begin{bmatrix} [0] & [0] \\ [0] & -[H] \end{bmatrix} \begin{Bmatrix} \{\overline{u}\} \\ \{\overline{p}\} \end{Bmatrix} + \begin{bmatrix} [K_T] & -[L] \\ -[L]^T & -[S] \end{bmatrix} \begin{Bmatrix} \dfrac{d\{\overline{u}\}}{dt} \\ \dfrac{d\{\overline{p}\}}{dt} \end{Bmatrix} = \begin{Bmatrix} \dfrac{d\{f\}}{dt} + A\dfrac{d\{\overline{T}\}}{dt} \\ W\dfrac{d\{\overline{T}\}}{dt} - \{fp\} + \dfrac{d\{g_p\}}{dt} \end{Bmatrix} \qquad [2.17]$$

Expressing equation [2.17] using the reverse finite difference method, the following equations are obtained:

$$\begin{bmatrix} [K_T] & -[L] \\ -[L]^T & -[S] - \Delta t[H] \end{bmatrix} \begin{Bmatrix} \{\Delta\overline{u}_n\} \\ \{\Delta\overline{p}_n\} \end{Bmatrix} = \begin{Bmatrix} \{\Delta f_n\} + [A]\{\Delta\overline{T}_n\} \\ -\Delta t\{fp_n\} + W\{\Delta\overline{T}_n\} + \{\Delta g_p\} + \Delta t[H]\{\overline{p}_{n-1}\} \end{Bmatrix} \qquad [2.18]$$

in which Δ express the each incremental vector between time step. One can obtain the solution of the displacement and pore water pressure with given initial condition and boundary conditions.

By obtaining the width of fracture process zone w_c or crack band width by the comparison of experiment and analysis, it is possible to identify the fracture energy for the concrete failing by the effective stresses.

It will be extended to deformational property of not only concrete but also general porous media system.

5. Analytical Estimation for Pore Water Experiment due to Lateral Water Pressure

Using the proposed analytical model and assuming aggregate as elastic body and cement paste as elasto-plastic body of which yield function is Drucker-Prager type[Wu 90], the experimental results are analyzed. A model for the analysis is 1/8 part of a specimen with the size of $15 \times 15 \times 30cm$, and it is divided into eight elements, as shown in figure 2.13. It should be noted here that though the experimental specimen is a cylinder, the numerical model is a rectangle block of

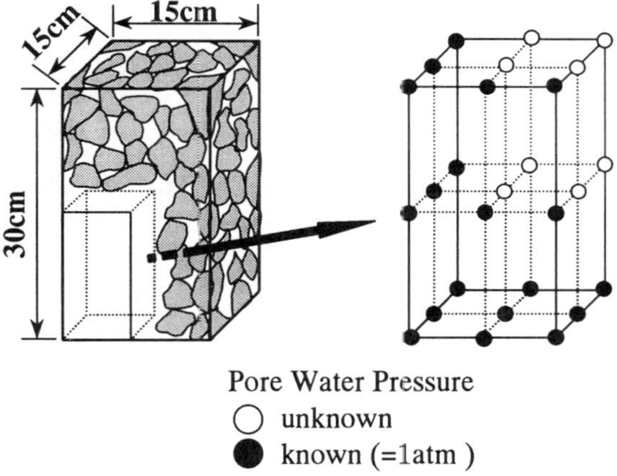

Figure 2.13. *Analytical Model*

Age (Days)	Young's Modulus (MPa)	Poisson's Ratio	Permeability (cm/sec)	Initial Cohesion (MPa)
1	8.32×10^{3}	0.15	3.33×10^{-8}	1.18
3	15.0×10^{3}	0.15	6.66×10^{-8}	1.86
Age (Days)	Frictional Angle(°)		Bulk Modulus of Water(MPa)	ω_c / l_0
	Initial	Final		
1	30	35	2000	0.6
3	30	35	2000	0.6

Table 2.1. *Material Parameters using in Analysis*

which sectional dimension is equal to the diameter of cylinder and the longer dimension of the rectangle is equal to that of the cylinder. Parameters used in analysis are listed in table 2.1. The bulk modulus of water is taken as $E_w = 2.0 \times 10^3\ N/mm^2$

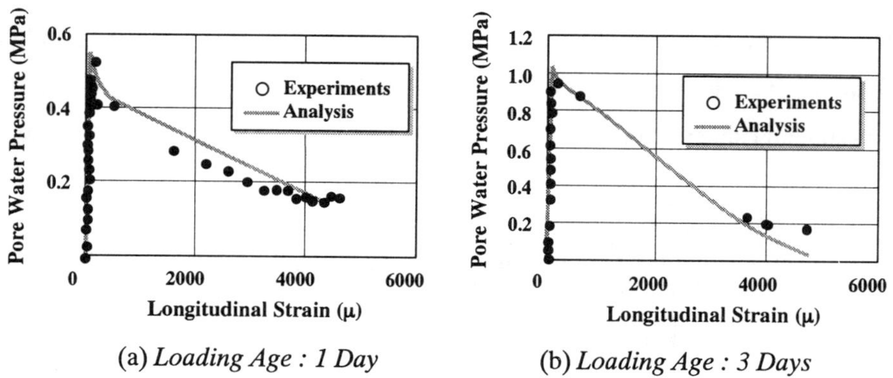

(a) *Loading Age : 1 Day* (b) *Loading Age : 3 Days*

Figure 2.14. *Analytical Simulation Results*

Figure 2.14(a) and (b) show the relationship between the pore water pressure and concrete strain in the longitudinal direction at the center of concrete specimens, respectively. In these figures, the solid line shows the analytical result and the symbol ● shows the experimental result. Figure 2.14(a) shows the results for the specimen at the age of 1 day (i.e. a weaker skeletal strength specimen) and figure 2.14(b) shows the results for the specimen at the age of 3 days (stronger skeletal specimen).

It is noted that the analytical results show a good agreement with the experimental results and the maximum value of pore water pressure are 0.53 and 1.0MPa for the specimens at the age of 1 and 3 days respectively. They are about tenth of uniaxial compressive strength and nearly equal to the uniaxial tensile strength.

The fracture energy consumed in those specimens are identifiable by equation (2.4), i.e. by the product of fracture process zone width of ω_c and the area covered by the effective stress and the strain curve. The best fitting fracture process zone widths of

ω_c are found to 9cm for the specimen of the age of 1 day and 3days respectively and the fracture energy for those specimen were 108.0 and 288.0 N/m respectively. The fracture process zone widths found in these experiments are exceedingly large compared with fully hardened concrete. This may be due to both the concrete age effect and the fracturing by pore pressure which acted uniformly in the wide range of specimen. It is still to be investigated.

6. Conclusion

Following to the first part, the experiments to fracture concrete by pore pressure have been performed and it was confirmed that concrete fails by pore pressure alone. The measurement of pore water pressure occurring inside of concrete has also carried out. From those findings, it is necessary for pore pressure to be taken into consideration in fracture of concrete which contain high pore water. For the purpose, fracture energy based on effective stress is proposed and the analytical model to simulate fracturing of concrete is developed and shown to be capable of estimating fracture.

7. References

[Den 91] Denzil, R.L., Fracture Mechanics based on Analysis of Thermal Crack Propagation of Massive Concrete, Doctoral Thesis, University of Nagoya, 1991.

[Ino 90] Inoue, S., Nishibayashi, S. and Kumano, T.: Study on the Estimation of Fatigue Life of Reinforced Concrete Beams under Water Environment and its Applicability to Design, *Proc. of 45th JSCE Annual Conference*, pp.636-637, 1990. (in Japanese)

[JGS 89] JGS (Japanese Geotechnical Society) : References on Mechanical Engineering for Partially Saturated Soil, 1989

[М и х 74] М и х а й л о в, А. В. :Л р о ц н о с т ъ Б е т о н а в з а в и с и е г о в л а г о с о р ж а н ц я, Б е т о н и Ж е л е з о б е т о н ., p.19, 1974. (in Russian)

[Mor 93] Morimoto, H., Iwamoto, T., Kurihara, T. and Koyanagi, O. : Characteristics of Compressive and Tensile Creep, *Proc. of the Chubu Cement Concrete*, No.8, pp.17-20, 1993. (in Japanese)

[Osh 93] Oshita, H., Ishikawa, Y. and Tanabe, T. :Creep Mechanism of Early Age Concrete Modeling with Two Phase Porous Material, *Proc. of the Fifth International RILEM Symposium*, pp.465-470, 1993.

[Tan 93] Tanabe, T. and Ishikawa, Y. :Time-Dependent Behavior of Concrete at Early Ages and its Modeling, *Proc. of the Fifth International RILEM Symposium*, pp.435-452, 1993.

[Wu 90] Wu, Z. S. and Tanabe, T. : A Hardening-softening Model of Concrete Subjected to Compressive Loading ,*Journal of Structural Engineering*, *Architectual Institute of Japan*, Vol.36B, pp.153-162, 1990.

Chapter 3

Computational Failure Analysis and Design

Splitting of Concrete Block Caused by Inside Pressure-Failure Mechanism and Size Effect

Josko Ozbolt — Jörg Asmus — Khalil Jebara

Institut für Werkstoffe im Bauwesen
University of Stuttgart
70550 Stuttgart, Germany
ozbolt@iwb-uni.stuttgart.de

ABSTRACT. *In the present paper the numerical results for the splitting failure mode which is typical for fastening elements that are used too close to the member edges or in a narrow concrete member are presented and discussed. To investigate the influence of the size and geometry on the splitting failure, three-dimensional FE analysis was carried out. In the analysis the non-local mixed formulation of the microplane model for concrete was used. The splitting fracture is generated by imposing controlled concentrated radial displacement in the hole of the anchorage element. Geometrical parameters of a concrete specimen (width and thickness), the size of the relatively small loading area (diameter and height) as well as the embedment depth of the fastening element are varied. The analysis shows a strong size effect on the ultimate pressure at splitting failure and good agreement with experimental observations.*

KEY WORDS: *Concrete, Splitting Failure, Fastening Technique, Microplane Model, Finite Elements, Size Effect.*

Introduction

Failure of fasteners can be caused by a rupture of steel, anchor pull-out or by a concrete cone failure. In the past, these failure modes have been intensively investigated and their failure load can be predicted using known design equations [ELI 97]. However, when the fasteners are pulled out from a concrete member in the anchorage zone, rather high radial splitting forces are generated. These forces may cause splitting failure. Splitting of concrete member may be expected if the member dimensions are relatively small or in large concrete specimens when the fasteners are installed near the edges or corners. The splitting failure load depends on the size and on the material properties of the concrete member as well as on the size of the area where the concentrated force is applied.

To avoid splitting failure several technical recommendations regarding minimum edge distance, minimum spacing and thickness of a concrete member exist [ELI 97]. These recommendations are based on several series of relevant experiments. In most of the experiments, a pull-out test which generates splitting failure was performed [ASM 97]. The ratio between tensile (pull-out) force and resulting radial force was varied by using different type of fasteners. However, it was not possible to obtain a direct relation between pull-out and splitting force since in the experiments it is difficult to control the size of the loading area related to the resulting radial (splitting) load. Therefore, to reduce further experimental work and to systematically investigate the influences of geometrical parameters such as width-b of a concrete member, member thickness-h, embedment depth-h_{ef}, edge distance-c, spacing-s (for multiple fastenings) and drilled hole diameter-d_B, numerical studies were performed [OZB 96a]. These studies were carried out using of the mixed constrained non-local microplane model for concrete (finite element code *MASA3*; [OZB 97]). In the present paper, a part of the numerical results for a single fastener are presented and discussed.

1. Numerical Analysis

1.1. *Geometry, Material Properties and FE Discretization*

The aim of the FE analysis was to study the influence of concrete member thickness, edge distance with a constant drilled hole diameter (d_B = 18 mm) and with a constant height of the load transfer area (h_{LE} = 10 mm) on the splitting failure. Furthermore, the influence of the size of the loading area was investigated as well. The study was carried out for a double symmetric concrete strip with a drilled hole in the mid of it (see Figure 1). The specimen was loaded by controlling the radial displacement applied over a relatively small load transfer zone of the fastener.

The analysis was performed using of the eight node solid finite elements based on the linear strain field assumption (eight integration points). The mesh size and the form for different FE models in the crack initiation zone were of the same shape and size. The volume close to the load transfer zone is modeled by a finer mesh (element size of approximately 2.5-3 mm). The size of the elements increased moving away from the area of the load transfer zone. The typical finite element mesh used in the analysis is shown in Figure 2.

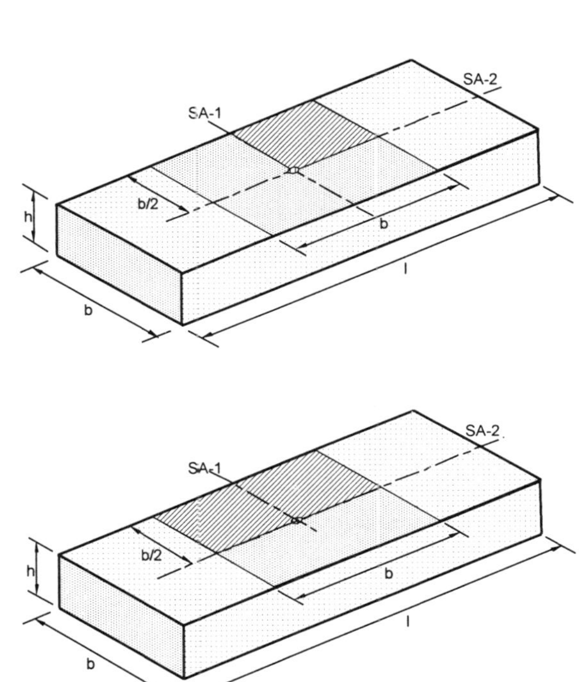

Figure 1. *Geometry of the specimen: a) fastening in the middle of a concrete strip (quarter, model A) and b) fastening in the middle of a concrete strip (half plate, model B)*

The following material properties for concrete were used: uniaxial tensile strength $f_t = 2.2$ MPa, uniaxial compressive strength $f_c = 34$ MPa, fracture energy $G_F = 0.05$ N/mm, initial modulus of elasticity $E = 25500$ MPa and Poisson's ratio $\nu = 0.18$. The local constitutive laws resulting from the microplane model are plotted in Figure 3.

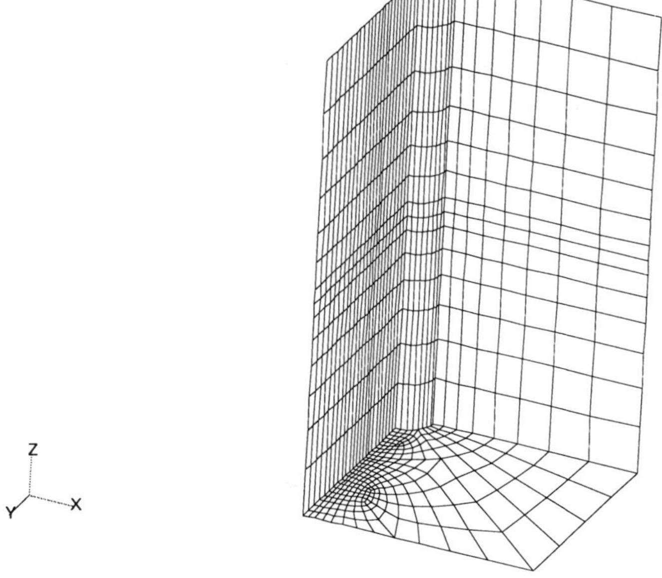

Figure 2. *The typical FE mesh for a single fastener placed in the middle of a concrete member - two symmetry planes (model A)*

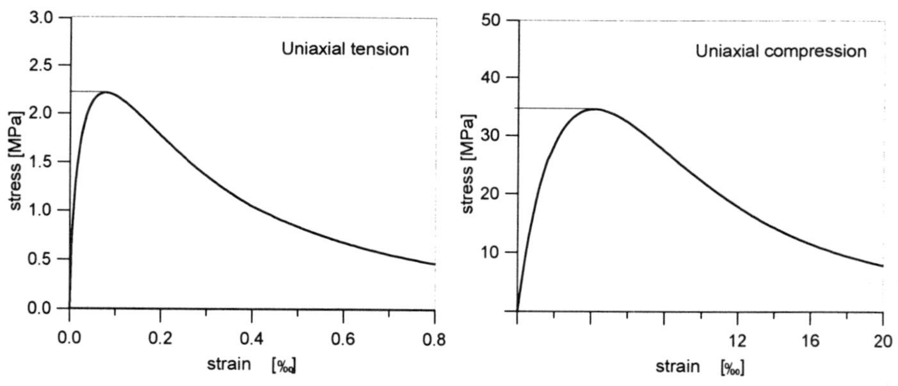

Figure 3. *Constitutive laws for: a) uniaxial tension and b) uniaxial compression*

To assure the objectivity of the analysis with respect to the mesh size and orientation, the non-local integral approach (microcrack interaction approach; [OZB 96b]) was employed. Besides the concrete constitutive law, the governing parameter in the non-local analysis is the so-called characteristic length [OZB 96a]. In the present study the characteristic length was set at $l_{ch} = 5$ mm.

All numerical results presented here are obtained by modeling only a quarter of the concrete plate, i.e., two symmetry planes are assumed (Model A, see Figure 1a). Two additional calculations using the model where only one plane symmetry is considered (model B, see Figure1b) served as a comparison to see whether the model with two symmetry planes could realistically predict a splitting failure mode. No doubt, the model B is closer to reality, however, from the computational point of view it is more time consuming.

1.2. *Comparison between Two Different Discretizations*

To check whether in this study both symmetry planes can be exploited without running into danger that the results could possibly be completely wrong, two models are first compared for examples in which two different member width were used (b = 160 and 640 mm) and all other parameters kept constant.

The calculated load-displacement (L-D) curves resulting from models A and B (member width b = 160 and 640 mm) are plotted and compared in Figure 4. The splitting force is calculated as:

$$S_p = \int_A p\,dA \qquad [1]$$

were p= pressure over the loading area and A= area of the loading zone. Note that this force is actually the total radial force which generates splitting failure of the specimen. For b = 160 mm the L-D curves as well as the ultimate loads agree well. Both models (A and B) predicted the splitting failure. However, for the strip width of 640 mm the L-D curves show a different behaviour (compare Figures 4a and 4b). The reason for this is due to different failure modes. The failure mode of model B was not the splitting but rather a local concrete compression failure in the loading zone. Consequently, for the member width b = 640 mm the results of both models are not comparable. However, for a member width b = 160 mm both models predicted the same failure mode and similar failure loads. The failure modes after the ultimate load are plotted in Figure 5 in terms of the maximal principal strains. The dark zones indicate damage (cracks).

For the model with two symmetry planes the crack pattern is shown in Figure 5a. Measuring the direction of the crack in relation to the SA-1 axis (see Figure 1), the splitting crack propagates at approximately 40° (see Figure 5a). The observed direction in tests as well as in the calculations according to the model B (see Figure 5b) indicates that the splitting crack propagates approximately perpendicular to the longitudinal edges of the concrete members. Considering the applied boundary conditions for the model A (double symmetry) the crack should theoretically run with a 45° angle to the edge, which agrees well with the numerical results. Note that the mesh refinement close to one symmetry plane did not influence the crack direction (see Figure 5a).

Obviously, both models correctly predicted the splitting of the concrete block. The model A with two planes of symmetry seems to be reliable and therefore it was used in all subsequent calculations.

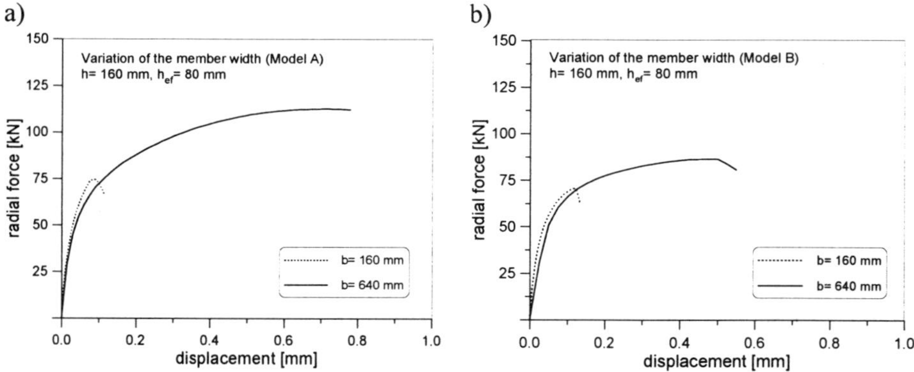

Figure 4. *Calculated load-displacement curves for different member widths obtained using: a) model A and b) model B*

1.3. *Influence of the Member Width*

To find the influence of the member width on the splitting failure load, altogether three calculations were carried out. The member height (h = 160 mm) and the embedment depth (h_{ef} = 80 mm) were kept constant, while the member width was varied (b = 160, 320 and 640 mm). In the case of b = 160 mm the used finite element mesh was unsymmetric, i.e., the mesh was along the symmetry axis SA-1 refined in order to check the influence of the mesh size on the localization of the splitting crack. In the other two cases only a light non-symmetry of the mesh was introduced.

The L-D curves for the three investigated cases (b = 160, 320 and 640 mm) are shown in Figure 6. For all three cases the failure is due to splitting of concrete. When small concrete members split, the typical failure is brittle and there is no indication or warning before it happens. On the contrary, the failure of larger concrete members is more ductile. The reason for this is the higher local damage of concrete in the zone where the load is applied since in larger specimen the total radial force is also larger.

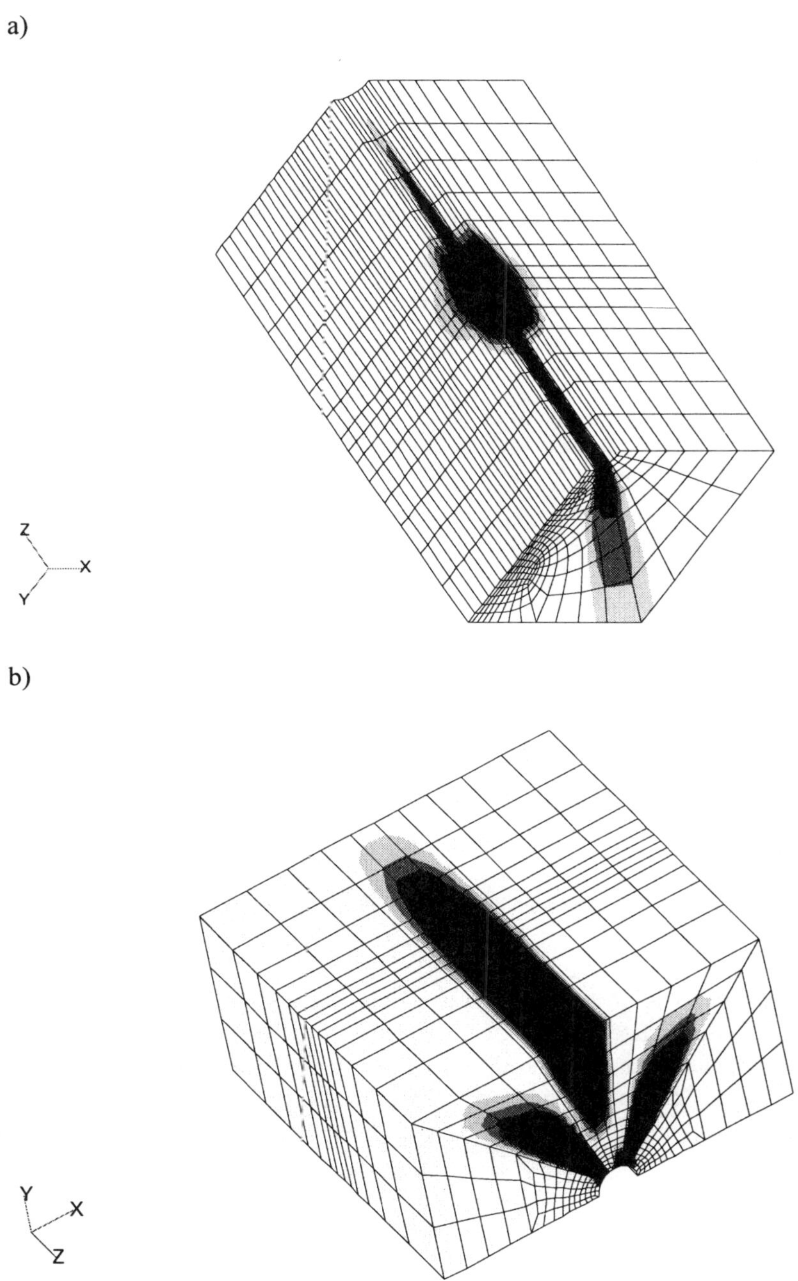

Figure 5. *Crack pattern in terms of total principal strains close after the peak load: a) model A and b) model B*

In Figure 7a, the relationships between the average ultimate pressure ($p = S_p/A$) normalized to the uniaxial compressive concrete strength are plotted as a function of the member width. As expected, when the member width increases the ultimate splitting strength increases as well and reaches for $b = 640$ mm the value that is approximately $p_{crit} = 7f_c$. Figure 7b shows the nominal splitting strength calculated as the ultimate splitting force divided by the cross section area of the concrete member ($S_p/(bh)$). The increase of the specimen width leads to the decrease of the nominal splitting strength.

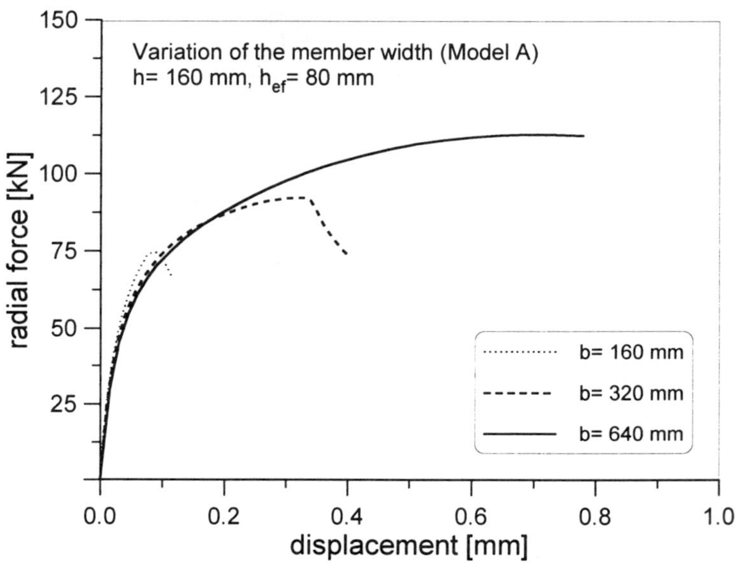

Figure 6. *Load-Displacement curves for three different member widths*

A similar trend is observed for the relative crack length w_r at peak load (see Figure 8). It is calculated as $w_r = w_a/l$ where w_a = the distance of the crack front measured perpendicular to the load surface area and l = the length of the splitting crack at failure. Obviously, with increase of the specimen width the critical crack length decreases, i.e., the cracking is less stable and consequently the nominal strength decreases.

a)

b)

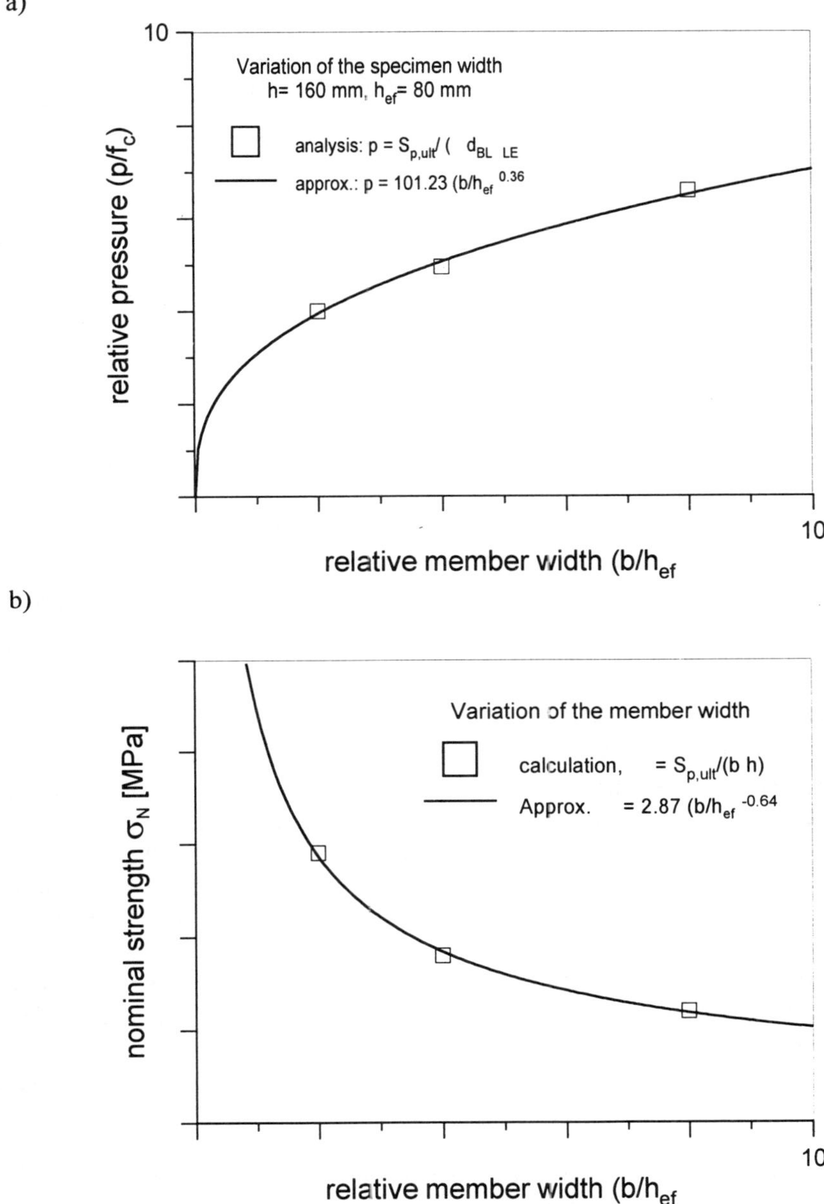

Figure 7. *Influence of the width of the concrete member: a) relative pressure and b) nominal splitting strength*

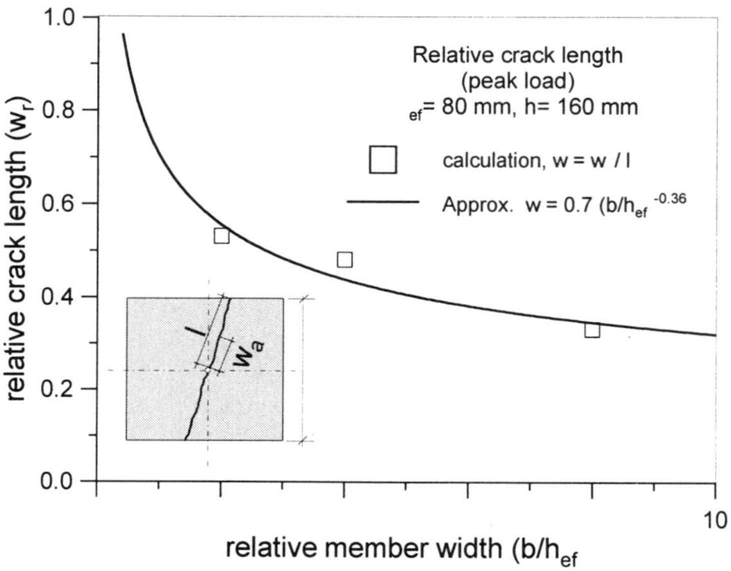

Figure 8. *Relative crack length at peak load as a function of the relative member width*

Before the splitting failure is reached, a small volume of concrete close to the loading zone is in a high 3D compressive state. This is due to the relatively small area of the load transfer zone and the strong lateral confinement. Above and under this compressed volume zone, high tensile strains (damage zones) localize in the radial as well as in the splitting direction (splitting crack). This can be seen from Figure 9 which shows the splitting crack front over the specimen depth at peak load in terms of principal stresses. The dark zones indicate the crack front. In front of this zone there is a hardening of the concrete and behind it the softening (cracked area). With a further increase of radial displacement (loading) the softening zones above and below the compressed concrete zone come together which results in a splitting failure. The splitting crack localizes in the weakest vertical plane which propagates along an angle of approximately 45^0 measured to the longitudinal axis of the test specimen (see Figure 5a).

The damage zones are the result of the non-homogeneous strain field which develops as a consequence of the non-uniform boundary conditions. The relative non-homogeneity of smaller concrete specimen is higher which results in a more brittle failure. This strain non-homogeneity decreases with increase of the specimen size. As a result, the behaviour under ultimate conditions for wider plates is more ductile since it is much harder to produce the non-homogeneity of strains that would result in splitting failure.

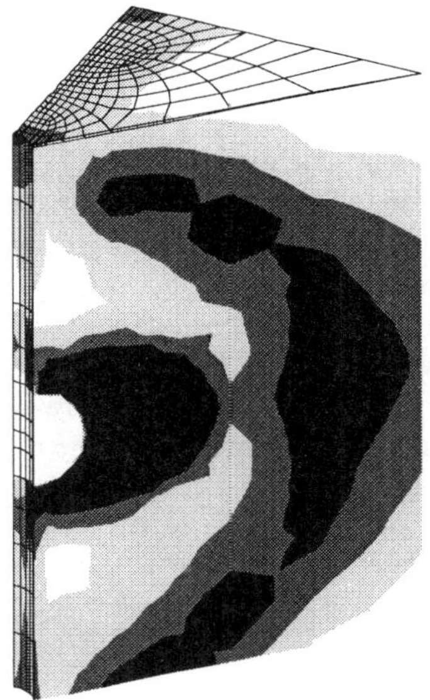

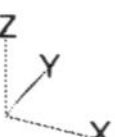

Figure 9. *The front of splitting crack in terms of max. principal stresses*

1.4. *Influence of the Concrete Member Height*

A series of FE calculations using a constant member width ($b = 320$ mm), member length ($l = 640$ mm) and embedment depth ($h_{ef} = 80$ mm) with a variable member height of $h = 120$, 160 and 240 mm were carried out to investigate the influences of the member height on the splitting failure load.

The resulting L-D curves are similar to those obtained for the variation of the specimen width, i.e., thinner members show a more brittle behaviour. As expected, the numerical analysis shows that with an increase of the member height the ultimate pressure increases as well (see Figure 10). However, if the member is thicker than $h = 2h_{ef}$ only a minor increase of the ultimate relative pressure is observed, i.e., for an increase of the embedment depth from $h = 2h_{ef}$ to $h = 3h_{ef}$ the pressure increases only for about 6%. Only two of the investigated cases ($b = h$ and $2h$) showed a clear splitting failure with a localized splitting crack. For $h > 2h_{ef}$ the specimen tends to fail not in the splitting failure mode but rather in a combined failure where the side concrete cone failure tends to dominate.

1.5. *Influence of the Embedment Depth*

To investigate the influence of the embedment depth on the splitting failure load, a series of calculations were carried out by varying the embedment depth between

h_{ef} = 40, 80, 160 and 320 mm. The ratios h/h_{ef} = 2 and b/h_{ef} = 4 were kept constant. Furthermore, the size of the hole diameter as well as the height of the loading zone were kept constant as well (d_B = 18 mm and h_{LE} = 10 mm).

Similar as in previous calculations the L-D curves indicate more brittle response for concrete specimen with smaller embedment depths. For embedment depths h_{ef} = 40, 80 and 160 mm the resulted failure mode was splitting. In the case of h_{ef} = 320 mm, a local compression failure took place. In Figure 11 the relationships between embedment depth and the relative ultimate pressure is shown. It can be seen that up to approximately h_{ef} = 200 mm the ultimate pressure increases proportionally with the embedment depth. However, for larger embedment depth a local concrete compression failure occurs. The relative pressure at which this takes place was observed to be about $9f_c$.

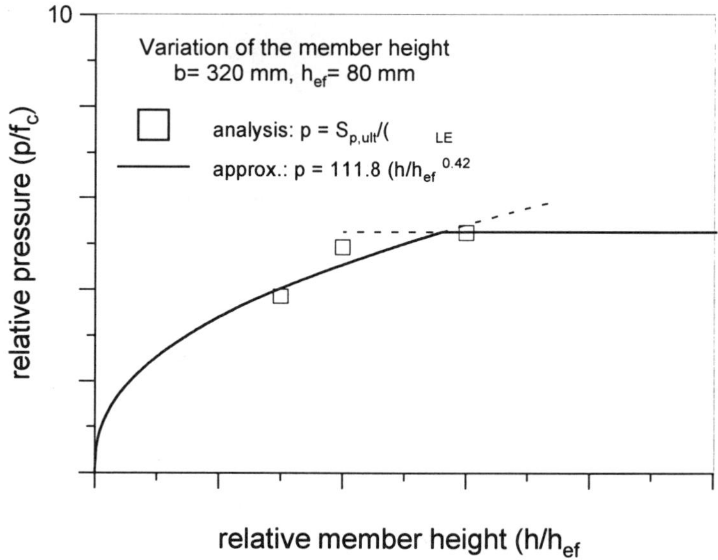

Figure 10. *Relative pressure as a function of the relative member height*

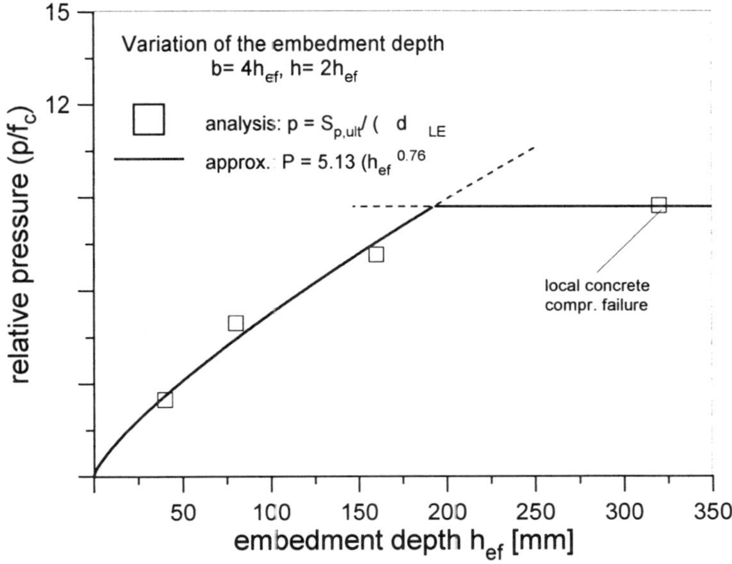

Figure 11. *Relative pressure as a function of the embedment depth*

1.6. *Influence of the Height of the Load Transfer Zone*

For all the results shown above the diameter of the drilled hole and the height of the load transfer zone were kept constant. Experimental investigations showed that the height of the loading zone influences the splitting failure load [ASM 97]. Therefore, to investigate this in more detail a series of calculations for a concrete member of constant size (h_{ef} = 160 mm, $h = 2h_{ef}$ and $b = 4h_{ef}$) were carried out. In these calculations, the diameter as well as the height of the loading zone were scaled proportionally with the specimen size (diameter: d_B = 18 and 72 mm; the height of the load transfer zone: h_{LE} = 10 and 40 mm).

The analysis principally shows that the failure load increases with increase of the loading area. Furthermore, smaller loading areas promote a more ductile behaviour. For all calculated cases a splitting failure was observed. Figure 12a shows the relation between the relative pressure at the peak load and the height of the loading zone for two different hole diameters. It can be seen that the ultimate pressure decreases with the increase of the height of the loading zone (see Figure 12a). The same tendency is observed when the hole diameter increases and the height of the loading zone is kept constant (see Figure 12b). Obviously, the size of the loading area strongly influences the ultimate pressure.

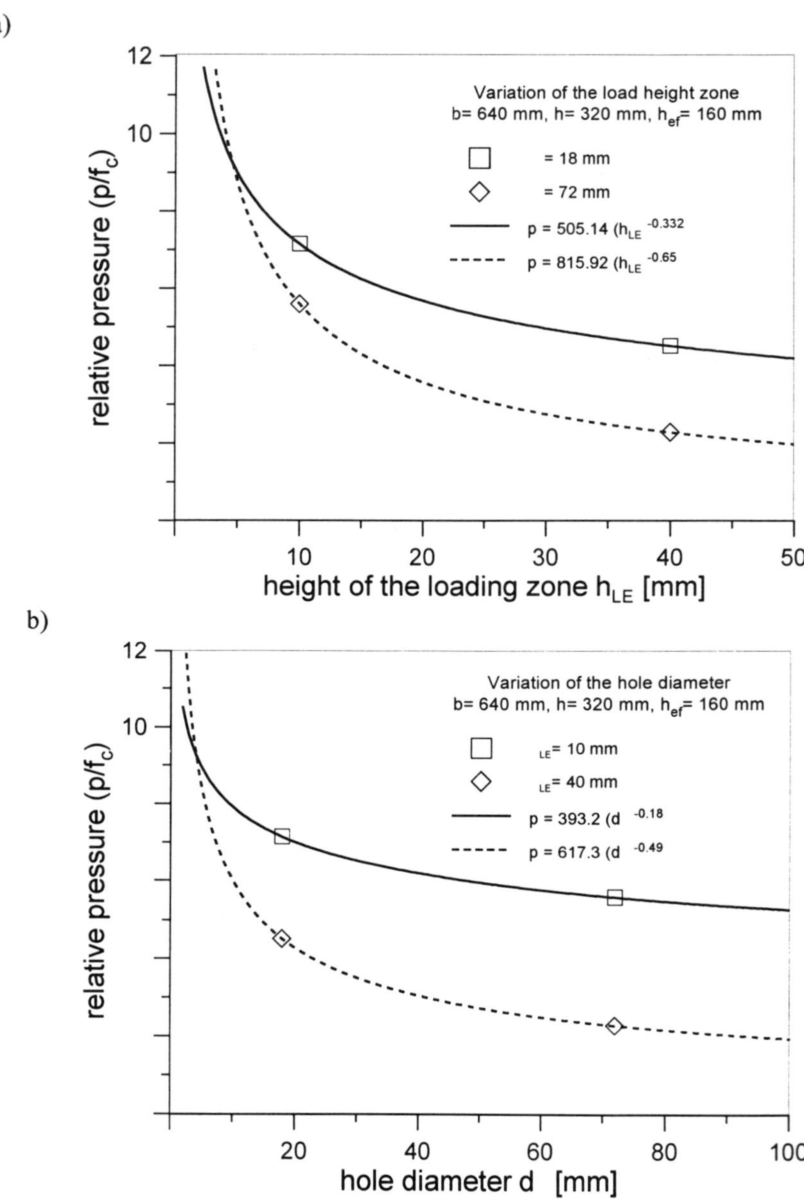

Figure 12. *Relative pressure as a function of: a) height of the loading zone and b) hole diameter*

2. Discussion of the Numerical Results

The presented results show that the splitting failure load of a concrete block subjected to inside pressure depends on the size and geometry of the concrete specimen. In Figure 13 the calculated results are summarized. The failure load for different geometrical configurations is plotted as a function of the embedment depth (characteristic size). The largest increase of the ultimate load is observed when the geometry, including the hole diameter and the height of the loading zone, is scaled proportionally. For this case the ultimate load is approximately proportional to the square of the embedment depth, i.e., there is no size effect on the ultimate load. In contrast to this, the lowest increase of the ultimate load is observed for the case where the specimen geometry is scaled proportionally but the size of the loading area is kept constant.

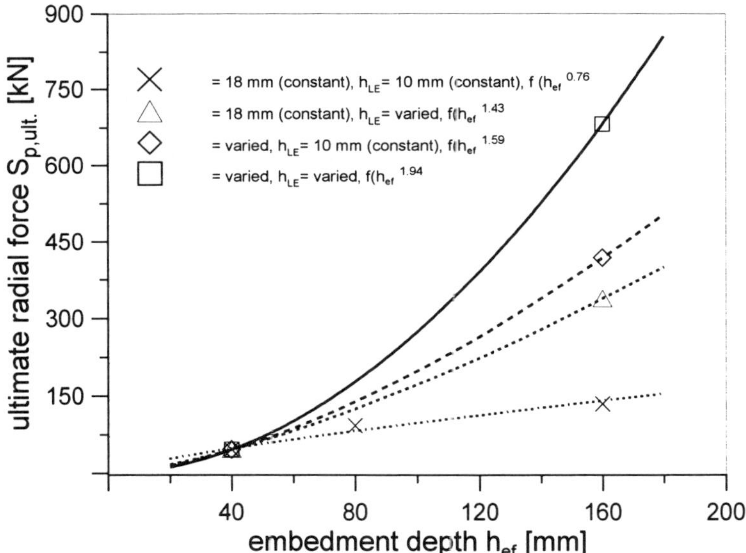

Figure 13. *Ultimate load for different geometrical configurations as a function of embedment depth*

When plotting the above calculated data as a relative pressure (p/f_c) versus the relative load area (d = loading area divided by the splitting failure area), as shown in Figure 14, it can be seen that for the size range of practical importance all calculated data are well fitted by the Bazant size effect formula [BAZ 84]. Furthermore, a fit of the calculated data by the power function yields an exponent of $n = -0.42$. The so called characteristic size resulting from the Bazant's size effect formula is very small ($d_0 = 0.03$). This shows that the size effect on the relative ultimate pressure is strong and close to that predicted by the linear elastic fracture mechanics size effect formula. The theoretical limit for the relative pressure in the case of $d \to 0$ is approximately $30f_c$. The numerical results are in good agreement with the experimental experience for partly loaded concrete areas [LIE 87].

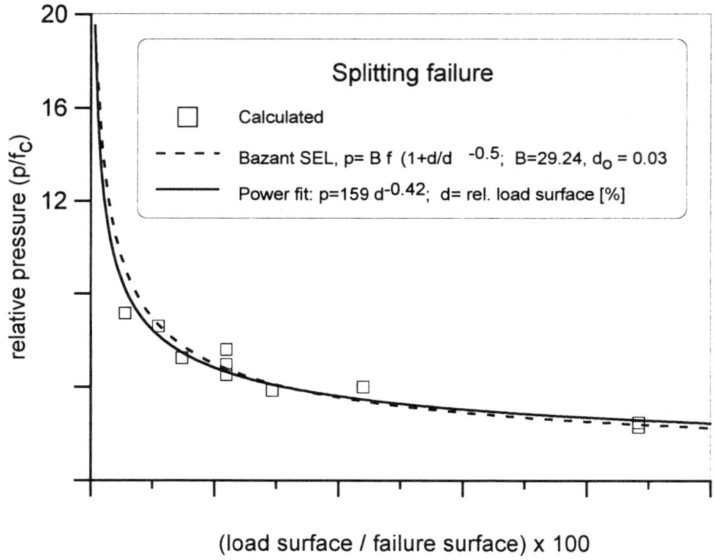

Figure 14. *Relative pressure as a function of the relative load area – size range of practical interest*

The extrapolation of the calculated results to the maximal size of $d = 100$ should theoretically yield the relative strength of one ($p/f_c = 1$), i.e., the compressive splitting strength should coincide with the uniaxial concrete compressive strength. In Figure 15, this extrapolation is plotted for the Bazant size effect formula and for the power function from Figure 14. Having in mind that both curves are obtained by fitting the calculated data of a very limited size range, the agreement with the lower theoretical limit is relatively good. This confirms that the numerical results are reliable.

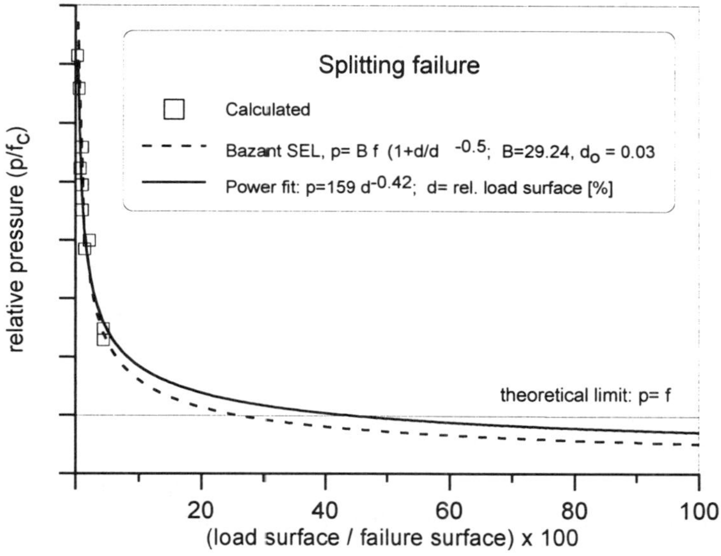

Figure 15. *Relative pressure as a function of the relative load area – full size range*

3. Conclusions

In the present paper the numerical results of the 3D finite element study for the splitting problem of a concrete block caused by a concentrated internal pressure are presented and discussed. The aim of the study was to better understand the failure mechanism as well as to investigate the influence of the member geometry on the ultimate splitting failure load.

The results show that the ultimate pressure at splitting failure strongly depends on the size and geometry of the specimen as well as on the size of the loading area. When the structure geometry and the loading area is scaled proportionally, the ultimate load increases somewhat proportionally as well, i.e., no size effect on the failure load is observed. However, if the structure size is scaled proportionally but the size of the loading area is kept constant, there is a strong size effect on the ultimate load. This is due to the localization of the damage and consequently leads to the decrease of the peak resistance by the increase of the size.

The size effect on the average ultimate pressure related to the relative loading area is strong and close to that predicted by linear elastic fracture mechanics. When the relative loading area takes the maximal possible value (one), the extrapolation of the ultimate pressure leads approximately to the uniaxial concrete compressive strength.

The numerical results are in good agreement with the experimental observations as well as with the theoretical and experimental studies for concrete members loaded by locally applied compressive forces.

4. References

[ASM 97] ASMUS J., Bemessung von zugbeanspruchten Befestigungen bei der Versagensart Spalten. Dissertation in preparation, Universität Stuttgart, Germany, 1997.

[BAZ 84] BAZANT Z.P., "Size Effect in Blunt Fracture: Concrete, Rock and Metal", *JEM, ASCE*, 110(4), p. 518-535, 1984. [ELI 97] ELIGEHAUSEN R., MALLEE R., REHM G., *Befestigungstechnik*, Ernst & Sohn, Berlin, Germany, 1997.

[LIE 87] LIEBERUM K.-H., Das Tragverhalten von Beton bei extremer Teilflächenbelastung. Dissertation, Technische Hochschule Darmstadt, Germany, 1987.

[OZB 96a] OZBOLT J., ASMUS J., JEBARA K., Dreidimensionale-Finite-Elemente-Analyse zur Versagensart Spalten durch Befestigungsmittel. Report Nr. 16/22-97/23, IWB, Universität Stuttgart, Germany, 1996.

[OZB 96b] OZBOLT J., BAZANT Z.P., "Numerical Smeared Fracture Analysis: Nonlocal Microcrack Interaction Approach", *IJNME*, 39(4), p. 635-661, 1996.

[OZB 97] OZBOLT J., LI Y.-J., KOZAR I., "Microplane Model for Concrete – Mixed Approach", submitted for publication in *IJSS*, 1997.

Failure Analysis of Quasi-Brittle Materials Using Interface Elements

Ignacio Carol — Carlos López

ETSECCPB, School of Civil Engineering
UPC, Technical University of Catalonia
E-08034 Barcelona, Spain

{carol, dlopezg}@etseccpb.upc.es

ABSTRACT. *A fracture-based normal/shear constitutive model for interfaces, which is based on the theory of plasticity with work-hardening, has been developed and validated with numerical and experimental results. This formulation can be interpreted as a generalization of Hillerborg's Fictitious Crack Model to mixed-mode loading. The model is implemented into a general-purpose finite element code with zero-thickness interface elements and advanced iterative strategies. Three examples of application taken from on-going research at ETSECCPB-UPC are also briefly discussed, i.e., microstructural analysis of concrete considered as a two-phase particle composite, microstructural analysis of cancellous (porous) bone, considered as a matrix with voids and the pull-out of rock bolts. In the first two examples, interfaces are inserted in between continuum elements all over the mesh representing potential crack trajectories, and the continuum elements themselves are considered linear elastic. For the pull-out analysis, interfaces are inserted along expected crack trajectories. All results show a good qualitative agreement with experimental observations and illustrate the capabilities of the model.*

KEY WORDS: *Fracture Mechanics, Interface Elements, Microstructural Analysis, Concrete, Cancellous Bone, Pull-out Test.*

1. Introduction

Since it was first proposed, the Fictitious Crack Model (FCM) [HIL 76] has been successfully applied to mode I cracking of specimens and structures. In its original form, (rigid-softening, only normal stress and opening displacement), however, it was not easy to implement the model into standard Finite Element calculation, and to use it in the presence of shear stress, inevitably present in general geometric and loading conditions.

As recognized by several authors, [ROT 88], [STA 90], [VON 92], zero-thickness interface elements with appropriate constitutive laws formulated in terms of shear and normal stress and the corresponding relative displacements provide the means of using the FCM concept in a modern numerical analysis of fracture.

2. Interface Constitutive Model

Interface behavior is formulated in terms of the normal and shear components of stresses (tractions) on the interface plane, $\sigma = [\sigma_N, \sigma_T]^t$, and corresponding relative displacements $\mathbf{u} = [u_N, u_T]^t$ (^{t} = transposed). The constitutive model is analogous to that used for each potential crack plane in the multicrack model [CAR 90], [CAR 91], [PRA 92], CAR 93], [CAR 95]; it has been recently described in detail and compared to other existing interface models [CAR 97], and its main features and verification are summarized in the following.

The constitutive formulation conforms to work-softening elasto-plasticity, in which relative plastic displacements can be identified with crack openings. The main features of the plastic model are represented in Figure 1. The initial loading (failure) surface $F = 0$ is given as a three-parameter hyperbola (tensile strength χ, c and $\tan\phi$; Figure 1a). Classic Mode I fracture occurs in pure tension. A second Mode IIa is defined under shear and high compression, with no dilatancy allowed (Figure 1b). The fracture energies G^I_f and G^{IIa}_f are two model parameters. After initial cracking, c and χ decrease (Figure 1c), and the loading surface shrinks, degenerating in the limit case into a pair of straight lines representing pure friction (Figure 1d). The process is driven by the energy spent in the fracture process, W^{cr}, the increments of which are taken equal to the increments of plastic work, less the frictional work in compression. Total exhaustion of tensile strength ($\chi = 0$) is reached for $W^{cr} = G^I_f$, and the residual friction ($c = 0$) is reached for $W^{cr} = G^{IIa}_f$.

Additional parameters α_χ and α_c allow for different shapes of the softening laws (linear decay for $\alpha_\chi = \alpha_c = 0$). The plastic model is associated in tension ($Q = F$), but not in compression, where dilatancy vanishes progressively for $\sigma_N \rightarrow \sigma^{dil}$. Dilatancy is also decreased as the fracture process progresses, so that it vanishes for $W^{cr} = G^{IIa}_f$. The dilatancy decay functions also include shape parameters α^{dil}_σ and α^{dil}_c (also linear decay for zero values of shape parameters). The elastic stiffness matrix is diagonal with constant K_N and K_T, that can be regarded simply as penalty coefficients. Some aspects of the constitutive implementation in a computer subroutine are described in Section 3.

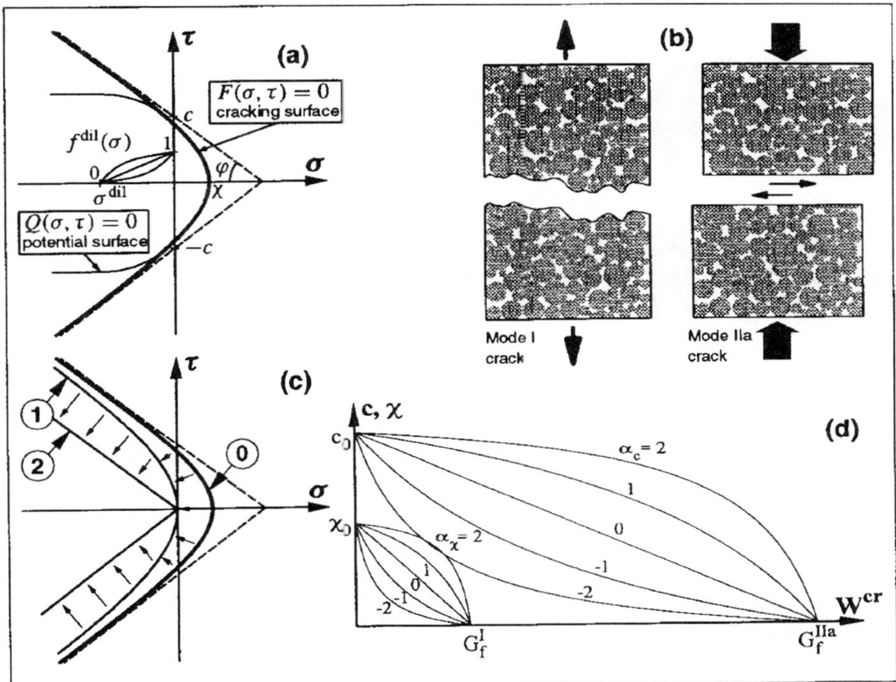

Figure 1. *Crack laws: (a) hyperbolic cracking surface F and plastic potential Q; (b) fundamental modes of fracture; (c) evolution of cracking surface; (d) softening laws for χ and c*

Three verification examples of the constitutive model are presented. The first one is a numerical test in pure tension. The resulting σ-u curves for various values of the fracture energy parameter G^I_f (other relevant parameters are K_N = 1,000 Mpa/mm, tensile strength χ_0 = 3 MPa and all shape parameters equal to zero) are represented in Figure 2. Note that, even with a zero shape parameter (linear softening function in terms of W^{cr}), the resulting softening curve in terms of crack opening is of the exponential type, with total area under each curve equal to the prescribed value of G^I_f.

The second example is a numerical shear test. First, normal compression of a prescribed value is applied. Then, the shear relative displacement is increased progressively with a constant normal stress, until a residual state is reached. The parameters used are $K_N = K_T = 25{,}000$ MPa/mm, $\tan\phi = 0.8785$, $\chi_0 = 3$ MPa, $c_0 = 4.5$ MPa, $G^I_f = 0.03$ N/mm, $G^{IIa}_f = 0.06$ MPa, $\sigma^{dil}_N = 30$ MPa, $\alpha^{dil}_\sigma = 2$ and all other shape parameters equal zero. The results are depicted in Figures 3 and 4. In Figure 3, shear stresses are represented against relative shear displacement for various values of normal compression. Note that, after the peak, all curves tend toward a residual value of the shear stress, which corresponds to basic friction times the normal stress. In Figure 4, dilatancy is represented by relevant normal displacements against relative shear.

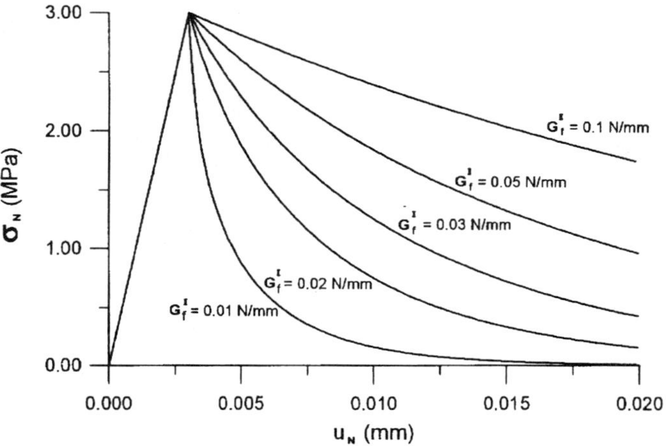

Figure 2. *Pure tension: normal stress versus relative displacement for different values of fracture energy*

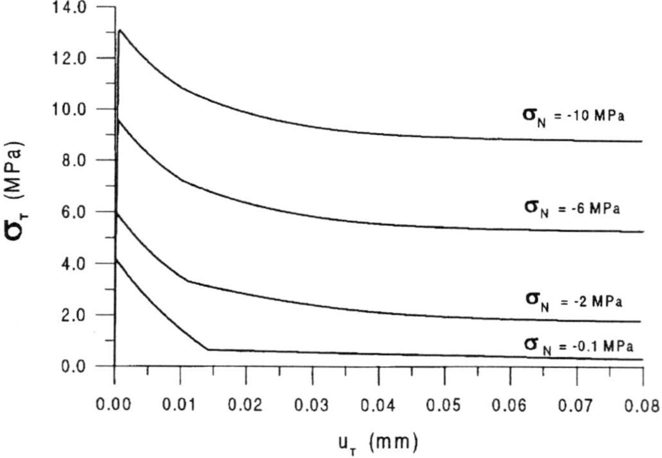

Figure 3. *Shear under constant compression: shear stress versus relative shear displacement for different values of normal stress*

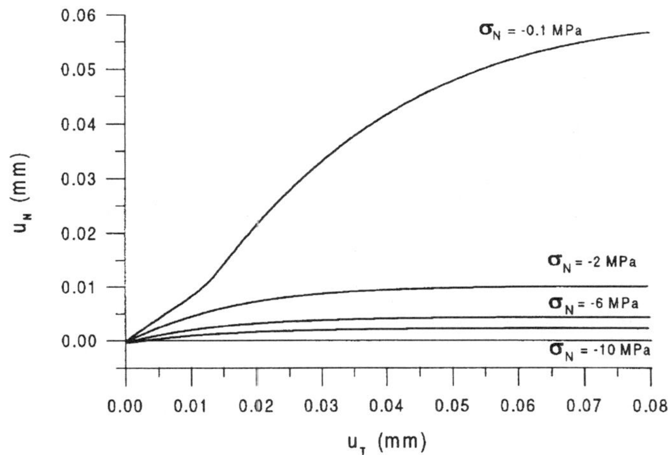

Figure 4. *Shear under constant compression: evolution of dilatancy for different values of normal stress*

The third constitutive example corresponds to an experimental test by Hassanzadeh [HAS 90]. Relative normal and shear displacements are prescribed to an interface in two steps. First, only a normal (opening) is applied till the peak stress is reached. Then, shear displacement is applied simultaneously with the normal displacement in a fixed proportion characterized by the angle $\tan\theta = u_N/u_T$.

Tests were run for various θ, starting with the limit case θ = 90° which actually corresponds to the test in pure tension. The parameter values used are $K_N = K_T =$ 200 MPa/m, $\tan\phi = 0.9$, $\chi_0 = 2.8$ MPa, $c_0 = 7$ MPa, $G^I_f = 0.1$ N/mm, $G^{IIa}_f = 1.0$ N/mm, $\sigma^{dil}_N = 56$ MPa, $\alpha_\chi = 0$, $\alpha_c = 1.5$, $\alpha^{dil}_\sigma = 2.7$, $\alpha^{dil}_c = 3$. The results obtained in the second part of the test are presented in Figures 5 and 6.

In Figure 5, normal stress is represented versus normal relative displacements. While for θ = 90° (pure tension), the usual exponential-type of decay is obtained, the imposition of a certain proportion of relative shear displacement causes the stresses to drop faster, to change the sign into compression, to reach a peak and then to vanish asymptotically. This is due to development of shear dilatancy that would exceed the prescribed normal opening rate.

In Figure 6, the shear stresses corresponding to the same tests are plotted against relative shear displacements. In both figures, numerical results (continuous lines) are represented together with experimental dots, showing how the proposed model not only gives reasonable numerical results, but is also capable of fitting experimental data obtained during non-trivial fully coupled normal/shear loading scenarios. Additional details of the constitutive formulation and test examples can be found in [CAR 97].

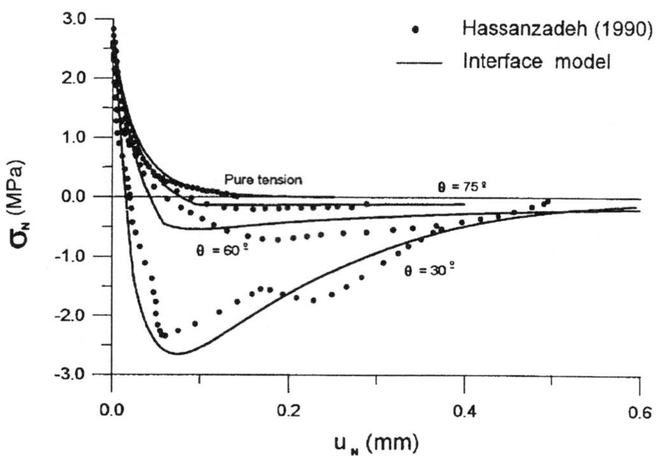

Figure 5. *Hassanzadeh's tests: normal stress versus relative displacement*

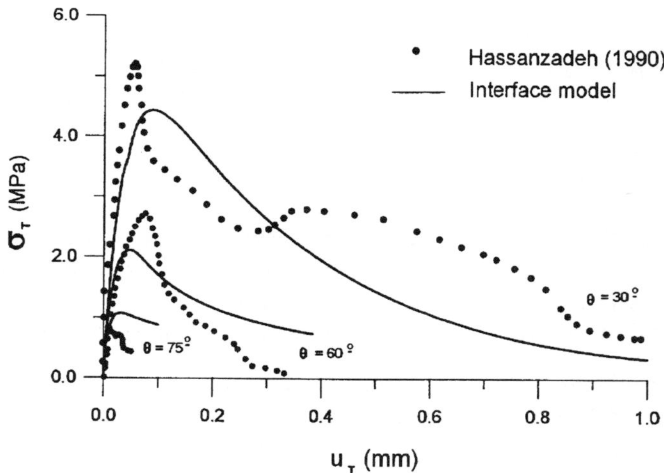

Figure 6. *Hassanzadeh's tests: shear stress versus relative displacement*

3. Numerical Implementation in the FE Code "DRAC"

The model has been implemented into a set of subroutines that have been added to the constitutive libraries of the FE code DRAC. A research-oriented geotechnical/structural FE program with 2D/3D capabilities, various element types including interfaces, post-processing module DRAC-VIU, etc., which have been entirely developed in-house in recent years at the Dept. of Geotechnical Engineering of ETSECCPB-UPC [PRA 93].

The constitutive subroutines perform relative displacement-to-stress calculations and include a substepping scheme to reduce integration errors. Substep size is determined depending on the type of prescribed relative displacement, and on variation of the flow rule direction [MAR 84], [CRI 91].

The interface elements used are "zero-thickness" isoparametric elements that can be inserted in between standard continuum finite elements. The nodes are grouped in pairs that match on each side those of the adjacent elements. The formulation follows the standard application of the Principle of Virtual Work, and the only special consideration refers to the integration rules which correspond to Newton-Cotes/Lobatto schemes (with integration points in between each pair of nodes) in order to avoid spurious oscillations in the resulting stress profiles. See Gens et al. [GEN 90] for more details.

The iterative strategy at the finite element level includes a linearized version of the arc-length standard procedure based on the norm of displacement increments of all nodes [RIK 72], [ROT 88]. This strategy works well in the type of microstructural calculations with interfaces along all possible crack paths, in which initially, cracks start opening all over the mesh and later most of them close, and deformations localize into one main crack.

4. Applications

4.1. *Microstructural Analysis of Concrete Specimens*

The first example of application is a study of fracture of concrete specimens considering a two-phase microstructure with large polygonal aggregates embedded on a mortar matrix. This is part of an on-going research effort initially motivated by concrete expansions in dam engineering [LOP 94]. More detailed results of the same study have been presented elsewhere [LOP 95], [LOP 96].

Square concrete specimens with 4x4 and 6x6 arrangements of aggregates are discretized as shown in Figures 7a-f. A number of interface elements are inserted along the aggregate-matrix interface, and also across the mortar matrix in order to allow most relevant failure mechanisms. Continuum elements between interfaces are considered linear elastic.

The geometry of the aggregates was taken from previous numerical work by Stankowski [STA 90], (Figure 7a,d). The mesh was rebuilt completely to provide straighter crack paths, following the ideas proposed by Vonk [VON 92], (Figure 7b,e). After introducing the interface paths, the resulting elastic cells were subdivided with triangular elements as needed to obtain a uniform element size (Figure 7c,f).

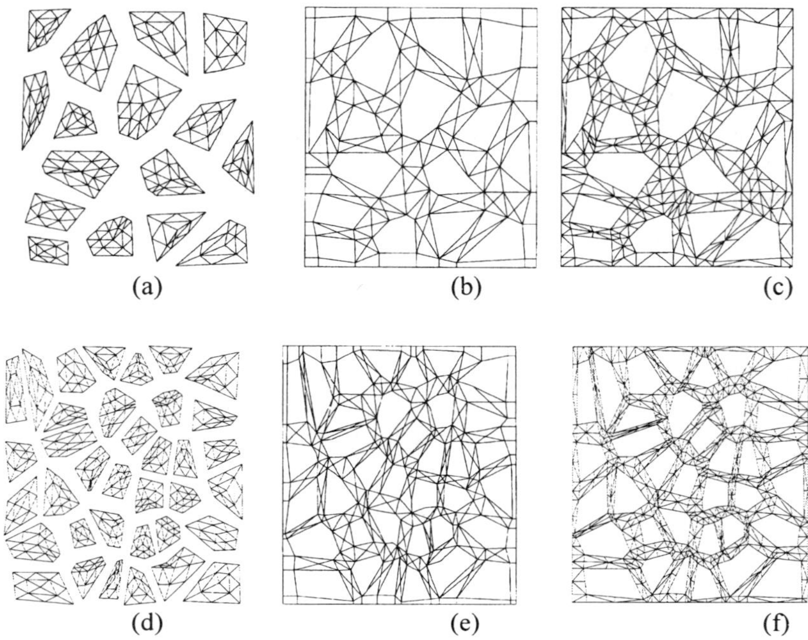

Figure 7. *FE discretization of the 4x4 and 6x6 arrangement: aggregates (left), interfaces inserted (center) and matrix (right)*

The load considered is pure tension, which has been applied in the form of a prescribed displacement on one side of the specimen, leaving the lateral displacement free in the transverse direction. Sum of the reactions divided by the size of the specimen gives an average stress. The calculations have been repeated with the load applied along the x and y directions for the 4x4 and 6x6 mesh and the resulting average stress-average strain curves are represented in Figure 8. It can be noted that all of them yield similar results in spite of the relatively different geometries involved.

Figure 9 includes results of the 4x4 specimen loaded vertically, at three different stages of the loading sequence indicated in Figure 8, as well as the final deformed mesh (magnification factor = 200). The thickness of the lines in color represents the amount of energy spent in the fracture process (W^{cr}) at each point of the interface, in red if the crack is active (plastic loading), or blue if the crack is arrested (elastic unloading). It is apparent that, initially, many cracks start developing, and at some point deformations localize in one or two cracks that develop while all other cracks unload. Figure 10 shows the final state in terms of energy and also of the deformed state (magnification factor = 200) at the end of the calculations for the other three cases analyzed. Well-known phenomena such as crack bridging and branching can be observed in the results.

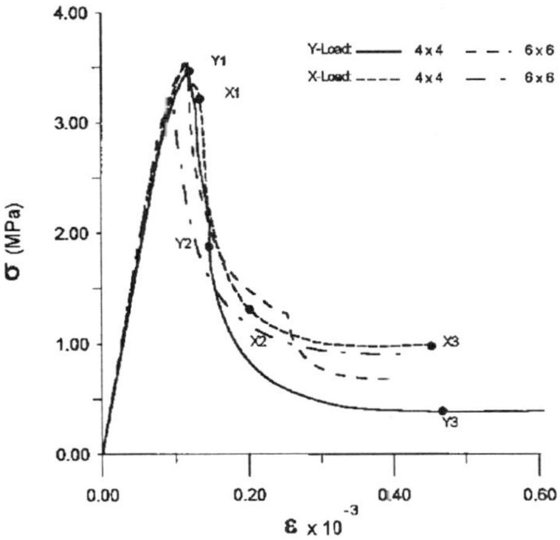

Figure 8. *Average stress-average strain curves*

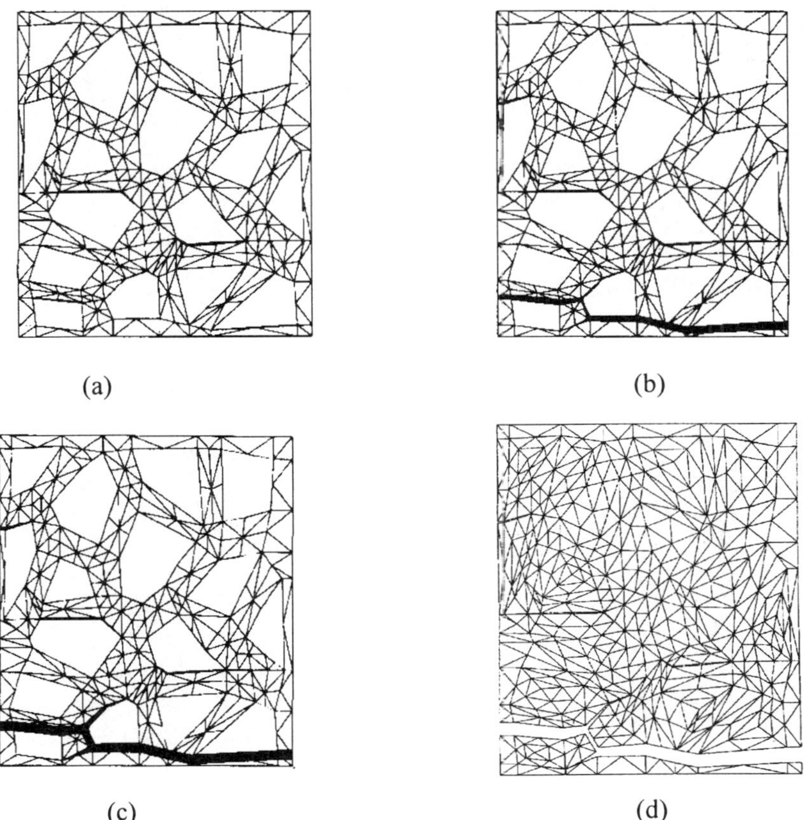

Figure 9. *Progressive cracking of the 4x4 mesh (a, b, c) represented by amount of energy spent and final deformed mesh (d) upon y loading*

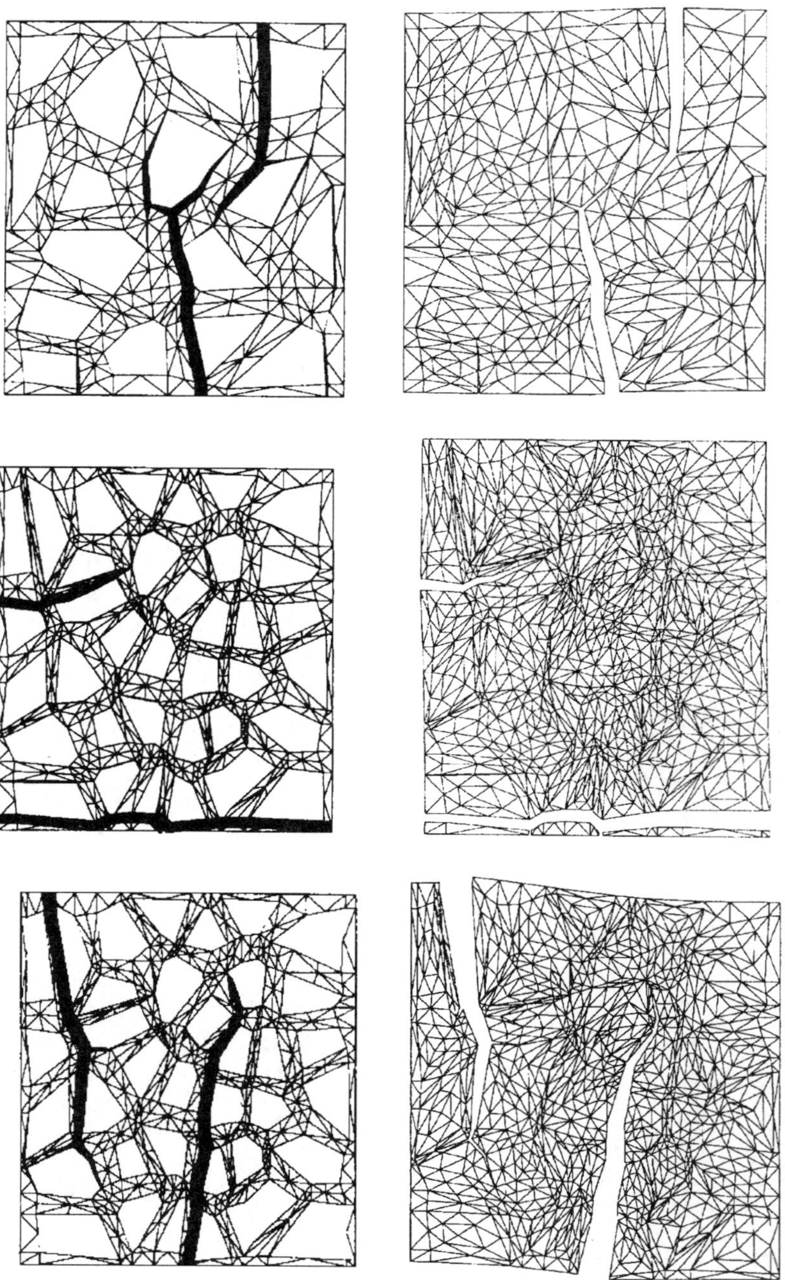

Figure 10. *Final state in terms of energy and also of the deformed state at the end of the calculations for the other three cases analyzed*

4.2. *Microstructural Analysis of Cancellous Bone*

The second example is taken from a study of fracture of trabecular (porous) bone, which is being carried out in collaboration with the Dept. of Structural Engineering of Politecnico di Milano. More details of this study can be found in [PIN 97].

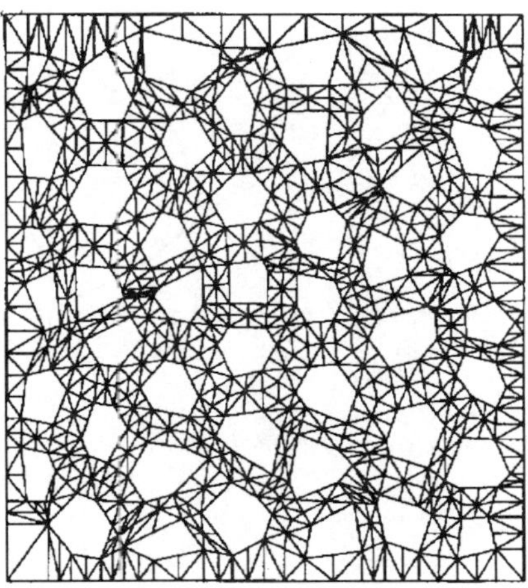

Figure 11. *FE discretization of trabecular bone architecture mesh*

The trabecular microstructure exhibits a geometry similar to the concrete, but with voids instead of aggregates. One of the various meshes used is represented in Figure 11.

This mesh was subjected to pure tension in the form of prescribed uniform displacements of the nodes on one side of the sample, and the sum of the reactions divided by the specimen size was taken as average stress. The results are represented in Figure 12 for two different sets of material parameters, together with an experimental curve for comparison.

Figure 13 depicts the energy spent and the deformed configuration at the final state of the calculations. Also in this case, interfaces open across the trabecular configuring one main macrocrack across the specimen.

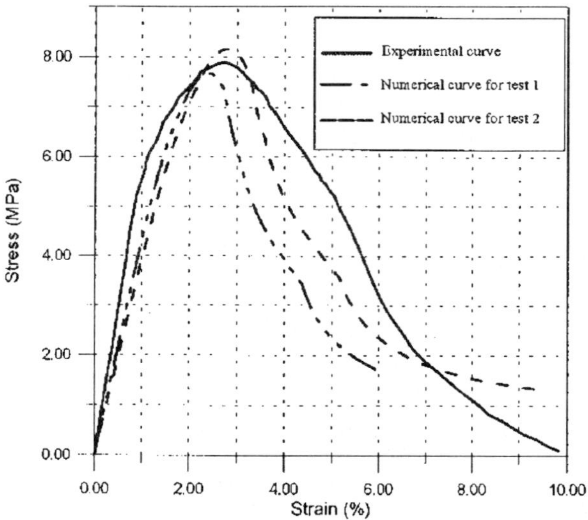

Figure 12. *Average stress-average strain curves*

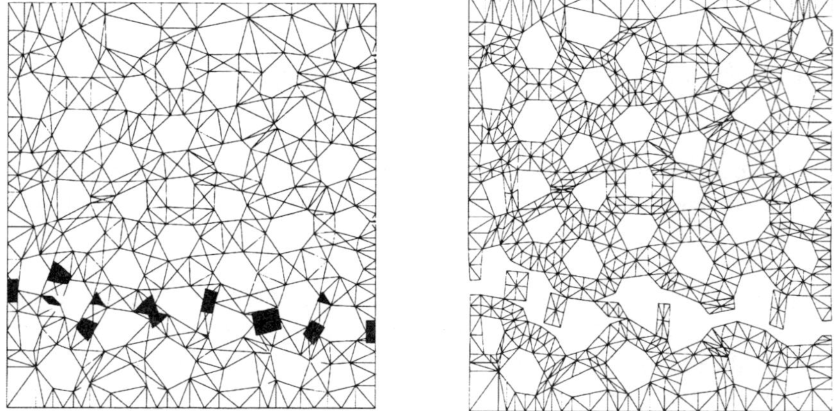

Figure 13. *Final state in terms of energy and deformed state*

4.3. Pull-out of Rock Bolts

The third example is taken from a study on pull-out of rock bolts carried out at the Department of Geotechnical Engineering ETSECCPB-UPC [ALO 96]. The rock bolt and surrounding material, shown in Figure 14, has been discretized with 1741 continuum elements and 1968 nodes. The mesh includes 122 interface elements along the rock bolt interface. The analysis is axisymmetric, and the parameters are, for the elastic continuum E = 1000MPa (soil), E = 30000MPa (mortar), E = 210000MPa (cable), ν = 0.30 (soil and cable), ν = 0.20 (mortar); for the interface: $K_N = K_T$ = 10 MPa/m, $\chi_0 = 0$, c_0 = 0.001MPa, $\tan\phi$ = 0.577 (ϕ = 30°), G^I_f = 30Nm/m^2, $G^{IIa}_f = 10G^I_f$, σ^{dil}= 0.0001MPa, and shape coefficients all equal zero.

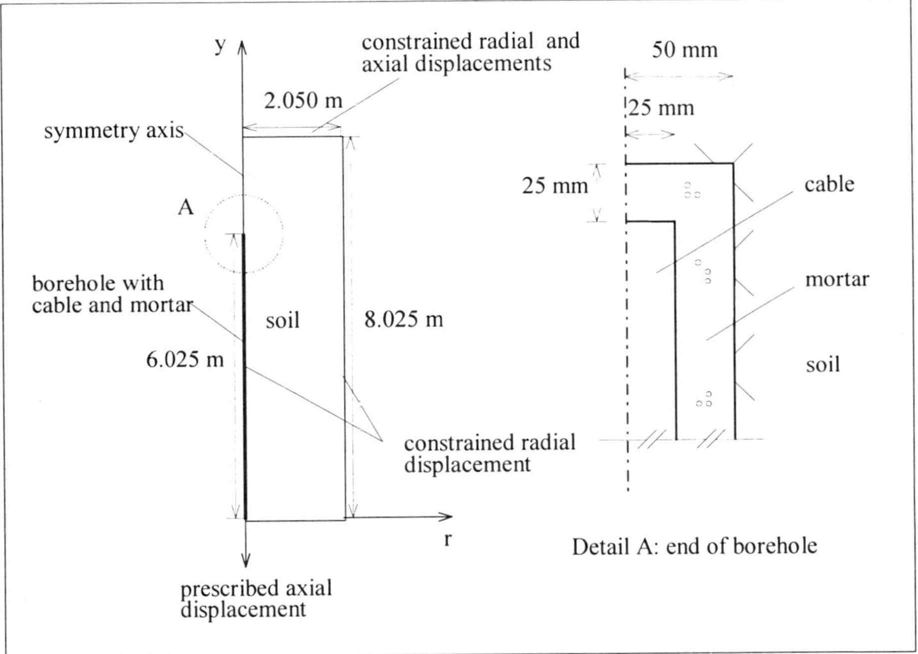

Figure 14. *Geometry of modeled pull-out*

A progressively increasing displacement is prescribed at the lower end of the rock bolt. Some results of the analysis are shown in Figures. 15, 16, and 17. In Figure 15 the total pull-out force (sum of reactions) is plotted against the prescribed displacement, and in Figures 16 and 17 the profiles of normal and shear stresses along the interface are represented at various stages of the prescribed pull-out displacement. The peak and softening response of this kind of tests are clearly reproduced in terms of F-δ, as well as the progressive failure shown in terms of stress profiles.

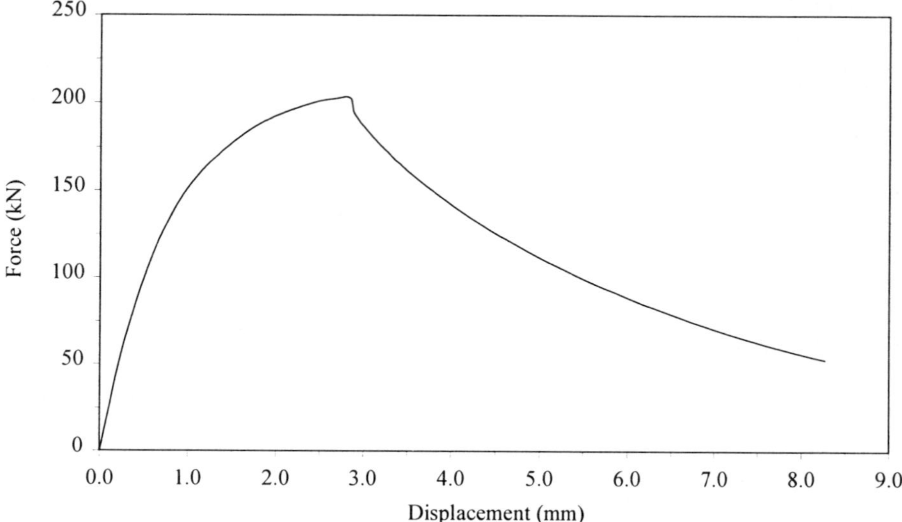

Figure 15. *Force-displacement curve for the rock bolt*

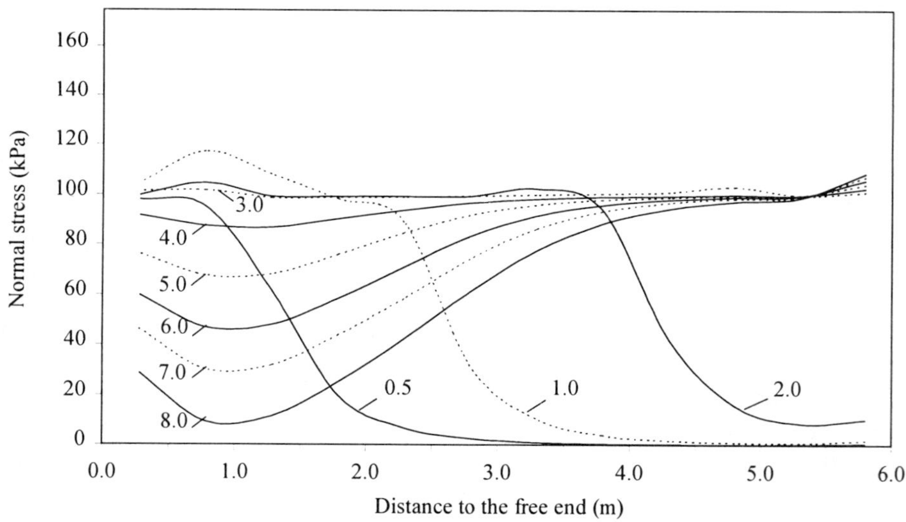

Figure 16. *Distribution of normal stresses for various prescribed displacements (in mm)*

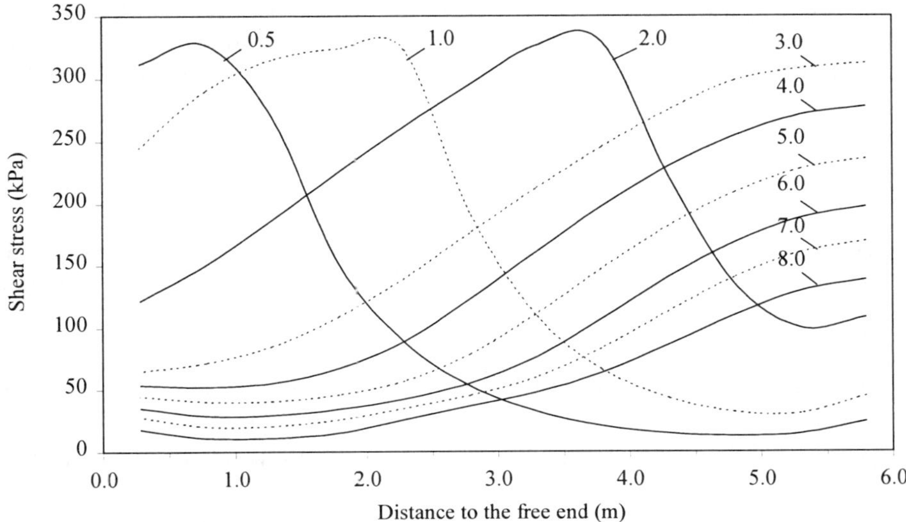

Figure 17. *Distribution of shear stresses for various prescribed displacements (in mm)*

5. Concluding Remarks

The interface model presented has been verified numerically and compared with experimental data, exhibiting meaningful results in all situations and the capability to capture non-trivial shear/normal coupled behavior. The interface has been implemented into a general-purpose finite element code and has been used in the numerical analysis of fracture of various materials. Two examples of application to microstructural analysis of normal concrete and cancellous bone subjected to tensile loading have been presented. Those results correspond to studies currently under development at ETSECCPB-UPC. Promising results are also being obtained in pure compression, tension/compression and brazilian tests. The concrete study is also being extended to high-strength concrete, in which interfaces are also included within the aggregate particles. In the third example, the model has also been used successfully for the fracture in the pull-out of injected rock bolts.

Acknowledgements

Partial support from DGICYT-MEC (Madrid, Spain) under grants PB95-0771 and PB96-0500 is gratefully acknowledged. The authors also wish to acknowledge the contributions of their colleagues who participated in the specific studies from which the examples of verification and application have been extracted: Pere Prat from Dept. of Geotechnical Engineering ETSECCPB-UPC, Antonio Aguado from

Dept. of Construction Engineering at ETSECCPB-UPC, Roberto Contro and Monica Pini at Dip. di Ingegneria Strutturale at Politecnico di Milano (Italy); and Eduardo Alonso from Dept. of Geotechnical Engineering also at ETSECCPB-UPC.

References

[ALO 96] ALONSO E., CASANOVAS J., ALCOVERRO J.,"Anclajes", Simposio sobre estructuras de contención de terrenos, Sociedad Española de Mecánica del Suelo y Cimentaciones, pp. 279-330, Santander, Spain, 1996.

[CAR 90] CAROL I., PRAT P.C., "A statically constrained microplane for the smeared analysis of concrete cracking", In Bicanic and Mang, editors, *Computer aided analysis and design of concrete structures*, Vol. 2, pp. 919-930, Zell-am-See, Austria, Pineridge Press, 1990.

[CAR 91] CAROL I., PRAT P.C., "Smeared analysis of concrete fracture using a microplane based multicrack model with static constraint", In J.G.M. van Mier, J. G. Rots, and A. Bakker, editors, *Fracture processes in concrete, rock and ceramics*, pp. 619-628, Noordwijk, The Netherlands. E & FN SPON, London, 1991.

[CAR 93] CAROL I., PRAT P.C., GETTU R., "Numerical analysis of mixed-mode fracture of quasi-brittle materials using a multicrack constitutive model", In H. P. Rossmanith and K. J. Miller, editors, *Mixed-mode fatigue and fracture*, pp. 319-332. Mechanical Engineering Publications Ltd., London. ESIS Publication 14, 1993.

[CAR 95] CAROL I., PRAT P. C., "A multicrack model based on the theory of multisurface plasticity and two fracture energies", In Owen, D.R.J., Oñate, E., and Hinton, E., editors, *Computational plasticity (COMPLAS IV)*, Vol. 2 pp.1583-1594, Barcelona, Pineridge Press, 1995.

[CAR 97] CAROL I., PRAT P.C., LÓPEZ C.M., "A normal/shear cracking model. Application to discrete crack analysis", ASCE *J. of Engineering Mechanics*. 123 (8): 1-9, 1997.

[CRI 91] CRISFIELD M., *Non-linear finite element analysis of solids and structures*, Volume 1, 1991.

[GEN 90] GENS A., CAROL I., ALONSO E., "A constitutive model for rock joints. Formulation and numerical implementation", *Computers and Geotechnics*, 9:3-20, 1990.

[HAS 90] HASSANZADEH M., "Determination of fracture zone properties in mixed mode I and II", *Engineering Fracture Mechanics*, 35 (4/5): 845-853, 1990.

[HIL 76] HILLERBORG A., MODÉER M., PETERSSON P.E., "Analysis of crack formation and crack growth in concrete by means of Fracture Mechanics and Finite Elements", *Cement and Concrete Research*, 6(6):773-781, 1976.

[LOP 94] LÓPEZ C.M., AGUADO A., CAROL I., "Numerical studies of two gravity dams subjected to differential expansions of the microstructure", In E. Bourdarot, J. Mazars, and V. Saouma, editors, *Dam fracture and damage*, pp. 163-168, Balkema, 1994.

[LOP 95] LÓPEZ C.M., CAROL I., AGUADO A., "Fracture of microstructural concrete: a numerical study using interface elements", In Batra, R., editor, *Contemporary Research in Engineering Science*, Proceedings, Eringen Medal Symposium honoring S.N. Atluri, pp. 55-65, New Orleans, Society of Engineering Science, 32nd Annual Meeting, Springer Verlag, October, 1995.

[LOP 96] LÓPEZ C.M., CAROL I., AGUADO A., "New results in fracture analysis of concrete microstructure using interface elements", *Anales de Mecánica de la Fractura*, 13: 92-97, 1996.

[MAR 84] MARQUES J., "Stress computation in elastoplasticity", *Engng Comput.*, 1: 42-51 1984.

[PIN 97] PINI M., LÓPEZ C.M., CAROL I., CONTRO R., "Microstructural analysis of cancellous bone using interface elements", In Rossmanith, H.P., editor, *Damage and Failure of Interfaces-(DFI-1)*, Vienna (Austria), 1997.

[PRA 92] PRAT P.C., CAROL I., GETTU R., "Numerical analysis of mixed-mode fracture processes", *Anales de Mecánica de la Fractura*, 9:75-80, 1992.

[PRA 93] PRAT P.C., GENS A., CAROL I., LEDESMA A., GILI J.A., "DRAC: A computer software for the analysis of rock mechanics problems", In H. Liu, editor, *Application of computer methods in rock mechanics*, Vol. 2, pp. 1361-1368, Xian, China, 1993.

[RIK 72] RIKS E., "The application of Newton's method to the problem of elastic stability", *J. Appl. Mech.*, 39: 1060-1066, 1972.

[ROT 88] ROTS J.G., Computational modelling of concrete fracture, PhD thesis, Delft University of Technology, The Netherlands, 1988.

[STA 90] STANKOWSKI,T., Numerical simulation of progressive failure in particle composites, PhD thesis, Dept. CEAE, University of Colorado, Boulder, CO 80309-0428, USA, 1990.

[VON 92] VONK R., Softening of concrete loaded in compression, PhD thesis, Technische Universiteit Eindhoven, Postbus 513, 5600 MB Eindhoven, The Netherlands, 1992.

Modeling Material Failure as a Strong Discontinuity with the Material Point Method

Howard L. Schreyer[*] — Deborah L. Sulsky[**] — S.-J. Zhou[*]

** Department of Mechanical Engineering*
*** Department of Mathematics and Statistics*
University of New Mexico
Albuquerque, NM USA 87131

ABSTRACT. *A discrete constitutive equation for modeling material failure as a decohesion or separation of material to form two free surfaces is a relatively simple approach. However, numerical simulations based on such a model involve considerable complexity including remeshing if the finite element approach is used. Here, a basic formulation involving the simultaneous application of the continuum and decohesion constitutive equations is described together with a numerical approach based on the material point method. Preliminary results indicate that failure propagation can be predicted at an arbitrary angle without the dispersion of the crack front that is often observed with conventional finite elements.*

KEY WORDS: *Decohesion, Localized Deformation, Softening, Material Point Method, Strong Discontinuity.*

1. Introduction

The propagation of cracks through concrete is just a manifestation of material failure. Numerous models of failure have been proposed together with numerical procedures for obtaining solutions to the governing boundary value problem. Unfortunately, the material model for failure is often entwined with the numerical approach so that it is difficult to determine which aspect is the limiting component if predictions do not match experimental data. Here, we attempt to carefully differentiate these two essential components by first concentrating on a basic failure model, and then proposing the use of a relatively new numerical procedure, the material point method. Preliminary results indicate that the difficulties associated with the finite element method are not present.

There are many criteria for material failure but the definitions of failure are often vague or defined implicitly through each criterion. We take material failure to mean the process by which two new free surfaces are formed, with brittle fracture as an obvious example. However, there are other forms of material failure as exemplified by ductile rupture, delamination, the breaking of grain boundaries and the pullout of reinforcing rods or fibers. Our interest is to represent all of these phenomena with a single model that incorporates the essential features of the state of stress or strain at which failure initiates and predicts the correct energy dissipated. Our focus is not on replicating the details of failure, although this can be done in some cases, but on predicting the effect of failure on the far-field stress distribution and on structural response as reflected, for example, by a force-deflection curve. The proposed approach is based on a constitutive equation that describes decohesion. When used with the material point method, which is a relatively new computational method that is particularly robust for problems with large deformations, the proposed approach has a simple structure in that the decohesion comes into the analysis through the constitutive equation only. There is no attempt to enforce the geometrical continuity of a crack. Instead, compatibility is enforced in an averaged sense.

Failure modeling involves both theoretical formulations of constitutive equations and numerical simulations, and the two aspects should be carefully delineated. However, the finite element method has become the method of choice for the majority of engineering applications so that the formulation of the constitutive equations is often tailored for use by finite elements; conversely, limitations imposed by the finite element method are often interpreted unjustly as a limitation of the theoretical approach. In the following brief survey, we attempt to keep the discussion of the two phases distinct, if at all possible.

A large number of papers related to failure have been based on a zone of softening with an assumed width in which a continuum constitutive equation continues to be used [BAZ 84], [BOR 87], [ROT 87], [OLI 89], [DAH 90], [WIE 98]. The theoretical difficulty with such an approach is the possible loss of ellipticity and material stability within the band. When used with finite elements, the band width is associated with the size of the elements and the accuracy is then limited when the elements become highly deformed.

An alternative (discrete) approach is to consider material failure as a strong discontinuity in displacement with traction related to the discontinuity. There is a long history in which discrete constitutive equations are postulated directly as reflected by Barenblatt [BAR 59], and Hillerborg et al. [HIL 76]. Feenstra et al. [FEE 91] and Corigliano [COR 93] provide a nice summary of previous models and describe numerical methods based on the use of interface elements. The use of discrete constitutive equations has not met with complete favor partially because strong discontinuities are difficult to handle numerically and convergence with mesh refinement and mesh insensitivity is difficult to show. The use of interface elements may require frequent remeshing if the crack surface propagates in a curved manner, and double nodes which separate with the evolution of decohesion [SCH 92]. However, Dvorkin et al. [DVO 90] provide a nice approach that overcomes many of these objections by handling discontinuities at the element level rather than enforcing discontinuities to be along element boundaries.

A fundamentally different approach is described in more recent work by Simo et al. [SIM 93] in which the continuum constitutive equation is extended beyond the loss of ellipticity condition into the softening regime. They argue that this extension should be accompanied by distribution theory which, in effect, leads to a strong discontinuity. The theory has since been extensively developed by Oliver [OLI 89], Simo and Oliver [SIM 94], Armero and Garikipati [ARM 95], Larsson and Runesson [LAR 96], Oliver [OLI 96] and Armero [ARM 97]. The final result is discrete constitutive equations relating stress to the discontinuity in displacement, and here also the discontinuity is handled at the element or constitutive level.

We have opted for a particular combination of these ideas in an attempt to provide an approach that is as simple and as straightforward as possible. First, we propose the direct introduction of discrete constitutive equations with the thought that they should be introduced when ellipticity is lost, although a direct failure initiation criterion can be used. No attempt is made to model the post-crack frictional effects that may occur with surfaces with rough cracks [FEE 91] although such features can be added. Second, the discontinuity is considered to be part of the constitutive equation and is applied in a manner analogous to that of Dvorkin [DVO 91], Oliver [OLI 96a], [OLI 96b] and Armero [ARM 97]. A point that is undergoing failure is also considered to be a material point in the continuum so that the decphesion and continuum constiutive equations must be simultaneously satisfied subject to the restriction of traction equilibrium. Third, we invoke the constitutive equation in the material point method. The arguments for the direct calculation of the strong discontinuity in displacement, which we also call decohesion, and the use of the material point method are summarized as follows:

(i) We retain the conceptual simplicity inherent with the discrete constitutive approach that material failure does not happen abruptly but occurs smoothly with a gradual reduction in traction as the displacement discontinuity increases.

(ii) We believe it is extremely difficult to evaluate properties of any constitutive equation in the failure regime. However, it is probably easier to select material parameters for a discrete constitutive equation than for a continuum model extended into the softening regime.

(iii) The discrete equation can be applied, if desired, at the instance when ellipticity is lost so that there is a high probability that well posedness can be retained although a stability analysis must be performed [SUO 92].

(iv) The essential aspects of prescribed stress at the initiation of failure and prescribed energy dissipation at the end of failure are automatically included in this model.

(v) Once decohesion is initiated on a surface of discontinuity, the adjacent continuum tends to unload into the elastic regime, so the computational simplicity of only needing to combine decohesion with elasticity covers the vast majority of practical cases.

(vi) The decohesion constitutive equation can be developed in a thermodynamical setting, in concert with many current continuum models, and can include plasticity, damage, viscoelastic and viscoplastic features that are associated strictly with the decohesion.

(vii) The application of decohesion constitutive equations in the material point method retains the simplicity of current applications of strong discontinuities at the element level in the finite element method. However, double nodes or interface elements are not needed and there is the additional potential advantage that mesh orientation and mesh distortion are not factors that need to be considered.

(viii) Following the method outlined by Allix and Corigliano [ALL 95], [ALL 96] there is the potential of relating the decohesion constitutive equation to mixed-mode fracture.

(ix) Finally, the use of a discrete constitutive equation may still be a suitable model for diffuse failure if the primary objective is to obtain an efficient solution for the region away from the failure zone.

The next section provides only a brief description of the material point method since the method has been fully described in previous papers. Section 3 describes the basic structure of the decohesion model used in our analysis. Analytical and numerical solutions to model problems [ZHO 98] including a convergence study are given in Section 4, which is then followed by conclusions concerning the general applicability of the method for material failure in general including delamination.

2. The Material Point Method

The material point method [SUL 94], [SUL 95], [SUL 96] discretizes a solid body by marking a set of material points in the original configuration that are tracked throughout the deformation process. Let $\mathbf{x}_p^n$, $p = 1,...,N_p$ denote the current position of material point p at time t^n, n = 0, 1, 2, ... These material points provide a Lagrangian description of the solid body that is not subject to mesh tangling. Each point at time t^n has an associated mass, m_p, density, ρ_n^p, velocity, $\mathbf{v}_p^n$, Cauchy stress tensor, $\boldsymbol{\sigma}_p^n$, strain, $\mathbf{e}_n^p$, and any other internal variables necessary for the constitutive model. If temperature changes are important, internal energy or temperature may also be ascribed to the material points. The material point mass is constant in time, insuring that the continuity equation is satisfied. Other variables must be updated with reference to conservation of momentum, conservation of energy, or from the constitutive model.

To make the computations tractable, at each timestep of a dynamic algorithm, information from the material points is interpolated to a background computational mesh. This mesh covers the computational domain and is chosen for computational convenience. A particularly simple choice is a regular rectangular grid. After information is interpolated to the grid, equations of motion are solved on this mesh which is considered to be an updated Lagrangian frame. For example, to solve the momentum equation on the grid using an explicit FE algorithm, one must know the value of the momentum at the beginning of the timestep at the nodal positions. The nodal momentum, $m_i^n \mathbf{v}_i^n$, is the product of the nodal mass and nodal velocity, and each is determined by interpolation,

$$m_i^n = \sum_{p=1}^{N_p} m_p N_i(\mathbf{x}_p^n)$$
$$m_i^n \mathbf{v}_i^n = \sum_{p=1}^{N_p} m_p \mathbf{v}_p^n N_i(\mathbf{x}_p^n) \qquad [1]$$

In the above, $N_i(x)$ is the nodal basis function associated with node I. In this paper, $N_i(x)$ are the tensor products of piecewise linear functions. The internal forces are determined from the particle stresses according to

$$f_i^{int} = -\sum_{p=1}^{N_p} G_{ip}^n \boldsymbol{\sigma}_p^n m_p / \rho_p^n \qquad [2]$$

The quantity G_{ip}^n is the gradient of the nodal basis function evaluated at the material point position, $G_{ip}^n = \nabla N_i(\mathbf{x})|_{\mathbf{x}_p^n}$. The momentum equation is solved with the nodes considered to be moving with the deformation to give nodal velocities, $\mathbf{v}_i^L$, at the end of this Lagrangian timestep of size Δt,

$$m_i^n \frac{\mathbf{v}_i^L - \mathbf{v}_i^n}{\Delta t} = \mathbf{f}_i^{int}. \qquad [3]$$

At the end of this Lagrangian step, the new nodal values of velocity are used to update the material points. The material points move along with the nodes according to the solution given throughout the elements by the nodal basis functions

$$\mathbf{x}_p^{n+1} = \mathbf{x}_p^n + \Delta t \sum_{i=1}^{N_n} \mathbf{v}_i^L N_i(\mathbf{x}_p^n). \qquad [4]$$

Similarly, the material point velocity is updated via

$$\mathbf{v}_p^{n+1} = \mathbf{v}_p^n + \sum_{i=1}^{N_n} (\mathbf{v}_i^L - \mathbf{v}_i^n) N_i(\mathbf{x}_p^n). \qquad [5]$$

The sums in these last two equations extend from 1 to N_n where N_n is the number of nodes in the computational mesh.

A strain increment for each material point is determined using the gradient of the nodal basis function,

$$\Delta \mathbf{e}_p = \frac{\Delta t}{2} \sum_{i=1}^{N_n} [G_{ip}^n \mathbf{v}_i^L + (G_{ip}^n \mathbf{v}_i^L)^T]. \qquad [6]$$

This strain increment is then used in an appropriate constitutive equation for the material being modeled to update the stress at the material point. Any internal variables necessary in the constitutive model can also be assigned to the material points and transported along with them. Once the material points have been completely updated, the computational mesh may be discarded and a new mesh defined, if desired, and then starts the next timestep.

The material point method has several advantages. The Lagrangian description provided by the material points can undergo large deformations without mesh tangling. Since the computational mesh is under user control, it can be chosen so that reasonable timesteps may be taken in this Lagrangian frame. Usually, the timestep is restricted by the CFL condition for an explicit algorithm, where the critical timestep is the ratio of the mesh size to the wave speed. Note that this condition depends on the more favorable mesh spacing, not on the material point spacing. Since equations are solved in an updated Lagrangian frame on the FE mesh, the nonlinear convective terms, troublesome in Eulerian formulations, are not an issue. Finally, the material points transport material properties and internal variables without error.

3. Discrete Constitutive Equation For Decohesion

3.1. *The Theoretical Model*

We define the initiation of material failure as the time when a material point first experiences a discontinuity in displacement but continues to function as a point in a solid continuum. A collection of such points in a neighborhood defines a failure surface, Γ. Although the material manifestation is a single surface, one observes spatially two surfaces, Γ_U and Γ_L, as sketched in Figure 1. Each dotted line illustrates points in space identified with a single material point which can be considered associated with any one of the spatial points on the line. The sketch illustrates the material surface as a thick line between the two spatial surfaces but if a Lagrangian description is used, the material surface may be at a totally different location. Failure is said to be complete when traction can no longer be sustained on the material surface, i.e., the spatial surfaces no longer have any ligaments connecting them even though one point on each surface is identified as a single material point. The discontinuity in displacement is called decohesion.

Here, we present a development of discrete constitutive equations using thermodynamics as a framework with the result that the dissipation inequality is automatically satisfied. The approach entails two essential assumptions consisting of (1) the form of the free energy, and (2) the form of the evolution equations. Each assumption leads to a different model which can only be tested by solving a problem for which either qualitative or quantitative data exist. To allow for the presentation of different models in a convenient manner, we present the general framework first, and then show the implications inherent in specific assumptions.

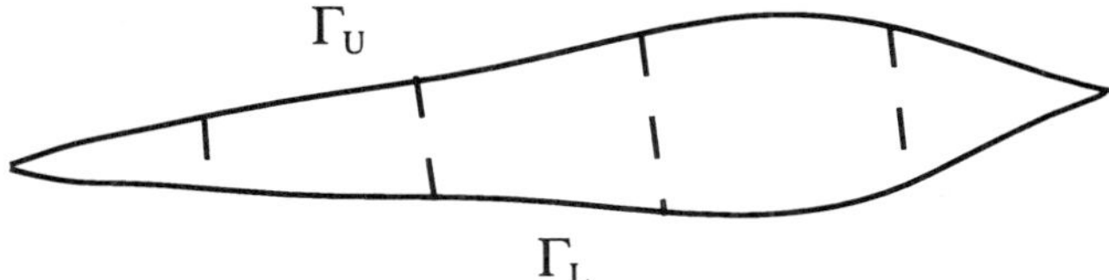

(a) Decohesion in spatial configuration - two surfaces.

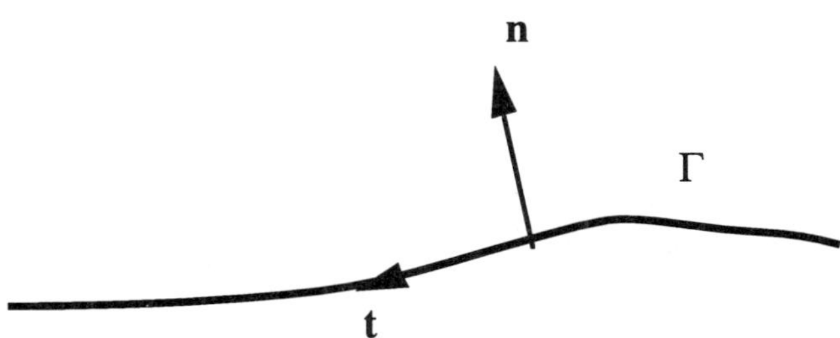

(b) Decohesion in material configuration - surface with discontinuities

Figure 1. *Material failure as represented by two spatial surfaces, Γ_U and Γ_L, or one material surface, Γ*

The approach is analogous to what one might use for a rigid-plastic continuum for which the elastic part of the response is ignored, i.e., the total strain and the plastic strain are identical. The internal strain energy does not exist and the stress must be provided by the solution to a boundary value problem. However, there remains a contribution to the free energy associated with hardening and evolution equations for plasticity variables.

Consider a situation where loads are applied to a body which is continuous except on a material failure surface, Γ, which displays a strong discontinuity, or decohesion, $u^d = [u]$. Any point on the suface is also a point in the continuum which is assumed to be governed by linear elasticity so that the stress, σ, and strain, e, are linearly related by the elasticity tensor, E:

$$\sigma = E{:}e \qquad [7]$$

If n denotes the normal to the surface, then the traction, τ, is given by $\boldsymbol{\tau} = \boldsymbol{\sigma} \cdot \mathbf{n}$. The rate at which power is being added to the surface by this traction is $\boldsymbol{\tau} \cdot \dot{\mathbf{u}}^d$ in which a superposed dot denotes a derivative with respect to time. We postulate that the free energy per unit surface area consists of an initial energy, U_0, due to residual stresses that resulted from the curing process, and a term, U_d, which represents the effect of decohesion:

$$U = U_0 - U_d(\bar{u}) \qquad [8]$$

The use of the negative sign is meant to suggest that normally energy is provided by the original material to the decohesion process. The parameter, $\bar{u}$, is a scalar representation of the state of decohesion. The specific choice for U_d (positive) is part of a particular model The decohesion, u^p, is viewed as "permanent" decohesion and is introduced similarly to plastic strain in elasto-plastic continuum models. In the absense of elasticity, $u^p = u^d$.

If only u^p and $\bar{u}$ are considered to be the primary variables describing decohesion, the dissipation rate is

$$D_s = \boldsymbol{\tau} \cdot \dot{\mathbf{u}}^d - \dot{U} = \boldsymbol{\tau} \cdot \dot{\mathbf{u}}^p + \bar{\tau}\dot{\bar{u}} \qquad \text{with} \qquad \bar{\tau} = \frac{\partial U_d}{\partial \bar{u}} \qquad [9]$$

The generalized traction, $\bar{\tau}$, is conjugate to $\bar{u}$. Instead of the traction starting at zero, as it does for some existing discrete models [NEE 87, COR 93], we visualize that when the failure process starts, $u^d = 0$, $u^p = 0$, and the traction is the initial vector, τ_0 which depends on the path.

We parameterize the development of decohesion through a single, dimensionless monotonically increasing variable, λ, and the evolution equations

$$\dot{\mathbf{u}}^p = \dot{\lambda}\mathbf{m}^e \qquad \dot{\bar{u}} = \dot{\lambda} m^e \qquad [10]$$

in which m^e and m^e denote evolution functions that depend on τ and $\bar{\tau}$. m^e is a vector whose inner product with τ is assumed to be positive, semi-definite. If we introduce an effective traction

$$\tau^e = \boldsymbol{\tau} \cdot \mathbf{m}^e \qquad [11]$$

then the dissipation rate becomes

$$D_s = \dot{\lambda}[\tau^e + \bar{\tau} m^e] \qquad [12]$$

To ensure that the dissipation is positive, define a decohesion function as follows:

$$F_d = \tau^e + \bar{\tau} m^e - F_0 \qquad F_0 > 0 \qquad [13]$$

The function has been constructed in the usual manner so that F_d is negative when the tractions are zero. We assume decohesion does not occur unless $F_d = 0$ in which case the dissipation rate becomes $D_s = \dot{\lambda} F_0$, a positive scalar (and $F_d > 0$ is not permitted). The evolution equations can now be interpreted as parameterized in terms of dissipation which is a monotonically increasing parameter. The total dissipated energy is simply $D = \lambda F_0$ which depends only on the value of λ and not on the path followed to achieve a given value of λ.

Next, we consider the decohesion condition, $F_d = 0$. At the initiation of decohesion, $\tau = \tau_0$ and $\mathbf{m}^e = \mathbf{m}_0^e$. If we assume $\bar{\tau} = 0$ at the initiation, then

$$F_0 = \tau_0^e \equiv \boldsymbol{\tau}_0 \cdot \mathbf{m}_0^e \qquad [14]$$

and the decohesion condition reduces to

$$\tau^e = \tau_0^e - \bar{\tau} m^e \qquad [15]$$

Typically, $\bar{\tau}$ increases to the point where τ^e goes to zero and $\bar{\tau} = \bar{\tau}_s \equiv \tau_0^e / m_s^e$ which is defined to be separation. The values of $\mathbf{u}^p$ and $\bar{u}$ at separation are denoted by $\mathbf{u}_s^p$ and $\bar{u}_s$ respectively. Unless there is a load reversal which brings the two spatial surfaces back into contact, it is assumed that τ^e remains zero after separation.

With the use of [8], the stored surface energy lost due to separation is

$$\Phi_U = -U_d(\bar{u}_s) \qquad [16]$$

The total energy per unit area that must be provided to cause total separation is variously called the fracture energy, or the energy of separation, Φ_F, and consists of the sum of the stored energy and the dissipated energy:

$$\Phi_F = \Phi_U + \Phi_D \qquad [17]$$

Because Φ_U is negative, the fracture energy is less than the dissipation.

In summary, the resulting set of constitutive equations in rate form becomes

$$\begin{aligned}
&\text{(i)} \quad \dot{\boldsymbol{\sigma}} = \mathbf{E}:\dot{\mathbf{e}} && \text{Continuum elasticity} \\
&\text{(ii)} \quad \dot{\boldsymbol{\tau}} = \dot{\boldsymbol{\sigma}} \cdot \mathbf{n} && \text{Traction equilibrium} \\
&\text{(iii)} \quad \dot{\mathbf{u}}^p = \dot{\lambda} \mathbf{m}^e \quad \dot{\bar{u}} = \dot{\lambda} m^e && \text{Evolution equations} \\
&\text{(iv)} \quad \bar{\tau} = \frac{\partial U_d}{\partial \bar{u}} && \text{Constitutive equation} \\
&\text{(v)} \quad F_d \equiv \tau^e - (\tau_0^e - \bar{\tau} m^e) = 0 && \text{Consistency} \\
&\text{(vi)} \quad \tau^e = \boldsymbol{\tau} \cdot \mathbf{m}^e \ \text{ and } \ \tau_0^e = \boldsymbol{\tau}_0 \cdot \mathbf{m}_0^e
\end{aligned} \qquad [18]$$

Two additional assumptions remain to completely formulate constitutive equations; (i) the form of the evolution equations for the evolution functions m^e and m^e, and (ii) the form of the function U_d which provides a constitutive relation between $\bar{\tau}$ and $\bar{u}$. Even slight changes in the forms of these assumptions can have significant effects on predictions. Since the only possible way to evaluate the suitability of decohesive constitutive equations is indirectly through comparisons of solutions to problems with features provided by experimental data, we consider [18] to be the basic format and provide different models based on plausible assumptions. The results of choosing specific forms for m^e, m^e and U_d are given next.

3.2. *Model 1: Associated Evolution Equations*

Here we are more specific in our formulation of the decohesion model. In the theoretical formulation, it is most convenient to use dimensional parameters so that physical interpretations can be easily made; conversely, for numerical implementation of the theory, dimensionless variables should be used. With these objectives in mind, we choose λ to be dimensionless and consider $\bar{u}$ to have the dimension of length (in analogy with the decohesion u^d). We choose the evolution functions and the decohesion energy to be of the following forms:

$$\mathbf{m}^e = \bar{u}_0 \frac{\mathbf{A}_d \cdot \boldsymbol{\tau}}{(\boldsymbol{\tau} \cdot \mathbf{A}_d \cdot \boldsymbol{\tau})^{1/2}} \qquad m^e = \bar{u}_0 \qquad U_d = U_0 \frac{(\bar{u} / \bar{u}_0)^{q+1}}{(q+1)} \qquad [19]$$

in which A_d is taken to be a positive definite (dimensionless) tensor whose components are material parameters, as is $q \geq 0$. Additional material parameters (constants) are the reference decohesion scalar, $\bar{u}_0$, and the reference surface energy, U_0. We define a reference scalar traction, $\bar{\tau}_0$, by the relation $U_0 = \bar{u}_0 \bar{\tau}_0$. The immediate result of these choices is that

$$\tau^e = \bar{u}_0 (\boldsymbol{\tau} \cdot \mathbf{A}_d \cdot \boldsymbol{\tau})^{1/2} \qquad \bar{u} = \lambda \bar{u}_0 \qquad \bar{\tau} = \bar{\tau}_0 (\lambda)^q \qquad [20]$$

and the evolution functions are obtained as derivatives of the damage function with respect to the corresponding conjugate variables, i.e., the evolution functions are said to be "associated":

$$\mathbf{m}^e = \frac{\partial F_d}{\partial \boldsymbol{\tau}} \qquad m^e = \frac{\partial F_d}{\partial \bar{\tau}} \qquad [21]$$

At this point, we consider a two-dimensional formulation with n denoting the normal to the failure surface and t a unit tangent vector as indicated in Figure 1. Corresponding components of τ are τ_n and τ_t respectively. If the traction consists only of a normal component, specify the failure initiation value as τ_{nf} and, similarly, let τ_{tf} denote the failure initiation value for a purely shear case. One approach for incorporating these failure initiation conditions is to choose the components of A_d with respect to this local basis as follows:

$$[A_d] = \bar{\tau}_0^2 \begin{bmatrix} \frac{1}{(\tau_{nf})^2} & 0 \\ 0 & \frac{1}{(\tau_{tf})^2} \end{bmatrix} \qquad [22]$$

The consequences of this choice are

$$D_s = \dot{\lambda} U_0 \qquad \bar{\tau}^* = \lambda^q \qquad F_d = U_0 F_d^* \qquad F_d^* = \tau^{e*} - (1 - \bar{\tau}^*)$$
$$\tau^e = U_0 \tau^{e*} \qquad \tau^{e*} = \frac{(\boldsymbol{\tau} \cdot \mathbf{A}_d \cdot \boldsymbol{\tau})^{1/2}}{\bar{\tau}_0} = (\frac{\tau_n^2}{\tau_{nf}^2} + \frac{\tau_t^2}{\tau_{tf}^2}) \qquad [23]$$

where τ^{e*} is a dimensionless effective traction, $\bar{\tau}^*$ is a dimensionless form of $\bar{\tau}$, and F_d^* is a dimensionless decohesion function. In deriving these equations, the identities $F_0 = \tau_0^e = U_0$ have been used. The identities follow from the conditions at the initiation of decohesion that $\bar{u} = 0$ and $\tau^{e*} = 1$.

We note that the decohesion condition $(F_d^* = 0)$ reduces to $\tau^{e*} = 1 - \bar{\tau}^*$. As decohesion occurs, $\bar{u}$ increases and τ^{e*} decreases to zero when $\bar{u} = \bar{u}_0$. Therefore $\bar{u}_0$ can be interpreted as the value of $\bar{u}$ at which separation occurs. In the post separation regime, $\bar{u} > \bar{u}_0$, the decohesion condition is $\tau^{e*} = 0$. Finally, we give an alternative form for the decohesion evolution function:

$$\mathbf{m}^e = \frac{U_0}{\tau^{e*}} (\frac{\tau_n}{\tau_{nf}^2} \mathbf{n} + \frac{\tau_t}{\tau_{tf}^2} \mathbf{t}) \qquad [24]$$

Since the dissipation rate is $\dot{\lambda} U_0$, the energy dissipated per unit surface area at any moment is simply λU_0. From [20], $\lambda = 1$ at separation so the maximum energy dissipated is U_0 which provides a physical interpretation and a method for determining this particular parameter. The formulation implies that the dissipated energy is independent of path which is generally not representative of real materials. The final value of the stored energy is obtained by substituting $\bar{u} = \bar{u}_0$ in [19] and using [16]. In summary, the final dissipated, stored and total failure energies are:

$$\Phi_D = U_0 \qquad \Phi_U = -\frac{1}{(q+1)} U_0 \qquad \Phi_F = \frac{q}{(q+1)} U_0 \qquad [25]$$

Suppose a pure opening-mode path is followed, or $\tau_t = 0$. Then it is easily shown that $u_n = \bar{u}(\bar{\tau}_0 / \tau_{nf})$. If $\bar{\tau}_0$ is chosen to equal τ_{nf}, then $\bar{u}$ equals u_n for Mode I. Experimental data obtained from a pure shear mode test can then be used to assess the adequacy of the model in a process similar to that used to evaluate Mises plasticity. The limitation of a single value of dissipated energy can be circumvented by using a multifunctioned decohesion surface. The development of such a surface is beyond the scope of what we wish to achieve. For the given model, the required data are τ_{nf}, τ_{tf}, U_0 and q. We can choose $\bar{\tau}_0 = \tau_{nf}$ and then $\bar{u}_0 = U_0 / \tau_{nf}$ to provide values for all of the parameters.

Sometimes nonassociated models are required to provide a better fit with experimental data including observations of the mode of failure. Next we give an example of how a particular nonassociated model can be constructed.

3.3. *Model 2: Non-associated Evolution Functions*

Suppose we retain all aspects of the previous model with the exception that the evolution equation for the permanent decohesion is in the normal, or opening direction irrespective of the state of traction:

$$\dot{u}_n^p = \dot{\lambda} m_n^e \qquad \dot{u}_t^p = 0 \qquad m_n^e = \mathbf{m}^e \cdot \mathbf{n} \qquad [26]$$

We retain the previous expression for τ^e and the decohesion function. Therefore, the dissipation rate for normal mode decohesion must be evaluated specifically from the following equation:

$$D_{s,n} = \boldsymbol{\tau} \cdot \dot{\mathbf{u}}^p + \bar{\tau}\dot{\bar{u}} = \dot{\lambda}[\tau_n \dot{u}_n^p + \bar{\tau}\,\bar{u}_0] \qquad [27]$$

which will be less than that obtained with the associated rule (sometimes called the Principle of Maximum Dissipation) if τ_t is not zero for at least part of the decohesion.

A corresponding development for Mode II (pure shear) evolution can be obtained by merely replacing normal components of traction with shear components.

4. Numerical Application

4.1. *Incorporation with the Material Point Method*

In general, there is no need to determine the actual shape of the deformed material element associated with each material point. However, when material separation occurs, there is a need to consider the effect on the strain field over the material element (compatibility). For small deformations, which would be the case normally for quasibrittle materials and for small rotations, the original configuration can be used. Typically, each cell with the material point method is chosen to be a square element with each side of length h and the element associated with each material point can also be chosen similarly to be a square of size $h_p = h / \sqrt{n_p}$ where n_p is the number of material points per cell. Over each material element, the increment in strain, $\Delta\mathbf{e}$, is assumed to be constant as is the increment in decohesion, $\Delta\mathbf{u}^d$, as indicated in Figure 4 which also shows the unit vector $\mathbf{m} = \Delta\mathbf{u}^d / |\Delta\mathbf{u}^d|$. For future use, define the opening, $\mathbf{M}_n$, shear, $\mathbf{M}_t$, and failure, $\mathbf{M}_m$, tensor modes as follows:

$$\mathbf{M}_n = \mathbf{n} \otimes \mathbf{n} \qquad \mathbf{M}_t = \frac{1}{2}(\mathbf{n} \otimes \mathbf{t} + \mathbf{t} \otimes \mathbf{n}) \qquad \mathbf{M}_m = \frac{1}{2}(\mathbf{n} \otimes \mathbf{m} + \mathbf{m} \otimes \mathbf{n}) \qquad [28]$$

We note that $\mathbf{M}_m$ reduces to $\mathbf{M}_n$ and $\mathbf{M}_t$ when $\mathbf{m} = \mathbf{n}$ and $\mathbf{m} = \mathbf{t}$, respectively.

For a given time increment, if the total (average) strain increment, $\Delta\mathbf{e}$, is considered fixed, the result of the decohesion is that the effective strain increment in the remaining part of the material in the element must be reduced (relaxed) by what might be called a decohesion strain increment, $\Delta\mathbf{e}^d$, which satisfies a weak form of the compatibility condition

$$\int_{\Omega_p} \Delta\mathbf{e}^d dV = \int_{\partial\Omega_d} \Delta u^d \mathbf{M}_m dA \qquad \Delta\mathbf{u}^d = \Delta u^d \mathbf{m} \qquad [29]$$

in which dV and dA denote differentials of volume on the material element, Ω_p, and of area on the decohesion surface, $\partial\Omega_d$, respectively. The magnitude of the decohesion increment, which is in the direction of $\mathbf{m}$ by definition, is Δu^d. With the assumptions that the decohesion and strain are constant over each material element, the result is the following expression relating the "relaxation" or "decohesion" strain increment to the increment in decohesion:

$$\Delta\mathbf{e}^d = \frac{\Delta u^d}{L_e}\mathbf{M}_m \qquad L_e = \frac{V_p}{A_d} \qquad [30]$$

The effective length, L_e, is merely the ratio of the element volume to the area of the decohesion surface within that element. For the two-dimensional case illustrated in Figure 2, the effective length is

$$L_e = \frac{h_p}{\cos\alpha_p} \qquad 0 \le \alpha_p \le \frac{\pi}{4} \qquad [31]$$

with a corresponding formula for an angle measured with respect to the other side of the material element if $\alpha_p > \pi/4$.

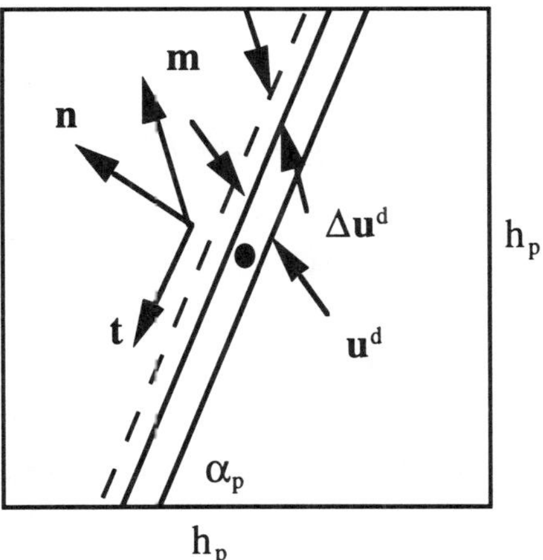

Figure 2. *A typical material element with decohesion*

4.2. *Solution Algorithm*

The constitutive equations subroutine is invoked with the total strain increment, $\Delta\mathbf{e}$, prescribed, with the total decohesion equal to the plastic decohesion, and with $\mathbf{n}$ given and assumed fixed. It is computationally more efficient to define an alternative mode vector, $\mathbf{m}^*$, from which the mode vector, $\mathbf{m}^e$, and an alternative tensor mode, $\mathbf{M}^*$, are easily determined. Let $\boldsymbol{\sigma}_{pr}$, $\mathbf{u}^p_{pr}$ and λ_{pr} denote the values of $\boldsymbol{\sigma}$, $\mathbf{u}^p$ and λ respectively, at the end of the previous step. The requirement is to solve the following set of nonlinear equations:

$$\begin{array}{ll}
\Delta\boldsymbol{\sigma} = \Delta\boldsymbol{\sigma}^{tr} - \mathbf{E}:\Delta\mathbf{e}^d & \Delta\boldsymbol{\sigma}^{tr} = \mathbf{E}:\Delta\mathbf{e} \\
\boldsymbol{\tau}_{pr} = \boldsymbol{\sigma}_{pr}\cdot\mathbf{n} & \Delta\boldsymbol{\tau}^{tr} = \Delta\boldsymbol{\sigma}^{tr}\cdot\mathbf{n} \\
\mathbf{m}^* = \dfrac{\mathbf{A}_d\cdot\boldsymbol{\tau}}{\bar{\tau}_0\tau^{e*}} & \mathbf{m}^e = \bar{u}_0\mathbf{m}^* \\
\boldsymbol{\sigma}_m = \mathbf{E}:\mathbf{M}^* & \boldsymbol{\tau}_m = \boldsymbol{\sigma}_m\cdot\mathbf{n} \\
\Delta\mathbf{e}^d = \Delta\lambda\mathbf{M}^* & \Delta\mathbf{u}^p = \Delta\lambda\mathbf{m}^e \\
\Delta\boldsymbol{\tau} = \Delta\boldsymbol{\tau}^{tr} - \Delta\lambda\boldsymbol{\tau}_m & \boldsymbol{\tau} = \boldsymbol{\tau}_{pr} + \Delta\boldsymbol{\tau} \\
\mathbf{u}^p = \mathbf{u}^p_{pr} + \Delta\mathbf{u}^p & \lambda = \lambda_{pr} + \Delta\lambda \\
\bar{\tau}^* = \lambda^q & \tau^{e*} = \left(\dfrac{\boldsymbol{\tau}\cdot\mathbf{A}_d\cdot\boldsymbol{\tau}}{\bar{\tau}_0}\right)^{1/2}
\end{array} \qquad [32]$$

$$\mathbf{M}^* = \frac{1}{2}\frac{\bar{u}_0}{L_e}(\mathbf{n}\otimes\mathbf{m}^* + \mathbf{m}^*\otimes\mathbf{n})$$

$$F_d^* \equiv \tau^{e*} - (1-\bar{\tau}^*) = 0$$

$$\boldsymbol{\sigma} = \boldsymbol{\sigma}^{tr} - \Delta\lambda\boldsymbol{\sigma}_m$$

The first step is to assume that no decohesion occurs in order to obtain a trial stress and traction:

$$\boldsymbol{\sigma}^{tr} = \boldsymbol{\sigma}_{pr} + \Delta\boldsymbol{\sigma}^{tr} \qquad \boldsymbol{\tau}^{tr} = \boldsymbol{\sigma}^{tr}\cdot\mathbf{n} \qquad [33]$$

and then determine the value of the damage function, F_d^{*tr}, for this trial traction and the existing value λ_{pr}. If $F_d^{*tr} \le \varepsilon$, the step is purely elastic with no additional decohesion and no further action is required. If the inequality is not satisfied, the decohesion variables must be incremented.

Next we describe a one-step algorithm which enforces the requirement $F_d = 0$ to order $(\Delta\lambda)^3$. Perform a Taylor expansion of F_d about the trial state:

$$F_d = a(\Delta\lambda)^2 + 2b\Delta\lambda + c + O(\Delta\lambda)^3 \qquad [34]$$

in which the last term indicates the order of the remainder and

$$a = \frac{1}{2}\frac{\partial^2 F_d}{\partial\lambda^2}\bigg|_{tr} \qquad 2b = \frac{\partial F_d}{\partial\lambda}\bigg|_{tr} \qquad c = F_d^{tr} \qquad [35]$$

We choose $\Delta\lambda = \Delta\lambda_1$ and $\Delta\lambda = \Delta\lambda_2$ to be the solutions to the first-order and second-order equations respectively; i.e., $2b\Delta\lambda_1 + c = 0$ and $a(\Delta\lambda_2)^2 + 2b\Delta\lambda_2 + c = 0$ or

$$\Delta\lambda_1 = -\frac{c}{2b} \qquad \Delta\lambda_2 = \frac{-b + \sqrt{b^2 - ac}}{a} \qquad [36]$$

with the sign chosen so that in the limit of infinitesimal $\Delta\lambda$ we have $\Delta\lambda_2 = \Delta\lambda_1$.

Consider the case when the model choice of [20] is used and the Taylor expansion is applied to the dimensionless damage function F_d^*. It follows from [35] that

$$\begin{aligned} c &= (\tau^{e*})^{tr} - (1 - \lambda_{pr}^q) \\ 2b &= \frac{\partial\tau^{e*}}{\partial\boldsymbol{\tau}} \cdot \frac{\partial\boldsymbol{\tau}}{\partial\lambda} + q\lambda_{pr}^{q-1} \\ 2a &= \frac{\partial\boldsymbol{\tau}}{\partial\lambda} \cdot \frac{\partial^2\tau^{e*}}{\partial\boldsymbol{\tau}^2} \cdot \frac{\partial\boldsymbol{\tau}}{\partial\lambda} + \frac{\partial\tau^{e*}}{\partial\boldsymbol{\tau}} \cdot \frac{\partial^2\boldsymbol{\tau}}{\partial\lambda^2} + q(q-1)\lambda_{pr}^{q-2} \end{aligned} \qquad [37]$$

in which all terms are to be evaluated at λ_{pr} and the trial value of the traction. Note that when $q = 1$, the last term in the expression for a is zero. With a modest amount of algebra involving [32], it follows that

$$\frac{\partial\boldsymbol{\tau}}{\partial\lambda} = -\boldsymbol{\tau}_m \qquad \frac{\partial\tau^{e*}}{\partial\boldsymbol{\tau}} = \frac{\mathbf{m}^*}{\bar{\tau}_0} \qquad [38]$$

and we immediately have the relation

$$2b = -\frac{\mathbf{m}^* \cdot \boldsymbol{\tau}_m}{\bar{\tau}_0} + q\lambda_{pr}^{q-1} \qquad [39]$$

Next, we proceed to obtain the coefficient a in [37]. We utilize [32] and [38] to obtain

$$\frac{\partial^2 \tau^{e*}}{\partial \boldsymbol{\tau}^2} = \frac{1}{\bar{\tau}_0^2\ \tau^{e*}}[\mathbf{A}_d - \mathbf{m}^* \otimes \mathbf{m}^*] \qquad [40]$$

which leads to

$$\frac{\partial \boldsymbol{\tau}}{\partial \lambda} \cdot \frac{\partial^2 \tau^{e*}}{\partial \boldsymbol{\tau}^2} \cdot \frac{\partial \boldsymbol{\tau}}{\partial \lambda} = \frac{[(\boldsymbol{\tau}_m \cdot \mathbf{A}_d \cdot \boldsymbol{\tau}_m) - (\boldsymbol{\tau}_m \cdot \mathbf{m}^*)^2]}{\bar{\tau}_0^2\ \tau^{e*}} \qquad [41]$$

Finally

$$\frac{\partial^2 \boldsymbol{\tau}}{\partial \lambda^2} = -\frac{\partial \boldsymbol{\tau}_m}{\partial \lambda} = -\mathbf{n} \cdot \mathbf{E}: \frac{\partial \mathbf{M}^*}{\partial \lambda} = -\mathbf{n} \cdot \mathbf{E}: \frac{\bar{u}_0}{L_e}(\mathbf{n} \otimes \frac{\partial \mathbf{m}^*}{\partial \lambda}) \qquad [42]$$

in which the minor symmetry of **E** has been used. We note that

$$\frac{\partial \mathbf{m}^*}{\partial \lambda} = \frac{\partial \mathbf{m}^*}{\partial \boldsymbol{\tau}} \cdot \frac{\partial \boldsymbol{\tau}}{\partial \lambda} = -\frac{1}{\bar{\tau}_0} \frac{\partial^2 \tau^{e*}}{\partial \boldsymbol{\tau}^2} \cdot \boldsymbol{\tau}_m = -\frac{1}{\bar{\tau}_0}[\mathbf{A}_d - \mathbf{m}^* \otimes \mathbf{m}^*] \cdot \boldsymbol{\tau}_m$$

$$\frac{\partial \tau^{e*}}{\partial \boldsymbol{\tau}} \cdot (-\mathbf{n} \cdot \mathbf{E} \cdot \frac{\bar{u}_0}{L_e} \mathbf{n}) = -\frac{1}{\bar{\tau}_0}(\mathbf{m}^* \otimes \mathbf{n}):\mathbf{E} \cdot \frac{\bar{u}_0}{L_e} \mathbf{n} = -\frac{\boldsymbol{\tau}_m}{\bar{\tau}_0} \qquad [43]$$

with the result

$$\frac{\partial \tau^{e*}}{\partial \boldsymbol{\tau}} \cdot \frac{\partial^2 \boldsymbol{\tau}}{\partial \lambda^2} = \frac{[(\boldsymbol{\tau}_m \cdot \mathbf{A}_d \cdot \boldsymbol{\tau}_m) - (\boldsymbol{\tau}_m \cdot \mathbf{m}^*)^2]}{\bar{\tau}_0^2\ \tau^{e*}} \qquad [44]$$

a result identical to that of [41]. Therefore

$$a = \frac{[(\boldsymbol{\tau}_m \cdot \mathbf{A}_d \cdot \boldsymbol{\tau}_m) - (\boldsymbol{\tau}_m \cdot \mathbf{m}^*)^2]}{\bar{\tau}_0^2\ \tau^{e*}} + \frac{q(q-1)}{2} \lambda_{pr}^{q-2} \qquad [45]$$

Even with this rather simple form, preliminary numerical results indicate that the use of the first-order equation for $\Delta\lambda_1$ in [36] is sufficiently accurate and the extra computations required to get a is not justified.

4.3. *Separation*

The procedure outlined above holds until separation occurs as indicated by $\lambda \geq 1$. The given algorithm can be used also for separation with the revised damage function $F_d^* = \tau^{e*}$ so that $F_d^* = 0$ enforces the separation condition that $\tau = \mathbf{0}$ and $\Delta\tau = \mathbf{0}$. To prevent the possibility of numerical problems when F_d^* is close to zero, the decohesion criterion is applied for $F_d^* \geq \varepsilon$ and the separation condition is used when $F_d^* < \varepsilon$ where ε is a small positive number (typically 0.01). The value of the stress component, $\sigma_{tt} = \mathbf{t} \cdot \boldsymbol{\sigma} \cdot \mathbf{t}$, is automatically adjusted through the equilibrium equations.

4.4. *Example Solution*

The application given in this section is restricted to a case for which (i) the initiation of failure is given directly by a failure criterion rather than by a discontinuous bifurcation analysis, and (ii) the orientations of the surfaces of decohesion are known *a priori*. The results of these numerical analyses are to suggest the following: (a) To illustrate that the use of jump in displacement as an internal variable together with a weak implementation of compatibility provides a simple and useful algorithm in the MPM, (b) to show that the MPM does not exhibit the finite element pathologies associated with distorted meshes and instabilities with the result that additional features such as enhanced strains are not required, and (c) to illustrate through an example that the material point method does not exhibit the orientation effect often seen with finite elements when discontinuities are allowed to propagate at various angles to the mesh sides.

In this example, a material with isotropic elastic properties but anisotropic thermal properties is cooled from room temperature (20° C) to -50° C, at which point the temperature is held fixed. The cooling rate is -10° degs C/μs. The coefficient of thermal expansion is $2.25\text{x}10^{-4}$/deg C along the 1-1 material axis but an order of magnitude smaller, $2.1\text{x}10^{-5}$/deg C, along the 2-2 material axis. Several grains of this material constitute a composite material where the grains are assembled into the composite with random orientation.

Figure 3 shows a 1 mm square sample of grains marked by material points. An arrow indicates the 1-1 material direction of each grain. The grains have isotropic elastic properties, with Young's modulus 3.28 GPa, Poisson's ratio 0.3, and density 1.9 kg/m^3. The interstitial region has the same material properties and is also represented by material points (Figure 4), however, these points are allowed to follow a non-associated decohesion constitutive model in which the mode is normal to the grain boundary. The decohesion constitutive properties are as follows: the peak normal traction is τ_{nf} = 5.5 Mpa, q =1, and U_o = 0.055 N-m/m^2. The peak stress in shear was not requireed because of the nonassociated evolution equation.

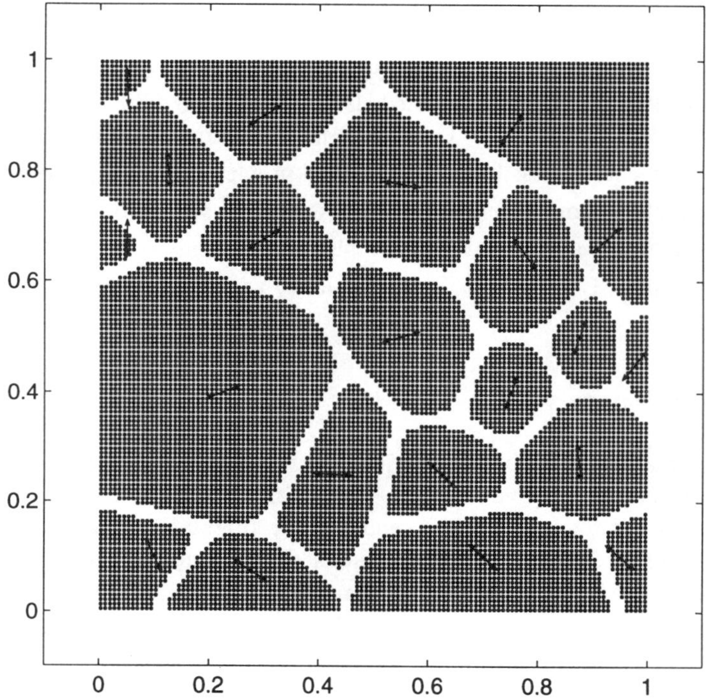

Figure 3. *Sample of anisotropic grains with the direction of the 1-1 material axis indicated with an arrow*

The normal to a grain boundary is computed by assigning a scalar color, with the value one, to the material points making up the grain, interpolating that color to the background mesh and then taking the gradient. The gradient gives the inward normal to the grain. Figure 5 shows an example of the computed normals for a grain. The computational mesh is square with side length of 25 microns.

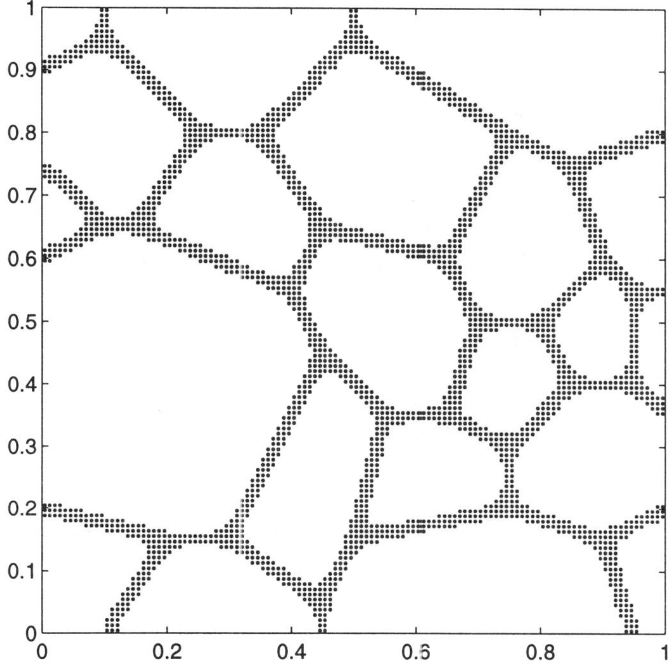

Figure 4. *Interstitial region is also discretized using material points*

As the material cools, the anisotropic thermal properties cause the contraction to be nonuniform. The effect of the variability in the stress field is seen in Figure 6 where contours of the jump in displacement are shown at various times. Decohesion starts in the upper right quadrant of the sample, and the location of the initial decohesion has the largest value throughout the computation. However, the decohesion does not propagate from the maximum all the way across the domain, instead a different path on the left side of the sample seems more likely to span the domain. This calculation indicates the complex interaction of geometry and material properties that make predicting the path of crack propagation so difficult.

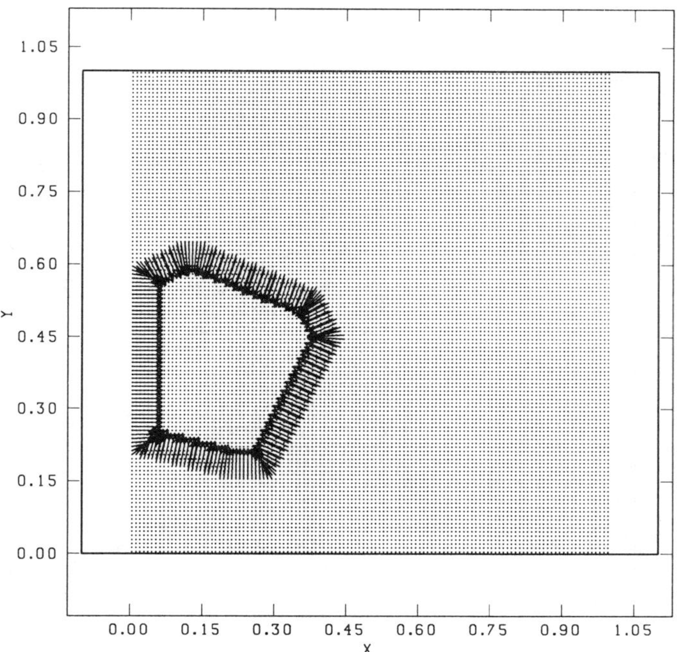

Figure 5. *The normal to the boundary is computed by taking the gradient of the color function*

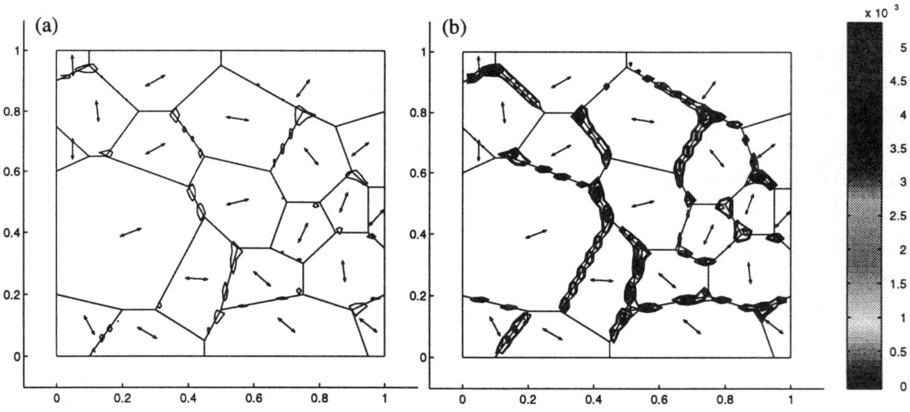

Figure 6. *Grayscale contour plot of the jump in displacement at time (a) 2μs and (b) 5μs*

5. Summary

A rigid, plastically softening decohesion model has been combined with continuum elasticity and traction continuity at a material failure surface to provide a relatively simple description for the evolution of material failure. When incorporated in the material point method, the result is a constitutive equation subroutine that is similar to softening plasticity. A length parameter associated with a material element provides a mechanism for ensuring convergence with mesh refinement. As the tip of a failure surface propagates through the mesh, the formulation inherent with the material point method appears to preclude the diffusion of the crack tip, a feature often seen with conventional finite elements. Further investigations involving the propagation of curved cracks are necessary to determine whether or not the proposed method is general and robust. Nevertheless, in that light of the long history of complex numerical analysis in connection with crack propagation, we believe that the simplicity of the decohesion formulation in the material point method holds considerable promise for development into a general method for predicting material failure.

Acknowledgment

This work was partially supported, at various times, by Los Alamos National Laboratory, Sandia National Laboratories and the Air Force Office of Scientific Research.

6. References

[All 95] ALLIX, O., et al., "Damage analysis of interlaminar fracture specimens", *Composite Structures,* Vol. 31, p. 61-74, 1995.

[ALL 96] ALLIX, O., CORIGLIANO, A., "Modeling and simulation of crack propagation in mixed-modes interlaminar fracture specimens", *International Journal of Fracture,* vol. 77, p. 111-140, 1996.

[ARM 97] ARMERO, F., "Large-scale modeling of localized dissipative mechanisms in a local continuum: applications to the numerical simulation of strain localization in rate-dependent inelastic solids", to appear in *Mechanics of Cohesive-Frictional Materials,* 1997.

[BAR 59] BARENBLATT, G.I., "The formation of equilibrium cracks during brittle fracture; general ideas and hypotheses: axially symmetric cracks", *J. Appl. Math. Mech.,* vol. 23, p. 622-636, 1959.

[BAZ 83] BAZANT, Z.P., OH, B.H., "Crack band theory for fracture of concrete", *Materials and Structures RILEM,* vol. 16, p. 155-177, 1983.

[BAZ 97] BAZANT, Z.P., LI, L.N., "Cohesive crack with rate-dependent opening and viscoelasticity - I. Mathematical model and scaling; II. Numerical algorithm, behavior and size effect", *Int. J. of Fracture, vol. 86(3),* p. 247-266 and 267- 279, 1997.

[BOR 87] BORST, R. DE, "Smeared cracking, plasticity, creep and thermal loading - A unified approach", *Computational Methods in Appl. Mech. Engrg.,* vol. 72, p. 89-110, 1987.

[DAH 90] DAHLBOM, O., OTTOSEN, N.S., "Smeared crack analysis using generalized fictitious crack model", *J. of Engineering Mechanics,* vol. 116, p. 55-76, 1990.

[COR 93] CORIGLIANO, A., "Formulation, identification and use of interface models in the numerical analysis of composite delamination", *Int. J. Solids Structures,* vol. 30(20), p. 2779-2811, 1993.

[DVO 90] DVORKIN, E. et al., "Finite Elements with Displacement Interpolated Embedded Localization Lines Insensitive to Mesh Size and Distortions", *Int. J. for Num. Methods in Eng.,* vol. 30, p. 541-564, 1990.

[FEE 91a, 91b] FEENSTRA, P.H., et al., "Numerical Study of Crack Dilatancy. I. Models and Stability Analysis", and "II. Applications", *J. of Engineering Mechanics,* vol. 117, p. 733-753 and p. 754-769, 1991.

[HIL 76] HILLERBORG, A., et al., "Analysis of Crack Formation and Crack Growth in Concrete by Means of Fracture Mechanics and Finite Elements", *Cement and Concrete Research,* vol. 6, p. 773-782, 1976.

[LAR 96] LARSSON, R., RUNESSON, K., "Element-embedded localization band based on regularized displacement discontinuity", *J. of Engineering Mechanics,* vol. 122(5), p. 402-411, 1996.

[OLI 89] OLIVER, J., "A consistent characteristic length for smeared cracking problems", *Int. J. Num. Meth. Eng., vol. 28,* p. 461-474, 1989.

[OLI 96a, 96b] OLIVER, J., "Modelling Strong Discontinuities in Solid Mechanics via Strain Softening Constitutive Equations. Part 1. Fundamentals", and "Part 2. Numerical Simulations", *Int. J. for Num. Methods in Eng.*, vol. 39, p. 3575-3600 and p. 3601-3623, 1996.

[NEE 87] NEEDLEMAN, A., "A continuum model for void nucleation by inclusion debonding", *J. of Applied Mechanics,* vol. 54, p. 525-531, 1987.

[ROT 87] ROTS, J.G., DE BORST, R., "Analysis of mixed-mode fracture in concrete", *J. of Engineering Mechanics,* vol. 113, p. 1739-1758, 1987.

[SCH 92] SCHELLEKENS, J.C.J., Computational Strategies for Composite Structures, Ph.D. Dissertation, Dept. of Civil Engineering, The Delft University of Technology, Delft, The Netherlands, 1992.

[SIM 93] SIMO, J.C., et al., "An analysis of strong discontinuities induced by strain-softening in rate-independent solids", *Computational Mech.,* vol. 12, p. 277-296, 1993.

[SIM 94] SIMO, J.C., OLIVER, J., "A new approach to the analysis and simulation of strain-softening in solids", *Fracture and Damage in quasi-Brittle Materials*, ed. by Z.P. Bazant et al., Workshop held in Prague, 1994

[SUO 92] SUO, Z., et al., "Stability of Solids with Interfaces", *J. Mech. Phys. Solids,* Vol. 40, No. 1, 613-640, 1992.

[SUL 94] SULSKY, D., et al., "A Particle Method for History-Dependent Materials", *Computer Methods in Applied Mechs. and Engineering,* vol. 118, p. 179-196, 1994.

[SUL 95] SULSKY, D., et al., "Application of a Particle-in-Cell Method to Solid Mechanics", *Computer Physics Communications,* vol. 87(1&2), p. 236-252, 1995.

[SUL 96] SULSKY, D., SCHREYER, H.L., "Axisymmetric Form of the Material Point Method with Applications to Upsetting and Taylor Impact Problems", *Comp. Methods in Applied Mechanics & Eng.,* vol. 139(*1-4*), p. 409-429, 1996.

[WIE 98] WIEHE, S., et al., "Classification of smeared crack models based on material and structural properties", *Int. J. Solids Structures,* vol. 35(12), p. 1289-1312, 1998.

[ZHO 98] ZHOU, S., The Numerical Prediction of Material Failure Based on the Material Point Method, Ph.D. Dissertation, Department of Mechanical Engineering, The University of New Mexico, Albuquerque, NM, 1998.

Implementation and Application of an Algorithm for Incremental Adaptive Finite Element Analysis of Concrete Plates

Thomas Huemer — Roman Lackner — Herbert A. Mang

Institute for Strength of Materials
Vienna University of Technology
Karlsplatz 131202, A-1040 Vienna

{Thomas.Huemer, Herbert.Mang, Roman.Lackner}@tuwien.ac.at

ABSTRACT. *Adaptive Finite Element (FE) analyses permit control of the discretization error. As regards adaptive non-linear FE-analysis, the error is estimated after each load or displacement step of an incremental calculation. Based on the result of the error estimation an improved FE-mesh is generated. Re-calculations are perfomed until the error no longer exceeds a userdefined tolerance. Based on the advancing front method, an automatic mesh generation algorithm for quadrilateral elements is developed. A given mesh density function is used for the mesh generation. Even if the current mesh density function is strongly graded, geometrically well-shaped elements are generated. The developed error estimator is based on the rate of the mechanical work. It permits consideration of elasto-plastic material behavior in conjunction with softening because of cracking of concrete. A new algorithm for incremental analysis of materially non-linear problems is presented. It is based on a technique for estimating the incremental error. The algorithm does not require re-calculations of the structural response from load steps preceding the current load step. Thus, the proposed procedure can be used effectively for the analysis of complicated structures with a large number of degrees of freedom. The usefulness of this procedure is demonstrated by studying an unreinforced concrete plate. A comparison of the obtained results with results from computations based on uniformly refined FE-meshes shows that the adaptive procedure gives a significant improvement in the predictive precision of the localized (crack) zone.*

KEY WORDS: *Finite Element Method, FEM, Non-Linear Adaptivity, Error Estimation, Mesh Generation, Fictitious Crack Concept.*

Introduction

The Finite Element (FE) Method has become the most powerful tool for structural analysis. Apart from research, this method is widely used in practical engineering. However, the FEM is often applied without good understanding of the method's background. Modern pre- and postprocessors facilitate the development of appropriate models for the analysis as well as the interpretation of the results by means of color plots of results. Nevertheless, powerful pre- and postprocessors will not be of much help if the underlying FE discretizations are inadequate. Poor discretization may lead to wrong results. In conjunction with other deficiencies, such as an inadequate design, this may cause the damage of a structure. As an example for a spectacular structural failure, the implosion of the offshore platform Sleipner A [SCH 93] [COL 97] may be mentioned. An inadequate arrangement of the reinforcement bars and a too coarse FE mesh, which was directly created out of a CAD model, caused the failure of one of the 32 tricells. The latter tricells are triangular voids formed by the interior walls between the 24 cells of the platform. As a result of the collapse, the platform sank. It imploded at the ground of the North Sea and destroyed a $180 million structure. Fortunately, nobody died.

One of the ways to prevent such a collapse is to check the quality of the analysis model and the obtained FE results. If the results do not meet the prespecified requirements of accuracy, a more accurate discretization must be generated, followed by a recalculation of the problem. The automated process of getting an improved approximation of the solution of the underlying mechanical problem is one of the goals of an adaptive finite element analysis.

The following ingredients are necessary for such an analysis: (1) a FEM program for the solution of the mechanical problem, (2) a reliable program for error estimation, and (3) a program for calculating the mesh density function. This function is derived from the distribution of the relative element error. It is used as the basis for mesh generation. Moreover, an efficient and robust mesh generator for the types of finite elements used is necessary as well as a program for bandwidth optimization, which, in general, is included in the FEM code [LEE 95]. A FEM code used for adaptive analysis has to meet special requirements. The input of displacements must be possible. For materially non-linear problems, the input of plastic strains and internal variables must also be possible.

This paper is organized as follows: Section 1 contains the definition of an incremental error estimator applicable to elasto-plastic material behavior. In Section 2, remarks concerning the mesh density function are made. Section 3 deals with mesh generation, using the Advancing Front Method (AFM) and specific strategies for the generation of well-shaped quadrilateral finite elements. In Section 4, an algorithm for incremental adaptive FE analysis is proposed. As a numerical example, an L-shaped plate made of plain concrete is investigated in Section 5. Possible remedies for problems which have occurred during this investigation are given in Section 6. Section 7 contains a summary and an outlook for further pertinent research work.

1. Error estimation

Non-linear FE analyses are usually performed in an incremental-iterative manner. Hence, it is useful to establish an error measure based on incremental quantities. This measure is related to a reference quantity, yielding a relative incremental global error, η, given as

$$\eta^2 = \frac{\Delta e^2}{\Delta u^{h,2}} \,. \qquad [1]$$

In Eq.[1], Δe^2 is the absolute incremental global error measure and $\Delta u^{h,2}$ is the aforementioned incremental reference quantity. These quantities are obtained by means of integration over the time interval $[t_n, t_{n+1}]$:

$$\Delta e^2 = \int\limits_{t_n}^{t_{n+1}} (\dot{e^2})\, d\tau \quad \text{and} \quad \Delta u^{h,2} = \int\limits_{t_n}^{t_{n+1}} (\dot{u^{h,2}})\, d\tau. \qquad [2]$$

The relative incremental global error η permits an assessment of the accuracy of the analysis over the entire spatial domain for the specified time (load) increment. It may be divided into an elastic and plastic part. The respective contributions to Δe^2 are

$$\Delta e^2 = \Delta e^{e,2} + \Delta e^{p,2} \qquad [3]$$

with the elastic part given as

$$\Delta e^{e,2} = \int\limits_{\Omega} \sum_{i,j=1}^{3} |(\sigma_{ij}^* - \sigma_{ij}^h) \sum_{k,l=1}^{3} [D_{ijkl}(\Delta\sigma_{kl}^* - \Delta\sigma_{kl}^h)]\, |\, d\Omega \qquad [4]$$

and the plastic part as

$$\Delta e^{p,2} = \int\limits_{\Omega} \sum_{i,j=1}^{3} |(\sigma_{ij}^* - \sigma_{ij}^h)(\Delta\varepsilon_{ij}^{p,*} - \Delta\varepsilon_{ij}^{p,h})|\, d\Omega \,. \qquad [5]$$

Quantities with the superscript "h", such as the stresses σ_{ij}^h and their increments $\Delta\sigma_{ij}^h$, are part of the original FE solution. The superscript "*" identifies quantities which are obtained by a recovery technique. This is a mode of post-processing of the original FE solution in order to get an improved solution. This solution is then used for the evaluation of the error measures. These measures are estimates. Knowledge of the true errors would require knowledge of the true solution. D_{ijkl} is the compliance tensor.

Based on the rates of the elastic and the plastic part of the mechanical work, $\dot{\mathcal{W}}^e = \boldsymbol{\sigma}^T \dot{\boldsymbol{\varepsilon}}^e$ and $\dot{\mathcal{W}}^p = \boldsymbol{\sigma}^T \dot{\boldsymbol{\varepsilon}}^p$, the increment of the reference quantity is defined as

$$\Delta u^{h,2} = \Delta u^{h,e,2} + \Delta u^{h,p,2} \qquad [6]$$

with

$$\Delta u^{h,e,2} = \int_{\Omega} \sum_{i,j=1}^{3} |\sigma_{ij}^{h} \sum_{k,l=1}^{3} \left[D_{ijkl} \Delta \sigma_{kl}^{h} \right] | d\Omega \qquad [7]$$

and

$$\Delta u^{h,p,2} = \int_{\Omega} \sum_{i,j=1}^{3} |\sigma_{ij}^{h} \Delta \varepsilon_{ij}^{p,h}| \, d\Omega \, . \qquad [8]$$

Again the superscript "h" refers to the finite element solution. The relative incremental local (element) error η_{τ} is obtained by restricting the integrals in Eqs.[4] and [5] to one element. Summation over all m_e elements of the FE mesh, yields the respective global error $\eta^2 = \sum_{\tau=1}^{m_e} \eta_{\tau}^2$.

The recovered quantities in Eq.[4], $\boldsymbol{\sigma}^*$ and $\Delta\boldsymbol{\sigma}^*$, are obtained by a modified stress recovery technique. The error in the incremental plastic strains, $\Delta\varepsilon^{p,*} - \Delta\varepsilon^{p,h}$, is evaluated by means of the so-called "plastic-strain recovery".

Modified stress recovery

The modified stress recovery is based on the superconvergent patch recovery (SCPR) technique proposed by Zienkiewicz and Zhu [ZIE 92]. This standard smoothing method exploits continuity conditions at the element edges, orginating from the equilibrium conditions. The least-squares method is used for the solution of the minimization problem concerning the difference between the discontinous FE stress distribution, $\boldsymbol{\sigma}^h$, and the continous recovered distribution, $\boldsymbol{\sigma}^*$.

The modified stress recovery [LAC 98B] allows for weak discontinuities. They are characterized by a jump in the strain distribution, $[\![\boldsymbol{\varepsilon}]\!] \neq \mathbf{0}$, but no jump in the displacement distribution, $[\![\mathbf{u}]\!] = \mathbf{0}$ [VAR 95]. These compatibility conditions lead to the modified constraint condition for the recovered tractions $\mathbf{t}^*$:

$$\mathbf{t}^*|_i^{\epsilon} + \mathbf{t}^*|_j^{\epsilon} = \mathbf{0} \, , \qquad [9]$$

where ϵ represents the common edge of two adjacent finite elements i and j. The condition in Eq.[9] allows jumps of the recovered stress component $\sigma_{||}^*|^{\epsilon}$ parallel to the element edge ϵ. The mechanical rationale for such jumps are situations arising in cracked concrete. The stress distribution from the FE solution of a patch consisting of four finite elements is shown in Fig.1a, whereas Fig.1b illustrates the stress distribution obtained from the modified stress recovery.

"Plastic strain recovery"

Since the evolution of plastic deformations is a local evolution, a smoothing technique using a patch of elements cannot be applied to evaluate $\Delta\varepsilon^{p,*} - \Delta\varepsilon^{p,h}$ in Eq.[5]. Therefore, the "plastic strain recovery" is performed as follows [LAC 98A]: (a) employing the recovered stress distribution $\Delta\boldsymbol{\sigma}^*$, a new trial state $\tilde{\boldsymbol{\sigma}}_{n+1}^{trial}$ is defined; (b) projection of $\tilde{\boldsymbol{\sigma}}_{n+1}^{trial}$ back to the yield surface $f(\boldsymbol{\sigma}, \boldsymbol{\alpha}_{n+1}^h)$ yields $\tilde{\boldsymbol{\sigma}}_{n+1}$; (c) the error measure for the increment of the plastic strains is obtained as

$$\Delta\varepsilon^{p,*} - \Delta\varepsilon^{p,h} := \Delta\tilde{\gamma} \frac{\partial f}{\partial \tilde{\boldsymbol{\sigma}}} \, . \qquad [10]$$

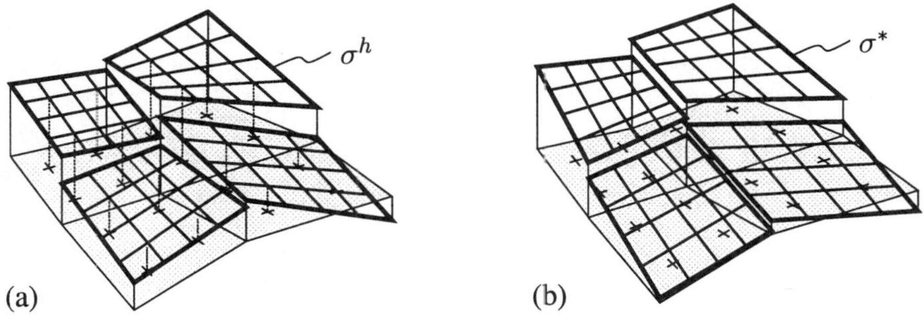

Figure 1. *Stress distribution obtained from (a) the finite element solution and (b) the modified patch recovery technique allowing jumps along the element edges*

Fig.2 illustrates that approach. It is noteable that the "plastic strain recovery" is activated only if plastic loading has occurred, that is

$$\Delta\alpha_{n+1}^h \neq 0\,, \qquad [11]$$

where α^h is the strain-like internal variable.

2. Mesh density function

If the relative global error η exceeds its prescribed value $\bar{\eta}$, $\eta > \bar{\eta}$, then the old mesh consisting of m_e finite elements is replaced by a new mesh with $\bar{m}_e$ elements. The refinement strategy according to Li and Bettess [LI 95] is employed, whereby the relative local (element) error of the new mesh is assumed to be distributed uniformly over all elements:

$$\bar{\eta}_1 = \bar{\eta}_2 = \ldots = \bar{\eta}_\tau = \ldots = \bar{\eta}_{\bar{m}_e}\,. \qquad [12]$$

The new number of elements of the mesh, $\bar{m}_e$, is obtained from

$$\bar{m}_e = \left(\sum_{\tau=1}^{m_e} \frac{\eta_\tau}{\bar{\eta}}\right)^2 . \qquad [13]$$

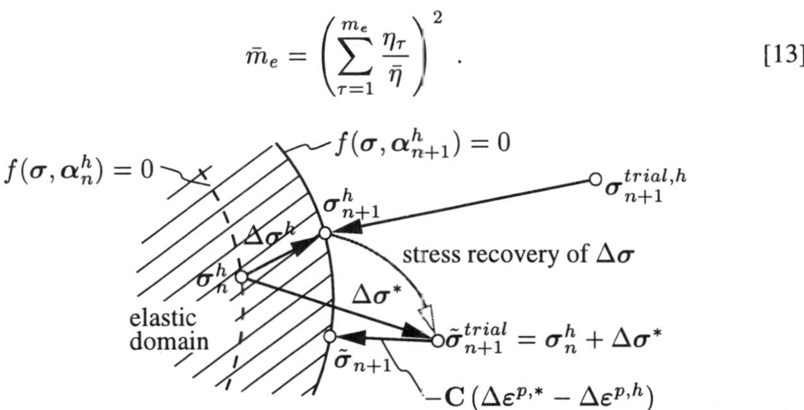

Figure 2. *On the evaluation of an estimate of the error of the incremental plastic strain*

Here, m_e is the number of elements of the old mesh and η_τ is the relative element error. The new characteristic element size $\bar{h}|_\tau$ follows from the element size h_τ and the relative element error η_τ of element τ of the old mesh [LAC 98B]

$$\bar{h}|_\tau = h_\tau \sqrt{\frac{\bar{\eta}}{\sqrt{\bar{m}_e}\eta_\tau}} \, . \qquad [14]$$

The mesh density function $\bar{h}|_\tau$ is then evaluated for all elements of the old mesh and transformed to nodal values. With the old mesh as a background grid for the mesh generation, the new mesh size can be obtained at every point of the domain.

3. Mesh generation

In the context of an adaptive FE analysis, a mesh generator should allow for:

- generation of meshes with quadrilateral elements
- variable element sizes over the whole domain
- fully automatic generation of the complete input of the FEM program
- simple description of the geometric domain and the mesh density
- generation of well-shaped elements even in case of a high gradation of the element size
- robust and efficient implementation.

The listed requirements exclude some of the traditional mesh generation methods, which were successfully used in the pre-processing step of a FE analysis. The widely used transformation methods can not deal with varying element sizes over the whole domain. Moreover, most of them are semi-automatic methods. They require manual subdivision of the domain in so called macro- or super-elements of well defined shape (i. e., four-sided patches). The triangulation methods are only useful for generation of meshes with triangular elements. Most of the grid-based approaches are only able to create meshes with elements of uniform size. Out of many different concepts for generation of FE meshes, the Advancing Front Method (AFM) fulfills all of the aforementioned requirements. Hence, it is well suited for use in adaptive FE analyses.

The AFM was first published by Peraire *et al.* [PER 87] for the generation of meshes with triangular elements. For the first time a background grid was used. It supplies properties such as the element size for the mesh generator over the whole domain (Sec.2). Zhu *et al.* [ZHU 91] extended the AFM of Peraire *et al.* [PER 87] for the generation of meshes consisting of quadrilateral elements. The method is based on the fact that ervery domain which is surrounded by a polygon with an even number of sides can be subdivided entirely into quadrilateral elements. The concept of a background grid is also used for supplying the element size.

Herein, a mesh generator based on the works of Peraire *et al.* [PER 87] and Zhu *et al.* [ZHU 91] is used. Additional mesh generation strategies were implemented

to enforce meshes with well-shaped elements even for the case of a relatively high gradation of the mesh density function [HUE 98].
The process of mesh generation can be summerized as follows:

1. *Geometric representation of the domain*: The boundary of a multiply connected region is discretized by means of so-called geometry elements. Geometry elements are straight lines or arcs; all of them are fixed in space by geometry nodes and their coordinates (Fig.3a).

2. *Generation of background grid*: The background grid provides the information about the element size of the elements of the mesh to be generated. Each node represents one discrete value of the mesh density function. If mesh density values between nodes are requested, an interpolation is performed. For the ease of interpolation, the background grid consists of triangular elements. In an adaptive analysis, the background grid is established by subdividing the quadrilateral elements of the old mesh along the shorter diagonal (Fig.3b).

3. The *generation of boundary nodes* is the first genuine step during mesh generation. The nodes are generated according to the element size supplied by the mesh density function. The lines between the boundary nodes form the first so-called active front. For the generation of a mesh consisting of quadrilateral elements, the active front must consist of an even number of active front sides (Fig.3c).

4. *Offset element generation*: This optional step of mesh generation is very useful for obtaining well-shaped quadrilateral elements along the boundary of the domain. There, one or more rows of elements are generated by moving the boundary by an offset distance according to the mesh density function to the interior of the domain (Fig.4a). The updated active front represents the interior boundary of the domain for the continuation of the mesh generation.

5. *Node and element generation*: During this main step of mesh generation, nodes and elements are created simultaneously, starting always from the shortest side

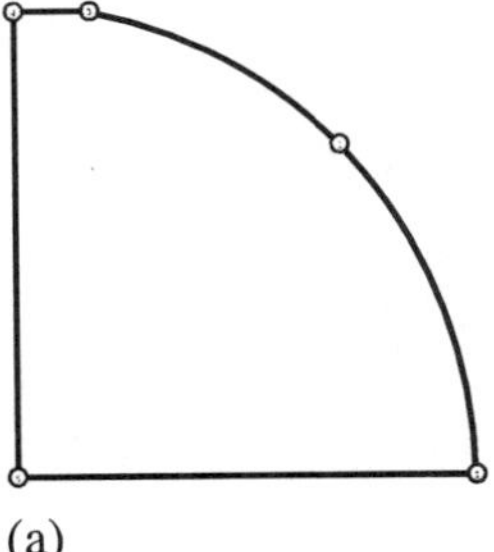

(a)

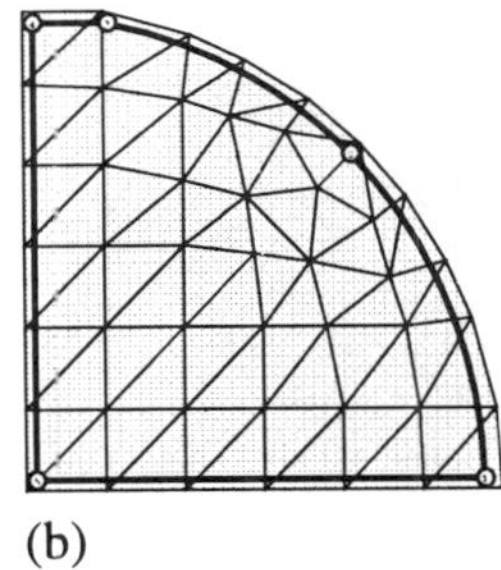

(b)

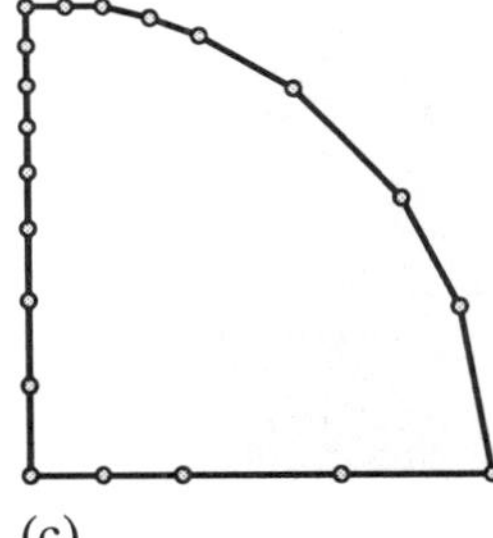

(c)

Figure 3. *Steps 1 to 3 of mesh generation: (a) geometric representation of the domain; (b) background grid; (c) boundary node generation*

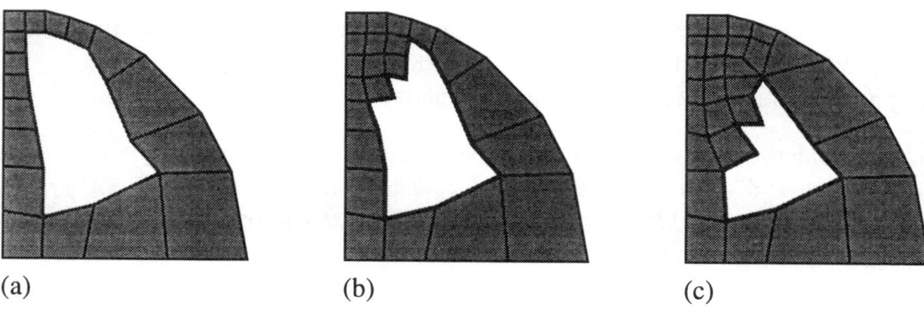

(a) (b) (c)

Figure 4. *Steps 4 and 5 of mesh generation: (a) offset element generation; (b) and (c) two different stages during mesh generation*

of the active front. In addition to node and element generation, the active front is being updated permanently. During the ongoing process of covering the domain with elements, additional active fronts may be generated. Each of these active fronts is treated in the same way. The generation of the mesh will terminate if all active fronts vanish, i. e., if the whole domain is covered with elements. Figures 4b and c refer to two different stages of the process of mesh generation.

6. *Mesh quality enhancement*: after generation of the FE mesh, some regions of the mesh may consist of elements with badly formed geometric shapes. These elements or groups of elements are deleted or replaced in this optional step. Fig.5a, shows elements which should be deleted from the mesh. In Zhu *et al.* [ZHU 91] other examples for an enhancement of the mesh quality are given.

7. A *mesh smoothing* step should always be the last step in the process of mesh generation. In an iterative procedure, the position of the nodal points is changed in order to get better-shaped elements. This process is terminated if the changes of all nodal positions are smaller than a prespecified threshold value. Fig.5b shows the mesh before and Fig.5c after some smoothing runs.

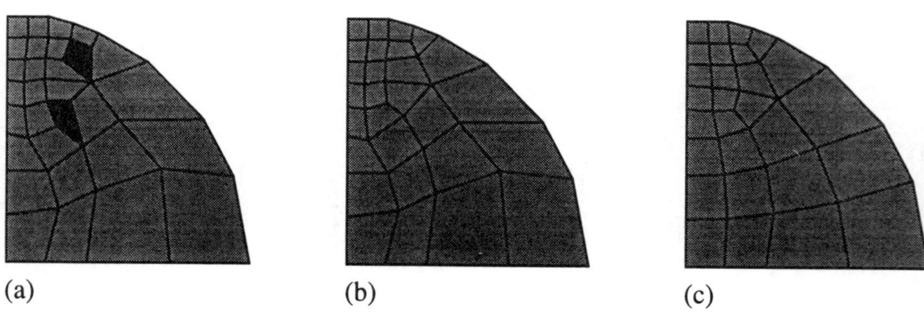

(a) (b) (c)

Figure 5. *Steps 6 and 7 of mesh generation: (a) and (b) enhancement of mesh quality; (c) smoothing of mesh*

4. Algorithm for adaptive analysis

In the algorithm for adaptive analysis, the previously described module (error estimator, program for calculating the mesh density function, mesh generator) is integrated in a FEM code to form an adaptive analysis tool. The FEM code has to meet the aforementioned requirements (see Introduction). It is used as an encapsulated calculation module. Here the calculations are performed with the commercial multi-purpose FE program MARC [MAR96].

The prescribed load or displacement is applied incrementally according to a predefined history. After each load increment the error is estimated. If the relative global error η is greater than the prescribed error tolerance $\bar{\eta}$, a new FE discretization is performed. As regards the restart of the calculation, two major strategies are possible:

(a) the calculation is restarted at the beginning of the load history (increment 0) or

(b) the calculation is restarted at the beginning of the actual load increment.

The first approach requires reconsideration of the entire history which would be prohibitively expensive for large-scale analyses. Hence, the second approach will be applied.

The algorithm proposed for adaptive FE analysis is an incremental procedure based on the Newton iteration method for solving non-linear problems. Fig.6 shows the load-displacement-diagram for the special case of a system with one degree of freedom. It is assumed that the system is in equilibrium after applying load increment n and that the relative global error η is smaller than $\bar{\eta}$.

The displacement increment resulting from the applied load increment $\Delta f_{ex,n+1}$ is calculated by means of the Newton iteration method. Then the actual relative incremental global error is estimated. If η is smaller than the prescribed relative global

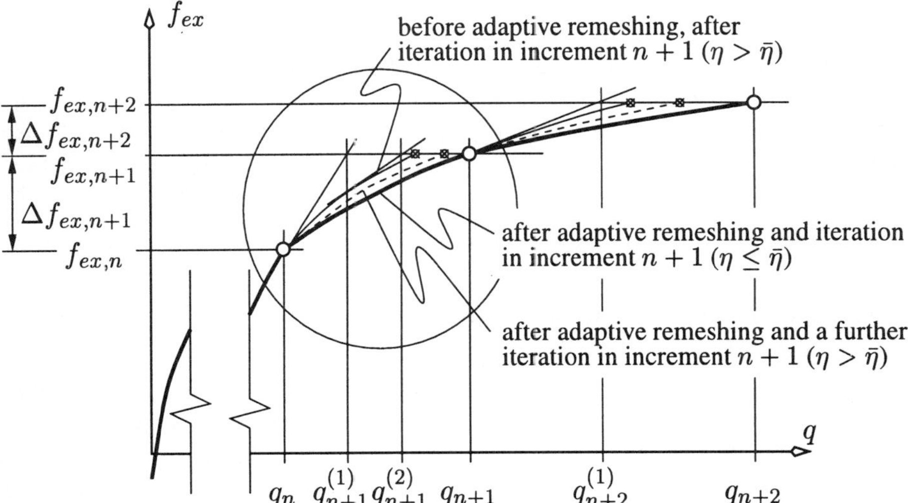

Figure 6. *Algorithm for adaptive analysis for a system with one degree of freedom*

error $\bar{\eta}$, the next load increment will be applied. Otherwise a new mesh is generated. It is based on the distribution of the actual relative local (element) error. According to the incremental restart strategy, only the displacement increment corresponding to the load increment $\Delta f_{ex,n+1}$ is recalculated. The previous history is accounted for by prescribing the complete set of accumulated state variables (displacements, plastic strain tensor and internal variables) as initial state variables. A special algorithm is used for the transfer of quantities from the old to the new mesh. Subsequently, the relative incremental global error η is estimated. If it is greater than $\bar{\eta}$, then the procedure will be continued. Otherwise the next load increment is applied.

Because of its time-efficiency, the algorithm is a powerful tool for large-scale analyses. Nevertheless, a transfer of the physical data from the old to the new mesh is required. In this work, a transfer scheme employing the displacements, the recovered stresses and the internal variables is used (see [LAC 98B]). The new displacements are obtained by interpolation. The remaining quantities are computed by averaging techniques.

Figure 7 shows the flowchart of the algorithm for adaptive FE analysis. After reading the input data (geometry definition, material properties, load history, parameter definitions for the error estimator, mesh generator and FEM code) the initial FE mesh is generated. The previously initialized increment counter is increased by 1 and a FE analysis is performed. If the iteration does not converge, it is assumed that the collapse load has been reached. In this case the adaptive analysis will be terminated. Otherwise, the relative incremental global error η is estimated. If it is greater than the user-defined error $\bar{\eta}$, a new mesh density function is determined. It is based on the distribution of the relative incremental local error (see Section 2).

Then, a new FE mesh is generated. It is based on the desired element sizes which are supplied by the mesh density function. In the next step, the displacements, plastic strains, and internal variables are transferred to the new mesh. The increment counter is decreased by 1. An equilibrium iteration is performed at the same load level, i. e., without application of a (new) load increment. After increase of the increment counter and application of the actual load increment, a FE calculation and an error estimation are performed. If the relative incremental global error is again greater than the user-defined relative incremental global error, this cycle is repeated. If this is not the case and if the full load history has not been applied yet, the next load increment is applied. Otherwise, the adaptive analysis is finished and post-processing can begin.

5. Example: L-shaped plate made of concrete

In order to verify the proposed modules and the implementation of the algorithm for adaptive finite element analysis, an L-shaped plate made of plain concrete has been analysed. In Figure 8, the geometric dimensions and the boundary conditions are shown. Moreover, the prescribed displacement and the material properties are given.

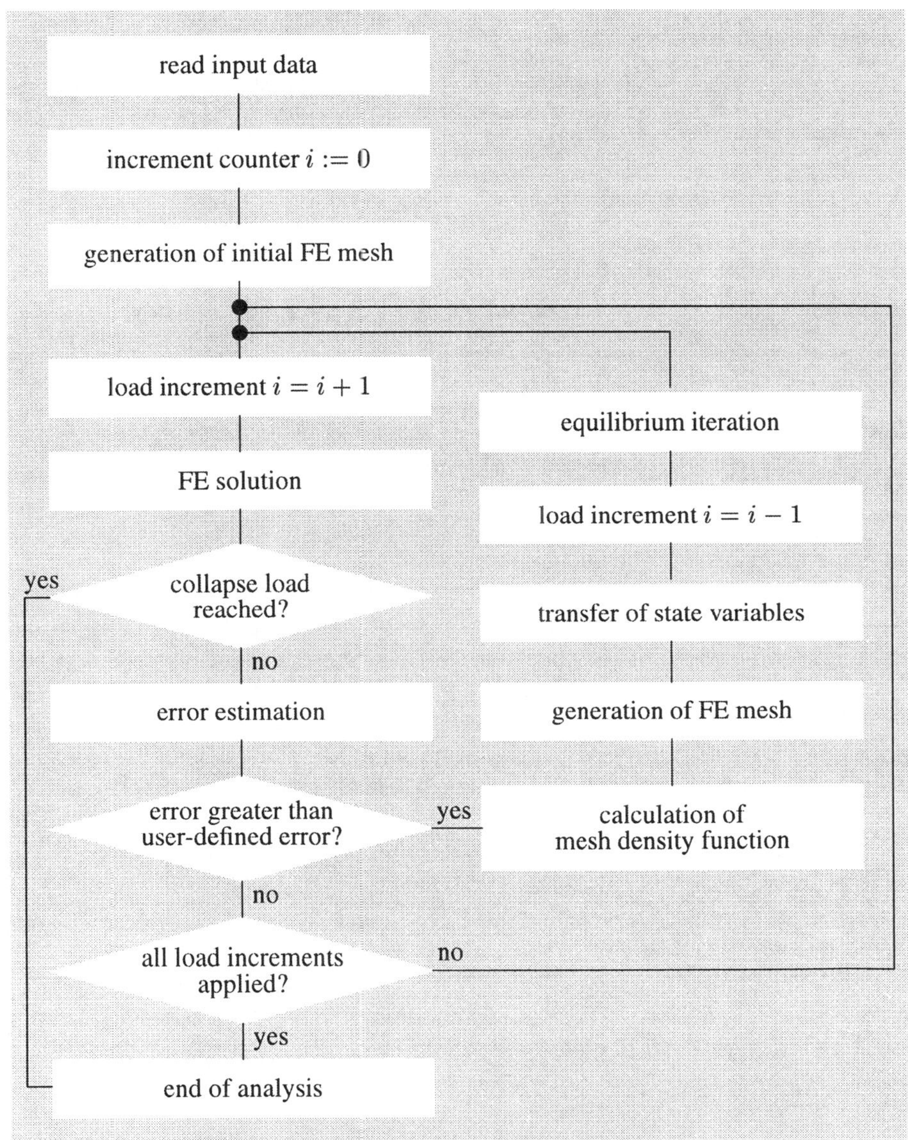

Figure 7. *Flowchart of the algorithm for adaptive FE analysis*

The prescribed displacement, $\bar{u} = 2.0$ mm, is applied incrementally. The respective displacement history consists of 170 increments. The first three increments are given as $\Delta u_1 = \Delta u_2 = \Delta u_3 = 0.10$ mm, resulting in a total of 0.30 mm. Then, 167 increments of $\Delta u_i = 0.01$ mm, $i = 4, \ldots, 170$, each, follow (Fig.9).

Four-node plane-stress elements are used. They are based on a displacement formulation. The material behavior of concrete is described by a 2D multi-surface plasticity model, representing a special case of the isotropic 3D model proposed in

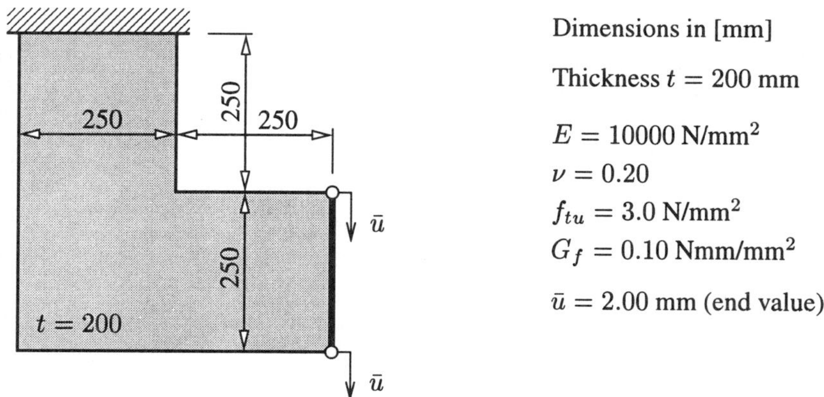

Figure 8. *L-shaped plate made of concrete: dimensions, boundary conditions, material properties and prescribed displacement*

[MES 97]. The constitutive behavior is governed by two independent hardening and softening mechanisms. The ductile material in the compression region is accounted for by an isotropic hardening Drucker-Prager model. The smeared crack approach is used for modelling the fracture of concrete. A Rankine criterion (based on the maximum tensile strength) in conjunction with an isotropic exponential softening law describes the brittle behavior of concrete. Use of the fictitious crack model with a constant characteristic length ℓ_c for each element serves the purpose of maintaining the objectivity of the analysis in case of strain softening. The calibration of the softening law is based on the characteristic length and on the tensile fracture energy G_f.

At first, a comparative study with four uniform FE meshes was performed in order to prove the objectivity of the employed material model with regards to different discretizations. Fig.10 shows the meshes I, II, III, and IV with 300, 1000, 2340, and 5400 elements respectively.

Fig.11 shows plots of diagrams of the displacement $\bar{u}$ versus the vertical reaction-force representing the integral of the vertical stresses along the clamped part of the boundary. Meshes I and II show nearly the same structural response. The results

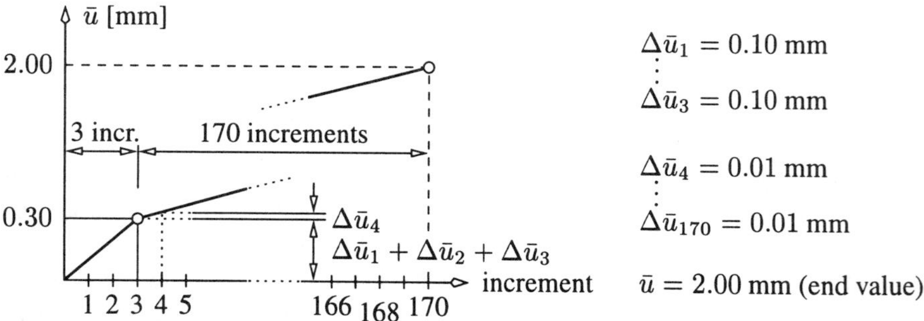

Figure 9. *Concrete L-shaped plate: displacement history*

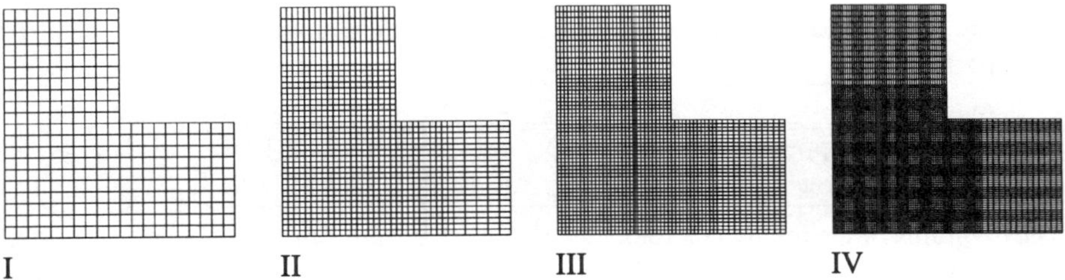

Figure 10. *Comparative study: uniform meshes I, II, III, and IV*

obtained with meshes III and IV deviate from the results computed using the meshes I and II. This seems to indicate that the results are not objective.

Figure 12 provides the explanation of the situation. The four illustrations refer to a part of the four meshes. The Rankine softening parameter α_R is shown at a prescribed displacement of $\bar{u} = 1.0$ mm. α_R is the measure for the tensile behavior . In regions with $\alpha_R = 0$ the tensile strength has not been reached so far. Regions with $\alpha_R > 0$ identify zones of cracked concrete. Using a fictitious crack model, a discrete crack is associated with the width of one finite element. Therefore, as can be seen in Fig.12, the extension of the crack shows a strong dependency on the FE discretization. In mesh I and in mesh II the crack follows the first horizontal row of elements above the level of the sharp corner of the L-shaped plate. The respective $\bar{u}$-R diagrams in Fig.11 agree very well.

With regards to the meshes III and IV, the locations of the crack differ significantly from the ones of the meshes I and II. The cracks start at the inbound corner of the plate at an angle of approximately 20°. At a certain distance from this corner, which depends on the discretization, the cracks follow a horizontal row of elements. Therefore the geometric dimensions of the intact parts of the plate differ. Consequently, the global structural response must be different.

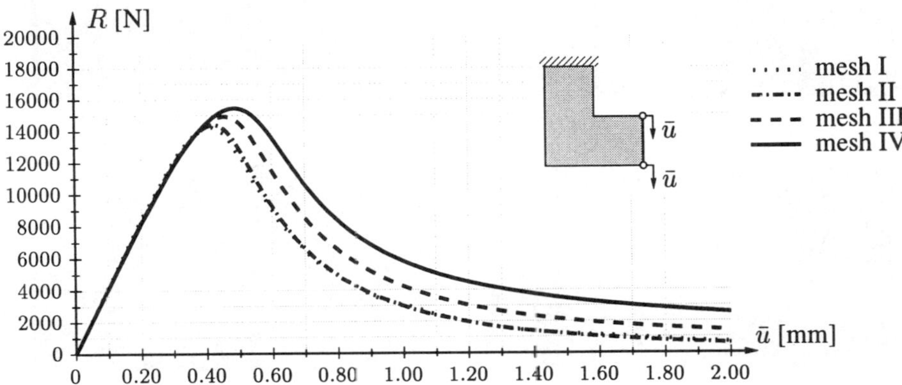

Figure 11. *Comparative study: displacement – reaction-force diagrams for different meshes*

Then, three adaptive analyses with different upper and lower bounds for the error were performed:

		A	B	C
Upper error limit:	$\bar{\eta}_{max}$	20 %	15 %	10 %
Lower error limit:	$\bar{\eta}_{min}$	16 %	12 %	8 %

The diagrams of the relative incremental global error were similar to the ones reported in [LAC 98B]. Fig.13 shows the distribution of the Rankine softening parameter α_R in a part of the investigated structure. These results were obtained from the adaptive analyses A, B, and C at a prescribed displacement of $\bar{u} = 1.0$ mm.

Figure 13 indicates a strong mesh gradation along the computed crack and in its vicinity, which clearly demonstrates the potentials of the proposed mesh generator. With decreasing error limits, the location of the crack is captured better. As regards the analysis C, a very small crack divides the two undamaged parts of the plate. Nevertheless, there *were* problems which were encountered in the adaptive analysis: the displacement–reaction-force diagram showed an incorrect, non-smooth behavior. A stiffening of the response was noticed at each transfer of state variables [HUE 98].

Therefore, a modified strategy was employed. Three different meshes, A*, B*, and C*, which were obtained from the adaptive analyses A, B and C at a prescribed displacement of $\bar{u} = 1.0$ mm, were used for a non-adaptive analysis (Fig.14). The entire displacement history (see Fig.9) was considered.

The displacement–reaction-force diagrams obtained from the modified analyses are shown in Figure 15. This figure also contains the respective diagrams obtained for the meshes I, II, III, and IV (see Fig.11). The agreement of the structural response

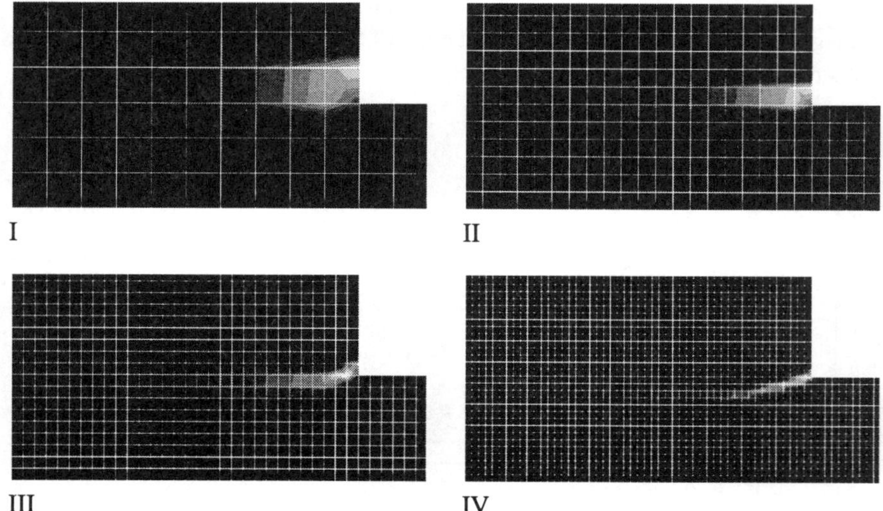

Figure 12. *Comparative study: parts from meshes I, II, III, and IV; Rankine softening parameter* α_R

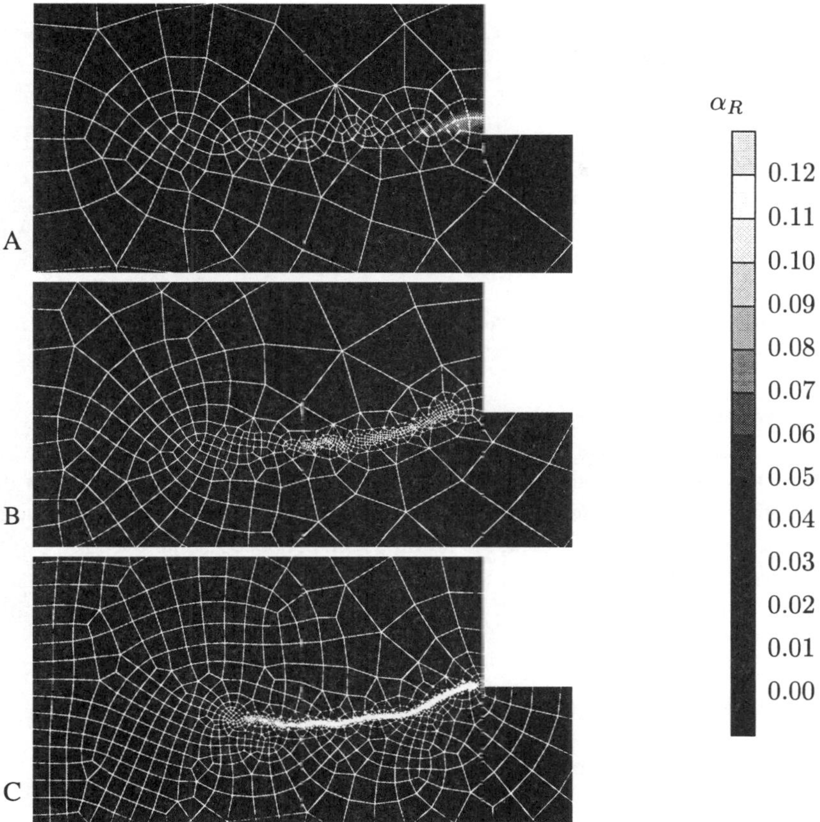

Figure 13. *Adaptive analyses A, B, and C: Rankine softening parameter α_R at a prescribed displacement of $\bar{u} = 1.0$ mm*

analysed with meshes A*, B*, and C* is rather good. This is evident from the good capturing of the location of the crack by using the the adaptive analysis.

6. Possible Remedies

The load-displacement curve is characterized by artificial stiffening after each refinement of the mesh and, hence, after each transfer of variables. Remedies for improving this unsatisfactory situation will be investigated in the following.

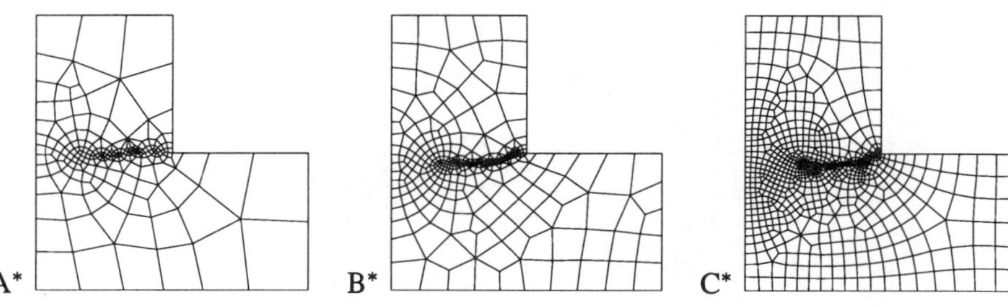

Figure 14. *Modified analyses: meshes A*, B*, and C* from adaptive analyses at a prescribed displacement of* $\bar{u} = 1.0$ *mm*

6.1. *Remedies within the framework of the FEM*

Crack is associated with strain localization. Preserving the objectivity of the results with respect to the element size in case of localization requires the introduction of a length-scale parameter. Herein, the fictitious crack concept, characterized by choosing a characteristic dimension of the element as this parameter, is employed. For an assessment of the performance of the FEM in case of cracking of concrete, different aspects such as:

- the FE formulation (displacement and mixed formulation, respectively),
- the FE discretization (non-aligned and aligned discretization, respectively), and
- computation of the characteristic length ℓ_c,

will be investigated. The following subsections contain brief descriptions of the aforementioned options.

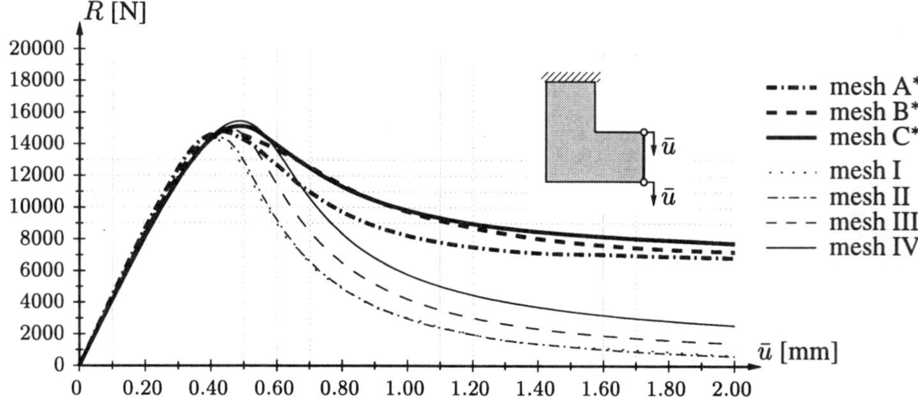

Figure 15. *Modified analyses: displacement – reaction-force diagrams*

6.1.1. *Displacement versus mixed formulation*

For the calculations reported in this work, a four-node quadrilateral element has been employed. When subjected to bending, this element exhibits shear locking. This effect is significantly reduced by treating the stresses and strains as independent variables (mixed formulation) and chosing their distributions over the element domain appropriately. The strain interpolation obtained from the displacement formulation and the chosen strain distribution for the mixed formulation are given by

$$\varepsilon_D = \left[\begin{array}{ccc|cc} 1 & & & \eta & \\ & 1 & & & \xi \\ & & 1 & \xi & \eta \end{array}\right] \mathbf{q}_\varepsilon \quad \text{and} \quad \varepsilon_{HW} = \left[\begin{array}{ccc|cc} 1 & & & \eta & \\ & 1 & & & \xi \\ & & 1 & & \end{array}\right] \mathbf{q}_\varepsilon\,, \qquad [15]$$

where the subscripts "D" and "HW" refer to the displacement formulation and the underlying variational principle (Hu-Washizu principle) of the mixed formulation respectively. $\mathbf{q}_\varepsilon$ denotes the vector containing the strain degrees of freedom, (ξ, η) are the local element coordinates.

6.1.2. *FE discretization*

For the assessment of the quality of crack simulations the academic example of a bar under tensile loading is chosen. Figure 16 contains the geometric dimensions, the material parameters and the chosen FE meshes.

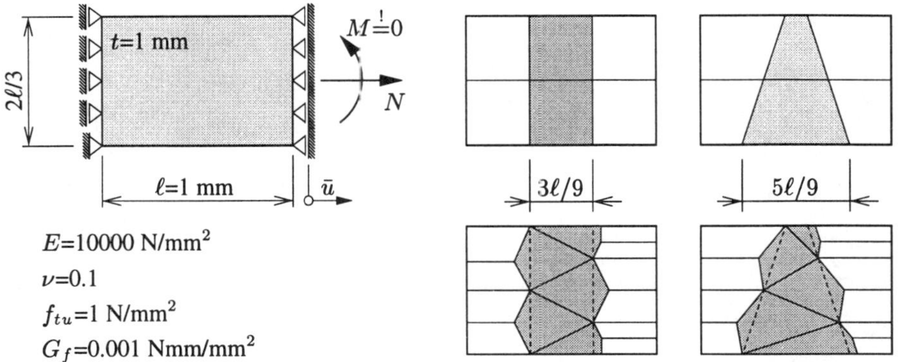

Figure 16. *Bar under tensile loading: geometric dimensions, material parameters and chosen FE meshes*

Localization is triggered by an artificial weakening of the elements located at the center line of the bar (grey-shaded elements in the four illustrations of FE meshes in Figure 16). These meshes are characterized by aligned and arbitrary (non-aligned) discretizations of the domain of localization as well as by constant and variable crack band widths.

6.1.3. *Determination of the characteristic length ℓ_c*

Several modes of determination of the characteristic length ℓ_c as a function of the element size were reported in the literature. They may be divided into two main groups:

- constant ℓ_c (independent of the state of loading),

  (i) which is the same at each integration point of an element ($\ell_c^{(i)}=\sqrt{A^e}$, where A^e represents the area of the finite element), and

  (ii) which is different at each integration point of an element ($\ell_c^{(ii)}=\omega\ell_c^{(i)}$, where ω is a weighting factor containing the proportionate area of the considered integration point A^{ip}: $\omega=\sqrt{4A^{ip}/A^e}$), and

- variable ℓ_c (depending on the state of loading), determined from

  (iii) the crack orientation ($\ell_c^{(iii)}=\omega(A^e/d)$) (see Figure 17),

  (iv) the direction of the relevant principal plastic strain ($\ell_c^{(iv)}=\omega d$) (see Figure 17), and

  (v) the layout of the singular band $\phi(\mathbf{x})$ which is set to zero behind the crack and to one in front of the crack [OLI 89] (see Figure 17) ($\ell_c^{(v)}=(\partial\phi(\mathbf{x})/\partial x_n|_{ip})^{-1}$, where x_n is the coordinate orthogonal to the crack; the crack orientation is computed as in (iii)).

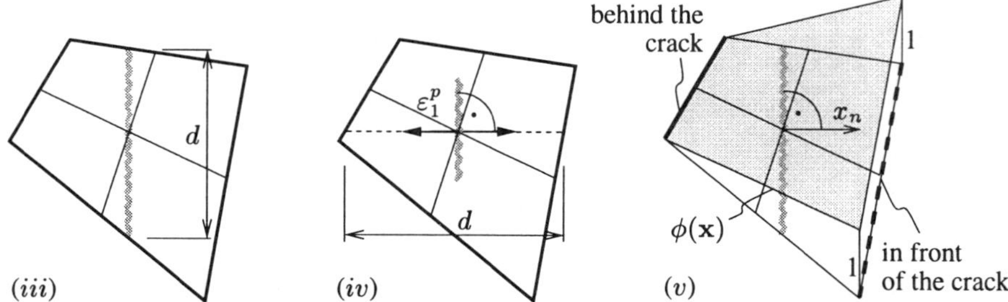

Figure 17. *On the evaluation of the characteristic length depending on the state of loading*

6.1.4. *Assessment of FE simulations of cracks*

For the assessment of the described options, the values of the axial force N of the bar subjected to axial tension via application of a prescribed displacement $\bar{u}$ are documented. Moreover, also the values of the bending moment M, which should be zero, are documented. Linear softening is assumed for the post-peak material domain.

Figure 18 contains the obtained evolutions of N and M for all meshes, based on the displacement formulation. It is seen that the use of variable characteristic lengths

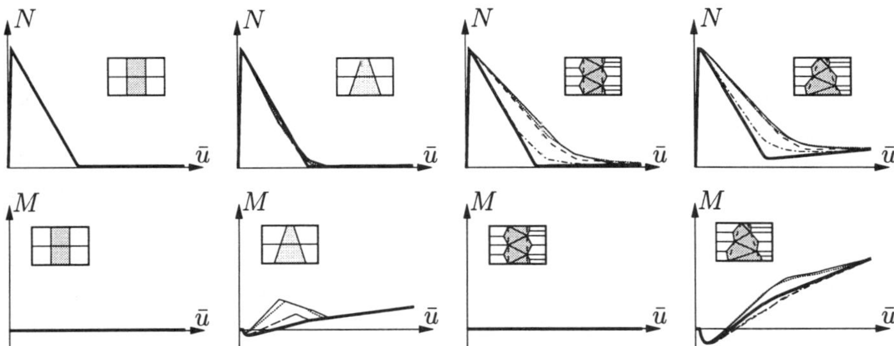

Figure 18. *Axial force N and bending moment M for different meshes based on the displacement formulation: Comparison of different ways of evaluating ℓ_c, (i) (——), (ii) (⋯⋯), (iii) (- - - - -), (iv) (-·-·-·-), (v) (——)*

ℓ_c results in an improvement of the solution. The characteristic length determined according to [OLI 89] ($\ell_c^{(v)}$) gives the best response. Mesh alignment is found to reduce the influence of locking on the results. This is reflected by a reduction of the error of M and by the correct response of the axial force (see Figure 18).

Figure 19 contains the evolutions of N and M as obtained from the displacement and the mixed FE formulation. A reduction of locking is observed for the mixed formulation. This reduction is larger for the smaller the value of M. Additional mesh alignment yields the correct solution, i.e., $M = 0$.
Our findings can be summerized as follows:

- the evaluation of the characteristic length according to [OLI 89] ($\ell_c^{(v)}$) gives the best structural response,
- the mixed formulation reduces locking of the four-node bilinear element but does not eliminate this effect, and
- mesh alignment reduces locking of this element.

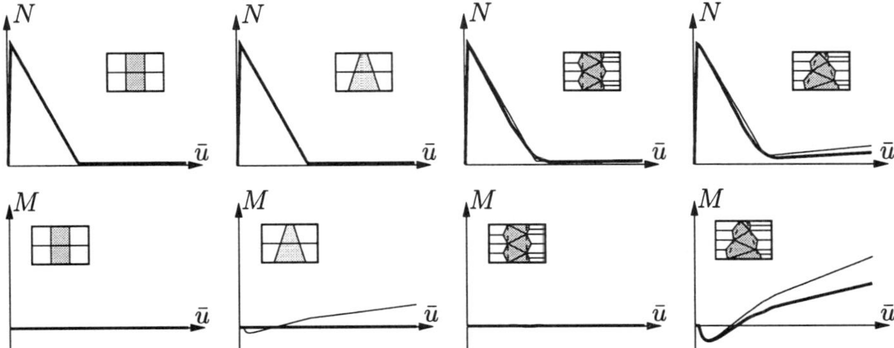

Figure 19. *Axial force N and bending moment M for different meshes based on the displacement formulation (——) and the mixed formulation (——) respectively*

6.2. *Remedies concerning the transfer scheme*

As mentioned previously, artificial stiffening was observed in the load-displacement diagram after each transfer of variables from the old to the new mesh.

An illustrative 1D example is used for the explanation of the reason for this deficiency (see Figure 20). The old mesh is assumed to consist of three linear truss elements (with one integration point each, located in the center of the element). The new mesh is assumed to consist of two finite elements. Localization is assumed to occur in the center element of the old mesh. The value of the internal variable is chosen as $\alpha = 2$. Transferring the internal variable[1] α to the new mesh by means of averaging [LAC 98B], the distribution of α, which was restricted to one element of the old mesh, extended to both elements of the new mesh. Obviously, an extension of the localization zone leads to a decrease of the value of the internal variable α. Hence, "artificial hardening" takes place. It results in stress points located within the increased elastic domain. This leads to a retrieval of elastic stiffness explaining the observed stiffening in the load-displacement diagram of the L-shaped plate.

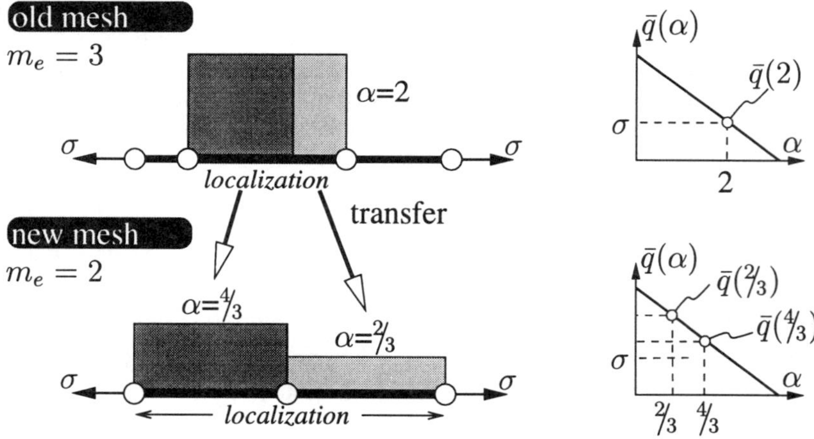

Figure 20. *Investigation of the nature of the deficiency of the transfer of the internal variable α by means of averaging of this quantity*

An improved version of the employed transfer scheme, see [LAC 98B], is reported in [LAC 98C]. It is characterized by an additional update of the internal variables after the transfer at material points located in the domain of plastic loading characterized by $f = 0$, with f denoting the yield function. The new value of α is computed from the transferred stresses and the yield condition, $f(\sigma, \alpha) = 0$ (for details see [LAC 98C]).

[1] In [LAC 98B] a mode of scaling of the internal variable α is reported, which involves a constant reference length. For the sake of simplification of the present example, for all elements of the old and the new mesh with localization, the same length was chosen. Hence, no scaling of α is required.

7. Summary and Conclusions

In this paper, the main ingredients of incremental adaptive finite element analysis of concrete plates were described. They are integrated in an efficient and robust algorithm for adaptive finite element analysis of non-linear problems with special emphasis on strain softening in the context of cracked concrete. These ingredients are:

- a projection-type error estimator for elasto-plastic material behavior based on incremental quantities,
- a program for calculation of the mesh density function from the distribution of the relative incremental element error and
- a mesh generator for discretizations with quadrilateral elements based on the Advancing Front Method.

An L-shaped plate made of plain concrete was chosen as a numerical example. As expected, material softening in consequence of concrete cracking caused serious problems, namely:

- element locking for both structured and unstructured meshes in case of crack propagation across element rows,
- an artificial increase of the structural resistance after each transfer of internal variables in the context of remeshing.

These problems were discussed *in extenso*. Possible remedies were proposed, namely, use

- of finite elements based on a mixed formulation in combination with mesh alignment in order to reduce locking, and
- of an improved transfer scheme resulting in a smooth structural response as regards the displacement – reactionforce diagram.

Bibliography

[COL 97] COLLINS M., VECCHIO F., SELBY R. and GUPTA P., The failure of an offshore platform . *Concrete International (ACI)*, vol. 19, num. 8, p. 28–35, 1997.

[HUE 98] HUEMER T., *Automatische Vernetzung und adaptive nichtlineare statische Berechnung von Flächentragwerken mittels vierknotiger finiter Elemente* . Dissertation, Vienna University of Technology, 1998. In German.

[LAC 98A] LACKNER R. and MANG H., Adaptive FE ultimate load analysis of reinforced concrete plates and shells . In *Proceedings of the Structural Engineering World Congress (SEWC)*, San Francisco, USA, 1998. Elsevier Science, Oxford. In print.

[LAC 98B] LACKNER R. and MANG H., Adaptive FEM for the analysis of concrete structures . In DE BORST R., BIĆANIĆ N. and MANG H., Eds., *Proceedings of the International Conference on Computational Modelling of Concrete Structures*, vol. 2, p. 897–919, Badgastein, Austria, 1998. Balkema Publishers, Rotterdam.

[LAC 98C] LACKNER R. and MANG H., Simulation of local failure of concrete plates on the basis of error control . In *Proceedings of the 3rd International Conference on Fracture Mechanics of Concrete and Concrete Structures*, Gifu, Japan, 1998. Aedificatio Publishers, Freiburg.

[LEE 95] LEE C. and LO S., An automatic adaptive refinement procedure using triangular and quadrilateral meshes . *Engineering Fracture Mechanics*, vol. 50, p. 671–686, 1995.

[LI 95] LI L.-Y. and BETTESS P., Notes on mesh optimal criteria in adaptive finite element computations . *Communications in Numerical Methods in Engineering*, vol. 11, p. 911–915, 1995.

[MAR96] MARC Analysis Research Coorporation, Bredewater 26, 2715 CA Zoetermeer, The Netherlands, *MARC/MENTAT* , 1996. Multi-Purpose Finite Element Package.

[MES 97] MESCHKE G., MANG H. and LACKNER R., Recent accomplishments and future research directions in computational plasticity of reinforced concrete structures . In OWEN D., OÑATE E. and HINTON E., Eds., *Computational Plasticity, Proceedings of 5th Int. Conf.*, p. 119–144. CIMNE, 1997.

[OLI 89] OLIVER J., A consistent characteristic length for smeared cracking models . *International Journal for Numerical Methods in Engineering*, vol. 28, p. 461–474, 1989.

[PER 87] PERAIRE J., VAHDATI M., MORGAN K. and ZIENKIEWICZ O., Adaptive remeshing for compressible flow computations . *Journal of Computational Physics*, vol. 72, p. 449–466, 1987.

[SCH 93] SCHLAICH J. and REINECK K.-H., Die Ursache für den Totalverlust der Betonplattform Sleipner A . *Beton- und Stahlbetonbau*, vol. 88, num. 1, p. 1–4, 1993. In German.

[VAR 95] VARDOULAKIS I. and SULEM J., *Bifurcation analysis in geomechanics*. Blackie Academic & Professional, Chapman & Hall, London, 1995.

[ZHU 91] ZHU J., ZIENKIEWICZ O., HINTON E. and WU J., A new approach to the development of automatic quadrilateral mesh generation . *International Journal for Numerical Methods in Engineering*, vol. 32, p. 849–866, 1991.

[ZIE 92] ZIENKIEWICZ O. and ZHU J., The superconvergent patch recovery and *a posteriori* error estimates. Part 1: The recovery technique . *International Journal for Numerical Methods in Engineering*, vol. 33, p. 1331–1364, 1992.

Error Indicator to Assess the Quality of a Simplified Finite Element Modelling Strategy

Shahrokh Ghavamian * — Gilles Pijaudier-Cabot *,**
Jacky Mazars *

* *LMT-Cachan, ENS Cachan, France*

** *Institut Universitaire de France*

ABSTRACT. *The purpose of this paper is to describe a tool developed in order to equip a simplified finite element software with quality assessment. These are error "estimators" or "indicators" whether they are applied to elastic linear or damage and plasticity analysis of R.C. elements described by the classical beam theory. A number of test cases are provided to demonstrate the capabilities of this tool. Both static and dynamic analyse s are performed, and the correlation between error measures and displacement or force results is shown.*
KEY WORDS: *Error Measure, Non Linear Analysis, Dynamic, Simplified Modelling, Damage Mechanics, Reinforced Concrete.*

1. Introduction

Advances achieved in construction engineering and growing architectural requirements constitute major needs in understanding well the material behaviour entering in the response of structural components. This is necessary to achieve more complex, safer and cheaper structures by optimising their functioning. During the past two decades, the continuous progress made in computer technology and performance improvements of numerical resolution techniques such as finite element methods (FEM), have greatly contributed to the popular use of computers in the construction industry.

In using these techniques, the validity of computational results and their efficiency are of great concern, and their inadequate use may produce unacceptable results having defaults, as this was the case in the Sleipner offshore platform which cost its failure [COL 97]. That is why disposing of any kind of indications on the quality of FE results is of great interest.

In literature, many studies concerning this issue can be found; these are based on commonly named a posteriori error calculation techniques. The basic concept is to determine the error which is due to the "discretisation" of any "continuum". The error calculation evaluates the difference between two solutions: the one concerning the reference problem (continuum formulation) which is a mathematical representation of the reality, and the discrete problem (FEM). Many parameters may be considered as error sources, such as: the geometrical representation, the time and space discretisation, the boundary and loading conditions, the time integration scheme, the non linear system resolution technique used for the integration of differential equations, and finally the accuracy used in storing numbers.

The aim of this paper is to present how the quality of a FEM computation may be estimated or measured, as a result of the space discretisation process only, which constitutes an important aspect among the above potential error sources.

From a literature review, we can conclude that error calculation techniques are mostly elaborated based on three different approaches:

Analysis of the stress field

Initially developped by Zienkiewicz and Zhu [ZIE 87; 92], it allows to estimate the error by comparing two set of results: one being discrete, the other smoothened. The latest one is built from the previous discrete solution field.

Residue of equilibrium equations

This method proposed by Babuska and Reinholdt [BAB 78; 82] is based on the direct estimation of errors by measuring the residual forces in the equation of equilibrium which is never totally balanced in standard FEM analyses.

Error in the constitutive relation

This method proposed by Ladevèze [LAD 91] consists in measuring the distance between two solution fields: one which is kinematically admissible, the other which is statically admissible. The distance can be viewed as a non satisfaction of the material constitutive relations.

The technique proposed here presents similarities to the one suggested in the last approach. It consists in measuring the difference between two stress fields, solutions of the FE model, one being KA (kinematically admissible) the other being SA (statically admissible). To demonstrate the method, we consider an elastic linear problem. The resolution of the problem consists in finding a displacement field U and a stress field σ so that:

- U satisfies kinematic boundary conditions,
- σ satisfies the equations of equilibrium,
- σ and $\varepsilon(U)$ satisfy the constitutive law.

Since the FEM is based on displacement methods [ZIE 88], the approximation applies on stress field, thus affecting its quality, since:

– equilibrium is not necessarily reached inside the elements (at nodes this is monitored by the required precision),

– the static boundary conditions are not satisfied point wise,

– the normal stress vector is not continuous across the interface between two adjacent elements.

We have applied and implemented this technique in a calculation program named EFICOS which serves for non linear static and dynamic analysis [OWE 80], [BAZ 86], [GHA 98], [MAZ 98] of reinforced concrete frames.

It is a finite element program, kinematically based on a "layered beam" approach. Beam elements possess two nodes, and are built by the superposition of layers with non linear stress-strain laws (Figure 1). The Euler-Bernoulli assumptions on the deflection of cross-sections allow the use of uniaxial stress-strain laws within each layer. These are damage based laws for concrete [LAB 91] and plasticity for steel.

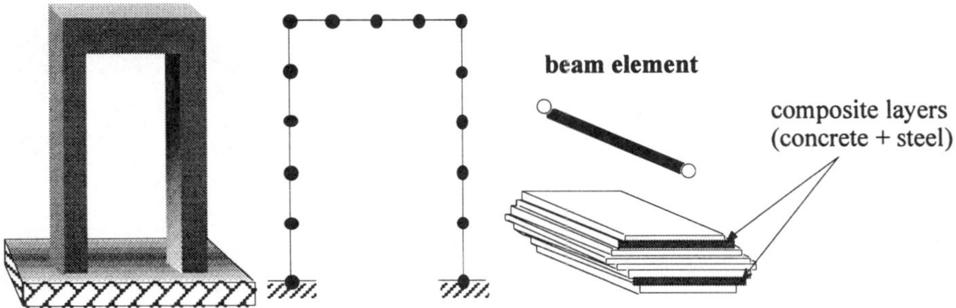

Figure 1. *Mesh discretisation by multilayered beam decomposition*

External loads as well as inertial forces are defined uniquely at nodes of the modelling. For dynamic problems, the part of error inherent to time discretisation as not considered in the error calculation technique. However, the Newmark's implicit scheme with constant acceleration, guarantees numerical stability within linear elastic behaviour at least.

It is very important to make a distinction between the error calculation technique is applied to linear elastic analysis within static loading, and the error calculation applied to any other type of non-linear or dynamic computation (non-linear static and dynamic). In the first case, the error measured can be considered as an "estimator" in the mathematical sense. In fact, there is an exact solution to the reference problem, which is described within the linear elasticity of classical beam theory and the error tends to zero when the numerical solution tends to the analytical one.

For any non linear analysis, the analytical solution does not exist and therefore the measured errors should be considered as error "indicators". While in the case of linear elasticity, low value of error indicated that the exact "reference" solution is almost reached, for non linear analysis it is only a stabilised solution which is obtained when error is reduced to its minimum because other sources of error such as time integration still exist.

2. Theoretical background

2.1. *Error measure in static*

For this type of analysis, the reference problem is constituted by a Euler-Bernoulli beam (no shear strain, small strains and small displacements) subject to forces applied upon the ends (Figure 2).

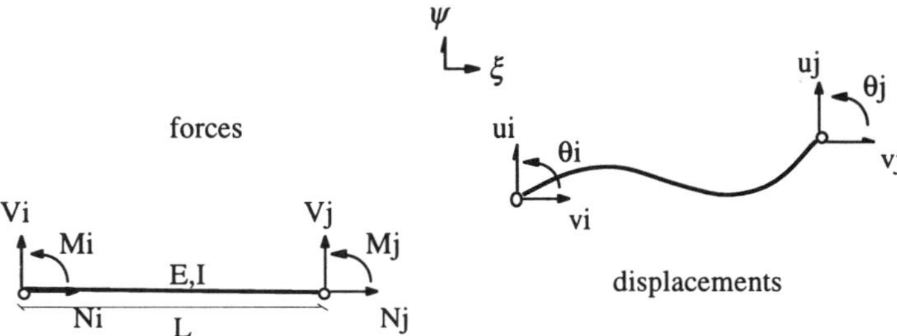

Figure 2. *Mechanical description of the static reference problem*

The formulation of the problem consists in finding a displacement field U and a generalised stress field Σ fulfilling the conditions:

– U verifies kinematic restrains at nodes,
– Σ verifies the following equilibrium equations,

$$\begin{cases} \dfrac{dN(\xi)}{d\xi} = 0 \\ \dfrac{dT(\xi)}{d\xi} = 0 \\ \dfrac{dM(\xi)}{d\xi} + T(\xi) = 0 \end{cases} \qquad [1]$$

– Σ and $\varepsilon(U)$ verify the constitutive law.

$$\sigma(\xi,\psi) = E(1-d)\,\varepsilon(\xi,\psi) \qquad [2]$$

The displacement solution to this problem is then:

$$u(\xi) = (1 - X)\, u_i + (X)\, u_j$$
$$v(\xi) = v_i + \left(3X^2 - 2X^3\right)\left(v_j - v_i\right) + \left(X - 2X^2 + X^3\right) L\theta_i + \left(-X^2 + X^3\right) L\,\theta_j \quad [3]$$
with $X = \xi/_L$

The finite element result provides nodal displacements. Nodal forces are then calculated by using the stiffness matrices of each element. These results are prolonged along each element using interpolation functions derived from equilibrium equations:

$$\begin{cases} \dfrac{dN(\xi)}{d\xi} = 0 \\ \dfrac{dT(\xi)}{d\xi} = 0 \\ \dfrac{dM(\xi)}{d\xi} + T(\xi) = 0 \end{cases}$$

The stress which derives from a kinematically admissible field is computed using the constitutive relation and strains calculated from displacements which are available at any point of a beam (longitudinally using Equation [3], and transversally using plane section assumption). Generalised KA stresses are then obtained by performing a numerical integration over the cross section (Figure 3).

The error is measured by calculating the difference between these two KA and SA fields along each element, by applying an Euclidean norm.

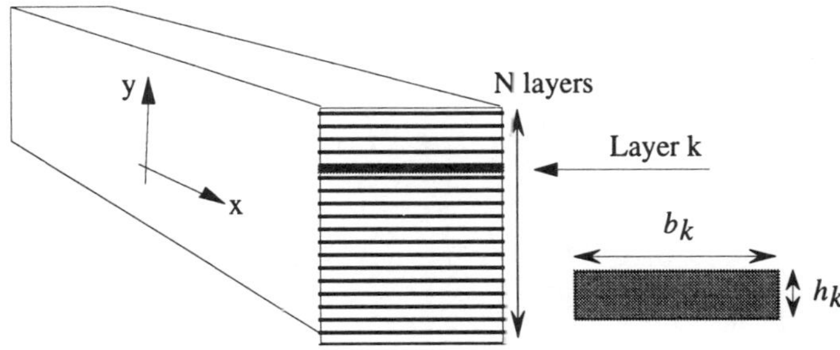

Figure 3. *Numerical integration over the cross section of a multilayered beam*

$$E^i_{L2} = \left| \Sigma_{KA} - \Sigma_{SA} \right|_{L2} = \left[\int_{L^i} \left(\Sigma_{KA} - \Sigma_{SA} \right)^2 d\xi \right]^{1/2} \quad [4]$$

E^i_{L2}: error measure using the L2 norm (Lebegue) over element i

Σ_{CA}: generalised stress (N or M), kinematically admissible

Σ_{SA}: generalised stress (N or M), statically admissible

To realise the integration, the Gauss method is used. Although SA fields are well defined (see later) in both elasticity and non-linearity, this is not the case for KA fields with non-linearity. Thus, the number of integration points necessary for that depends on the order of each generalised stress field functions, Σ_{SA} and Σ_{KA}. Considering the bending moment, this is as follows:

- *Static linear elastic behaviour*

$$\frac{dM(\xi)}{d\xi} + V(\xi) = 0 \quad \Rightarrow \quad M_{SA} \text{ polynomial of order 1}$$

$$M^i(\xi) = \sum_{k=1}^{n} b^i_k \, h^i_k \, \sigma^i_k(\xi) \, \psi^i_k \text{ where } \sigma^i_k(\xi) = E^i_k \left(\psi^i_k \frac{d^2 v(\xi)}{d\xi^2} + \frac{du(\xi)}{d\xi} \right) \qquad [5]$$

Thus, M_{KA} is polynomial of order 1, and 2 integration points would be sufficient.

$$E^i_{L2} = \left| M^i_{KA} - M^i_{SA} \right|_{L2} = \sqrt{\frac{L^i}{2}} \left[\left(M^i_{KA} - M^i_{SA} \right)^2_{\xi = {}^1\!/_2 (1 - {}^1\!/_{\sqrt{3}})} + \left(M^i_{KA} - M^i_{SA} \right)^2_{\xi = {}^1\!/_2 (1 + {}^1\!/_{\sqrt{3}})} \right]^{1/2}$$

- *Static non linear behaviour*

Considering the constitutive law of behaviour of constituents (concrete and steel), the variation of Young's modulus along an element within each layer is very complex. For the sake of simplicity we use a linear form, bearing in mind that the importance of this fact needs to be investigated more in detail. Under these circumstances the order of M_{KA} is 2, which thus needs 3 integration points.

$$E^i_{L2} = \left| M^i_{KA} - M^i_{SA} \right|_{L2} = \sqrt{\frac{L^i}{2}} \left[\frac{5}{9} \left(M^i_{KA} - M^i_{SA} \right)^2_{\xi = {}^1\!/_2 (1 - \sqrt{3/5})} + \frac{8}{9} \left(M^i_{KA} - M^i_{SA} \right)^2_{\xi = {}^1\!/_2} + \frac{5}{9} \left(M^i_{KA} - M^i_{SA} \right)^2_{x = {}^1\!/_2 (1 + \sqrt{3/5})} \right]^{1/2} \qquad [6]$$

2.2. *Error measure in dynamics*

The reference problem is considered to be quite similar to the static one, except that inertial forces are considered as a body force uniformly distributed along the beam. This implies a uniform constant distribution of the density. This time, the formulation of the problem would be:

– U verifies kinematic boundary and initial conditions
– Σ verifies equilibrium equations,

$$\begin{cases} \dfrac{\partial N(\xi)}{\partial \xi} + m(\xi)\,\dfrac{\partial^2 u(\xi,t)}{\partial t^2} = 0 \\ \dfrac{\partial V(\xi)}{\partial \xi} + m(\xi)\,\dfrac{\partial^2 v(\xi,t)}{\partial t^2} = 0 \\ \dfrac{\partial^2 M(\xi)}{\partial \xi^2} + m(\xi)\,\dfrac{\partial^2 v(\xi,t)}{\partial t^2} = 0 \end{cases} \qquad [7]$$

(rotational moment of inertia is neglected in the last equation)

– Σ and $\varepsilon(U)$ verify the constitutive law.

This time, the order of the polynomial $M(\xi)$ is 5, which undoubtedly poses the problem of calculation cost. In order to remain within acceptable computational costs we made the decision to modify the reference problem by considering inertial forces no longer as uniformly distributed body forces, but more as nodal forces which is the same as for the static case. Hence, the error calculation in dynamic analysis are performed in the same way as in static.

3. Different error quantity measures

At this point we can use the two different error measures that may be obtained by applying the above technique. One is called the "Relative Local Error" RLE, and the other the "Relative Global Error" RGE. These are totally complementary, and are useful for different purposes. While RGE provides the FE user with information concerning the overall quality of his results, RLE is useful in a mesh refinement process. These two quantities are obtained as follows:

$$RLE^i(t) = \frac{\underset{t\in[0,t]}{Sup}\sqrt{\int_{e^i}\left(\Sigma_{KA} - \Sigma_{SA}\right)^2 d\xi}}{\underset{t\in[0,T]}{Sup}\sqrt{\sum_e \int_{e^i}\left(\Sigma_{KA} + \Sigma_{SA}\right)^2 d\xi}} \qquad [8]$$

$$RGE^i(t) = \frac{\underset{t\in[0,t]}{Sup}\sqrt{\sum_e \int_{e^i}\left(\Sigma_{KA} - \Sigma_{SA}\right)^2 d\xi}}{\underset{t\in[0,T]}{Sup}\sqrt{\sum_e \int_{e^i}\left(\Sigma_{KA} - \Sigma_{SA}\right)^2 d\xi}} \qquad [9]$$

(T represents the total duration of the loading time, both in static and in dynamics).

One can notice that these two measures are related to each other:

$$RGE^{i}(t) = \sqrt{\sum_{e} \left(RLE^{i}(t)\right)^{2}} \qquad [10]$$

4. Applications

In this section, our goal is to demonstrate the potentiality of these error measures in several case studies. Throughout these examples, some explanations are provided in order to better understand the meaning and the evolution of the error measures with different discretisations.

4.1. *Case study n° 1*

We are concerned here with the elastic linear behaviour of an anchored cantilever rectangular beam subjected to a vertical deflection. Figure 4b illustrates the problem and provides the mechanical and geometrical specifications. The FE modelling is composed of 1 beam element containing 10 cross-sectional layers. This structure was chosen because the FE solution should be close to the exact solution; that means we should expect zero error in results. The only difference being the cross-sectional discretisation. However, an error of 0.5% was measured (RLE and RGE are the same when the modelling is composed of only one element). By investigating the effect of the number of layers on this error, we found the dependence of the error on the number of layers as shown in Figure 4a.

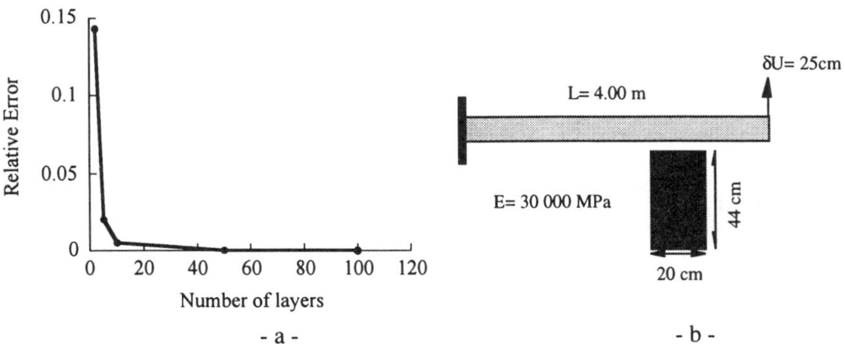

Figure 4. *Case study n°1. a- variation of the relative error with the number of layers, b- description of the problem*

In order to explain this observation, it is interesting to see how M_{KA} (kinematically admissible bending moment) is calculated:

$$M^i(\xi) = \sum_{k=1}^{n} b_k^i h_k^i \sigma_k^i(\xi) \psi_k^i \quad \text{avec} \quad \sigma_k^i(\xi) = E^i \left(\psi_k^i \frac{d^2 v(\xi)}{d\xi^2} + \frac{du(\xi)}{d\xi} \right) \qquad [11]$$

The moment is a polynomial of degree two with respect to the vertical direction. Since the position of the layers are not geometrically compatible with Gauss numerical integration points, a sufficient number of layers is necessary to approach the exact integration value. Consequently, we can conclude that at least ten layers are necessary if an acceptable quality in results is expected. This argument is not only valid for linear elasticity, but also for any other case.

4.2. *Case study n° 2*

Here we study the dynamic (seismic) non linear behaviour of the former problem. The beam is considered as reinforced concrete with 1.5% longitudinal steel rebars. The loading used is a synthetic accelerogram applied transversally at the anchorage. Table 1 indicates all material properties of the specimen. The non linear constitutive laws are: damage based model for concrete and elastoplastic for steel. The concrete model is able to describe the unilateral functioning of concrete by allowing crack opening and reclosure, and two damage scalars (compressive and tensile partitioning) monitoring stiffness and strength reduction. Also inelastic strains are generated upon non-linear range of functioning. For more details, the reader may refer to [LAB 91].

Three dicretisations with 1, 3 and 10 constant length elements are used. Each element is composed of 10 layers, containing the upper and lower steel rebars. This time the purpose was to demonstrate the mesh refinement dependency on the error measure.

Steel rebars	E (Young's modulus)	[MPa]	200
	f_e (yield stress)	[MPa]	400
	E_T (post elastic tangent stiffness)	[MPa]	20
Concrete	E (Young's modulus)	[MPa]	30 000
	f_y (peak compressive stress)	[MPa]	28
	f_t (peak tensile stress)	[MPa]	2
	Damage based model [LAB 91]		
Loading	Duration of the loading	[s]	20.0

Table 1. *Material properties of study case n° 2*

Figure 5 shows the deflection time history response and the RGE evolution. It is interesting to notice the difference in results for the three meshes, and the correlation with their respective RGE evolution. We should bear in mind that the error measures are maxima (sup) of the time history error values.

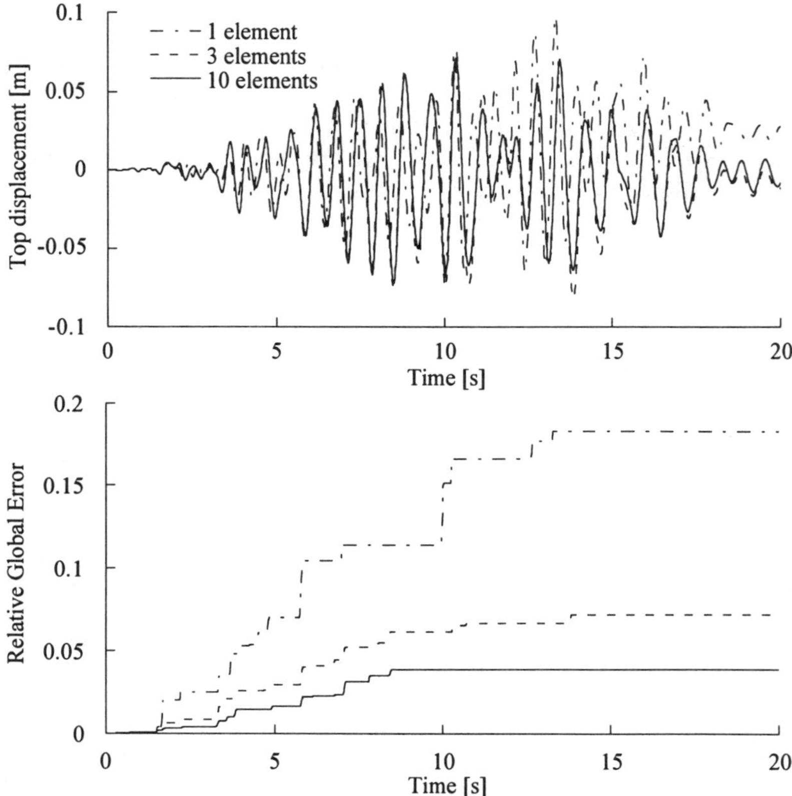

Figure 5. *Time histories.* TOP: *deflection.* BOTTOM: *Relative Global Error evolution*

4.3. *Case study n° 3*

This is the non linear analysis of a single-bay plane structure as shown in Figure 6, while material properties are the same as described in case study n° 2 (Table 1). The loading, 0.5 m horizontal top displacement, was applied through 100 increments. This time, four different meshes were used, by refining homogeneously the entire structure. Figure 8 illustrates the deflection shape, tensile damage mapping, rebar yielding and RLE distributions.

Figure 7 provides comparisons between the different discretisations: the evolution of the RGE, the shear reaction force at the left leg anchorage, and also the variation of the final RGE.

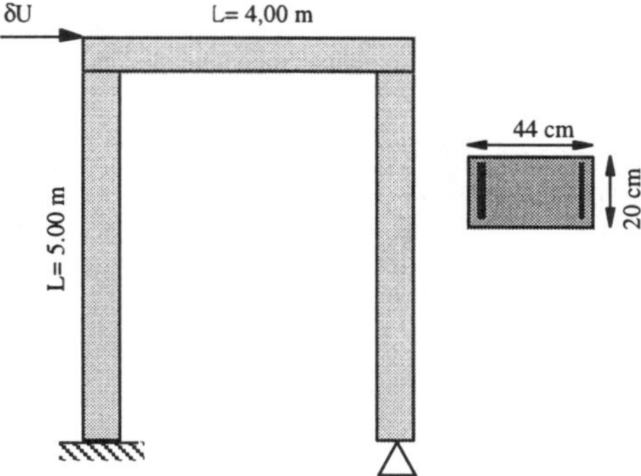

Figure 6. *Case study n° 3*

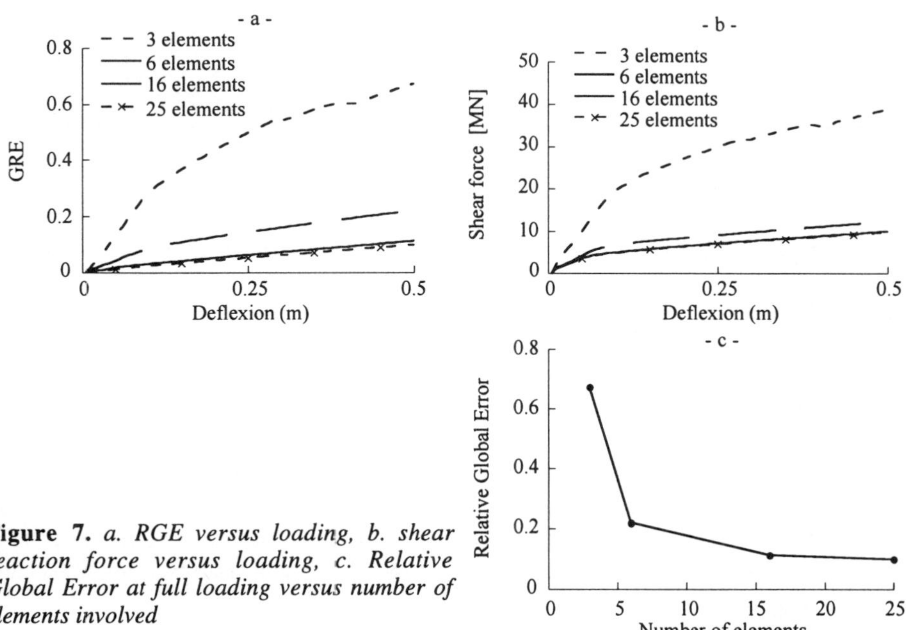

Figure 7. *a. RGE versus loading, b. shear reaction force versus loading, c. Relative Global Error at full loading versus number of elements involved*

When refining the discretisation, we used a homogeneous discretisation in order to show the influence of the number of elements. From the RGE reduction as a consequence of mesh refinement (Figure 8c) it is clear that there is no need to use many elements everywhere. However, the remaining RGE can be strongly reduced by using more than 10 layers per element. In most cases, the influence of layer numbers on RGE can be as high as the number of elements involved.

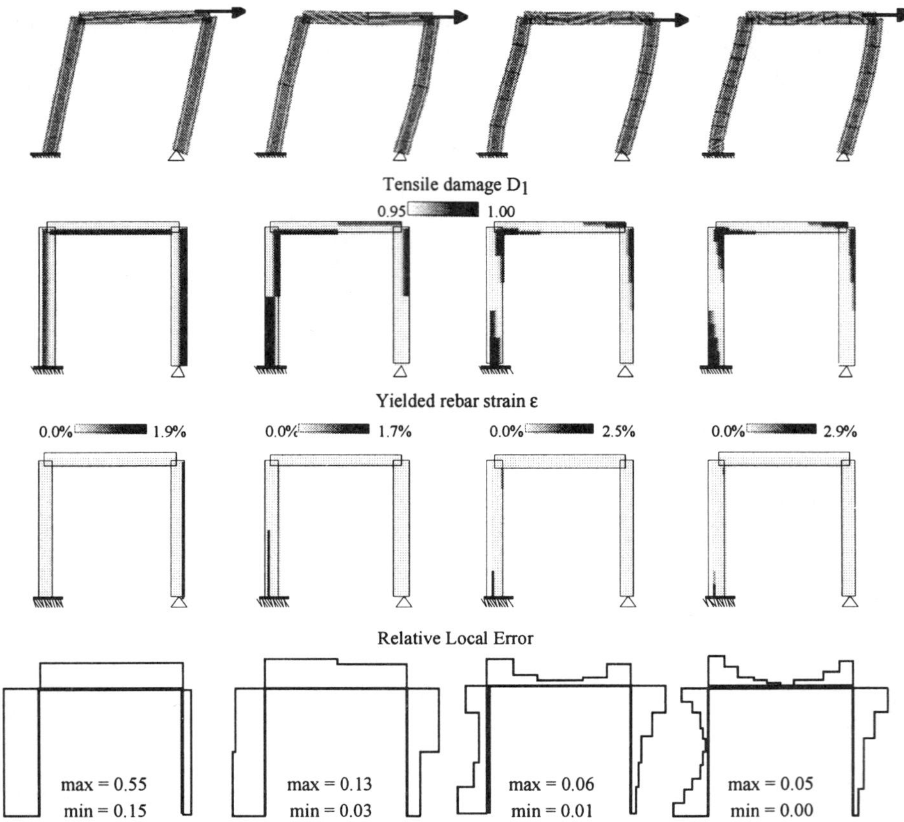

Figure 8. *Local results for 4 different mesh refinements*

6. Conclusion

The presented Relative Global Error allows to qualify the geometrical discretisation of a structure using multilayered beam elements. Meanwhile, the value of Relative Local Error per element appears to be of great use in optimising the *a posteriori* mesh refinement process (in terms of element and layer numbers, and their arrangement).

The usefulness of these measures were demonstrated by several examples, showing the strong correlation of the error with calculated forces, displacement and material non-linearities.

The search for better, optimised computational results requires the consideration of several parameters:

- number and size of elements, and their arrangement,
- number and size of layers, and their positioning.

The influence of each of these may become dominant on the error that is generated by the geometrical discretisation. That is why the mesh optimisation must become automatic by monitoring all these parameters. With a quality assessment based on such error indicator, the use of the Simplified Finite Element technique may reach its full capacity, specifically for computation time consuming analyses which are dynamic non-linear calculations.

8. References

[BAB 78] BABUSKA, I., RHEINHOLDT, W.C. (1978). Error estimates for adaptative finite element computation, *SIAM J. Num. Anal.*, vol. 15, n° 4, 736-754.

[BAB 82] BABUSKA, I., RHEINHOLDT, W.C. (1982). Computational error estimates and adaptative processes for some nonlinear structural problems. *Comp. Meth. In Applied Mech. And Engrg.* 34, 895-937.

[BAZ 86] BAZANT, Z.P., PAN, J., PIJAUDIER-CABOT, G. (1986). Softening in reinforced concrete beams and frames. *ASCE Journal of Structural Engineering*, vol. 113, n° 12.

[COL 97] COLLINS, M.P., VECCHIO, F.J. (1997). The failure of an offshore platform. Concrete International-ACI. August 1997.

[GHA 98] GHAVAMIAN, Sh. (1998). Méthode simplifiée pour la simulation du comportement sismique des structures en béton armé. Traitement des effets de l'élancement et estimateur d'erreurs. Thèse de doctorat ENS Cachan.

[LAB 91] LA BORDERIE, C.L., MAZARS, J., PIJAUDIER-CABOT, J. (1991). Response of plain and reinforced concrete structures under cyclic loading. ACI STP 134, W. Gerstle and Z.P. Bazant, eds, 147-172.

[LAD 91] LADEVÈZE, P., PELLE, J.P., ROUGEOT, P. (1991). Error estimation and mesh optimization for classical finite elements. *Engrg. Computation,* 8:69-80.

[MAZ 98] MAZARS, J., GHAVAMIAN, SH., RAGUENEAU, F. (1998). EFICOS: A finite element technique in predicting the behaviour of R.C. structures under severe loading. The French technology of concrete, 13TH FiP congress. Ed. AFPC-AFREM Paris. p. 59-72. Amsterdam.

[OWE 80] OWEN, D.R.J., HINTON, E. (1980). *Finite elements in plasticity: Theory and practice.* Pineridge Press Ltd., Swansea, England.

[ZIE 87] ZIENKIEWICZ, O.C., ZHU, J.Z. (1987). A simple error estimator and adaptative procedure for practical engineering analysis. *Int. J. for Num. Meth. In Engineering.,* vol. 24, pp. 337-357.

[ZIE 88] ZIENKIEWICZ, O.C., TAYLOR,(1988). The finite element method, Vol. 1. *Basic formulation and linear problems,* London, McGraw-Hill.

[ZIE 92] ZIENKIEWICZ, O.C., ZHU, J.Z. (1992). The superconvergent patch recovery and *a posteriori* error estimates, part 1. The recovery technique. I*nt. J. for Num. Meth. In Engineering,* vol. 33, pp. 1331-1364.

A Methodology for Discretisation Objective, Discrete, Dynamic Fracture

Paul Klerck* — Roger Owen* — Jianguo Yu* — Tony Crook**

**University of Wales, Swansea*
Department of Civil Engineering
SA2 8PP

***Rockfield Software Ltd.*
Innovation Centre
University of Wales, Swansea
SA2 8PP

ABSTRACT. *A discrete dynamic fracture methodology is proposed based on a combination of localisation and discrete finite element (DEM) concepts. Fracture is considered in the form of a rotating smeared crack model utilising tensile strain-softening to represent material degradation. Rate dependence is introduced to extirpate the ill-posedness of the hyperbolic, dynamic system by ensuring the dispersive character of the softening continuum and by on introducing a length scale to govern the width of the localisation zone. Discrete fracture is governed by a nodal averaging of a local damage measure with localisation envisaged a priori to discrete fracture insertion. The efficacy of the methodology is demonstrated by application to the common dynamic fracture manifestations of spalling and corner fracture.*

KEY WORDS: *Dynamic, Fracture, Discrete, Localisation, Rate Dependence, Smeared Crack.*

Introduction

The ability of a finite element methodology to accommodate the continuum-discrete transition is of paramount importance in the modeling of post failure interaction in dynamic fracture systems [MUN 95]. In the pursuit of objective discretisation in post failure analysis, it remains to ensure objectivity a priori to post failure interaction. A combination of localisation and discrete element concepts is proposed whereby discrete fracture is initiated after localisation of damage. A fracture model is implemented within an explicit, finite (discrete) element code (DEM) permitting local re-meshing following discrete crack insertion [MUN 95].

The smeared crack approach is adopted utilising strain-softening to model material de-cohesion and energy dissipation [BAZ 83], [DEB 85], [ROT 87], [WIL 86]. The dynamic manifestation of the inviscid strain-softening continuum is the loss of hyperbolicity and thus Lipschitz continuity of the governing differential equations [JEF 76]. The dynamic initial value problem is immediately ill-posed with the softening elliptic sub-domain rendering smooth solutions. No interaction occurs between the hyperbolic and the elliptic sub-domains resulting in pathological discretisation dependence [BAZ 85], [SLU 92a], [NEE 88], [BEL 87], [BEL 86], [WU 84].

The ill-posedness is extirpated by the inclusion of rate dependence in the strain-softening formulation [LOR 90], [PRE 90], [SLU 92a], [SLU 92d], [NEE 88], which is physically justified on the basis of a finite fracture time [WAS 94]. This renders the softening continuum dispersive and introduces a discretisation independent length scale that governs the width of the localisation zone [SLU 92a], [NEE 88]. Diagnostics of discontinuous and diffuse bifurcation [SOB 90], [PRE 90], [HIL 58], [DRU 59], derived from the consideration of weak discontinuities in hyperbolic systems [HIL 61], [JEF 76], [PRE 90], reveal the constraints of effective viscous regularisation [SIM 89], [OWE 82], [MIT 90], [OWE 80]. A rate dependent strain-softening formulation is implemented in an orthogonal, rotating crack model [DUC 72], [GUP 84], [BHA 94], [GUZ 95], [COP 80], [MIL 85], which is deemed to yield a more reliable structural response than the overly stiff fixed crack models [MIL 85], [BHA 94], [DEB 85], [GAJ 91], [FOS 96]. Discrete fracture is governed by the nodal averaging of a local damage measure [MUN 95].

Numerical examples considering the dynamic fracture manifestations of spalling and corner fractures [HOP 21], [JOH 72], [RIN 63], [KOL 53] illustrate the efficacy of the proposed fracture methodology. It should be noted that the present exposition concerns the attainment of discretisation objectivity with respect to discrete fracture distribution and energy dissipation and does not concern post failure interaction.

1. Smeared Cracking with an Inviscid, Tensile Strain-Softening Continuum

For the majority of structural solids including rock, the total energy required to break all the bonds on any cross section of material in tension does not differ widely from 1 Joule per square meter.[GOR 79]. Materials are not brittle because of their low tensile strength but because of the low energy required to propagate fractures. Unlike more ductile materials, which dissipate large amounts of energy in the dislocation mechanisms associated with plastic deformation, brittle materials do not possess effective energy absorbing mechanisms and the work done during fracturing in tension is virtually confined to that needed to break bonds. Thus, it is widely viewed that the tensile strain-softening observed in quasi-brittle materials is not a material property, but the manifestation of inhomogeneous deformation, geometry and boundary conditions. Read and Hegemier [REA 84] demonstrate that an apparent structural strain-softening response may be obtained by utilising actual cross sectional areas in the calculation of stress.

The inhomogeneous nature of materials such as concrete and rock exhibit dispersed process zones that dissipate energy in a myriad of micromechanical processes. Tensile strain-softening exists as a phenomenological description of quasi-brittle materials for scales far in excess of that at which the governing micromechanical processes dominate and provides an effective means of dissipating the total fracture energy.

1.1. *Motivation for a Rotating Crack Model*

The smeared crack concept envisages a cracked solid as an equivalent anisotropic continuum with degraded properties in the directions normal to the crack orientations [GUZ 95]. A crack is introduced at a Gauss point when the maximum principal stress surpasses the tensile yield strength of the material. The crack is deemed orthogonal to the maximum principal strain (and stress) so that the principal axes of orthotropy are aligned with the principal strain (and stress) directions.

The fixed crack model proposed by Bazant and Oh [BAZ 83] enforces fixed crack alignment with the incipient crack directions and utilises a constant shear retention factor. This model exhibits deficient softening in the post peak regime [ROT 87] with the constant shear retention factor permitting an indefinite increase in shear stress and thus prohibiting shear energy dissipation. In addition fixed crack models result in severe stress locking due to the zigzag propagation of crack profiles in continuous finite element meshes [BHA 94]. The (non-orthogonal) multiple fixed crack model proposed by de Borst and Nauta [DEB 85] does much to alleviate the stress locking phenomena but suffers form ill-conditioning of the tangent stiffness matrix at close angular spacing of cracks. The lack of coupling between multiple cracks also results in excess energy dissipation.

To ameliorate the crudeness of the constant shear retention factor Willam et al. [WIL 86] and Rots et al. [ROT 87] introduce a shear softening relation based on shear fracture energy. However, this additional softening does not provide a remedy to the overly stiff prediction of the fixed crack approach and reduces stability [GUZ 95]. It is noted that shear retention factors are difficult to determine, and the lack of coupling between the normal stress and the shear stress is questionable.

An alternative to the fixed crack model is the rotating crack model [DUC 72], [GUP 84], [BHA 94], [GUZ 95], [COP 80], [MIL 85], [ROT 89], which enforces coincident rotation of the principal axes of orthotropy and the principal strain (and stress) axes. Consequently, no shear stresses are realised parallel to the cracks and thus no shear retention factor needs to be stipulated. Bazant [BAZ 83] expresses concern that rotation of the principal axes of orthotropy is inadmissible as it implies the rotation of directional microstructural defects. This observation does not eviscerate the method and some reconciliation is possible if the orientation of the rotating crack is considered as an average direction. The general lack of alignment between the principal axes of stress and strain dictates the choice of the tangential shear modulus [BAZ 82], [GUZ 95]. It is also apparent that the rotating crack model is independent of the loading history.

Notwithstanding the above observations, the rotating crack model performs more effectively than all the variations of the fixed crack model in alleviating stress locking and consistently yields a softer, lower bound (more conservative) response [GUZ 95], [ROT 89], [WIL 89]. It is widely accepted that the rotating crack model provides a more reliable structural response prediction than the fixed crack models [MIL 85], [BHA 94], [DEB 85], [GAJ 91], [FOS 96].

1.2. *Implementation of an Explicit, Orthogonal Rotating Crack Model*

A rotating smeared crack model is implemented permitting concomitant cracking in both principal directions of orthotropy. The principal axes of orthotropy are aligned with the principal strain axes and co-axiality of the principal stress and strain is enforced by modifying the shear modulus. The onset of damage invokes the uncoupling of the (orthogonal) principal orthotropic directions and results in a constitutive relationship equivalent to the sum of two orthogonal (uncoupled) one-dimensional smeared crack equations.

The state variables pertinent to the stress update procedure at time t_n are the total strain vector $\boldsymbol{\varepsilon}_n = [\varepsilon_{xx\,n}, \varepsilon_{yy\,n}, \gamma_{xy\,n}]^T$ and the total strain increment vector $\Delta\boldsymbol{\varepsilon}_n = [\Delta\varepsilon_{xx\,n}, \Delta\varepsilon_{yy\,n}, \Delta\gamma_{xy\,n}]^T$. Prior to incipient cracking, the constitutive relationship is isotropic linear-elastic and the global stress vector $\boldsymbol{\sigma}_n$ at time t_n is given by,

$$\boldsymbol{\sigma}_n = \left[\sigma_{xx\,n}, \sigma_{yy\,n}, \tau_{xy\,n}\right]^T = \mathbf{C}\boldsymbol{\varepsilon}_n \qquad [1]$$

where $\mathbf{C}$ is the linear elastic modulus for plane strain. To enforce alignment of the principal stress and principal strain axes [GUZ 95], [BAZ 83] the shear modulus G is given by,

$$G = \frac{\left(\sigma_{xx\,n} - \sigma_{yy\,n}\right)}{2\left(\varepsilon_{xx\,n} - \varepsilon_{yy\,n}\right)} \qquad [2]$$

The angular displacement between the principal axes of orthotropy and the global Cartesian axes at time t_n is given by,

$$\theta_n = \frac{1}{2}\tan^{-1}\left(\frac{2\tau_{xy\,n}}{\sigma_{xx\,n} - \sigma_{yy\,n}}\right) = \frac{1}{2}\tan^{-1}\left(\frac{\gamma_{xy\,n}}{\varepsilon_{xx\,n} - \varepsilon_{yy\,n}}\right) \qquad [3]$$

The principal stresses at time t_n are given by,

$$\boldsymbol{\sigma}_n^{principal} = \left[\sigma_{1\,n}, \sigma_{2\,n}\right]^T = \mathbf{N}^T \boldsymbol{\sigma}_n \qquad [4]$$

$$\mathbf{N} = \begin{bmatrix} \cos^2\theta_n & \sin^2\theta_n \\ \sin^2\theta_n & \cos^2\theta_n \\ \frac{1}{2}\sin 2\theta_n & -\frac{1}{2}\sin 2\theta_n \end{bmatrix} \qquad [5]$$

Cracking is initiated if the first principal stress, defined in Equation [4], surpasses the virgin tensile strength of the material σ^y. The onset of cracking introduces an uncoupled one-dimensional equation in each of the principal orthotropic directions, and the principal stresses (Equation [4]) become the crack stresses $\boldsymbol{\sigma}_n^{cr}$. Unloading and reloading in tension occurs with the Young's modulus E. The remainder of the derivation considers a single principal orthotropic direction undergoing tensile loading.

The total strain increment vector, with respect to the principal axes of orthotropy at time t_n, is given by,

$$\Delta\boldsymbol{\varepsilon}_n^{principal} = \left[\Delta\varepsilon_{1\,n}, \Delta\varepsilon_{2\,n}\right]^T = \mathbf{N}^T \Delta\boldsymbol{\varepsilon}_n \qquad [6]$$

It is assumed that all the non-linear strain in each principal orthotropic direction is accounted for by the opening of the associated crack. A loading strain increment $\Delta\varepsilon_{i\,n}$ is decomposed into an elastic increment $\Delta\varepsilon_{i\,n}^{e}$ and a crack increment $\Delta\varepsilon_{i\,n}^{cr}$ [DEB 85].

$$\Delta\varepsilon_{i\,n} = \Delta\varepsilon_{i\,n}^{e} + \Delta\varepsilon_{i\,n}^{cr} \qquad [7]$$

The loading crack stress increment is given by,

$$\Delta\sigma_{i\,n}^{cr} = E\Delta\varepsilon_{i\,n}^{e} = E\left(\Delta\varepsilon_{i\,n} - \Delta\varepsilon_{i\,n}^{cr}\right) \qquad [8]$$

The one-dimensional strain-softening relationship renders a constant, negative elasto-plastic modulus $h = \partial\sigma_i^{cr} / \partial\varepsilon_i^{cr}$ such that,

$$\Delta\sigma_{i\,n}^{cr} = h\Delta\varepsilon_{i\,n}^{cr} \qquad [9]$$

where h is usually a function of fracture energy G_f [WIL 86], [FEE 93], [ROT 88]. Combining Equation [9] with Equation [8] and Equation [7] renders the increment in crack stress,

$$\Delta\sigma_{i\,n}^{cr} = \left(\frac{Eh}{E+h}\right)\Delta\varepsilon_{i\,n} = \overline{C}\Delta\varepsilon_{i\,n} \qquad [10]$$

where $\overline{C}$ is the one-dimensional elasto-plastic tangent modulus (negative for softening) subject to $-E \leq h < 0$. The global increment in stress is given by,

$$\Delta\boldsymbol{\sigma}_n = \mathbf{N}\Delta\boldsymbol{\sigma}_n^{cr} \qquad [11]$$

The constant softening modulus defined in Equation [9] is introduced to demonstrate the modus operandi of the proposed smeared-crack model. The remainder of the paper will be concerned with the motivation and subsequent implementation of a rate dependent crack law proposed by Sluys [SLU 92a]. The incorporation of any such model in the formulation derived above amounts to a substitution of an appropriate equation for Equation [10].

2. Dynamic Smeared Cracking with Inviscid and Viscous Regularised Tensile Strain-Softening Continua

2.1. *Dynamic Manifestations of the Inviscid Strain-Softening Continuum*

The dynamic equation governing one-dimensional, longitudinal wave propagation is given by,

$$\overline{C}\frac{\partial^2 u}{\partial x^2} - \rho\frac{\partial^2 u}{\partial t^2} = 0 \qquad \textbf{[12]}$$

where $\overline{C}$ is the one-dimensional tangent modulus, negative for strain-softening.

Considering a constant tangent modulus, a positive value of the discriminant, $\Delta_{discr} = \overline{C}/\rho$ of the second-order differential equation, renders the equation type *hyperbolic*. Discontinuities in solution derivatives across some surface or hyper-surface are transported along the two characteristic curves at the characteristic velocities. A discontinuity in the first solution derivative (velocity) is termed a shock wave, while a discontinuity in the second solution derivative (acceleration) is termed an acceleration wave. The characteristic velocities are positive and real for a positive tangent modulus and given by [LUB 90],

$$v^{char} = dx/dt = \pm\sqrt{\overline{C}/\rho} \qquad [13]$$

A negative tangent modulus yields a negative discriminant and indicates an *elliptic* governing equation. As widely quoted, the characteristic velocities are imaginary, although this statement is somewhat redundant as elliptic systems possess no characteristic curves! At best, the imaginary characteristic velocities indicate a change in type of the governing differential equation. The harmonic solutions to the elliptic equation are considered smooth or *analytic* with the solution at a point equal to its Taylor series expansion about the point. Discontinuities in the solution or any of its derivatives are not permitted which is characteristic of physical systems in equilibrium.

The numerical manifestations of the elliptic character of the governing differential equation at incipient softening has been the subject of much research [SLU 92a], [BAZ 85], [NEE 88], [BEL 86], [BEL 87], [WU 84]. The resultant pathological discretisation dependence stems from an inability of the hyperbolic and the elliptic sub-domains to interact. Strain-softening is confined to a cross-sectional plane of zero width which results in singularities in strain, wave trapping phenomena [WU 84] and instability. There is no length scale introduced by the formulation to set the minimum width of the softening (localisation) zone. In numerical solutions this manifests itself as the localisation into the narrowest band that the discretisation can resolve. Consequently, the dissipated energy tends to zero with mesh refinement.

2.2. *Acceleration Waves and Localisation Diagnostics*

The spurious dynamic strain localisation in an inviscid strain-softening continuum can be viewed as the occurrence of a stationary discontinuity. This phenomenon can be elucidated by considering acceleration waves in hyperbolic systems [HAD 03], [HIL 61], [LUB 90], [LOR 90]. Consider an acceleration wave with unit normal $\mathbf{n}$ moving in three-dimensional Euclidian space $\mathfrak{S}^3$ at speed c. The acceleration $\mathbf{a}$ is discontinuous across the wave front while the velocity $\mathbf{v}$, the Cauchy stress $\boldsymbol{\sigma}$ and the mass density ρ are all continuous. The jump operator $[\bullet]$ denotes the difference between the values of the enclosed at function an infinitesimal distance on either side of the wave front. The symbol ∇ will denote the spatial gradient operator. The *Hadamard compatibility condition* [HAD 03] yields the following jump relationships,

$$c[\nabla \mathbf{v}] = -[\mathbf{a}] \otimes \mathbf{n} \qquad [14]$$

$$c[\nabla \boldsymbol{\sigma}] = -[\dot{\boldsymbol{\sigma}}] \otimes \mathbf{n} \qquad [15]$$

where ⊗ denotes the dyadic product. The balance of linear momentum is given by,

$$\nabla \boldsymbol{\sigma} : \mathbf{I} + \rho(\mathbf{b} - \mathbf{a}) = 0 \qquad [16]$$

where **b** is the body force per unit mass assumed continuous and **I** is the second-order identity tensor. The constitutive relationship is given by,

$$\dot{\boldsymbol{\sigma}} = \overline{\mathbf{C}} : \nabla \mathbf{v} \qquad [17]$$

where the rate of deformation $\dot{\boldsymbol{\varepsilon}}$ is equivalent to the symmetric part of the velocity gradient $\nabla \mathbf{v}$. $\overline{\mathbf{C}}$ is the fourth-order, inviscid material tangent modulus depending on state variables and assumed symmetric. Combining Equation [14], Equation [15], Equation [16] and Equation [17] renders the jump in acceleration [**a**] as the result of the following eigenvalue problem,

$$(\mathbf{Q} - \rho c^2 \mathbf{I})[\mathbf{a}] = 0 \qquad [18]$$

where **Q** is the second-order *acoustic tensor* given by,

$$\mathbf{Q} = \mathbf{n} \cdot \overline{\mathbf{C}} \cdot \mathbf{n} \qquad [19]$$

The acoustic tensor emerges as a diagnostic of discontinuous bifurcation [SOB 90]. For a symmetric tangent modulus, the acoustic tensor is symmetric and all its eigenvalues are real. The positive quadratic form of the symmetric acoustic tensor renders positive eigenvalues equal to ρc^2 and thus real wave speeds. Hyperbolicity of the governing dynamic equations is ensured by a positive determinant of the acoustic tensor, $det\mathbf{Q} > 0$. A zero or negative eigenvalue renders the governing dynamic equations elliptic and exhibits a non-positive determinant of the acoustic tensor, $det\mathbf{Q} \leq 0$. A zero eigenvalue indicates a stationary discontinuity [HIL 61] (localisation) with the discontinuity occurring in the spatial velocity gradient and not the acceleration. Negative eigenvalues equal to ρc^2 result in imaginary wave speeds.

The diagnostics fo diffuse failure, indicating instability and loss of uniqueness coincide for a symmetric tangent modulus [SOB 90]. Inviscid tensile softening materials violate the stability postulates of Drucker [DRU 59] and Hill [HIL 58], [WIL 94]. Stability can be formulated in terms of second-order work d^2W,

$$d^2W = \int_V \dot{\varepsilon}^T \dot{\sigma} dV > 0 \qquad [20]$$

The stability criteria exists as a lower bound condition for the occurrence of limit points and localisation bands and incorporates the constraints on the acoustic tensor for real positive wave speeds.

2.3. *Motivation for Viscous Regularisation*

Rate dependence is well established in the regularisation of hyperbolic systems [JEF 76]. Viscosity introduces a length scale which is the spatial equivalent of the relaxation time. The inclusion of rate dependence into constitutive relationships ensures hyperbolicity of the governing dynamic equations and extirpates the pathological discretisation dependence [LOR 90], [PRE 90], [SLU 92a,b,c], [SLU 92d], [NEE 88], subject to constraints on the time increment size [SIM 89], [OWE 82] and the loading frequency [SLU 92a].

Notwithstanding the numerical convenience of rate dependence, the inclusion can also be justified on physical grounds for fracture modeling. The concept of finite fracture time is integrally linked to the perception of fracture as a nucleation and growth process. Reconciliation of the fracture propagation speed with the theoretical Raleigh wave speed is only approached in the case of perfect cleavage planes [WAS 94]. Viscosity accounts for the time lag between initiation of the fracture process zone and the realisation of discrete fracture.

2.4. *One-Dimensional Tensile Strain-Softening with Viscous Regularisation*

The one-dimensional, linear viscoplasticity model considered was proposed by Sluys [SLU 92a] and exists as a special case of power law Perzyna viscoplasticity [PER 66], [OWE 80], [MIT 90]. A closed form solution is derived for the increments in stress $\Delta\sigma$ and viscoplastic strain $\Delta\varepsilon^{vp}$. The incremental total strain decomposition in one-dimension is given by,

$$\Delta\varepsilon = \Delta\varepsilon^e + \Delta\varepsilon^{vp} \qquad [21]$$

where $\Delta\varepsilon^e$ is the increment in elastic strain and $\Delta\varepsilon^{vp}$ is the increment in viscoplastic strain. The stress rate equation in incremental form is given by,

$$\Delta\sigma = E\left(\Delta\varepsilon - \Delta\varepsilon^{vp}\right) \qquad [22]$$

The viscoplastic flow rule is given by,

$$\dot{\varepsilon}^{vp} = \left(\sigma - \sigma^{y} - h\varepsilon^{vp}\right)/m \qquad [23]$$

where, σ is the one-dimensional stress, σ^{y} is the incipient tensile strength, $h = \partial\sigma/\partial\varepsilon^{vp}$ is the scalar viscoplastic softening modulus ($h < 0$) and m is the material viscosity. The integration of the viscoplastic strains is over a time interval Δt_n from time t_{n-1} to t_n such that the increment in viscoplastic strain is given by,

$$\Delta\varepsilon_n^{vp} = \left[(1-\theta)\dot{\varepsilon}_{n-1}^{vp} + \theta\dot{\varepsilon}_n^{vp}\right]\Delta t_n \qquad [24]$$

where $\theta = 0$ defines an explicit integration and $0 < \theta \leq 1$ defines an implicit integration. Consider the limited Taylor series expansion of the viscoplastic strain rate given by,

$$\dot{\varepsilon}_n^{vp} = \dot{\varepsilon}_{n-1}^{vp} + \frac{\partial\dot{\varepsilon}^{vp}}{\partial\sigma}\Big|_{n-1}\Delta\sigma_n + \frac{\partial\dot{\varepsilon}^{vp}}{\partial\varepsilon^{vp}}\Big|_{n-1}\Delta\varepsilon_n^{vp} \qquad [25]$$

Combining Equation [22], Equation [23], Equation [24] and Equation [25] yields the increment in viscoplastic strain given by,

$$\Delta\varepsilon_n^{vp} = \frac{E\Delta\varepsilon_n + (1/\theta)m\dot{\varepsilon}_{n-1}^{vp}}{E + h + m/(\theta\Delta t_n)} \qquad [26]$$

Substituting Equation [26] into Equation [22] yields the increment in stress,

$$\Delta\sigma_n = \overline{C}_n\left(\Delta\varepsilon_n - \frac{m\dot{\varepsilon}_{n-1}^{vp}\Delta t_n}{m + \theta\Delta t_n h}\right) \qquad [27]$$

where the tangent modulus $\overline{C}_n = \partial\sigma_n/\partial\varepsilon_n$ is given by,

$$\overline{\mathbf{C}}_n = E\left(\frac{m/\theta\Delta t_n + h}{m/\theta\Delta t_n + h + E}\right) \qquad [28]$$

The tangent modulus Equation [28] approaches the inviscid elasto-plastic tangent modulus (negative for strain-softening) as the viscosity of the material m tends to zero. Hyperbolicity of the governing dynamic equations is ensured by enforcing the positive definiteness of the tangent modulus, which results in the following time increment constraint [OWE 80], [OWE 82], [SIM 89], [MIT 90],

$$\theta \Delta t_n < \frac{m}{\|h\|} \qquad [29]$$

The inclusion of the above formulation into the rotating crack model amounts to a substitution of Equation [27] for Equation [10] while recognizing the equivalence of the viscoplastic strain ε^{vp} and the crack strain ε^{cr}. Numerical applications have been considered by Sluys [SLU 92a,b,c], Loret and Prevost [LOR 90], [PRE 90], and Needleman [NEE 88].

2.5. *The Localisation Length Scale*

The ability of waves to disperse is of paramount importance if an initial value problem is to remain well posed during strain localisation. Propagation of stress waves through localisation zones is only possible if the waves are able to transform with changes in the wave numbers k of the Fourier components. Dispersive waves exhibit phase velocities which are functions of the wave number [WHI 74],

$$c_{phase} = \omega / k = \mathrm{W}(k) / k \qquad [30]$$

where ω is the angular frequency equal to the real *dispersive relationship* W(k), such as that for the one-dimensional case $\partial^2 \mathrm{W}(k) / \partial k^2 \neq 0$ [WHI 74]. The velocity with which the entire group of Fourier components travels is called the group velocity and is given by,

$$c_{group} = \partial \mathrm{W}(k) / \partial k = \partial \omega / \partial k \qquad [31]$$

Energy travels with the group velocity [WHI 74]. The dispersive relationship for the one-dimensional viscous regularised strain-softening model is given by [SLU 92a],

$$\left(\rho m \omega^2 - mEk^2 \omega\right) i - \rho (E + h) \omega^2 + hEk^2 = 0 \qquad [32]$$

where the wave number is complex $k = k_r + \alpha i$. The dispersive displacement solution is given by,

$$\mathrm{u}(x,t) = \mathrm{A}\mathrm{e}^{-\alpha x}\mathrm{e}^{i(k_r x - \omega t)} \qquad [33]$$

where A is the maximum displacement amplitude, α is the damping parameter and k_r is the real part of the wave number [SLU 92a],

$$k_r^2 = \frac{\rho\omega^2}{2E}\left(\frac{\left(m^2\omega^2 + h^2 + Eh\right) + \left(\left(m^2\omega^2 + h^2 + Eh\right)^2 + (mE\omega)^2\right)^{1/2}}{m^2\omega^2 + h^2}\right) \qquad [34]$$

$$\alpha^2 = \frac{\rho\omega^2}{2E}\left(\frac{-\left(m^2\omega^2 + h^2 + Eh\right) + \left(\left(m^2\omega^2 + h^2 + Eh\right)^2 + (mE\omega)^2\right)^{1/2}}{m^2\omega^2 + h^2}\right) \qquad [35]$$

From Equation [34] and Equation [35] it is observed that for increasing frequency transient loading, the group and phase velocities tend to the linear elastic longitudinal wave velocity, while spurious velocities tending to infinity result for quasi-static frequencies. This demonstrates the breakdown of the viscous regularisation effect as the quasi-static limit is approached and is equivalent to the invicid material response.

The damping coefficient α governs the spatial exponential attenuation of the solution amplitude as the wave propagates in the rate dependent region and thus defines the implicit length scale ℓ.

$$\lim_{\omega \to \infty} \alpha = \ell^{-1} \equiv \left(\frac{2\mathrm{mc_L}}{\mathrm{E}}\right)^{-1} \qquad [36]$$

Rate dependence introduces a length scale into the initial value problem which is independent of the finite element discretisation. Once again, the regularisation effect is observed to be lost for low frequency waves resulting in arbitrary length scales.

3. The numerical Model

The proposed softening model has been implemented within an explicit dynamic solution algorithm that permits effective simulation of the post-fracture response of brittle materials [MUN 95]. Within the algorithm, discrete fractures are modelled by splitting the initial finite element discretisation, allowing fragmentation and subsequent particle flow. This is achieved by constructing a global damage map based on the nodal averaging of a Gauss point damage measure and subsequently inserting discrete fractures with local re-meshing when appropriate. The local damage measure is defined as the fraction of total failure and is given by,

$$F_i^e = \frac{2\sigma^y}{\sigma_{in}^y + \sigma^y} - 1 \qquad [37]$$

where σ_{in}^y is the current tensile strength and σ^y is the virgin tensile strength. The fracture indicators are averaged at the nodes and indicate discrete fracture along a radiating element side when greater than unity (corresponding to sufficient energy release to propagate a fracture, with a length corresponding to the element characteristic length). The fracture indicators also act as weighting factors for the averaging of the fracture directions at each node.

4. Numerical Examples

The fracture response of a system subject to transient loading frequently bears little resemblance to its statically loaded counterpart. Constructive stress wave interference, the interaction with boundaries, and the volumetric expansions associated with diverging spherical waves are responsible for the majority of dynamic fracture phenomena. Corner fractures and spalling are two of the most common types of fractures that arise from interference between stress waves. The following two examples will model both these dynamic manifestations and demonstrate the ability of the proposed methodology to yield discretisation objective results with respect to dissipated fracture energy and discrete fracture distributions.

4.1. *Corner Fractures in a Hollow, Internally Loaded, Square Cylinder*

Consider a thick-walled, square cylinder with an explosively loaded (circular) cylindrical bore [JOH 72], [RIN 63]. A trapezoidal loading pulse is adopted as an approximation to the confined explosive pulse. The system is modeled in plane strain utilising quarter symmetry. Figure 1 defines the system graphically.

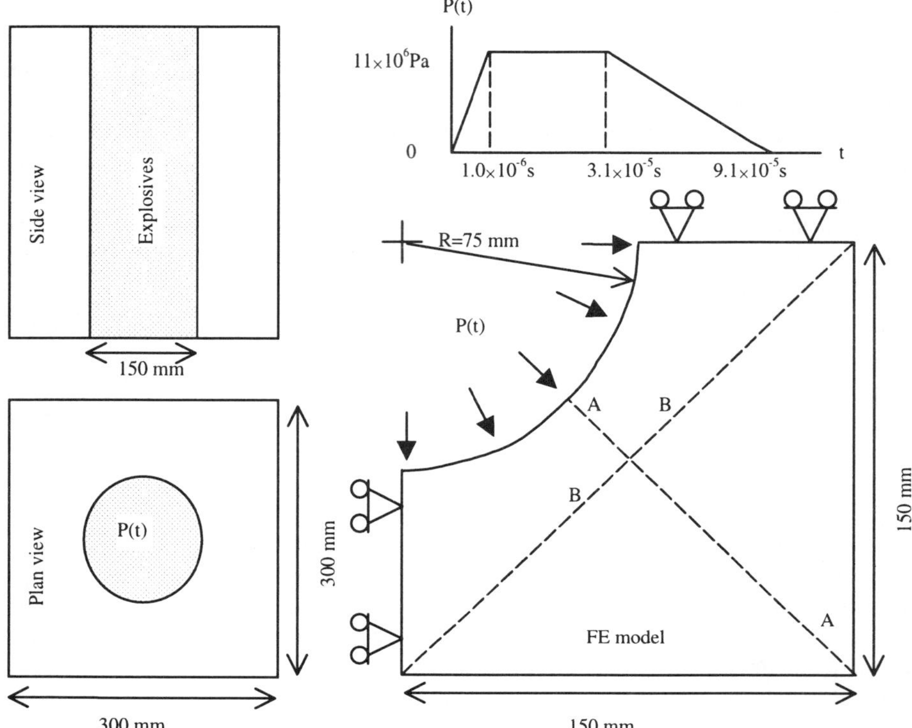

Figure 1. *Square cylinder with explosively loaded circular bore a) Side view b) Plan view c) Explosive loading pulse d) Finite element model*

The material considered is a quartzite rock with Young's modulus E=70×10^9Pa, mass density ρ = 2800kgm^{-3} and Poisson's ratio ν = 0.2. The dilatational (longitudinal) elastic wave speed is c_L = 5000ms^{-1}. The constant softening modulus is h = 150×10^9Pa, the tensile strength is σ^y = 6×10^6Pa and the viscosity parameter is m = 94000Nsm^{-2}. It should be noted that a reduced tensile strength is adopted to make allowance for viscous effects. The localisation length scale is given by (see Equation [36]),

$$\ell = \frac{2mc_L}{E} \cong 13mm \qquad [38]$$

Three uniform discretisations are considered, consisting of 1*mm*, 1.5*mm* and 2*mm* linear triangular elements respectively. Single point Gaussian quadrature is utilised with the timestep taken as 90% of the critical central difference timestep. Discrete fracturing is only permitted along element boundaries.

The posed problem is considered in many publications on dynamic fracture owing to its apparent simplicity in elucidating the evolution of corner fractures [JOH 72], [RIN 63]. The internal explosive loading imparts a diverging, cylindrical, compressive stress wave that propagates towards the unconstrained system boundaries. The reflected tensile waves from adjacent boundaries combine constructively along the diagonals (A-A in Figure 1*d*), which results in fracture if the tensile strength of the material is surpassed. Fractures are deemed to originate at the corners and to propagate along the diagonals. Figure 2 demonstrates the qualitative evolution of corner fractures graphically.

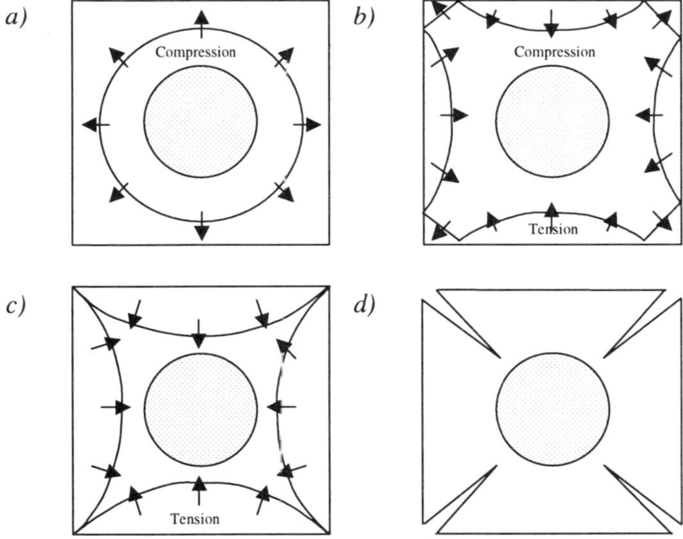

Figure 2. *a) Incident cylindrical compression front b) Tensile reflection from the unconstrained sides c) Tensile fronts combining along the diagonals d) Final discrete fracture distribution*

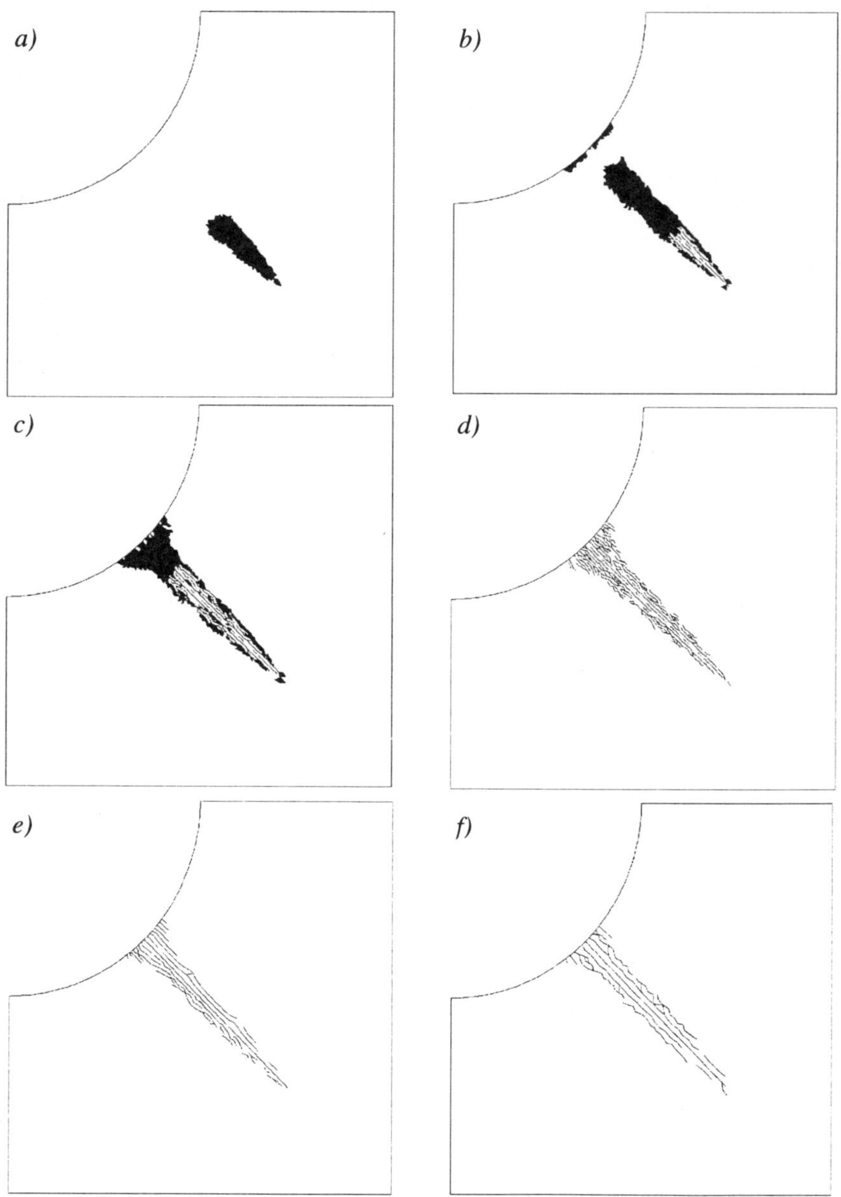

Figure 3. *a) b) c) Evolution of the fracture process zone for 1mm elements d) Discrete fracture distribution for 1mm elements, e) Discrete fracture distribution for 1.5mm elements, f) Discrete fracture distribution for 2mm elements*

The wave interaction and resultant fracture pattern discussed above and displayed in Figure 2 are based on a number of restrictive assumptions. It is assumed that the loading compression pulse is of zero width and experiences no attenuation of amplitude with divergence from the source. It is also assumed that the pulse is wholly reflected as a dilatational pulse from the unconstrained material boundary, irrespective of the angle of incidence. Physical loading pulses have a finite width and the resultant stress at any given time is the superposition of the incident and the reflected pulses. Amplitude attenuation is mandatory for a diverging cylindrical pulse, the amplitude decreasing inversely with the square root of the radius from the source. The attenuated compressive pulse will impinge obliquely on the unconstrained material boundary near the corners and will be reflected as both a distortional and dilatational components. The relative magnitudes of the reflected distortional and dilatational amplitudes exhibit surprising sensitivity to the angle of incidence and the Poisson's ratio of the material. It is observed that the incident dilatational pulse is almost wholly reflected as a distortional pulse near the corners. Physically, the combination of material parameters, loading and system geometry prevents tensile fracture commencing from the corners.

The numerical results are displayed in Figure 3, demonstrating the evolution of the fracture process zone and the resulting discrete fracture distributions for the different discretisations. The numerical results confirm that fractures are initiated some distance from the corners and propagate along the diagonals towards the center. The fracture process zone is observed to have a width set by the length scale (13*mm*) and is plotted in solid black in Figure 3. Discrete fracture is confined to the localisation zone.

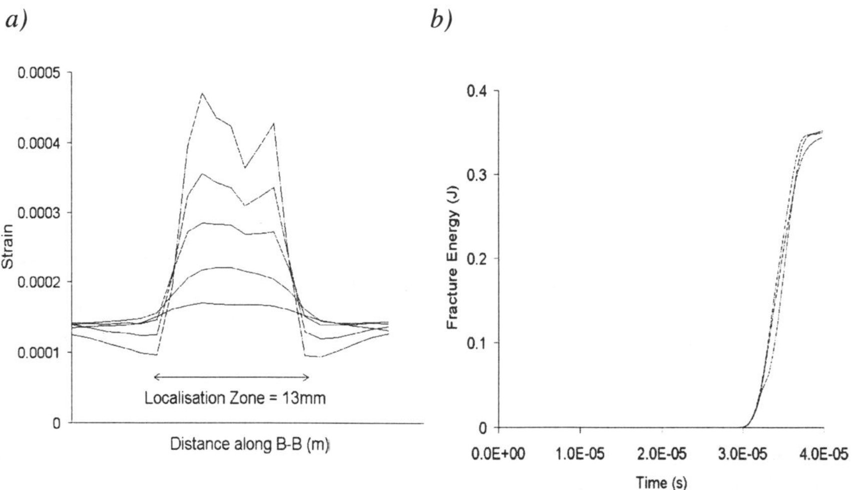

Figure 4. *a) Strain evolution in the localisation zone for 1mm elements b) The equivalence of fracture energy dissipation for all discretisations*

Figure 4 displays the evolution of strain in the localisation zone and verifies the equivalence of dissipated fracture energy for the different discretisations.

It is noted that fracture occurs through the thickest section of the square cylinder which is contrary to what is normally observed in quasi-static systems. Experimental results displaying fractures to the corners can usually be explained by the gas pressure from the bore forcing the initial fracture to open, thus propagating the fracture from the initiation point to the corner.

4.2. *Spalling in an Unsupported Rectangular Block*

Spalling or scabbing results from compressive dilatational waves reflecting off unconstrained material boundaries as tensile waves. This provides a mechanism for the proliferation of tensile fracture far removed from the origin of the incident compressive wave. Spalling and subsequent momentum entrapment may be one of the mechanisms for material ejection in mining excavations following impingement of seismic waves on the unrestrained boundaries. The theory of spalling and momentum entrapment is due to Hopkinson [HOP 21], [JOH 72], [RIN 63], [KOL 53], and is encompassed by the theory of the Hopkinson pressure bar. Fracture occurs when the sum of the incident and the reflected stress waves surpasses the tensile strength of the material. The critical factors governing the spall are thus the dynamic tensile strength and the shape of the incident pulse.

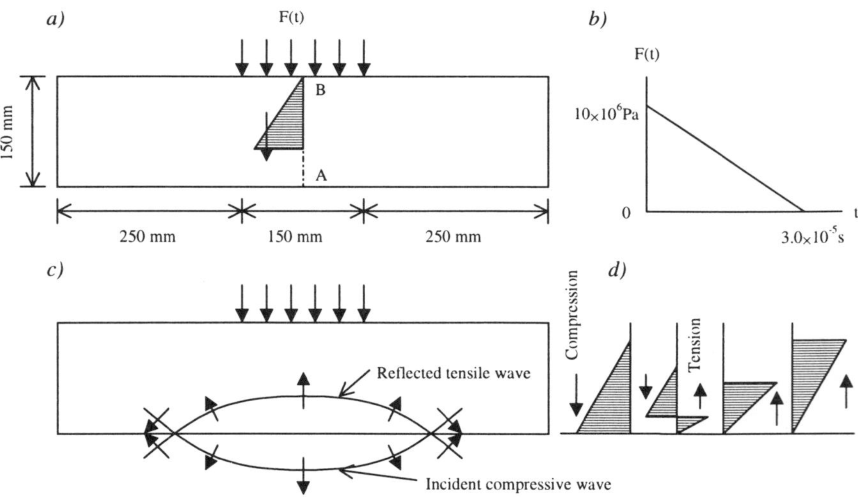

Figure 5. *a)Unsupported rectangular block subject to a distributed saw tooth loading pulse b) Saw tooth loading pulse c) Reflection of incident compression pulse from unconstrained bottom flange d) Net pulse shape during reflection*

The system considered is an unsupported rectangular block subject to a saw tooth pressure pulse applied over a finite width and is defined graphically in Figure 5. The distributed loading pulse generates a quasi-cylindrical compression wave of low curvature. The width of the model is chosen such that reflection from the block sides does not affect the spall formation .

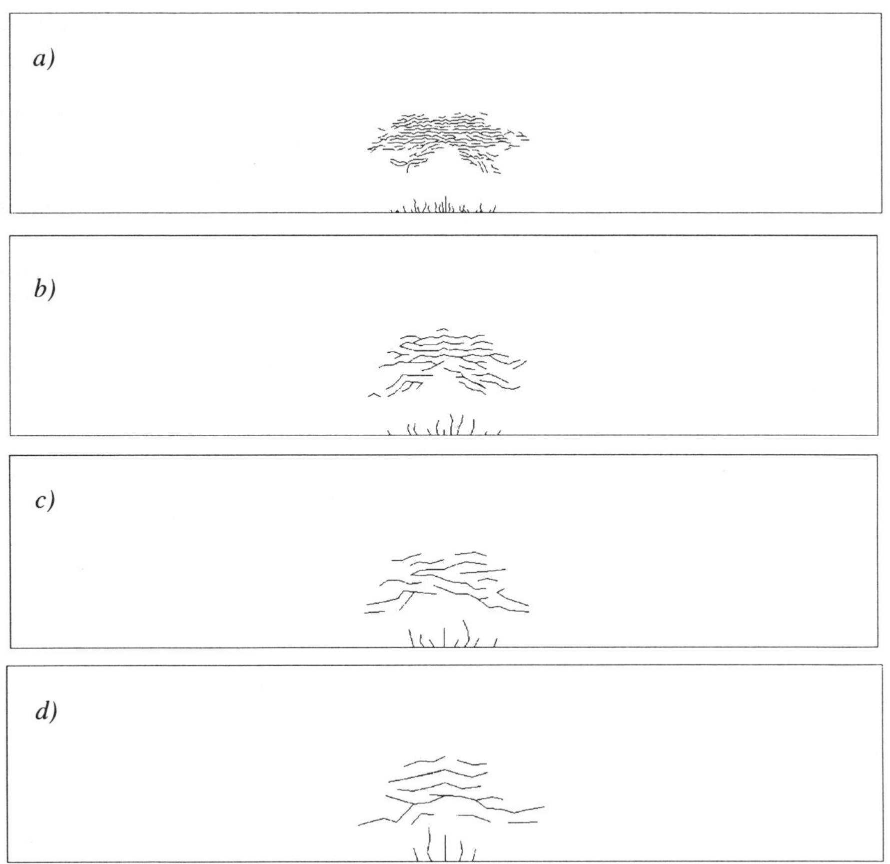

Figure 6. *Discrete fracture distributions for uniform discretisations of constant element size a) 2.5mm, b) 5mm, c) 7.5mm d) 10mm*

The material considered is a concrete with Young's modulus $E = 50\times10^9 Pa$, mass density $\rho = 2400 kgm^{-3}$ and Poisson's ratio $\nu = 0.2$. The dilatational (longitudinal) elastic wave speed is given by $c_L = 4564 ms^{-1}$ The constant softening modulus is taken as $h = 45\times10^9 Pa$, the tensile strength is $\sigma^y = 3.5\times10^6 Pa$ and the viscosity parameter is $m = 137\times10^3 Nsm^{-2}$. The length scale is given by,

$$\ell = \frac{2mc_L}{E} \cong 25mm \qquad [39]$$

Four uniform discretisations are considered, utilizing 2.5*mm*, 5*mm*, 7.5*mm* and 10*mm* linear triangular elements respectively. Single point Gaussian quadrature is utilized with the timestep taken as 90% of the critical central difference timestep. Discrete fracturing is only permitted along element boundaries.

The approximately horizontal incident pulse beneath the load is reflected from the unconstrained boundary as an essentially horizontal tensile pulse. Fracture appears at a finite distance from the unconstrained boundary once the tensile strength of the material has been surpassed. With increasing distance from the block center line, the oblique incidence of the compressive pulse renders dilatation and distortional reflected components. The reduced reflected dilatational tensile amplitude is insufficient to produce fracture. Figure 7 demonstrates the equivalence of the final discrete fracture distributions for the different discretisations. The fractures intersecting the unconstrained boundary perpendicularly can be defined as latent flexural fractures.

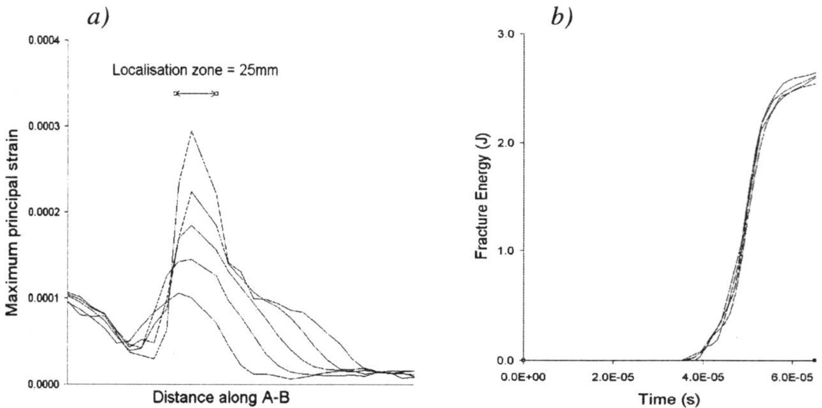

Figure 7. *a) Evolution of strain in the localisation zone b)Equivalence of dissipated fracture energy for all discretisations*

Figure 6*a* displays the evolution of strain along the center line A-B (see Figure 5) of the block for the uniform discretisation of 2.5*mm* elements. The strain is observed to localise into a zone of width set by the localisation length scale (25*mm*). The equivalence of dissipated fracture energy for the different discretisations is demonstrated in Figure 6*b*.

5. Conclusions

A combination of localisation and discrete element concepts has been proposed with localisation envisaged *a priori* to discrete fracture insertion. A smeared crack approach is adopted in the form of an orthogonal rotating-crack model utilising tensile strain-softening to model the material degradation. Rate dependence is introduced to ensure the dispersive character of the softening continuum and introduces a length scale that controls the width of the localisation zone. The numerical examples considered elucidate common manifestations of dynamic fracture and demonstrate the discretisation objectivity of the proposed methodology, with respect to the discrete fracture distribution and the dissipated fracture energy.

The regularisation effect of rate dependence is restricted to dynamic problems of high frequency and is subject to time increment constraints. It is noted, however, that the conditionally stable central difference integration scheme generally imposes more stringent constraints on the time increment than the viscous stability criteria. It is also apparent that the element size must be appropriate for the localisation zone resolution.

It is observed that a region which has completely failed will exhibit fractures along all element sides consistent with the prevailing fracture direction. Thus, within the localisation zone itself the discrete fracture spacing defaults to the element spacing and is thus arbitrary. For accurate physical modeling the need for a length scale governing crack spacing is acknowledged.

Further study includes the regularisation of the strain-softening continuum with gradient terms for the consideration of quasi-static analysis. Applications involving post failure interaction will also be considered.

6. References

[BAZ 83] BAZANT, Z.P., Comment on orthotropic models for concrete and geomaterials, *J. Engng. Mech.*, vol. 109, n° 3, p. 849-856, 1983.

[BAZ 85] BAZANT, Z.P., BELYTSCHKO, T.B., Wave propagation in a strain-softening bar: Exact solution, *J. Engng Mech. ASCE*, vol. 111, n° 3, p. 381-389,1985.

[BEL 86] BELYTSCHKO, T., BAZANT, Z.P., HYUN, Y., CHANG, T., Strain-softening materials and finite element solutions, *Computers & Struct.*, vol. 23, n° 2, p. 163-180, 1986.

[BEL 87] BELYTSCHKO, T., WANG, X., BAZANT, Z.P., HYUN, Y., Transient solutions for one-dimensional problems with strain softening, *J. Appl. Mech.*, vol. 54, p. 513-518, 1987.

[BHA 94] BHATTACHARJEE, S.S., LÉGER, P., Application of NLFM models to predict cracking in concrete gravity dams, *J. Struct. Engng.*, vol. 120, n° 4, p. 1255-1271, 1994.

[COP 80] COPE, R.J., RAO, P.V., CLARK, L.A., NORRIS, P., Modelling of reinforced concrete behaviour for finite element analysis of bridge slabs, Num. Meth. for nonlinear problems, Eds. Taylor et al., Pineridge Press Ltd., Swansea, vol. 1, p. 457-470, 1980.

[DEB 85] DE BORST, R., NAUTA, P., Non-orthogonal cracks in a smeared finite element model, *Engng. Comp.,* vol. 2, p. 35-46, 1985.

[DEB 86] DE BORST, R., Computational aspects of smeared crack analysis; Composite fracture model for strain softening and localised failure of concrete, Computational modeling of reinforced concrete structures, Eds. Hinton et al., Pineridge Press Ltd., p. 44-83, 1986.

[DRU 59] DRUCKER, D.C., A definition of stable inelastic material, *J. Appl. Mech. Trans.* ASME, March, p. 101-106, 1959

[DUC 72] DUCHON, N.B., Analysis of reinforced concrete membrane subjected to tension and shear, American Concrete Inst. J., p. 578-583, 1972.

[FEE 93] FEENSTRA, P.H., Computational aspects of biaxial stress in plain and reinforced concrete, Dissertation Delft University of Technology, Netherlands, 1993.

[FOS 96] FOSTER, S.J., BUDIONO, B., GILBERT, R.I., Rotating crack finite element model for reinforced concrete structures, *Computers & Struct.*, vol. 58, n° 1, p. 43-50, 1996.

[GAJ 91] GAJER, G., DUX, P.F., Simplified nonorthogonal crack model for concrete, *J. Struct. Engng.*, vol. 117, n° 1, p. 149-164, 1991.

[GOR 79] GORDON, J.E., *The new science of strong materials or Why things don't fall through the floor*, 2nd Ed., Penguin Books Ltd., 1979.

[GUP 84] GUPTA, A.K., AKBAR, H., Cracking in reinforced concrete analysis, *J. Struct. Engng.,* vol. 110, n° 8, p. 1735-1746, 1984.

[GUZ 95] GUZINA, B.B., RIZZI, E., WILLAM, K., PAK, R.Y.S., Failure prediction of smeared-crack formulations, *J. Engng. Mech.*, vol. 121, n° 1, p. 150-161, 1995.

[HAD 03] HADAMARD, J., *Leçons sur la propagation des ondes et les équations de l'hydrodynamique,* Hermann, Paris, 1903.

[HIL 58] HILL, R., A general theory of uniqueness and stability in elastic-plastic solids, *J. Mech. Phys. Solids*, vol. 6, p. 236-249, 1958.

[HIL 61] HILL, R., Acceleration waves in solids, *J. Mech. Phys. Solids*, vol. 10,p. 1-16, 1961.

[HOP 21] HOPKINSON, B., Collected Scientific papers, Cambridge Univ. Press, 1921.

[JEF 76] JEFFREY, A., *Quasilinear hyperbolic systems and waves,* Pitman Publishing Ltd., London, 1976.

[JOH 72] JOHNSON, W., *Impact strength of materials*, Edward Arnold (Publishers) Ltd., London, 1972.

[KOL 53] KOLSKY, H., *Stress waves in solids*, Clarendon Press, Oxford, 1953.

[LOR 90] LORET, B., PREVOST, J.H.,Dynamic strain localisation in elasto-(visco-)plastic solids Part 1. General formulation and one-dimensional examples, *Computer Meth. Appl. Mech. Engng.*, n° 83, p. 247-273, 1990.

[LUB 90] LUBLINER, J., *Plasticity theory*, Macmillan Publishing Co., New York, 1990.

[MIL 85] MILFORD, R.V., SCHNOBRICH, W.C., Application of the rotating crack models to R/C shells, *Computers and Struct.*, vol 20, p. 225-239, 1985.

[MIT 90] MITCHELL, G.P., Topics in the numerical analysis of inelastic solids, Dissertation University of Wales, Swansea, 1990.

[MUN 95] MUNJIZA, A., OWEN, D.R.J., BIĆANIĆ, N., A combined finite/discrete element method in transient dynamics of fracturing solids, *Engng. Computations*, vol. 12, n° 2, p. 145-174, 1995.

[NEE 88] NEEDLEMAN, A., Material rate dependence and mesh sensitivity in localisation, problems, *Computer Meth. Appl. Mech. Engng.*, n° 67, p. 69-85, 1988.

[OWE 80] OWEN, D.R.J., HINTON, E., *Finite elements in plasticity : theory and practice*, Pineridge Press Ltd., Swansea, UK, 1980.

[OWE 82] OWEN, D.R.J., DAMJANIĆ, F., Viscoplastic analysis of solids: stability considerations, Recent advances in non-linear computational mechanics, Eds. Hinton et al., Pineridge Press Ltd., Swansea, UK, p. 225-253, 1982.

[PER 66] PERZYNA, P., Fundamental problems in viscoplasticity, Advances in *Appl. Mech.*, n° 9, p. 243-377, 1966.

[PET 96] PETRINIĆ, N., Aspects of discrete element modelling involving facet-to-facet contact detection and interaction, Dissertation University of Wales, Swansea, 1996.

[PRE 90] PREVOST, J.H., LORET, B,Dynamic strain localisation in elasto-(visco-)plastic solids Part 2. Plane strain examples, *Computer Meth. Appl. Mech. Engng.*, n° 83, p. 275-294, 1990.

[REA 84] READ, H.E., HEGEMIER, G.A., Strain softening of rock, soil and concrete – A review article, *Mech. Mat.,* vol. 3, p. 271-294, 1984.

[RIN 63] RINEHART, J.S., PEARSON, J., *Explosive working of metals*, Pergamon Press Ltd., 1963.

[ROT 87] ROTS, J.G., DE BORST, R., Analysis of mixed-mode fracture in concrete, *J. Engng. Mech.,* vol. 113, n° 11, p. 1739-1758, 1987.

[ROT 88] ROTS, J.G., Computational modeling of concrete fracture, Dissertation Delft University of Technology, 1988.

[ROT 89] ROTS, J.G., BLAAUWENDRAAD, J., Crack models for concrete: discrete or smeared? Fixed, multi-directional or rotating? Heron, 34(1), 1989.

[SIM 89] SIMO, J.C., Strain softening and dissipation: A unification of approaches, Cracking and damage: Strain localization and size effect, Eds. Mazars et al., Elsevier Applied Science, London, p. 440-461, 1989.

[SLU 92a] SLUYS, L.J., Wave propagation, localisation and dispersion in softening solids, Dissertation Delft Inst. Tech., 1992.

[SLU 92b] SLUYS, L.J., DE BORST, R., Wave propagation and localisation in a rate-dependent cracked medium – model formulation and one-dimensional examples, *Int. J. Solids Struct.*, vol. 29, n° 23, p. 2945-2958, 1992.

[SLU 92c] SLUYS, L.J., DE BORST, R., Mesh sensitivity analysis of an impact test on a double-notched specimen, Rock Mechanics, Eds. Tillerson et al., Balkema, Rotterdam, p. 707-716, 1992.

[SLU 92d] SLUYS, L.J., BOLCK, J., DE BORST, R., Wave propagation and localisation in viscoplastic media, Proc. Complas. Int. Conf. (3rd: 1992), vol. 1, p. 539-550, 1992.

[SLU 96] SLUYS, L.J., DE BORST, R., Failure in plain and reinforced concrete - An analysis of crack width and crack spacing, *Int. J. Solids Struct.,* vol. 33, n° 20, p. 3257-3276, 1996.

[SOB 90] SOBH, N., SABBAN, S., STURE, S., WILLAM, K., Failure diagnostics of elasto-plastic operators, *Proc. Int. Conf. Numer. Meth. Engng. NUMETA 90*, Swansea, Eds. Pande et al., Elsevier Science Publishers Ltd., UK, vol. 2, p. 755-768, 1990.

[WAS 94] WASHABAUGH, P.D., KNAUSS, W.G., A reconciliation of dynamic crack velocity and Raleigh wave speed in isotropic brittle solids, *Int. J. Fract.,* n° 65, p. 97-114, 1994.

[WHI 74] WHITHAM, G.B., *Linear and nonlinear waves*, John Wiley & Sons Inc., US, 1974.

[WIL 86] WILLAM, K., BIĆANIĆ, N., STURE, S., Composite fracture model for strain softening and localised failure of concrete, Computational modelling of reinforced concrete structures, Eds. Hinton, Owen, Pineridge Press Ltd., Swansea, p. 122-153, 1986.

[WIL 89] WILLAM, K., PRAMONO, E., STURE, S., Fundamental issues of smeared crack models, Fracture of concrete and rock, Eds. Shah, Swartz, Springer-Verlag, NY, p. 192-207, 1989.

[WIL 94] WILLAM, K., MÜNZ, T., Failure conditions and localisation in concrete, Proc. EURO-C 1994 Int. Conf., vol. 1, p. 263-282, 1994.

[WU 84] WU, F.H., FREUND, L.B., Deformation trapping due to thermoplastic instability in one-dimensional wave propagation, *J. Mech. Phys. Solids*, vol. 32, n° 2, p. 119-132, 1984.

Restrained Cracking in Reinforced Concrete

Zdenek Bittnar — Petr Rericha

Czech Technical University
Faculty of Civil Engineering
Thakurova 7, 160 00 Prague 6
Czech Republic

ABSTRACT. *An analytical FEM model is used to simulate the crack development and coalescence in the vicinity of a reinforcement bar. The knowledge regained from the simulation should contribute to a prospective tension stiffening rule. A centrally reinforced cylinder is analyzed in axisymmetric geometry. The concrete model is a smeared crack elastic-plastic material with isotropic or directional strain softening. However, the crack coalescence cannot be assessed with sufficient accuracy by the local material model. A non-local material is being incorporated in the code. An attempt is made to establish a relation between the non-local continuum model and fracture mechanics parameters.*

KEY WORDS: *Reinforced Concrete, Stress, Strain, Degradation, Crack, Tension Stiffening, Strain Softening, Non-Local Material Model, Crack Pattern.*

Introduction

It is widely acknowledged that advances in the academic research of concrete do not adequately reflect in concrete design practice [HS90],[HM95]. Authors have pursued for several years the research in practice oriented material models for reinforced concrete. Recently, tension and cracking of this material received particular attention in the wake of advances in concrete fracture. Interaction of concrete and reinforcement bars is currently assessed using the tension stiffening concept. Several semi-empiric formulas were devised [BR66],[BH86],[CS78],[FM82], [GM90],[NO86]. The deficiency of the tension stiffening concept is well known. Replacement of this concept by some objective equivalent is of crucial importance for an effective and accurate analysis of reinforced concrete design.

There seems to be a single viable path to it i.e., to assume a cracking energy density distribution and crack pattern around reinforcement bars which are reasonably independent of reinforcement arrangement and boundary conditions. It is not easy to explore them either experimentally or analytically. Available experimental evidence is rather indirect (surface crack width and spacing). Direct observation of the crack development in the reinforced concrete body will hardly ever be possible. On the other hand, advanced material models of concrete allow the analytic simulations of the phenomenon. This is the prime issue of the present paper.

The first part features the material model of concrete. It is pointed out that nonlocal material model is necessary for this purpose. The weakest point of nonlocal models has been the calibration of their paramater values. In the second part, the crack development in an ideal continuum is explored. The objective is to establish the relations between the material constants of fracture mechanics and nonlocal continuum models. In the third part, FEM simulations of a particular problem are presented.

1. Nonlocal material model

It is known that reinforcement bars restrain the strain localization and crack development. In these conditions the strain localization zones interact with each other which makes the nonlocal material model indispensable. This phenomenon cannot be assessed by fracture mechanics models which implicitly assume an unconstrained crack development. Microcracks interactions, sorted by Bažant [BA94] are:

- Short-range – captured by the material model with characteristic length equal to the size of the local normalizing volume. This is a true material property.
- Long-range – captured by the FEM model provided a sufficiently fine element mesh is used. Bažant [BA94] incorporates even this long range interaction in the material model which, of course, makes the FEM solutions more robust. Nevertheless, it is not a true material property.

The present material model captures just the short range interactions. Two material constants are necessary as a minimum to define such a model (beyond the conventional local models constants like Young modulus and strength):

- g – cracking energy density which equals the area below the stress/strain diagram of the material.
- l – characteristic length which determines the short range interactions volume. This volume is the averaging volume of the nonlocal model.

Note that the above features are independent of the type of constitutive equations (plasticity, damage and others). Also, material constants of other local-

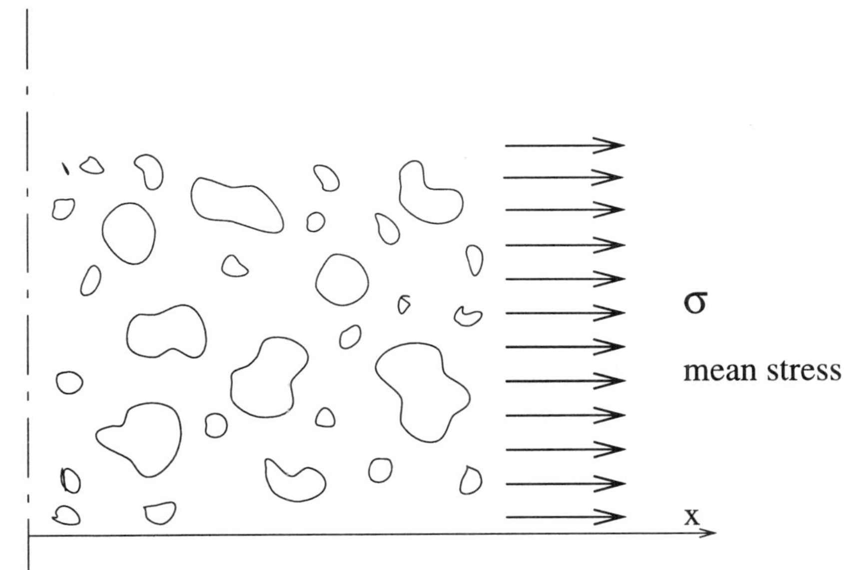

Figure 1: Uniaxial average stress state, coordinate x

ization limiters could be adjusted to model a physically equivalent material. These issues are beyond the scope of this paper.

Two material constants g and l define the strain localization behavior of the continuum model. They inevitably must have their counterparts in fracture mechanics since both material descriptions assess the same phenomenon. The unrestrained continuum is the medium where the relations between them can be established. Note that pure material properties are being explored and, consequently, homogeneous stress states are suitable for the purpose.

1. Analysis of crack development in continuum

A uniaxial tension stress state is considered with strain localization (arising macro crack) in the symmetry plane, see Fig. 1. It is obvious that the continuum material model is defined in terms of mean stresses and strains of the randomly inhomogeneous concrete. For definiteness, flow theory of plasticity is used with linear softening as shown in Fig. 2. The nonlocal degradation is governed by the nonlocal plastic strain $\bar{\varepsilon}^p(x)$

$$\bar{\varepsilon}^p(x) = \int_{-\infty}^{\infty} w(\zeta - x)\varepsilon^p(\zeta)\mathrm{d}\zeta \tag{1}$$

with linear weight function $w(x)$ as in Fig. 3. Stress and strain obviously

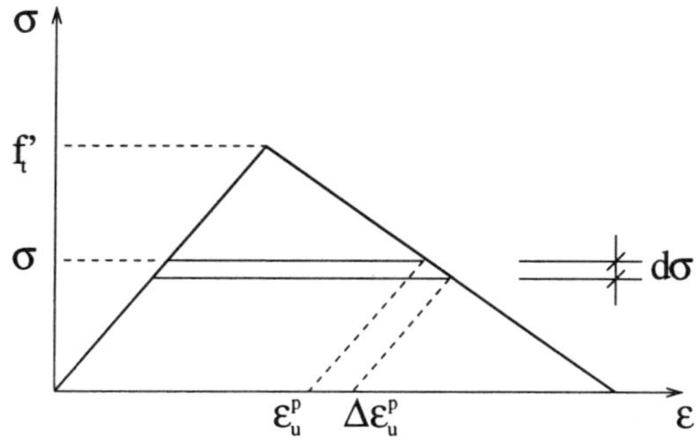

Figure 2: Elastic plastic material (flow theory) with linear softening

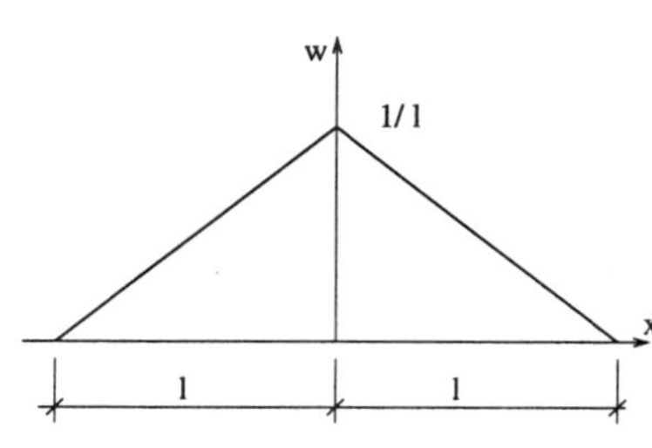

Figure 3: Weight function in unbounded continuum

vary with x only. Moreover, stress is constant everywhere by virtue of the elementary equilibrium. Displacement controlled loading is considered. The imposed displacement will be specified later on. The principle of virtual work applied to the work done by local and nonlocal strains yields the condition on the weight function:

$$\int_{-\infty}^{\infty} w(x)\mathrm{d}x = 1 \tag{2}$$

This condition may be used to modify the weight function near the boundary. The diagram in Fig. 3 may be clipped in its place in Fig. 1 and scaled up to meet equation 2. Analogous procedure is applicable in general biaxial and triaxial stress states too.

By virtue of the equilibrium bifurcation inherent to strain softening materials, infinitely many plastic strain distributions $\varepsilon^p(x)$ are possible for any intermediate stress state $\sigma(x) = \sigma$. Values of $\varepsilon^p(x)$ may fall anywhere between $0 < \varepsilon^p(x) < \varepsilon_u^p$, see Fig.2. One of these hypothetical strain distributions is shown in Fig.4. This is true for local and nonlocal strains as well. In spite of this ambivalence, unique strain distribution is obtained in experiments and numerical simulations. In order to derive it, a differential loading is considered from the state described in Figs 2 and 4. It is actuated by means of an imposed displacement du at the boundary of the elastic part $x = l_d$. The elongation is accompanied by a stress drop.

Consider first a local degradation material model. The stress decrease dσ must occur by elastic unloading (i.e., by decrease of ε) in all points of the volume V_u. The plastic strain may only increase in the volume V_l. It follows that the loading volume V_l may not grow throughout the loading history. When an infinitesimal initial V_l is assumed, the loading volume remains infinitesimal and plastic strain is totally localized

$$\varepsilon^p(x) = u \;\; \delta(x) \tag{3}$$

where $\delta(x)$ is the Dirac function. The asumption of an infinitely small initial volume V_l is consistent with the assumption of small random variations of the tension strength f_t'. Without loss of generality, the initial V_l is assumed to be located at $x = 0$. In a local material, localized strain implies total localization of degradation too. This distribution of the plastic strain does not allow for intermediate stress states at all; upon infinitesimal loading past f_t' the stress immediately drops to zero.

The degradation evolution in a nonlocal material can be analyzed in a similar manner. The hypothetical distribution of the plastic strain in Fig. 4 applies now to $\bar{\varepsilon}^p$ instead of ε^p. The averaged nonlocal plastic strain increment $\Delta\bar{\varepsilon}^p$ assumes the shape shown in Fig. 5. The total nonlocal plastic strain $\bar{\varepsilon}^p + \Delta\bar{\varepsilon}^p$ then exhibits a distribution with a maximum at $x = 0$, see Fig. 6. In the next loading increment the loading volume shrinks to a vanishing volume at $x = 0$ and any further local plastic increment is totally localized at this point. It may be concluded that for any initial distribution of ε^p, it becomes totally localized for any further loading. When the actual initial ε^p is localized at $x = 0$ it remains to be localized throughout the loading history, $\varepsilon^p = u\delta(x)$. The nonlocal plastic strain then is distributed according to the weight function:

$$\bar{\varepsilon}^p(x) = \begin{array}{ll} u\ w(x) & \text{for}\ \ \frac{u}{l} < \varepsilon_m \\ \varepsilon_m\ l\ w(x) & \text{for}\ \ \frac{u}{l} \geq \varepsilon_m \end{array} \tag{4}$$

Intermediate stress states exist and can thus be approximated by discretized solutions. The degradation zone remains distributed and limited to the characteristic length l. Physical meaning of the degragation zone implies that the characteristic length l of the nonlocal continuum model equals the crack band b_c of the fracture mechanics. Further, the energy dissipated in the process can be evaluated. Its density e in an elastic plastic material is a function of $\bar{\varepsilon}^p$

$$e(x) = \begin{array}{ll} g\frac{\bar{\varepsilon}^p(x)}{\varepsilon_m} & \text{for}\ \ \bar{\varepsilon}^p(x) < \varepsilon_m \\ g & \text{for}\ \ \bar{\varepsilon}^p(x) \geq \varepsilon_m \end{array}, \quad g = \frac{1}{2} f_t' \varepsilon_m \tag{5}$$

Substitution of Equation 4 and integration along x direction yields the expression for the energy E dissipated per unit area of the arising crack when $u \to \infty$:

$$E = \int_0^l e\ \mathrm{d}x = \int_0^l g\ l\ w(x)\ \mathrm{d}x = \frac{1}{2}\ l\ g \tag{6}$$

Since the fracture energy equals $2E$, a simple relation is obtained

$$G_f = l\ g = \frac{1}{2}\ l\ f_t'\ \varepsilon_m \tag{7}$$

Analogous argument leads to the same relations when deformation theory of plasticity or damage theory are adopted instead of the flow theory of plasticity used here. The crucial point of the deduction – elastic unloading in all points of volume V_u – remains valid. The validity can also be extended to other shapes of the softening branch in the stress strain diagram when proper

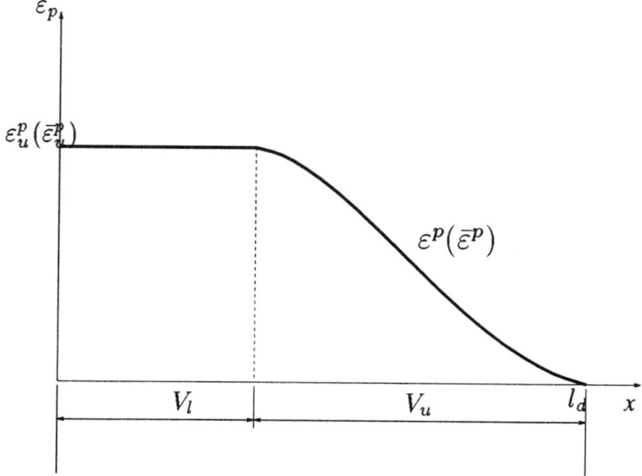

Figure 4: Hypothetical local plastic strain distribution

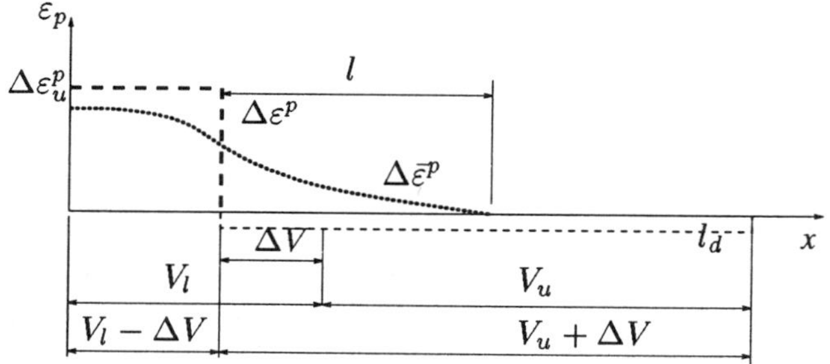

Figure 5: Increment of the local and nonlocal strains, decrease of the loading volume V_l

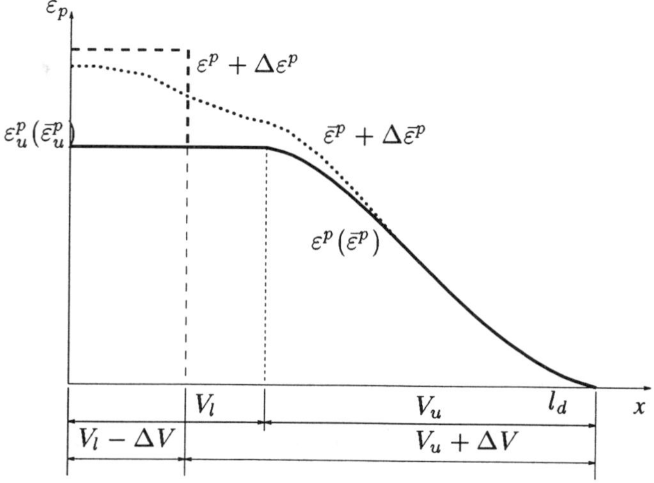

Figure 6: Total local and nonlocal strains after an increment, distribution of $\bar{\varepsilon}^p$ with maximum at $x = 0$

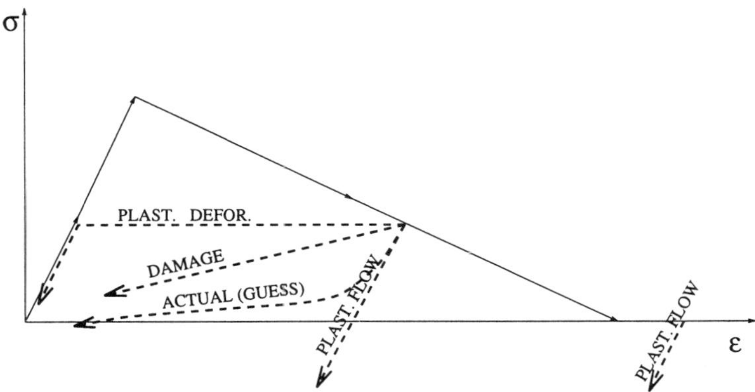

Figure 7: Unloading in plasticity and damage theories

dissipation energy density g is substituted. The conclusion of the section can be summarized in a chart:

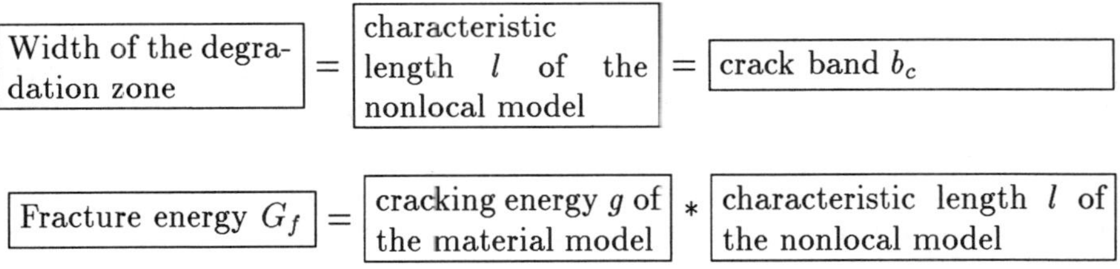

3. Material model in FEM simulations

The purpose of this section is to demonstrate the ability of the material model and its finite element implementation to solve actual boundary value problems and reflect the micro and macro crack interactions and coalescence. Graphics output is not yet available for solutions with nonlocal material model and the diagrams presented show the crack development in specimens with local models.

The local material model features an elastic plastic law, either flow or deformation theory. It is important to realize that quite general loading paths, including total unloading, take place in the major volume of the sample after local strain softening has occurred. Thus the two variants make a difference. The deformation theory requires the plastic strain to vanish before the stress starts to decrease. This may approximate closing of the cracks if the current strength and stress are very small. The loading path is shown in Fig. . It is suspected, however, that in the initial phase of the unloading the stress decreases according to the flow theory prediction and only after the stress has dropped to zero, the cracks start to close as indicated by the tentative curve 'actual' in Fig. . On the other hand, the flow theory fails when the stress reverses to negative (compression stresss occurs at positive strain) and thus in not suitable for general loading paths. In this particular problem there are never substantial compressive stresses in the axial direction and flow theory seems to be acceptable in this respect. The permanent plastic tension strain, however, obscures the strain field and its effect on the stress cannot be entirely neglected in complicated stress/strain distributions which occur in this problem. The issue is crucial, since degradation, localization and crack coalescence largely depend on the unloading behavior of the material. An automatic switch between the two theories or to damage-type constitutive law is being considered for future amendments. A similar approach was already used by Jirásek and Zimmermann [JZ97].

With respect to general loading paths, the issue of directional softening is important too. The up-to-now maximum tensile plastic strain tensor ε^p_{upt} tensor is stored and used to control tension softening. In each step, normal components of ε^p_{upt} are evaluated in the coordinate system that coincides with the principal directions of the current effective strain ε_{eff}. Strain ε_{eff} is defined as the strain that delivers the current stress $\boldsymbol{\sigma}$ when it is formally substituted into the rules of the deformation theory of plasticity. It obviously equals the current total strain ε in case the deformation theory of plasticity is used or it equals the difference $\varepsilon_{eff} = \varepsilon - \varepsilon^p$ in case the flow theory is adopted. Strain ε^p is the currrent plastic strain. The normal components of ε^p_{upt} are treated as its principal strains in the tension softening evaluation and updated by comparison with plastic parts of the principal components of ε_{eff}. This is not consistent and may be acceptable only when the the principal directions of ε_{eff} do not rotate much in each step. Stress $\boldsymbol{\sigma}$ is not stored at all. Two strain tensors, ε^p and ε^p_{upt}, are stored instead.

In comparison to the above issues, the selection of the yield surface (loading surface) seems to be of secondary importance. Rankine yield condition is used in the presented sample solutions. Its application is greatly facilitated by the fact that plastic strains are evaluated and tensor ε^p_{upt} is updated, in principal axes of tensors ε_{eff} and $\boldsymbol{\sigma}$.

Material parameters in the FEM simulations were determined in the following way. Fracture energy $G_f = 40$ N/m was determined from independent tests by the Hillerborg method. Tension strength $f'_t = 3.6$ MPa was measured directly by independent tension tests. The actual local material parameter $\varepsilon_m = 0.002$ was obtained from the maximum aggregate size 4 mm assuming $l = 12$ mm and using equation 7. This is possible owing to the conclusions of the previous section.

The computer code allows for the solution of two-dimensional problems, axisymmetric, plane strain and plain stress. Extension of a centrally reinforced cylinder has been simulated. Axial symmetry and plane symmetry are used in the analytical FEM model, see Fig.8. Because of the preliminary nature of

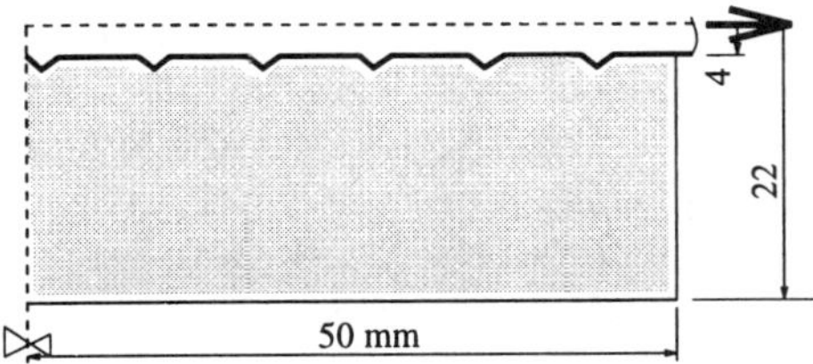

Figure 8: The specimen simulated, symmetry axis and plane

the present solutions, the reinforcement bar is not actually modelled by finite elements and loaded at the end as is the case in actual experiments, but a uniform axial strain is imposed on the bar/concrete boundary instead. This is equivalent to the limit case when the stiffness of the bar material grows to infinity and perfect bond exists. The bond/slip issue is still open on this scale. Whereas numerous semi-empiric formulas exist on the global force/slip basis, local bond/slip relations necessary for this type of analysis are lacking. The authors believe that the slip on this scale may be assessed by concrete failure in the adjacent concrete layer. Apparently, the presence of ribs prevents the free slip dislocations of the two surfaces by kinematic reasons and the error introduced by a perfect bond assumtion remains limited. There is, of course, the question whether the global slip can be assessed by the actual concrete material model and element mesh. As far as the present analyses are concerned, in the early phase of the crack development, not shown here, the degradation concentrates around the bar in a layer and the bar macroscopically is pull out of the concrete cylinder. Later on, the cracks rotate gradually to directions normal to the bar and coalesce into the principal cracks clearly indentifiable in the figures. This development suggests that the perfect bond assumption could be admissible. A number of aspects are yet to be clarified before the final conclusion is drawn on this issue.

4. FEM solutions

In the results presented here, the effect of various local material models is explored on the final crack pattern. The gradual crack development cannot be shown because of limited space. Final crack patterns are presented for the average strain $\varepsilon = 0.0024$ when the crack pattern does not change appreciably. Needless to say, the scaled plastic tension strain, not the actual crack width is shown in Figures 9 to 11. Nevertheless, its localization and direction is believed to represent the actual crack pattern.

In Fig. 9, the resulting crack pattern is shown for flow theory and directional softening. This material model should best approximate actual softening behavior of concrete in these conditions. The crack arresting effect of the reinforcement bar prevents localization in spite of the local material model being adopted. It is interesting to observe the difference when deformation theory of plasticity is used, as shown in Fig. 10. Numerous cracks have closed in the deformation theory solution. As expected, the unloading behavior affects the crack pattern considerably. Unfortunately, there is little experimental evidence on unloading of concrete in tension. Direct tests are impossible because of strain localization. Analytic simulation of tests with restrained strain localization, similar to the present one, can contribute to the topic.

The difference between isotropic and directional softening does not seem to affect the resulting crack pattern considerably as the comparison of Figs 10 and 11 shows. In this context, it is important to remark that very small or no degradation occurred in all sample solutions in the circumferential direction due to the hoop strain alone. In fact, it is the necessary condition for the axisymmetric model to be valid at all. If there were radial cracks and corresponding material degradation, their localisation would violate the axial symmetry. Nevertheless, in the isotropic degradation model the hoop stress is relaxed due to the degradation too, and the overall behavior should be softer. This is not observed in the results. This might be due to strong localization of the degradation in the two prinicpal cracks (the tension stresses vanish here entirely) whereas in the rest of the volume the hoop stress exists.

A rather unusal method was adopted for the solution of nonlinear equilibrium equations. Instead of the more or less standard loading path finding algorithms (arc-length and others), the explicit transient dynamic solution is used with the minimization of the inertia forces. The optimum load time history for this purpose was derived in [Ř86]. The solution features finite equilibrium accuracy with respect to the perfect quasistatic solution (some inertia forces are present) but provides a continuous history of the loading process instead. At discrete points, the iteration to exact equilibrium is possible via dynamic relaxation. Iterations are not performed in the basic run and the solver is thus robust and fast. It is worth mentioning that very small increments of strains and stresses are inherent to the method which facilitates material law implementation (recall the requirement of small step-to-step rotations of the principal axes of tensors ε_{eff} and $\boldsymbol{\sigma}$ in the previous section). In order to assess the effect of the 'parasitic' inertia forces in this particular problem, the solution in Fig.9 is recomputed with effectively doubled inertia forces. This is simply achieved by halving the time interval available to the solver to reach the same load level (average extension). The result in Fig. compares well with the reference solution in Fig. 9 except for the region at the intersection of the symmetry axis and symmetry plane. For the final purpose – the research in the cracking energy distribution around a reinforcement bar – the method appears to be sufficient.

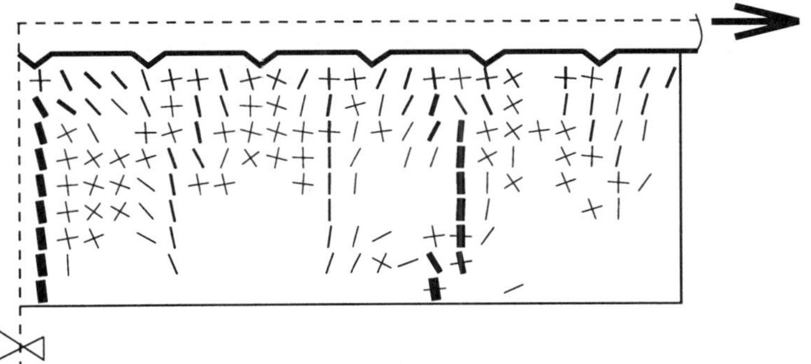

Figure 9: Flow theory, directional softening

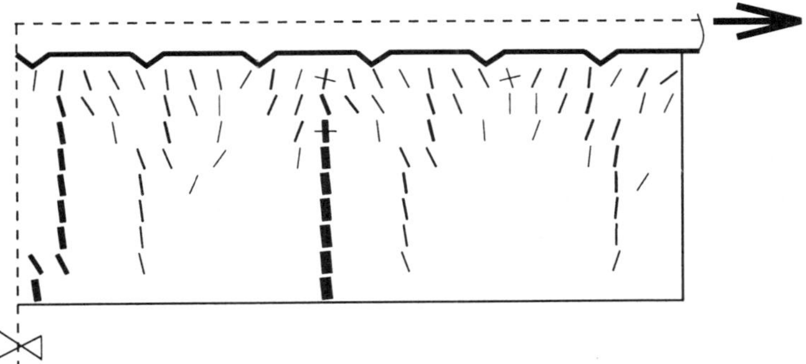

Figure 10: Deformation theory, directional softening

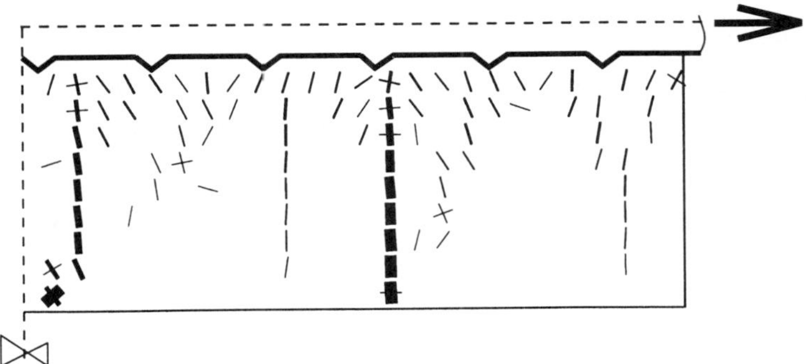

Figure 11: Deformation theory, isotropic softening

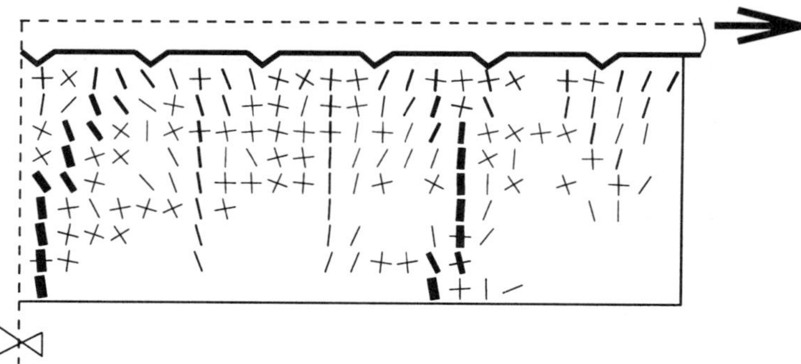

Figure 12: Flow theory, directional softening, lower presicion solution with about doubled inertia forces as compared to Fig. 9

5. Conclusions

1. Relations are derived between the two basic material constants of a non-local continuum and the two constants of the fracture mechanics. The relations hold for elastic plastic and damage type constitutive equations.

2. Local material model FEM simulations of the extension of a reinforced concrete cylinder exhibit considerable effect of the unloading concept on the final crack pattern. Classic plasticity theory options do not seem to describe the closing of cracks adequately and refinement is desirable.

3. The directional and isotropic degradations deliver similar crack patterns. For other specimen dimensions, the effect of this option is yet to be explored.

4. The quasistatic crack development can be analyzed with sufficient accuracy by the explicit dynamic simulation with minimized inertia forces.

6. References

[B94] Z.P. Bažant. Nonlocal damage theory based on micromechanics of crack interactions. *J. Engng Mechanics ASCE*, 120(3):593–617, 1994.

[BH86] S.B. Bhide. *Reinforced concrete elements in shear and tension.* Phd thesis, Univ. of Toronto, 1986.

[BR66] G.D. Base, J.B. Read, A.W. Beeby, and H.P.J. Taylor. Investigation of the crack control characteristics of various types of bars in reinforced concrete beams. Technical Report 18, Cement and concrete assoc., 1966.

[CS78] L.A. Clark and D.M. Speirs. Tension stiffening in reinforced concrete beams and slabs under short term load. Technical rep. 42.521, Cement and Concrete Assoc., London, 1978.

[FM82] H. Floegel and H.A. Mang. A tension stiffening concept for rc panels based on bond slip. *Proc. ASCE, J. Struct. Div.*, 108(12):2681–2701, 1982.

[GM90] A.K. Gupta and S.R. Maestrini. Tension stiffness model for reinforced concrete bars. *J.Struct. Engrg ASCE*, 116(3):769–790, 1990.

[HM95] G. Hofstetter and H.A. Mang. *Computational mechanics of reinforced and prestressed concrete structures.* Vieweg, 1995.

[HS90] H.T. Hu and W.C. Schnobrich. Nonlinear analysis of cracked reinforced concrete. *ACI Structural Journal*, 87(2):199–207, 1990.

[JZ97] M. Jirásek and T. Zimmermann. Nonlocal rotating crack model with transition to scalar damage. In D.J.R. Owen, E. Oñate, and E. Hinton, editors. *Computational plasticity.* CIMNE, 1997.

[NO86] P. Noakowski. Mitwirkungsgesetze zur Ermittlung der Verformungen und der Zwangsbeanspruchung bei gleichzeitiger Lastbeanspruchung. *Beton und Stahlbetonbau*, (12):230–236, 1986.

[R86] P. Řeřicha. Optimum load time history for non-linear analysis using dynamic relaxation. *Int. J. Numerical Methods in Engineering*, 23:2313–2324, 1986.

Failure of Concrete Beams Strengthened with Fiber Reinforced Plastic Laminates

Oral Buyukozturk — Brian Hearing — Oguz Gunes

Department of Civil and Environmental Engineering
Massachusetts Institute of Technology
Cambridge, MA 02139 USA

obuyuk@mit.edu

ABSTRACT. *Fiber reinforced plastics have been shown to be an effective choice of materials for improving the load capacity and stiffness of concrete structures retrofitted by bonding these materials on damaged components. However, failures of the retrofitted systems may become brittle and the bonded plastic sheets are susceptible to delamination from the concrete substrate. The influence of design and existing damage parameters on the failure processes of retrofitted concrete needs to be fully understood for reliable solutions to infrastructure renewal. This paper investigates brittle delamination failures of retrofitted reinforced concrete beams through combined experimental and analytical studies on laboratory beam specimens with varying lengths of bonded fiber reinforced plastic laminates. The addition of the laminate to the concrete beams is shown to increase load capacity and stiffness, but the specimens are observed to fail through debonding in the concrete substrate initiating at the ends of the laminate. Longer laminate lengths are found to debond at higher load levels, but the failures become more brittle with the longer lengths. Crack initiation and propagation is monitored using ultrasound based nondestructive experiments on the laboratory beams. An energy criterion governing the delamination process is developed incorporating design parameters of the retrofitted system including laminate reinforcement ratio and length. The use of a critical energy release rate as a failure criterion for these systems is explored, and an analytical technique for the application of this criterion is presented.*

KEY WORDS: *FRP Composite Laminate, Retrofit, Damage, Cracking, Concrete.*

Introduction

High-strength fiber-reinforced plastics (FRP) offer great potential for lightweight, cost-effective retrofitting of concrete infrastructure through external bonding to concrete members to increase strengths and stiffnesses. The use of FRP bonded to deteriorated, deficient, and damaged reinforced concrete structures has gained popularity in Europe, Japan, and North America. Fiber reinforced plastics have been used to retrofit concrete members such as columns, slabs, beams, and girders in structures including bridges, parking decks, smoke stacks, and buildings.

The application of FRP as external reinforcement to concrete structures has been studied by many groups. Theoretical gains in flexural strength using this method can be significant; however, researchers have also observed new types of failures that can limit these gains. These failures are often brittle, involving delamination of the FRP, debonding of concrete layers, and shear collapse, and can occur at loads significatly lower than the theoretical strength of the retrofit system. Among these failure modes, debonding or peeling of the FRP from the concrete surface is a major concern, yet little is known about the fracture processes and characteristics of these mechanisms. Thus, there is a need for improved knowledge of the delamination and peeling failure processes.

Researchers have indicated that failure criteria for laminated systems need to be established before confident application of FRP to strengthening concrete systems is possible [TRI 91], [MEI 92a]. For this, a fundamental understanding of influences and mechanisms of debonding is necessary. This paper investigates delamination failures of retrofitted reinforced concrete beams through experimental and analytical studies on laboratory beam specimens retrofitted with FRP laminate. First, available information on delamination failures of FRP retrofitted concrete systems are reviewed. Then, an experimental program conducted to further investigate the local delamination process using initially notched laboratory specimens is presented. Finally, criteria governing the local debonding process are examined, and the application of an energy-based failure criterion for these systems is explored.

1. Failure of Retrofitted Beams by Delamination

FRP retrofitted reinforced concrete has been observed to fail through a variety of modes, which can be grouped into distinct categories including flexural failures, shear failures, and debonding failures [BUY 98]. Debonding can occur through failure of any constituent material in the system, but a common debonding mode is delamination failure of the laminate, adherent, and a thin layer of concrete substrate peeling off of the concrete structure. Delamination in actual rehabilitated structures has been reported [KAR 97], [BUY 95], as well as in research programs incorporating slabs, beams, joints, and columns [MEI 92a], [CHA 94], [ARD 97]. For these reasons, interest in delamination processes, criteria, and characterization methodologies has increased.

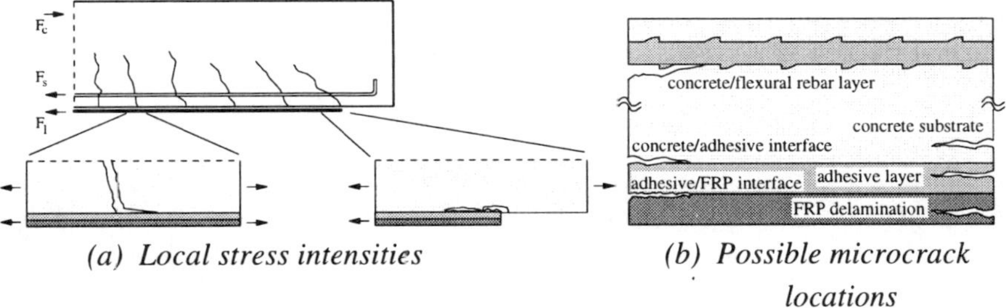

(a) Local stress intensities *(b) Possible microcrack locations*

Figure 1. *Local stress intensities and microcrack initiation scenarios*

Delamination can be caused by a number of reasons, including relative displacements of concrete in the vicinity of existing cracks, imperfect bonding between the composite and the concrete, cyclic loading induced debonding, environmental degradation, and other application and design specific flaws. Among these causes, delamination initiating from existing cracks or the end of the laminate has been concluded to be highly influential in retrofitted beam members [MEI 92b]. These failure processes are caused by local stress intensities in the concrete beam-adhesive-laminate interfacial region, as illustrated in Figure 1(a) [HAN 97], [TAL 97]. These intensities can initiate microcracking in the system at early load levels. Microcracks can form in any of the constituent materials or their interfaces, such as the laminate/adhesive interface, the adhesive/concrete interface, or the concrete/flexural steel layer, as shown in Figure 1(b). Upon further loading, these microcracks propagate and coalesce, ultimately forming macrocracks that can result in system failure.

This process has been studied by researchers using a variety of retrofitted system models. Studies using reinforced concrete beams retrofitted with externally bonded plates have been used to investigate global beam behavior during delamination failures. Theoretical peeling stresses in the adhesive layer of a beam with a bonded strengthening plate were derived by neglecting damage and cracks in the beam [TAL 97]. However, high speed video clips of plate end zone failures clearly show that cracks in the existing beam play a critical role in the peeling process, as shown in Figure 2 [HOL 97]. Even with the aid of high speed video, though, the exact local failure mechanism and direction of peeling of the steel plate delaminating is still unclear. The presence of shear cracks near the end of the laminate make it difficult to determine if local debonding initiated at the crack mouths or at the end of the laminate. Peeling failures of this type have also been observed in studies on the behavior of the retrofitted system with different bonded steel plate lengths [JAN 97]; it was concluded that unplated length influenced the peeling mechanisms. Again, however, the direction of peeling and exact local mechanism responsible for the system failure is indeterminable due to the flexural and shear cracks observed in the laboratory specimens. These and other reasons have lead researchers to use other specimens and scenarios to study the delamination process.

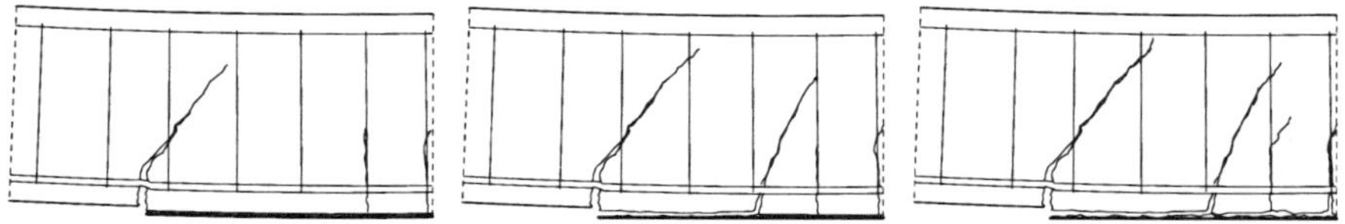

Figure 2. *High speed video clips of plate end zone failures, after [HOL 95]*

A variety of simplified models have been used to isolate the peeling failure process and study the bonding interaction between the concrete surface and external reinforcement. Experimental observations during lap shear pull-off tests have indicated that progressive bond failure is initiated by microcracks in the substrate at peak bonding stress followed by coalescence into debonding macrocracks [HAN 97]. Other studies have included theoretical models of unreinforced open sandwich specimens [ANA 85], [HAM 90] and peel tests [KAR 97] where local fracture under varying external peel loadings was studied through inclined lap-type specimens. Changes in peeling angles were concluded to affect the fracture morphology; higher angles of FRP peeling off the stationary slab were shown to shift the delamination process deeper into the mortar substrate. However, the application of this knowledge to laminated reinforced concrete flexural members is limited; the conditions of the concrete substrate in these studies may not have developed the complete microcrack initiation and propagation process found in components with conventional reinforcement subjected to generalized flexure and shear loadings. Thus, there exists a need for an isolated study of local delamination mechanisms of reinforced members under flexural beam loading.

2. Experimental Program

Specialized delamination beam specimens were developed and tested in an experimental program to study local delamination fracture processes. The objectives of the program were to isolate and monitor local delamination crack development and propagation, and to investigate influences of bonded laminate length on the delamination process. To meet these objectives, delamination specimens containing initial delamination notches were created with various lengths of bonded FRP laminate.

2.1. *Specimen*

To isolate local delamination mechanisms from other failure processes, a four-point delamination beam specimen was designed with over-reinforced shear capacity, as shown in Figure 3. The beam specimens were designed to be retrofitted with carbon FRP (CFRP) laminate that would theoretically increase the flexural capacity by 40% through tensile laminate rupture. To reduce the influence of shear cracks on the delamination process, the beams were over-reinforced with closely spaced stirrups that increased the shear strength to 175% of the flexural retrofitted capacity. Delamination was studied by propagation from the end of the laminate towards the center of the specimen. To simulate steady-state conditions, initial delamination cracks were introduced with a thin sheet of plastic placed between the adhesive and the concrete as the beam was retrofitted.

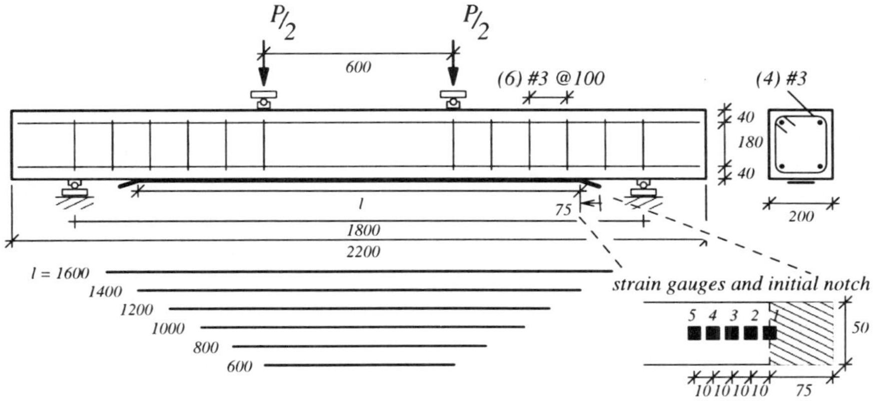

Figure 3. *Specialized delamination test specimen (dimensions in mm)*

Material	Tensile strength [MPa]	Compressive strength [MPa]	Elastic modulus [GPa]
Concrete	2.8	20.5	25.4
Steel	418.0	-	210.0
Adhesive	24.8	-	2.7
CFRP laminate	2100.0	-	155.0

Table 1. *Properties of materials used in the experimental program*

2.2. *Materials*

The concrete beam specimens were created with 7-day normal strength concrete with 10 *mm* maximum aggregate size and a *w/c* ratio of 0.5. In addition to compressive and tensile strength testing, the fracture energy of the concrete was measured at 42.6 J/m^2 using the RILEM Technical Committee 89-FMT suggested three point testing specimen. The internal reinforcement in the beam specimens consisted of (2) #3 ($\phi = 9.5$ *mm*) reinforcing bars for both tensile and compressive

2.3. *Test Procedure*

The specimens were loaded in four point bending in load steps of 4.5 *kN* until system failure. Midspan deflections and strain gauge signals were monitored continuously. The anchorage zones were inspected for delamination initiation at each load level using a magnifying scope and an ultrasound based NDT technique. For ultrasonic inspection, a commercially available ultrasonic transducer operating at 100 KHz was used with a pulser/receiver and digital oscilloscope.

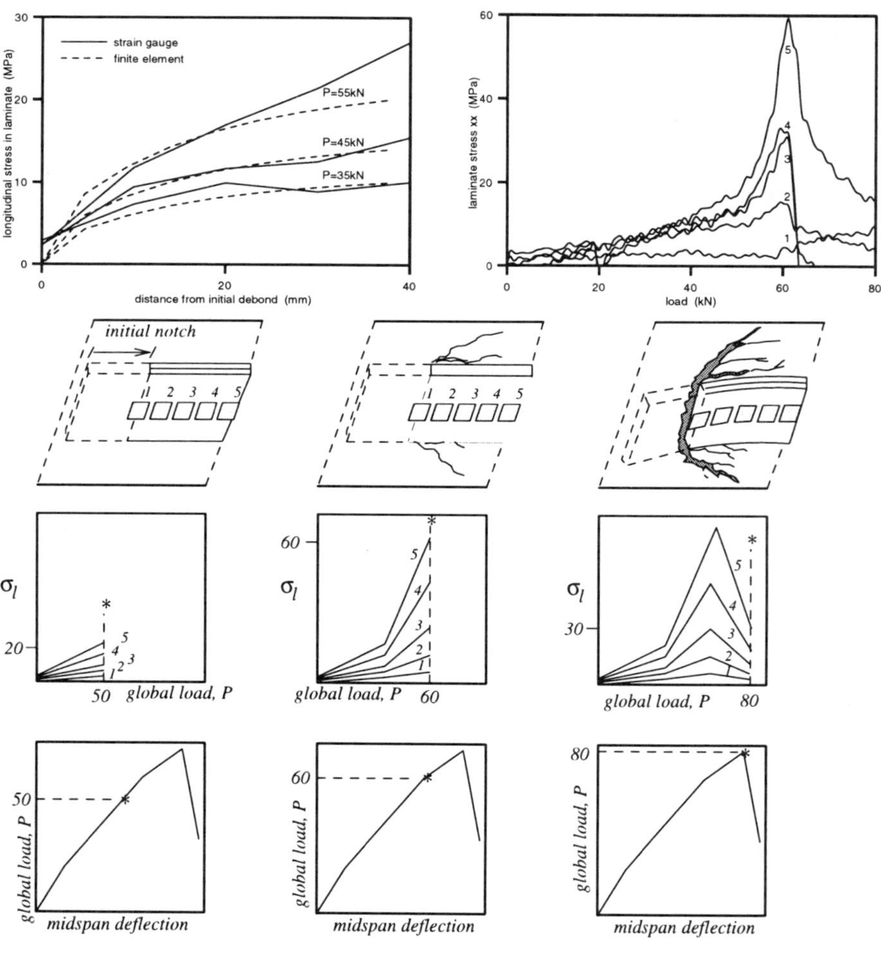

(a) Development of laminate stresses in anchorage zone *(b) Laminate stresses with load history*

Figure 5. *Results from the strain gauges on the beam with 1.4 m bonded length*

2.4. *Experimental Results*

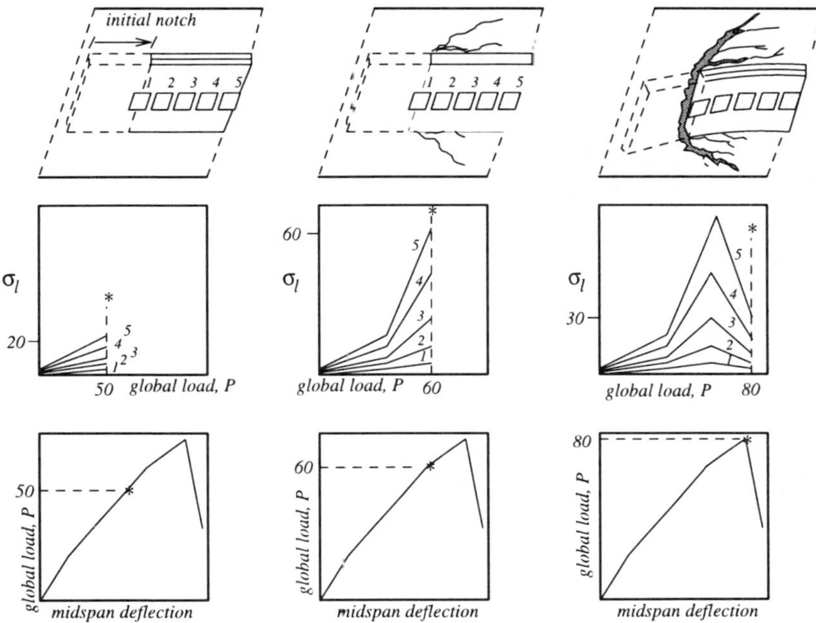

Figure 6. *Delamination process zone development with idealizations of Figures 5(b) and 4(a) for the 1.4 m laminate length beam*

First, unlaminated control beams were tested. The midspan deflection curves demonstrate traditional nonlinearities at cracking of the concrete and plasticity of the steel, as shown in Figure 4(a). Typical cracking patterns are illustrated in Figure 4(b), and it is shown that concrete cracking was largely limited to flexural cracks in the moment span. Then, laminated beams were tested; midspan deflection curves are also shown in Figure 4(a), and typical cracking patterns are illustrated in Figure 4(b). Ultimate loads from the testing program are presented in Table 2. The beams exhibited concrete cracking at levels higher than the unlaminated beams, with flexural cracking confined mostly to the constant moment span. Stiffnesses were observed to increase with the addition of the laminate. Strain gauge results indicated a development of laminate stresses in the anchorage region, as illustrated with different load levels in Figure 5(a).

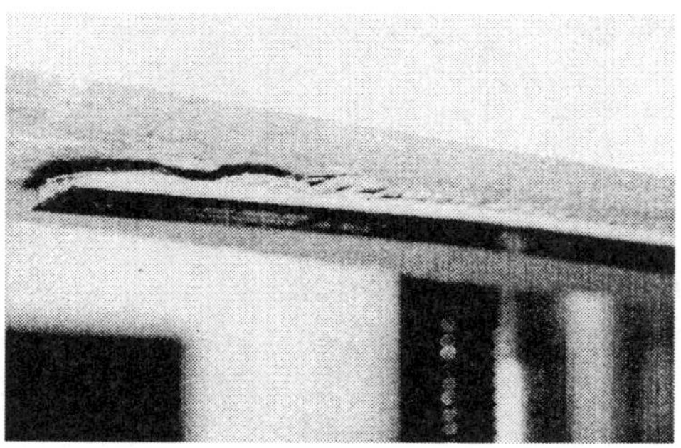

Figure 7. *Sample delaminated FRP with thin layer of concrete substrate*

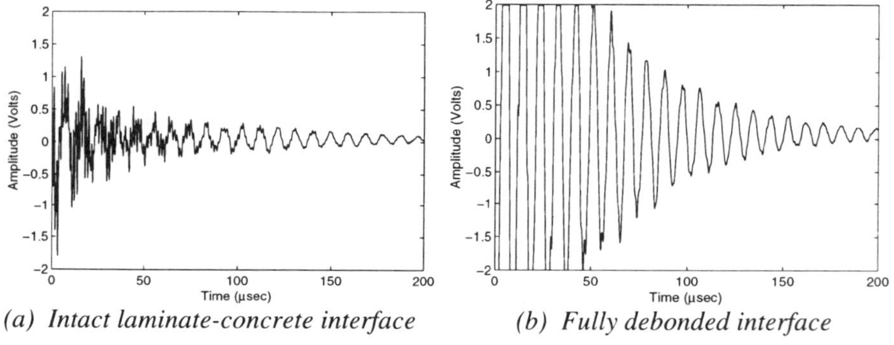

(a) Intact laminate-concrete interface *(b) Fully debonded interface*

Figure 8. *Time-amplitude responses obtained from ultrasonic inspection of retrofitted beams*

All retrofitted specimens failed through delamination in the concrete, leaving a thin layer of concrete substrate bonded to the delaminated FRP. Initial cracking in the concrete substrate at the delamination/anchorage zone was typically observed after flexural cracking, and before yielding of the steel or final delamination. These initial cracks were often accompanied by audible noises, and changes in stiffness were also observed in the midspan deflection curve. The strain gauge readings indicated that strains in the laminate reached their peak at crack initiation, and decreased upon further loading, as shown for the beam with 1.4 *m* bonded length in Figure 5(b). This indicates the initiation of the delamination process (at 60 *kN* for the beam shown), even though in this case the beam continued to resist more load. Upon additional loading the initial crack continued to widen until the ultimate load of the beam was reached (at 90 *kN* for the beam shown), where unstable delamination occurred resulting in peeling of the laminate, adhesive, and a thin layer of concrete. This process is illustrated in Figure 6 with idealizations of behavior demonstrated in Figure 5(b) and 4(a), and pictured in Figure 7.

Exploratory experiments were conducted using ultrasonic nondestructive testing. Figure 8 illustrates differences in the response recorded at the receiver for intact concrete-laminate interface (a) and fully debonded (b) interface respectively. The acoustic impedance of air is much higher than that of concrete, thus, the impedance mismatch at the laminate-concrete interface increases upon formation of debonding cracks, resulting in higher wave reflection and greater response amplitude. The responses obtained from ultrasonic inspection of in-service retrofitted beams gradually shifts from (a) to (b) with increasing debonding. Based on this, the measured responses during routine inspections can be used as a quantitative damage index to measure the integrity of bonding between the laminate and the substrate. The ultrasonic inspection technique is found to have a significant potential in providing quantitative information on damage accumulation at the beam-laminate interfacial region.

2.5. *Effect of Bonded Length*

From the results listed in Table 2, it is shown that loads at initiation of the delamination process and ultimate failure loads increased with longer laminate lengths. Beam stiffnesses were also found to increase, as shown in Figure 4(a). Beams with longer laminate lengths demonstrated more brittle failure, with sudden delamination along the entire shear span, as shown in Figure 4(b). These longer laminates also had less delay between first signs of delamination and ultimate failure. For example, the specimen with the longest laminate, 1.6 *m*, exhibited virtually no signs of crack initiation before sudden delamination and system failure. Shorter laminate lengths were found to fail at lower loads with less brittleness. Delamination crack initiation was observed early, and there was a larger load delay between initiation and ultimate failure than with the longer laminates. Based on observations and the results of the strain gauges, it is concluded that the delamination process is usually initiated before the ultimate system failure. The bonded laminate length was found to have significant influence on the load at which the delamination process is initiated. Causes of this process were further investigated in an analytical procedure.

3. Analytical Procedure

In the experimental procedure it was established that a delamination process zone develops at locations of stress intensities in the concrete-adherent-laminate region before ultimate delamination and failure of the system. The development of this process and stress field information were further investigated in an analytical procedure incorporating a finite element model.

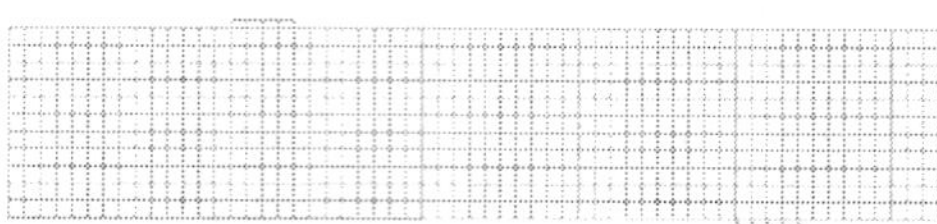

(a)Refined finite element mesh

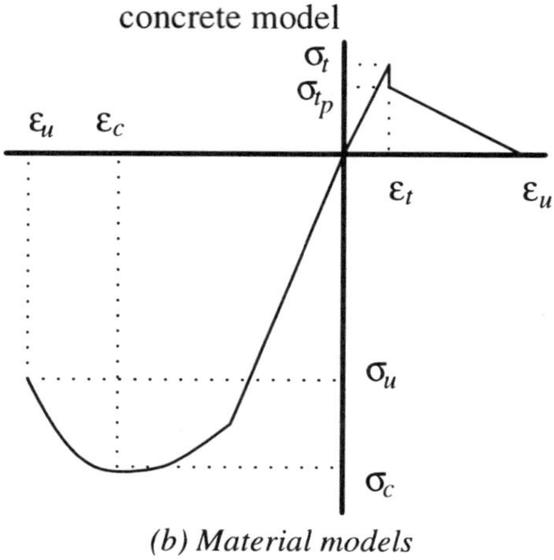

(b) Material models

Figure 9. *Models used in the finite element investigation*

3.1. *Finite Element Investigation*

Numerical simulations of the initially notched delamination specimen were conducted using a nonlinear finite element method incorporating a smeared crack concrete material model. The beams strengthened with CFRP plates were symmetrically modeled with 18.75 *mm* square 8-node 2D plane strain elements and 3-node truss members for reinforcing steel. The mesh, shown in Figure 9(a), assumes perfect bond at the adhesive, CFRP, and concrete interfaces to discount other failure modes such as interfacial debonding. The mechanical properties measured in the experimental procedure were used in the nonlinear material models. Concrete tensile softening was modeled through a bilinear curve, and the reinforcing steel, epoxy, and CFRP materials were assumed to have bilinear plasticity, as illustrated in Figure 9(b). Strains in the FRP plate in the anchorage region are plotted in Figure 6(a). During the simulations, a crack band originating from the initial notch was observed. The collapse load was achieved through nonsingular matrix formation at a concrete node in the anchorage region for all simulations conducted.

Cracking in the anchorage region under progressive loads and the development of the delamination crack band illustrated that the ultimate tensile strength of the concrete at the tip of the initial notch was reached at relatively early loads. The collapse load, indicating failure, was reached after the ultimate tensile strain in the concrete was exceeded. This was achieved after a significant crack band had developed from the anchorage zone, indicating that delamination failures can occur when both the strength and ultimate strain are exceeded. These two criteria have been used in a two-parameter approach investigating failure of pre-tensioned FRP reinforced concrete [LAD 81], [TRI 92]. However, the ultimate strain and peak stress used in the formulation are not well defined or readily measurable. Additionally, the use of softening stress-strain relationships have been shown to lead to mathematical results inconsistent with experiments (e.g., [BAZ 86]). The failure behavior at this scale may be better characterized through a fracture energy based model for quasi-brittle materials.

Fracture energy of quasi-brittle materials is defined by the area under the softening curve which represents the energy consumption per unit area of crack extension. An unambiguous definition of fracture energy in concrete where fracture is preceded by a process zone has been given [BAZ 87]. This definition of fracture energy can be used to analyze the effects of size dependency, fracture process zones of variable size, and a measure of material brittleness. The application of this criterion has been incorporated in analyses of critical fracture applications such as fiber and bar pullout tests, among others [BAZ 88], [BAZ 94]. Fracture energy release rates have also been investigated as criteria in shear-lap pull-off tests [TAL 96], [THE 86], unreinforced concrete sandwich specimens [ANA 85], [HAM 90], and peel tests [KAR 97]. Failure of specimens tested in these investigations were found to occur when the potential energy release of the specimen exceeded the constituent material's critical energy release rate. It has been concluded that this type of approach may also be applicable to reinforced concrete beams [ANA 85]. Based on these conclusions, energy release rates for the reinforced beam specimens used in this study were investigated using an energy based analytical technique.

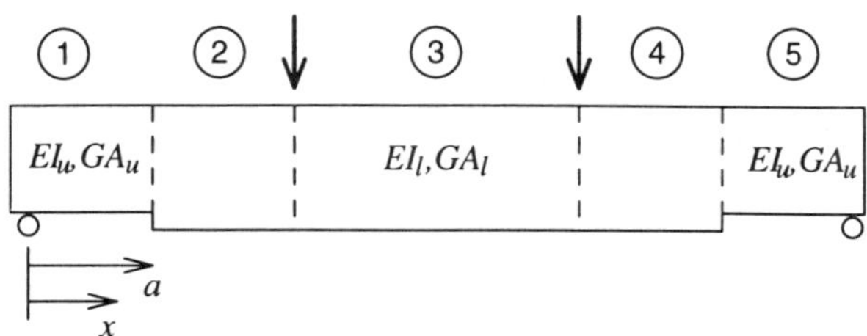

Figure 11. *Idealization of laboratory specimen into discrete cross sections*

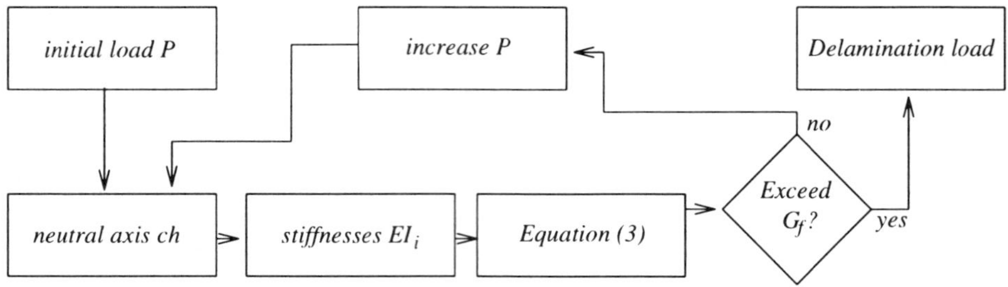

Figure 12. *Flowchart of nonlinear procedure used to calculate delamination load*

3.2. *Energy Based Criterion*

This section presents an analytical technique useful for applying the energy release rate criteria to predict the initiation of the delamination process. The Griffith energy criteria for fracture states that crack growth can occur if the energy required to form additional crack length can be delivered by the system. This condition for crack growth is

$$\frac{dW}{da} = \frac{d}{da}(F - U) \qquad [1]$$

where U is the elastic energy contained in the system, F is the work provided by external forces, and dW/da is the crack resistance. The elastic energy contained in a beam structure under bending and shear forces M and V can be estimated by integrating the normal and shear stresses over the cross sectional area. The beams tested in this investigation can be discretized into five divisions of similar cross sections and applied loading, as illustrated in Figure 11. The delamination process in the tested beams is simulated by advancing the delamination location of a by da, and transforming the laminated section (EI_l and GA_l) to equivalent unlaminated section (EI_u and GA_u) where EI_i and GA_i are the discrete section bending and shearing stiffnesses dependent on the integral load level, respectively

$$\frac{dW}{da} = \int_0^F \left[\frac{M_a^2}{2b}\left(\frac{1}{EI_u} - \frac{1}{EI_l}\right) + \frac{V_a^2}{2b}\left(\frac{k_u}{GA_u} - \frac{k_l}{GA_l}\right)\right] \qquad [2]$$

where k_u and k_l are constants in the calculation of shearing stresses of the cross section and b is the width of the beam [THE 86], [TOY 91].

Most delamination cracks in the tested laboratory specimens were observed to initiate before plasticity in the reinforcing steel. Thus, minimal error will be introduced by ignoring the nonlinear terms in Equation [2], reducing the expression to

$$\frac{dW}{da} = \frac{M_a^2}{2b}\left(\frac{1}{EI_u} - \frac{1}{EI_l}\right) + \frac{V_a^2}{2b}\left(\frac{k_u}{GA_u} - \frac{k_l}{GA_l}\right) \quad [3]$$

Stiffnesses can be estimated using conventional methods such as a strain compatibility and force equilibrium [SHA 94, PLE 94] to find neutral axis depth fraction c of beam height h

$$c^2[h] + c[h\eta_s\rho_s' + 2h\eta_s\rho_s + h\eta_l\rho_l] - 2[d'\eta_s\rho_s' + d\eta_s\rho_s + h\eta_l\rho_l] = 0 \quad [4]$$

where η_s and η_l are stiffness ratios of the reinforcing steel and laminate to the concrete and ρ_s, ρ'_s, and ρ_l are reinforcement ratios of the tensile steel, compressive steel, and laminate respectively. Stiffnesses can be estimated for a given load by assuming that concrete cannot sustain tension over its rupture strength. The influence of the thin FRP laminate on the shear capacity of the beam is unclear [JAN 97]; in this analysis, the shear components in Equation [3] are neglected. Taking moments of areas about the neutral axis ch, the stiffness EI is estimated by

$$EI = \frac{E_c b_c (ch)^3}{3} + A_s E_s (d_s - ch)^2 + E_{FRP} A_{FRP} h^2 (1-c)^2 \quad [5]$$

where E_i is the elastic modulus of material i (=c for concrete, s for steel, and FRP for FRP laminate), b_c is the width of the concrete, A_s and A_{FRP} are the areas of the reinforcing steel and FRP laminate, and d_s is the depth to the reinforcing steel.

Energy release rates were calculated using this analytical procedure and the results from the experimental program, as shown in Table 2. The critical energy release rates are shown to be similar and independent of laminate length. Differences in data are found in the shortest (*0.6 m*) and longest (*1.6 m*) laminate lengths tested; these differences are attributed to flexural steel yielding in the specimen with the short laminate and shear cracking near the support and laminate end of the specimen with the long laminate. The procedure was repeated in reverse, using the fracture energy of the concrete (G_f = 42.6 J/m^2) as the delamination criteria dW/da. Because the moment of inertia calculations are dependent on the applied loading, an incremental load loop was repeated using increasing load levels in Equations [4] and [5] until the applied moment and moments of inertia exceeded the concrete's fracture energy through Equation [3], as illustrated in Figure 12. These calculated delamination loads are plotted in Figure 13 with the results from the experimental program and finite element investigation. It is shown that this procedure approximates the delamination load trend reasonably well, and that the load increases with longer laminate lengths. This trend indicates a critical laminate length where the tensile strength of the laminate will be exceeded before delamination occurs. This type of procedure could lead to a criteria governing a minimum laminate length to prevent debonding from the ends of the laminate. Effects of relative laminate reinforcement ratio can also be examined through Equations [4] and [5]. Work is currently being conducted to apply an energy based criteria to scenarios where peeling initiates from existing cracks in the shear span of the beam.

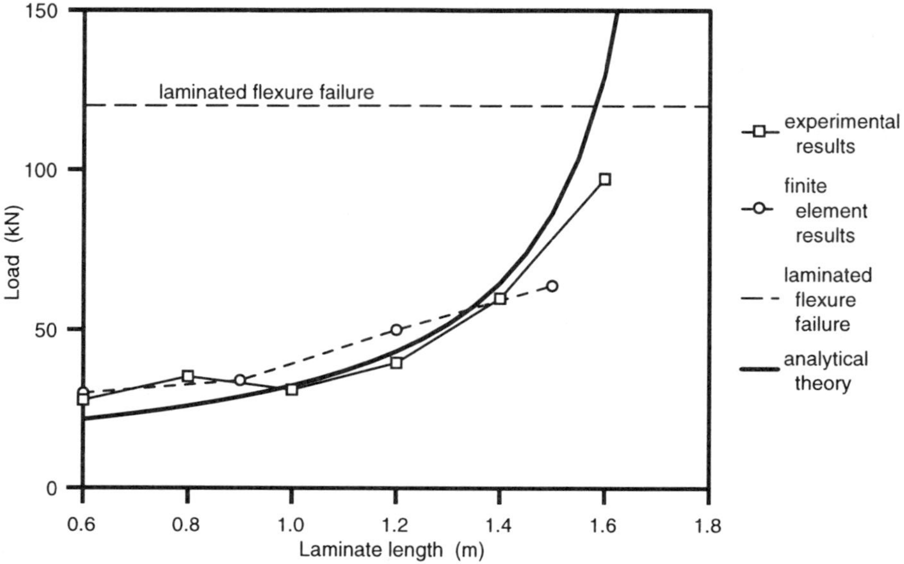

Figure 13. *Delamination loads predicted analytically with experimental and finite element results*

Conclusion

In this paper, delamination of reinforced concrete beams retrofitted with fiber-reinforced plastic laminates is investigated. Local failure processes are studied through laboratory tests on initially notched beam specimens retrofitted with various lengths of CFRP laminate. The beams were observed to fail through delamination in the concrete substrate near the laminate at ultimate loads that increased with longer laminate lengths. Observations of the failures indicated the development of a delamination process zone before unstable failure of the system. This mechanism is further investigated using a finite element model, and failure was shown to occur after ultimate tensile stresses and strains of the concrete were exceeded in the process zone. Based on this, the use of a fracture energy criterion as a governing measure of this process is explored. An analytical technique is used to calculate energy release rates of the experimental beams, and it was shown that the critical energy remained relatively constant for all the tested beams. The procedure is then conducted in reverse using the fracture energy of the concrete as the failure criteria, and the predicted loads are shown to match the experimental results. From this, it is concluded that the critical energy release rate of the concrete may be applicable as a criterion for local delamination. Furthermore, the influence of design parameters such as laminate reinforcement ratio and laminate length can be examined using this technique.

Acknowledgment

This work is supported by the National Science Foundation. The cognizant official is Dr. Ken Chong.

References

[ANA 85] ANANDARAJAH, A., VARDY, A. E. (1985). A theoretical investigation of the failure of open sandwich beam due to interfacial shear fracture. *The Structural Engineer*, 63B(3):85-92.

[ARD 97] ARDUINI, M., DI TOMMASO, A., NANNI, A. (1997). Brittle failure in FRP plate and sheet bonded beams. *ACI Structural Journal*, 94(4):363-371.

[BAZ 84] BAZANT, Z. P. (1984) Size effect in blunt fracture: concrete, rock, metal. *ASCE Journal of Engineering Mechanics*, 110(4):518-535.

[BAZ 86] BAZANT, Z. (1986) Mechanics of distributed cracking. *Applied Mechanics Review*, 39:675-705.

[BAZ 87b] BAZANT, Z. P., PHEIFFER, P. A. (1987) Determination of fracture energy from size effect and brittleness number, *ACI Materials Journal*, 84(6):463-480.

[BAZ 88] BAZANT, Z. P., SENER, S. (1988) Size effect in pullout tests. *ACI Materials Journal*, 85:347-351.

[BAZ 94] BAZANT, Z. P., DESMORAT, R. (1994) Size effect in fiber or bar pullout with interface fracture and softening slip. *ASCE Journal of Engineering Mechanics*, 120(9):1945-1962.

[BUY 95] BUYUKOZTURK, O. (1995) Private communication with Dr. J. Vielhaber, Director, Structural Research, Federal Laboratory of Material Testing (BAM). Berlin, Germany.

[BUY 98] BUYUKOZTURK, O., HEARING, B. (1998) Failure behavior of precracked concrete beams retrofitted with FRP. *to appear on ASCE Journal of Composites in Construction*.

[CHA 94] CHAJES, M., THOMSON, T., JANUSZKA, T., FINCH, W. (1994) Flexural strengthening of concrete beams using externally bonded composite materials. *Construction and Building Materials*, 8(3):191-201.

[HAM 90] HAMOUSH, S. A., AHMAD, S. H. (1990) Debonding of steel plate-strengthened concrete beams. *ASCE Journal of Structural Engineering*, 116(2):356-371.

[HAN 97] HANKERS, C. (1997) Zum verbundtragverhalten laschencerstarkter benbauteile unter night vorviegen ruhender beanspruchung. Technical report, Beuth Verlag GmbH, Berlin.

[HOL 97] HOLZENKAMPFER, P. (1997) Ingenieurmodelle des Verbunds geklebter Bewehrung für Betonbauteile. Technical report, Beuth Verlag GmbH, Berlin.

[JAN 97] JANSZE, W. (1997) Strengthening of Reinforced Concrete Members in Bending by Externally Bonded Steel Plates. PhD thesis, Delft University of Technology.

[KAR 97] KARBHARI, V. M., ENGINEER, M., ECKEL II, D. A. (1997) On the durability of composite rehabilitation schemes for concrete: Use of a peel test. *Journal of Material Science*, 32(1):147-156.

[LAD 81] LADNER, M., WEDER, C. (1981) Concrete structures with bonded external reinforcement (in German). Technical Report 705, EMPA.

[MEI 92a] MEIER, U. (1992) Carbon fiber-reinforced polymers: Model materials in bridge engineering. *Structural Engineering International*, 2(1):7-12.

[MEI 92b] MEIER, U., DEURING, M., MEIER, H., SCHWEGLER, G. (1992) Strengthening of structures with CFRP laminates: Research and applications in Switzerland. In *Advanced Composite Materials in Bridges and Structures*, pages 243-251, Canadian Society for Civil Engineering.

[PLE 94] PLEVRIS, N., TRIANTAFILLOU, T. (1994) Time-dependent behavior of RC members strengthened with FRP laminates. *ASCE Journal of Structural Engineering*, 120(3):1016-1041.

[SHA 94] SHARIF, A., AL-SULAIMANI, G. J., BASUNBUL, I. A., BALUCH, M. H., GHALEB, B. N. (1994) Strengthening of initially loaded reinforced concrete beams using FRP plates. *ACI Structural Journal*, 91(2):160-168.

[TAL 96] TALJSTEN, B. (1996) Strengthening of concrete prisms using the plate-bonding technique. *International Journal of Fracture*, 82(3):253-266.

[TAL 97] TALJSTEN, B. (1997) Strengthening of beams by plate bonding technique. *ASCE Journal of Materials in Civil Engineering*, 9(4):206-213.

[THE 86] THEILLOUT, J. N. (1986) Repair and strengthening of bridges by means of bonded plates. In Sasse, H. R., editor, *Adhesion Between Polymers and Concrete*, pages 601-621.

[TOY 91] TOYA, M., ONO, T., MIYAWAKI, T., KIKIOKA, K. (1991) Analyses of delamination of laminated beams subjected to three and four-point bending. *Proceedings of the Japanese Society of Mechanical Engineers*, 8:136-142.

[TRI 91] TRIANTAFILLOU, T., DESKOVIC, N. (1991) Innovative prestressing with FRP sheets: Mechanics of short-term behavior. *Journal of Engineering Mechanics*, 117(7):1652-1672.

[TRI 92] TRIANTAFILLOU, T., DESKOVIC, N., DEURING, N. (1992) Strengthening of concrete structures with prestressed fiber reinforced plastic sheets. *ACI Structural Journal*, 89(3):235-244.

Appendix

Summary of the Discussions During the Workshop

Discussion on Mechanics of Material Failure
M. JIRASEK

Discussion on Durability Mechanics (I)
F.-J. ULM

Discussion on Durability Mechanics (II)
Y. XI

Discussion on Computational Failure Analysis and Design
F.-J. ULM

List of Workshop Participants

Mechanics of Material Failure

Milan Jirásek

Laboratory of Structural and Continuum Mechanics
Swiss Federal Institute of Technology
LSC-DGC, EPFL, CH-1015 Lausanne, Switzerland

ABSTRACT. *This paper summarizes the discussion of five lectures presented in two workshop sessions on Friday morning, March 27, 1998.*

KEY WORDS: *Non-Local Continuum, Mesh-Induced Bias, Steel-Concrete Bond, Rate Effect, Fracture Energy, Concrete, Masonry.*

1. J. Ožbolt, J. Asmus, and K. Jebara: "Splitting of concrete block caused by inside pressure—failure mechanism and size effect"

1.1. *Determination of characteristic length of a nonlocal continuum*

Ožbolt explained that the parameters of his model are determined by running a two-dimensional or three-dimensional analysis of the uniaxial compression test and fitting the peak and the post-peak behavior observed in experiments. The mesh must be sufficiently fine to make nonlocal interaction among individual Gauss points possible. This condition is in the present application satisfied only in the region around the anchor. Farther away, the mesh is rather coarse and the nonlocal behavior might not be captured properly.

Bažant pointed out that the determination of the characteristic length from one load-displacement diagram is an ill-conditioned problem. Physically there should be only one characteristic length but good fits can be obtained with different values of this parameter if the local stress-strain diagram is adjusted accordingly.

De Borst proposed to determine the characteristic length from detailed measurements of strain distribution in the process zone; then the identification process is well-conditioned. According to him, the length scale cannot be inferred from the size effect law.

Pijaudier-Cabot argued in favor of identification from the size effect law.

He said that the load-displacement diagram for one size can be fitted with different values of characteristic length but proper scaling is obtained only with the correct one. Bažant later expressed his support. The material length may be defined as the effective length of the fracture process zone, which can be determined approximately as $c_f = D_0 g(\alpha)/g'(\alpha)$. The nondimensional function $g(\alpha)$ characterizes the geometry of the specimen and the type of loading, and parameter D_0 can be obtained solely by fitting the maximum load data; see [BK90] for more details.

According to de Borst, the results in a certain range of sizes can still be fitted with an incorrect internal length parameter. The available experimental results do not cover a sufficiently wide range of sizes. Pijaudier-Cabot specified that experiments can cover sizes in the range 1:10.

In response to additional questions, Ožbolt explained that in the example from his paper the characteristic length was 10 mm and the maximum aggregate size was 12 mm. It is optimal to work with a characteristic length approximately equal to the aggregate size but for large structures, e.g. dams, this is not always possible. If the characteristic length was changed to 20 or 30 mm the results would be qualitatively the same (the failure mechanism would not be affected) but quantitatively the error is difficult to estimate. Based on experimental data, Ožbolt feels that the error of his numerically computed results does not exceed 20%.

1.2. *Effect of mesh-induced directional bias and model symmetry on the simulated crack pattern*

De Borst questioned the objectivity of the results with respect to the mesh orientation. The cracks propagate in the radial direction, i.e. along the mesh lines. He would be convinced if the same pattern were obtained on a rectangular mesh. Ožbolt replied that he had run similar simulations on an approximately regular mesh and had not observed any significant difference. The pattern with three macroscopic cracks was the same.

Krysl pointed out that three cracks form on the model of 1/2 of the specimen but only one crack forms on the model of 1/4 of the specimen. The crack that would be expected on the boundary of the 1/4 model does not appear. Ožbolt confirmed that the results were affected by the assumption of four-fold symmetry. De Borst added that by making such an assumption we exclude bifurcations into nonsymmetric modes and overpredict the dissipated energy.

1.3. *Modeling of boundary conditions*

Gérard questioned the description of loading in the presented example. Ožbolt admitted that the boundary conditions are simplified. He took into account only the radial pressure but in reality there is an additional force acting in the axial direction. It is impossible to measure the ratio between the pullout and the splitting forces. The problem is very local and it is difficult to estimate the local characteristics.

1.4. *Role of compression failure*

Bažant wanted to know how important the effect of compressive failure was and whether realistic triaxial laws were required to model the failure process. Ožbolt replied that compression failure played an important role in this particular example and that the failure mechanism could not be simulated correctly with a model that does not properly capture dilatancy.

2. V. Červenka: "Applied brittle analysis"

2.1. *Possible improvements of the model*

In response to Ulm's question, Červenka mentioned the need for a three-dimensional extension of his model and for an improved modeling of compression. Gérard added that it would be highly desirable to simulate also the transport of water and gas.

Subsequent discussion was devoted to the comparison of Červenka's two-dimensional smeared crack model and Ožbolt's version of the three-dimensional microplane model. Červenka plans to handle the 3D extension by a plasticity-type approach rather than by the microplane model because he prefers simple models with a small number of material parameters.

2.2. *Geometric nonlinearity due to finite displacements*

Tanabe stressed the importance of geometric nonlinearity. He said that in the post-peak range this effect changes the results by 5-10%. Červenka agreed. He always runs his simulations with the finite displacement option on (updated Lagrangean formulation).

2.3. *Modeling of masonry*

In response to Maier's question, Červenka said that he models masonry on the macroscopic scale as a homogeneous and isotropic material. Ulm was suprised by the assumption of isotropy. Červenka explained that in large-scale analyses (three entire buildings in this case) it is impossible to model individual bricks. His model is initially isotropic and anisotropy is induced by cracking.

2.4. *Legal aspects and reliability of results*

In response to Ulm's question, Červenka said that the results of finite element analyses are accepted by the codes and by the official *Prüfingenieure* who approve designs in Germany or Switzerland; but some legislative problems still persist. In his opinion, the computed crack opening (in masonry) depends on the elastic properties of the material rather than fracture properties.

2.5. *Dependence of crack spacing on the mesh size*

Gérard wanted to know whether the crack spacing would depend on the mesh size. Červenka replied that when the cracking is diffuse there is one computational crack per integration point, and the spacing of these points of course depends on the element size. After localization, the distance between the major cracks that keep opening should be mesh-independent, provided that the crack band theory is implemented correctly. This has been confirmed by his numerous examples.

2.6. *Fixed versus rotating cracks*

In response to Ožbolt's question, Červenka explained that he used both the fixed and the rotating crack approaches. Models with rotating cracks are softer, the spacing between cracks is larger, and the cracks open more widely. In his opinion the two approaches provide bounds on the actual solution. Ožbolt had doubts that such a simulation tool could be used as a black box by an average user without previous experience in this area. Červenka said that some of the users of his program never request any assistance but still produce satisfactory results. He further explained the basic idea of the crack band theory (adjustment of the softening modulus as a function of the element size controling the width of the localized crack band).

2.7. *Modeling of bond*

In response to Cedolin's question, Červenka said that he did not explicitly model the debonding between concrete and steel. If the concrete is reinforced by ribbed bars, fracture usually happens in concrete and not directly at the interface. This is reflected by inclined cracks that form in a band of elements around the reinforcing bar. Ožbolt stressed that the bond properties would depend on the size of elements, and Červenka agreed.

3. P. Řeřicha and Z. Bittnar: "Restrained cracking in reinforced concrete"

3.1. *Nonlocal formulation of softening plasticity*

Jirásek wanted to know which particular version of nonlocal plasticity had been used in the present study. He pointed out that certain nonlocal formulations exhibit pathological behavior and generate spurious stresses [Jir98]. The explanation was adjourned to bilateral discussion.

3.2. *Parameters affecting the spacing between major cracks*

Ulm pointed out that the model is not suitable for cyclic loading. He was also surprised to see that the simulated spacing of principal cracks is completely different from the experimentally observed one, as reported for example by Goto [Got71] who observed cracks originating at the ribs. Řeřicha explained that he had not so far modeled the reinforcement bar with ribs by finite elements but by a uniform strain imposed on the steel/concrete boundary, which corresponds to an infinitely stiff bar and a perfect bond.

Ulm also criticized another aspect: If the results are scaled such that the effective concrete cover becomes e.g. 20 times the diameter of the reinforcing bar, the crack spacing is not realistic. Řeřicha argued that the simulated spacing is not scalable. According to empirical formulas, the visible crack spacing and width primarily depend on the cover rather than on the bar diameter. Some formulas even use a linear dependence between the cover and visible crack spacing, which would correspond to his results.

Ulm stated that for a cover larger than 5 to 7 times the bar diameter Řeřicha's model would not produce through cracks. Řeřicha explained that slip may occur in the concrete material, which could be captured by the model if a sufficiently fine mesh were used.

4. L. Cedolin, P. Bianchi, and A. Ratti: "A visco-damage model for the tensile behavior of concrete at moderately high strain-rates"

4.1. *Relation to models for metals at high strain rates*

Schreyer was impressed by Cedolin's work. He pointed out that the results appear to be similar to experimental data often seen in certain metals subjected to high strain rates. In this field some success has been achieved with models incorporating the inelastic strain rate (not just the inelastic strain) into an evolution equation for a state variable. An example is the Mechanical Threshold Stress model [FHG90] developed by Follansbee and colleagues at Los Alamos National Laboratory. This approach automatically takes into account the history of the inelastic strain rate. Cedolin said he had looked at empirical expressions in which the rate dependence is expressed as a function of the instantaneous strain rate, without being able to fit the experimental results.

4.2. *Physical sources of rate dependence*

In response to Schreyer's next question, Cedolin mentioned the Stéfan's effect as a possible source of rate dependence in the dynamic range.

Pijaudier-Cabot added that at high speeds (higher than those considered in Cedolin's paper) the rate effect could be attributed to the confining effect of inertia, which is then much more important than viscosity. According to his experience, if the viscosity is increased by a factor of 10 the numerical results do not change at all. Schreyer stated that this would be true only for very fast rates and Pijaudier-Cabot agreed.

Later in the discussion, Ulm confirmed that up to the rates considered by Cedolin viscous effects are dominant. He described the technique used by him and Rossi at LCPC, and he mentioned simulations of impact of a nuclear waste container. Viscosity is not the main factor at the impact point but it affects the propagation of stress waves through the container. Bažant said that the threshold should not be directly related to the strain rate but to an internal variable. Pijaudier-Cabot agreed, saying that the right mechanism we need is humidity. Ulm referred to the characteristic time of the diffusion process. At very high rates, water in the micropores changes its properties. We cannot consider the CSH molecules as staying quiet. They move, and inertia effects come into play. There are also other ranges, but between 10^{-1}/sec and 10/sec this phenomenon is dominant.

4.3. *Processes taking place in the lateral direction*

In response to Tanabe's question, Cedolin said that in his experiments the lateral strains remained very small, which made one-dimensional modeling possible. The stress distribution in the cross section was not exactly uniform but close to uniform.

4.4. *Alternative models*

Bažant suggested another way of modeling a wide range of rates. He described the rate-dependent version of his microplane model, in which the stress on a microplane is a function of both the strain and the strain rate, in the spirit of rate-dependent plasticity motivated by the activation energy model. He reported good fits for both static tests and missile impact simulations.

5. J. G. M. van Mier: "Towards a universal theory for fracture of concrete"

5.1. *Objective definition and evaluation of fracture energy*

Ulm raised the question whether the fracture energy in tension is an intrinsic parameter independent of the boundary conditions. De Borst pointed out the effect of the level of observation. The fictitious (cohesive) crack model assumes a planar crack and lumps all the tortuosity into a line. For different boundary conditions in the direct tension test, crack patterns which have a different total length are observed. The question is whether the increase of energy dissipation is exclusively due to the increase of the crack length or whether the fracture energy is affected by fractality. The data we have are not discriminative. The easiest solution would be to translate the effect of boundary conditions into the change of the crack length. Van Mier mentioned diffuse cracking in the prepeak range and the need to separate its contribution to energy dissipation from the fracture energy.

In response to Dougill's question, van Mier said that close observations of the cracks at the surface do not indicate that the cracks could carry substantial shear stresses. He observed that the crack surfaces sometimes touch but almost immediately the material parts rotate away from each other. Thus, there does not seem to be much chance of developing substantial friction at these contact points. The situation might be different inside the body where the material is more confined.

5.2. *Dependence of fracture energy on the stress state*

Barr drew attention to compressive failure, which consumes much more energy. Bažant pointed out that the least one can do is to distinguish fracture energy in tension from that in compression. However, energy dissipation in compression strongly depends on lateral strains, and the fracture energy might not have a unique value. It should be taken into account that at early stages of compressive softening there are many axial splitting cracks but no shear slip yet.

Van Mier described his observations [vM84] of a discontinuity band propagating from the corner of a cube specimen subjected to compression with a low confinement. In his opinion, this band does not develop from microscopic axial splitting cracks. In uniaxial compression, localization of deformations occurs, too, but in that case the development of a shear band is not seen so clearly. In reference to confined tests he mentioned the recent experimental work of van Geel [vG98] who reported the growth of a discontinuity band from the corners of prismatic specimens. Finally, van Mier agreed to Bažant's statement that slip starts only at later stages of compressive softening, and so the maximum load cannot be inferred from an assumed collapse pattern with Coulomb friction on inclined discontinuity surfaces.

After the workshop, van Mier added the following comments: The formation of a shear crack from an array of tensile cracks is a hypothetical mechanism. Gramberg (a mining engineer) was perhaps the first to recognize this in his thesis in 1970 dealing with the fracture of rocks [Gra70]. Stroeven [Str73] used this mechanism to explain failure of concrete, in particular the development of the so-called shear cones above large aggregates, based on the assumption that the aggregates have a larger Young's modulus than the surrounding cement matrix. The formation of a continuous inclined crack (often referred to as a shear crack) from an array of tensile splitting cracks has never been observed under compressive or shear load in plain concrete. However, in SIFCON with a high percentage of steel fibers, arrays of inclined cracks may develop under a global shear stress [vM92].

Acknowledgment

This discussion record has been prepared with the kind assistance of Z. P. Bažant, L. Cedolin, V. Červenka, J. G. M. van Mier, P. Řeřicha, H. L. Schreyer, and F.-J. Ulm.

References

[BK90] Z. P. Bažant and M. T. Kazemi. Determination of fracture energy, process zone length and brittleness number from size effect, with application to rock and concrete. *International Journal of Fracture*, 44:111–131, 1990.

[FHG90] P. S. Follansbee, J. C. Huang, and G. T. Gray. Low-temperature and high-strain-rate deformation of nickel and nickel- carbon alloys and analysis of the constitutive behavior according to an internal state variable model. *Acta metallurgica et materialia*, 38(7):1241–1254, 1990.

[Got71] Y. Goto. Cracks formed in concrete around deformed tension bars. *ACI Journal*, pages 244–251, 1971.

[Gra70] J. Gramberg. *Clastic and anti-clastic processes and their implications for rock mechanics.* PhD thesis, Delft University of Technology, Delft, The Netherlands, 1970.

[Jir98] M. Jirásek. Nonlocal models for damage and fracture: Comparison of approaches. *International Journal of Solids and Structures*, 35(31-32):4133–4145, 1998.

[Str73] P. Stroeven. *Some aspects of the micromechanics of concrete.* PhD thesis, Delft University of Technology, Delft, The Netherlands, 1973.

[vG98] E. van Geel. *Concrete behaviour in multiaxial compression—experimental research.* PhD thesis, Eindhoven University of Technology, Eindhoven, The Netherlands, 1998.

[vM84] J. G. M. van Mier. *Strain softening of concrete under multiaxial loading conditions.* PhD thesis, Eindhoven University of Technology, Eindhoven, The Netherlands, 1984.

[vM92] J. G. M. van Mier. Shear fracture in slurry infiltrated fibre concrete (SIFCON). In H. W. Reinhardt and A. E. Naaman, editors, *Higher Performance Fibre Reinforced Cement Composites*, pages 348–360. E&FN Spon, London, 1992.

Durability Mechanics (I)

Franz-Josef Ulm

Laboratoire Central des Ponts et Chaussées
58, boulevard Lefebvre, F-75732 Paris cedex 15

The following is the summary of the discussion in the session "Environmental and Ageing Issues 1". The reader can refer to the full papers included in the volume for more details about the authors' presentations.

The 'Chunnel' fire: thermal spalling due to chemoplastic softening in rapidly heated concrete — *F.-J. Ulm, O. Coussy, P. Acker*

Ulm presented a chemoplastic model for the behaviour of rapidly heated concrete, which he had applied for the expertise on the fire in the channel tunnel. In contrast to standard thermoplasticity approaches, the concrete material properties (Young's modulus, strength) depend here on the hydration degree. Furthermore, by applying the model for the expertise on the fire in the channel tunnel, which had devastated a part of concrete liners by thermal spalling, he showed that the thermal spalling was found to be initiated by a biaxial compressive stress clog at the hot side created by hindered thermal dilatation. By way of conclusion, he confirmed Bazant's analysis that internal pressure seemed not to be the cause of thermal spalling.

In reply to Heinfling, Ulm confirmed that there was sufficient physical evidence to consider that the dehydration was one of the main mechanism governing the thermal softening and thermal damage behaviour of rapidly heated concrete. This is well known from the cement chemistry. Additional thermal damage and softening may arise from the difference in thermal dilatation coefficient of the cement paste and the (siliceous) aggregates. However, this was not the case of the Chunnel concrete with calcareous aggregates.

Pijaudier-Cabot mentioned localisation problems which may arise for the coupled heat-diffusion-dehydration-mechanical softening problem, similar to the one he had studied for the leaching problem. Ulm agreed that the numerics needed further investigation, in particular for the case of dehydration for which the time scale of the dehydration kinetics is small with respect to the time scale of the structural heat conduction.

Contribution to the multiscale modelling of the behaviour of concrete and RC structures at elevated temperatures — *G. Heinfling, J.-M. Reynouard*

Heinfling presented some results concerning the modelling of concrete at elevated temperature with a thermoplasticity model for structural analysis of RC-structures, and some improvements of the model by considering the pore pressure and the effective stress concept. In his thermoplastic model the fracture energy G_f depends on the temperature. He showed the domain of application of the model and discussed the role of the pore pressure on the thermal spalling behaviour, of which the order of magnitude can be considered to drive internal cracking.

In reply to some questions, Heinfling confirmed that the difference in the fracture energy function $G_f = G_f(T)$ found in literature (Bazant and Kaplan; Baker) was due to the difference in the applied testing method, in particular with respect to confinement effects. Further discussion mainly concerned the role of the pore pressure on the thermal spalling. Bazant pointed out that he considers that the pore pressure was not at the origin of the thermal spalling, but could trigger the explosive fracture in rapidly heated concrete.

Durability Mechanics (II)

Yunping Xi

Department of Civil, Environmental and Architectural Engineering
University of Colorado, Boulder, CO 80309, USA

The following is the summary of the discussion in the session "Environmental and Ageing Issues 2". Question is from audience of the session; Answer is from speaker of the paper; and Comment is from the writer. Due to the length limitation, the materials presented in the session are not repeated in the discussion, and readers may need to refer to a specific paper for details.

Fracture of Concrete Saturated by Water — *T. Tanabe, H. Ohshita*

Question: Why the pressure measured inside of the structure is higher than the applied pressure at the boundary of the structure?

Answer: As the lateral liquid pressure is applied, there is an axial load applied at the same time in the vertical direction on the structure. The combination of the applied liquid pressure and the vertical load may be responsible for the high pressure measured inside the structure.

Question: The measured pore pressure is the pressure of the pore water? or the pressure in the pore wall?

Answer: Probably the combination of the two.

Comment: Measurement of pore pressure in concrete is extremely difficult. There have been two types of procedures used for measuring internal pore pressure of concrete. One procedure uses the sensors embedded in the concrete, and the readings are recorded from outside. The other one uses a thin tube connecting a small hole located inside of concrete and the pressure meter outside of the specimen. Both procedures do not produce reliable results. The capillary pores inside concrete is very small, in the range of micrometer, and on the other hand, the sensors available for measuring the pressure is relatively large, in the range of centimetres. Probably, the results obtained from the embedded sensor is a combined effect of the two, namely, the solid load carrying frame (i.e. the cement paste microstructure) and the liquid or vapour pressure in the pore. The main problem with this procedure is how to exclude the effect of pore wall. The problem of the second procedure is that the connecting tube cannot maintain the same pressurised environmental as in the concrete, and thus, the measured pressures tend to be lower than the real pressure.

A comprehensive Numerical Model for Concrete Bridge Deck Deterioration
V.E. Saouma, E. Hansen

Question: Finite element method was used for fracture analysis around a reinforcing bar. In the analysis, the cracks propagate along radial direction of the reinforcing bar is assumed due mainly to accumulation of the rust. What is the boundary condition used along the perimeter of the reinforcing bar?

Answer: Prescribed displacement conditions were used as boundary conditions. The amount of displacement was determined by the amount of the rust accumulation.

Question: The fracture propagation due to corrosion of reinforcing bars in concrete is driven mainly by the pressure at the interface between the reinforcing bar and the surrounding concrete. The pressure is developed during the process of rust accumulation, and the pressure pushes the concrete out along the radial direction. The displacement, the interface pressure, the rust accumulation, and the confining force of the surrounding concrete are all connected together. This is a coupled problem involving mass transfer, mismatch of two phases (steel bar with the rust and concrete) and fracture propagation. The prescribed displacement condition might not characterise correctly what is really happening in the interface.

Comment: A similar problem has been studied by the next speaker. Instead of interface pressure built up due to rust accumulation, the interface pressure is developed during alkali-silica reaction (ASR) in which an expansive ASR gel formed in the interface between the aggregate and the surrounding cement paste, and the pressure results in fracture propagation in the radial direction of the aggregate. The problem was solved by considering chemo-mechanical coupling, including ASR kinetics, ASR gel flow and pressure built up in a composite media.

Question: What was the expansion coefficient used for rust formation in the analysis?

Answer: The ratio of rust volume to iron volume varies between 1.5 to 4 depending the iron oxides formed in the electrochemical reaction.

Question: The title of the paper is related to deterioration of concrete bridge decks. In your analysis, has certain type of truck load been combined with the attack of steel corrosion?

Answer: At present stage of the analysis, the truck load was not included. Our focus is to establish a complete program that can handle the diffusion of chloride and moisture, electrochemical reaction of steel corrosion in concrete, the temperature effect, and fracture development due to rust accumulation. Truck load, however, will be included in the future.

Testing and Modelling Alkali-Silica Reaction and the Associated Expansion of Concrete — *Y. Xi, A. Suwito*

Question: ASR expansion does not occur immediately after the exposure of the concrete to humid environment, how to characterise the delayed expansion in the analysis?

Answer: The delayed expansion can be explained by the diffusion processes involved in ASR. In general, ASR expansion is a diffusion controlled problem. There are basically two levels of diffusion processes involved. One is the macrodiffusion at the boundary and inside of structural members, in which the moisture and aggressive chemicals (calcium and sodium ions) penetrate into the concrete member. This is a very slow process taking usually several years and even several decades. Typical examples are bridge piers and concrete gravity dams. The other one is microdiffusion at the boundary and inside of each aggregate, in which the alkali solution in capillary pores reacts with siliceous component in the aggregate. This is also a diffusion controlled process. But, it is not as slow as the macrodiffusion since the size of aggregate ranges only from micrometer to centimetre.

Question: During the presentation of the paper, it was mentioned that the model developed for ASR of concrete could also be used for other engineering problems, steel corrosion in concrete, for example. Can you elaborate how this can be done?

Answer: In the current version of the model, the centre core is an aggregate and the surrounding matrix is the effective media (concrete), and the ASR gel is considered to be a part of the aggregate (on the aggregate surface) which causes the expansion of aggregate particles. In the case of steel corrosion in concrete, the centre core will be the steel bar in stead of an aggregate, and the surrounding matrix will be the concrete, and the ASR gel is replaced by the rust. The whole process of steel corrosion in concrete is a diffusion controlled process involving chloride and moisture diffusions from the environment into the concrete as well as rust deposit in the interface transition zone between the aggregate and the concrete. The rust formation depends on kinetics of the electrochemical process. One important point is that ASR expansion is formulated in the present model by using a 3D composite model, while steel bars must be treated as a 2D problem. Therefore, some details and specific 2D formulation are apparently needed to be worked out.

Shrinkage and Weight Loss Studied in Normal and High Strength Concrete
B. Barr, A. El-Baden, A. Arslan

Question: A linear relationship between drying shrinkage and weight loss of concrete was adopted in the study. Is this true for the humidity range from 0% to 100%?

Answer: The test results obtained by the present study showed a very good linear relationship. But the range of the relative humidity used in the test is around 60%, that is, the concrete samples were taken out from the fog room and tested under 60% relative humidity. The effect of other humidity levels were not examined in this study.

Comment: For regular concrete, the relationship between drying shrinkage and weight loss exhibits linear behaviours, but with different slopes in different ranges of humidity. There are basically three or four ranges: below 11%, between 11% and 30%, between 30 and 70%, and above 70%. The change of the slope is due mainly to the fact that within each relative humidity range, there is a dominant shrinkage mechanism. For instance, at relative humidity ranging below 11%, the dominant shrinkage mechanism is the loss of interface water in C-S-H, which causes the collapse of layered C-S-H microstructure and thus results in large drying shrinkage.

Question: B3 model cannot predict very well the shrinkage of silica fume concrete and what can we do to improve the model?

Answer: Bazant suggested that the shrinkage half time in B3 model may be adjusted. This may improve model to certain extent.

Question: The coupling between creep and shrinkage is not considered in the study, which may be very important for high strength concrete.

Answer: The coupling problem was not in the scope of the present study.

Comment: The coupling between creep and shrinkage of regular concrete has been a very important research topic. In 1940s, Picket found that basic creep of concrete (loading without drying) plus drying shrinkage (drying without loading) is less than drying creep (loading and drying simultaneously). The extra deformation is due to the coupling of basic creep and drying shrinkage and is called Pickett effect. There are some fundamental studies on the mechanisms responsible for Pickett effect observed in regular concrete. The magnitude of the Picket effect, sometimes, is in the same order as basic creep and drying shrinkage. But, to the best of the writer's knowledge, so far, there have not been any research on Picket effect of high strength concrete.

Time Dependent Softening of Cracked and Ageing Concrete — *B.L. Karihaloo, S. Santhikumar*

Question: Is temperature effect being considered in the study?

Answer: Temperature effect was not taken into account in the present study. However, temperature does have significant effect on fracture and ageing of concrete. There are basically two types of temperature effects. One is the external thermal effect of environmental temperature fluctuation, and the other one is the internal thermal effect from the heat of hydration.

Question: Is there any direct physical proof on the friction effect in the tail of bridging stress vs. crack opening diagram?

Answer: There is not, as far as I know.

Computational Failure Analysis and Design

Franz-Josef Ulm

Laboratoire Central des Ponts et Chaussées
58, boulevard Lefebvre, F-75732 Paris cedex 15

The following is the summary of the discussion in the session "Computational Failure Mechanics". The reader can refer to the full papers included in the volume for more details about the authors' presentations.

Comments on microplane theory — *M. Jirasek*

Jirasek gave an overview of different microplane models with a focus on the microplane version of the Cordebois-Sidoroff approach (anisotropic damage model based on energy equivalence and working with both effective stress and effective strain) with kinematical and static constraints and proposed a finite strain generalisation of the theory. Jirasek showed that this kind of microplane model may be classified into a stiffness version and a compliance version. The stiffness version projects total strain and effective stress, while the compliance version projects total stress and effective strain.

Bazant agreed with the presented framework, and with the extension of the microplane theory to finite strain. As for this extension, he pointed out that a volumetric-deviator split can provide further simplification. Jirasek explained that the presented framework is an extension of the work presented by Carol and Bazant, and he agreed that the volume-deviator split is important for materials with internal friction such as concrete.

In reply to additional comments, Jirasek explained that the static constraint is also applied in single crystal plasticity (as well as in some of its generalisations to polycrystalline materials). In contrast to this static constraint, the stiffness version of the presented microplane model uses the static constraint on effective stresses, and not on total ones. In return, the key issue in the finite strain extension of the microplane theory was the proper choice of the strain quantities. However, from a computational point of view, the compliance version seems more robust than the stiffness version, and leads to more reasonable stress-strain diagrams.

Some questions arose concerning the projection of the normal vectors that enter the stress evaluation on the microplanes. Jirasek explained that they were not normal vectors, but material vectors (i.e., attached to the matter).

On regularising fields for non-standard-continuum models — *R. de Borst, G. Borino*

De Borst gave an overview of the main lines of modelling concrete fracture, starting from the classical separation in discrete and smeared crack models. As for local smeared crack models, he discussed advantage and disadvantage of (elasticity based) standard isotropic continuum damage models and of anisotropic continuum damage models which include the fixed crack model, as well as, according to de Borst, the Microplane models. Concerning non-local and enhanced smeared crack models, the nonlocal continuum damage model was mentioned as less computational efficient then gradient enhanced continuum damage models (isotropic and anisotropic).

In response to de Borst's statement that smeared crack models include the microplane model, it was pointed out that microplane models could also be seen as a generalisation of smeared crack models. Earlier, Jirasek tried to draw attention to the fact that multiple crack models are similar to statically constrained microplane models but quite different from kinematically constrained microplane models.

In the discussion on the evolution of the localisation zone, Jirasek's stated that nonlocal damage models produce a progressively shrinking zone of growing local strain. De Borst replied that if the softening curve has an exponential tail, this zone first shrinks but at late stages of the degradation process it expands.

Modelling material failure as a strong discontinuity with the material point method — *H.L. Schreyer, D.L. Sulsky, S. Zhou*

Schreyer discussed theoretical and computational aspects of the material point method (MPM) for the modelling of material failure. With respect to conventional methods (Eulerian, Lagrangian, etc.) this new numerical algorithm consists in transferring the (mechanical) information between a fixed background mesh and the material points.

Gradient operations related to the equations of kinematic and motion are treated with the mesh, while all constitutive equations are treated at the material points. This method can be applied to the modelling of discontinuities by increasing the number of material points for a fixed background mesh. In reply to a question, Schreyer agreed that one can consider this approach as an updated Lagrangian one, for which the updated reference configuration is defined by the updated system of material points.

Quasi brittle fracture analysis by the symmetric Galerkin boundary element method: recent developments — *A. Frangi, G. Maier*

Maier gave an overview of recent developments and applications of the Symmetric Galerkin Boundary Element Method (SGBEM) to the modelling of the non-linear behaviour of materials, with a focus on fracture. He explained that the method allows for the simulations of the fracture process, and was valid for both isotropic and anisotropic formulations.

During the discussion, Maier further explained that the BEM was not suitable for the modelling of spread non-linearities as those generated by viscoelastic behaviour, but showed rather good performance in the case of fracture.

Finite element analysis of fragmentation with continuous remeshing — *P. Krysl, M. Ortiz*

Krysl showed some results of recent developments in continuous remeshing applied to the simulation of strong non-linear behaviour of solids including surfaces of discontinuities such as discrete cracks.

List of Workshop Participants

BARR Ben
University of Wales, United Kingdom

BITTNAR Zdenek
Czech Technical University
Czech Republic

DE BORST René
Delft University of Technology
The Netherlands

BUYUKOZTURK Oral
Mass. Institute of Technology, USA

CAROL Ignacio
Technical Univ. of Catalunia, Spain

CEDOLIN Luigi
Politecnico di Milano, Italia

CERVENKA Vladimir
Cervenka Consulting, Czech Republic

CHAVANT Clément
Electricité de France, Research Division
France

DOUGILL John W.
The Institution of Structural Engineers
United Kingdom

GAMBAROVA Pietro
Politecnico di Milano, Italia

GERARD Bruno
Electricité de France, Research Division
France

HEINFLING Gregory
Electricité de France, Research Division
France

HUET Christian
EPFL-LMC, Switzerland

JIRASEK Milan
EPFL-LSC-DGC, Switzerland

KARIHALOO Bushan
University of Wales, United Kingdom

KRYSL Petr
California Institute of Technology, USA

LASNE Marc
Electricité de France, Research Division
France

LE BELLEGO Caroline
Electricité de France, Research Division
France

MAIER Giulio
Politecnico di Milano, Italia

MANG Herbert
Universität Wien, Austria

MAZARS Jacky
ENS de Cachan, France

OZBOLT Josko
Universität Stuttgart, Germany

PIJAUDIER-CABOT Gilles
ENS Cachan & Institut Universitaire de France, France

REYNOUARD Jean-Marie
INSA de Lyon, France

SAOUMA Victor
University of Colorado, USA

SCHREYER Howard
University of New Mexico, USA

TANABE Tada-aki
Nagoya University, Japan

ULM Franz
LCPC, France

VAN MIER Jan
Delft University of Technology
The Netherlands

WILLAM Kaspar
University of Colorado, USA

XI Yunping
University of Colorado, USA

CET OUVRAGE A ÉTÉ REPRODUIT
ET ACHEVÉ D'IMPRIMER PAR
L'IMPRIMERIE FLOCH À MAYENNE
EN JANVIER 1999.

DÉPÔT LÉGAL : JANVIER 1999.
N° D'IMPRIMEUR : 45074.
Imprimé en France